Understanding Marine Life

Understanding Marine Life

Edited by Percy Hugo

SYRAWOOD
PUBLISHING HOUSE
New York

Published by Syrawood Publishing House,
750 Third Avenue, 9th Floor,
New York, NY 10017, USA
www.syrawoodpublishinghouse.com

Understanding Marine Life
Edited by Percy Hugo

International Standard Book Number: 978-1-68286-681-8 (Hardback)

Cataloging-in-Publication Data

Understanding marine life / edited by Percy Hugo.
p. cm.
Includes bibliographical references and index.
ISBN 978-1-68286-681-8
1. Marine ecology. 2. Marine biology. 3. Marine sciences. I. Hugo, Percy.
QH541.5.S3 U53 2019
577.7--dc23

Printed in China.

TABLE OF CONTENTS

PREFACE

Every book is initially just a concept; it takes months of research and hard work to give it the final shape in which the readers receive it. In its early stages, this book also went through rigorous reviewing. The notable contributions made by experts from across the globe were first molded into patterned chapters and then arranged in a sensibly sequential manner to bring out the best results.

Marine life comprises of the plants, animals and other organisms that inhabit the seas, oceans or estuaries around the world. The marine ecosystem is very diverse and includes over 230,000 species and it is estimated that more than a million are yet to be discovered. The study of marine life is very important because marine organisms are the chief source of oxygen and also responsible for maintaining and shaping coastlines across the world. The study of this field branches out into marine conservation, marine biology, oceanography and hydrology. This book aims to equip students and experts with the modern evaluations and methodologies to achieve a multidisciplinary approach to the study of marine life. This book is a valuable compilation of topics, ranging from the basic to the most complex advancements in the field of marine life and their ecology, biology and conservation. It is a vital tool for all researching and studying this field.

It has been my immense pleasure to be a part of this project and to contribute my years of learning in such a meaningful form. I would like to take this opportunity to thank all the people who have been associated with the completion of this book at any step.

Editor

1

Application of Spirulina for Feeding Larvae of Green Tiger Shrimp, *Peneaus semisulcatus*

Ghaeni M[1]* and Matinfar A[2]

[1]*Department of Fisheries, Science and Research Branch, Islamic Azad University, Tehran, Iran*
[2]*Department of Aquaculture, Iranian Fisheries Research Institute, Tehran, Iran*

Abstract

This experiment was carried out in Bandargah Station in Bushehr province. *Peneaus semisulcatus* larvae were fed on *Spirulina platensis* as supplementation microalgae. In order to use *Spirulina platensis* in this study, we have to produce biomass and dry production of this microalga.

Determination of length and survival rate of *Peneaus semisulcatus* larvae was carried out as the objective of the study. Treatments including Z plus (as a supplementation without spirulina), M.C.F (as an imported enriched spirulina), the powder combination of *Spirulina platensis* with Z plus and *Spirulina platensis* as live food were compared to *Cheatocerus* microalgae as a control. The experiment was carried out from nauplii stage to early post larvae.

The results of the experiment indicated that the powder combination of *Spirulina platensis* with Z plus treatment has the most survival rate (76.5%). The maximum of length (4.3 mm) observed in control (*Cheatocerus spp*). The maximum mortality was observed in zoa stage in *Spirulina* microalgae treatment, because the spiral filamentous of *Spirulina sp.* was large.

Keywords: *Spirulina platensis; Peneaus semisulcatus;* Length; Survival

Introduction

Proper nutrition has long been recognized as a critical factor in promoting normal growth and sustaining shrimp health. The development of high-quality diets is a factor that has significantly contributed to explosive expansion of shrimp farming in the last two decades. Larval phase is the most sensitive stage in shrimp life, especially when the yolk sac is absorbed. In this stage, suitable food is necessary to inhibit mortality and provide normal metabolism [1]. Shrimp larval growth in aquaculture has traditionally been dependent on live food such as *Chaetoceros, Tetraselmis, Skeletonema* and *Chlorella [2,3]*. Culturing of live food is, however, expensive and complicated. Prepared diets, on the other hand, not only provide the essential nutrients that are required for normal physiological functioning, but also may greatly contribute to reducing the cost of production by eliminating or, at least, minimizing reliance on the live food and its often expensive associate infrastructure.

Lots of works have recently been carried out to find appropriate prepared diets as alternative sources to live food, among which *Spirulina* Powder (SPM), encapsulated *Spirulina* and Z plus supplement are worth mentioning. *Spirulina platensis* is a fast growing cyanobacter of large size (0.5 mm) and capable of growing in high salinity and alkalinity. The biomass of *S. platensis* has been recognized to be a "wonderful health food" for having high protein and various compounds, such as essential fatty acids (linoleic and gamma linolenic acids), essential amino acids, B-complex vitamins and biopigments [4-7]. Several studies have been conducted using dried *Spirulina* as a supplement in diets of crustaceans. *S. platensis* meal is available at a commercial scale; thus its use in feeds for aquaculture is possible [3]. *Spirulina* have already been tested as a substitute protein source for *Cyprinus carpio*, where equal or even higher growth rates were obtained by diets containing 25% algae meal, replacing 80% of the dietary fishmeal [8]. Addition of Spirulina in the diet of giant freshwater prawn (*Macrobrachium rosenbergii*) significantly improved growth, survival and feed utilization regardless of supplementation level in range of 5-20% [9]. Partial replacement of fish meal with *S. platensis* has also been evaluated in juvenile Pacific white shrimp, *Litopenaeus vannamei*, with promising result [10].

With a wideeographic distribution from the south and east Africa to India and Sri Lanka, including the Red Sea, Persian Gulf [11], the green tiger prawn, *P. semisulcatus*, is one of the most common and commercially important marine shrimp in the Indo-West Pacific region and enjoys a good aquaculture potential background. In recent years, there has been increasing interest in the culture of *P. semisulcatus* in some countries of the Middle East, including Kuwait and other Gulf countries [12]; Egypt [13] and Turkey [14]. Although a considerable range of information on the dietary and nutritional requirements of a number of penaeid shrimp, including *Penaeus japonicas, P. monodon*, and *Litopenaeus vannamei*, have been studied thoroughly for their nutritional requirements [15], not much information is available for *P. semisulcatus* [16].

The present work has focused on the application of *Spirulina* powder, produced in Iran as a new and alternative source to live feed in aquaculture industry, and its effect on the growth and survival rates of the larval green tiger shrimp, *P. semisulcatus*, as an endemic species to the Persian Gulf have been compared with other common food in hatcheries.

***Corresponding author:** Ghaeni M, Department of Fisheries, Science and Research Branch, Islamic Azad University, Tehran, Iran, E-mail: gonomwo@yahoo.com

Materials and Methods

This research was carried out in Bandargah Station of shrimp research center in Bushehr. The pure stock of *Spirulina platensis* was imported from Indonesia, in order to produce biomass and powder of *Spirulina, Spirulina* was cultured in laboratory condition and biomass of *Spirulina* was produced to for powder according to Jourdan procedure [17].

This study was conducted in Bandargah Station of Bushehr and the required nauplii of the project were prepared from cultured ablated brood stocks. Our study was designed as a follows: Four treatments from no. 1 up to no. 4 were *Spirulina platensis* powder that produced in Iran, *Spirulina platensis* powder and Z plus combination, Z plus powder and imported concentration of enriched *Spirulina* (M.C.F), respectively and the control was *Chaetoceros sp.* algae (Table 1). All treatments and control were prepared in triplicate. Larvae were randomly introduced in polyethylene containers. All containers were cleaned by detergents and disinfected by formalin and washed by tap water. Finally one hundred nauplii were introduced in each container (treatments and control). The containers were fully filled with aerated sea water with 30°C temperature and 36 ppt in salinity. A plan which was carried out four times a day was started from the stage of fifth nauplii to early post larvae. The powder solution with equal amount in each meal was 5 mg/l.

Chaetoceros sp. as a live food and Z plus powder as a supplement was used in Bandargah hatchery. Imported concentrations of enriched *Spirulina* (M.C.F) were also uses in private hatcheries in Bushehr. For these reasons, our treatments consist of M.C.F and Z plus powder and *Chaetoceros sp.* was used as a control of experiment. The chemical composition of the produced *Spirulina platensis* powder in Iran and Z plus powder have been shown in Table 2.

At the first stage of mysis, 0.5 gr/l (20 *Artemia nauplii*) of frozen *Artemia salina nauplii* was added for each container. So, we had to reduce the amount of algae with the ratio of 25% algae plus 75% *Artemia nauplii.* At three stages of zoa I, zoa III and mysis III, the average number of larva were counted three simultaneous times and the survival rate was measured. The total length of larvae was measured every day by an optical microscope.

Statistical Analysis

An ANOVA test was applied to compare the effects of different feeds on survival rate and length.

Results

The survival rates of *P. semisulcatus* larvae in 3 stages have been presented in Table 3 and Figure 1.

Treatment 1	Treatment 2	Treatment 3	Treatment 4	Control
Produced *Spirulina platensis* powder in Iran	*Spirulina platensis* powder and Z plus	Z plus powder	Imported concentration of enriched spirulina (M.C.F)	*Chaetoceros sp.*

Table 1: Experimental diet for larvae.

Food	Raw protein	Raw fat	Raw fiber	Moisture	ash
Z plus powder	50	15	2	12	8
Produced *Spirulina platensis* powder in Iran	46.6	0.1	-	9.46	0.7

Table 2: The chemical composition of produced *Spirulina platensis* powder in Iran and Z plus powder.

In the same column, different superscripts show significant differences.

The average length of *P. semisulcatus* larvae is shown in Figure 2 and 3 in all treatments, and in the control from zoa to mysis III.

Discussion

In the zoa stage, larvae start to be active for feeding by catching marine microalgae activity.

The existence of suitable floating food is necessary in the farm tank. *Skeletonema sp., Chaetoseros sp., Tetraselmis sp. and Chlorella sp.* microalgae are usually used in the zoa stage. Other foods, such as egg vitellus, egg and yeast are used as supplement food some times, but the water pollution in the tank will increase [2]. Also, application of *spirulina sp.* individually or in combination with other foods was used for species of shrimp larvae like *Litopenaeus vannamei* [10], *L. schmitti* [3,18], *Peneaus monodon* in different percents [19], while there is not enough information about the special species of *P. semisulcatus.*

In this research, *Spirulina* and other foods were used in hatcheries individually or in combination to obtain the best growth and survival

Stage Treatments	Zoa I	Zoa III	Mysis III
1	100	61.6^{a}	56^{c}
2	100	88.8^{b}	76.5^{a}
3	100	100^{c}	7^{d}
4	100	83.3^{d}	59.6^{bc}
control	100	83.3^{d}	68.5^{ab}

In the same column different superscripts show significant differences

Table 3: Survival rates of P. *semisulcatus* larvae in 3 stages.

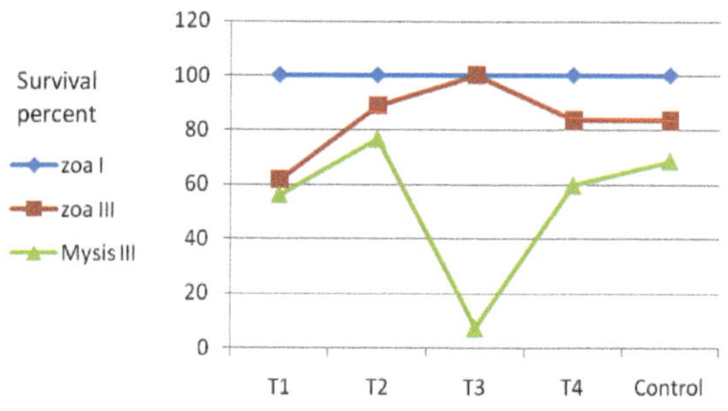

Figure 1: Survival rates of P. semisulcatus larvae in 3 stages in two modes.

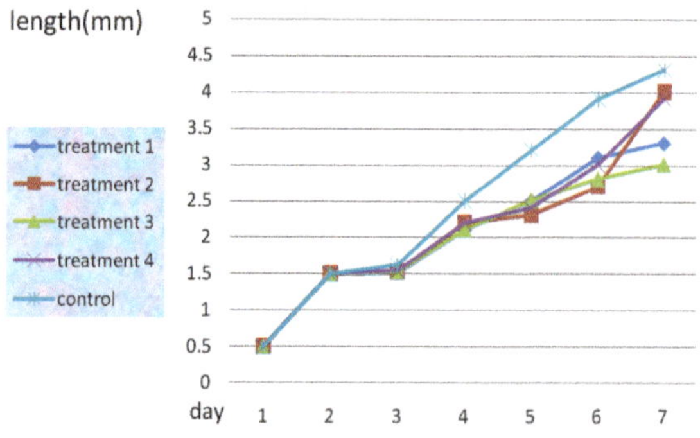

Figure 2: Length of *P. semisulcatus* larvae from zoa to mysis III during 7 days.

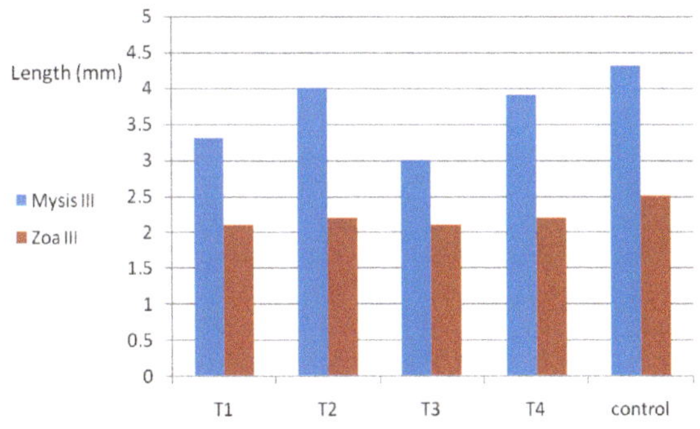

Figure 3: Length of *P. semisulcatus* larvae in zoaIII and mysis III during 7 days.

rates of *P. semisulcatus* larvae. Statistical analysis shows that there isn`t any significant difference in the growth factor but the difference between treatments and control was significant (P<0.001). Another report showed that the average growth of *Litopenaeus vannamei* juvenile fed by lyophilized powder of *Spirulina platensis* was 86 mg/day over the experimental period [10]. Also, the study of using *Spirulina platensis* in different dry weights for nursing black tiger shrimp post larvae 10 showed that the growth of shrimp larvae fed by using 5% of *Spirulina platensis* was significantly higher (14.25 mm) than that of shrimp larvae in control, and the survival rates for treatment groups were also higher (69.5%) than that of control [20].

Among other stages, the survival rate in mysis III was significant between the control and the combination of Z plus and the produced *Spirulina* powder in Iran (treatment 2, with the best survival rate) (P<0.001). While the survival rate of *P. semisulcatus* larvae fed only with *Chaetoceros* and *Tetraselmis* in mysis II was 85-96% [21], the survival rate of *P. orientalis* larvae fed on *S. platensis* was only 8.7%; those fed on *C. muelleri*, 28.7% and those fed on both algae, 17.3% for mysis [22].

At the stage of mysis III, the survival rate of the produced Spirulina powder in Iran was 56%. Some of the problems in this treatment were related to the difficult solvability of Spirulina powder in water, and the existence of bubbles on the surface of the container.

Ingthamjitr [19] evaluated the effect of spirulina and *Chaetoceros* on *Peneaus monodon* larvae from nauplii to the post larvae. The best growth was achieved with feeding by *Chaetoceros sp*. and *Spirulina* in 25:75 ratio. While in our project, there was not any difference between produced *Spirulina* and imported *Spirulina* in growth and survival rates. Their results are the same as our research. The survival rate for their combination of *Chaetoceros* and imported or produced *Spirulina* is more than separate treatments. Whiles in this research, treatment 2 with 1:1 ratio of the produced *Spirulina* and Z plus supplement had the most survival rate (67.5%) in mysis III stage, because Z plus supplement was improved by additive nutrients and after control, the maximum growth was seen in this treatment. Z plus had 15 % of unsaturated fatty acids which was prepared from marine resources.

Jaime-Ceballos et al. [18] evaluated *Spirulina platensis* in *Litopenaeus schmitti* larvae diet from mysis I to post larvae. The survival rate was about 80% in each treatment. In contrast to control, the size of post larvae was so small. Whiles in this research, the survival rate in all treatments was less than 80% dependent on the amount and times of the feeding (4 times in contrast to 6 times feeding), that *P. semisulcatus* larvae are more sensitive than other species. In the work of Jaime-Ceballos et al. [3], *Spirulina platensis* powder is replaced by *Chaetoceros mulleri* in *Litopenaeus schmitti* larvae diet, and then the length of larvae are compared with each other.

There was significant relation between the level of replacement and the length of larvae, the maximum length was acquired in 30%, but in this research, the combination ratio for Z plus and produced *Spirulina* powder was 50% (1:1) and the best survival rate was observed with this ratio (P<0.001).

Treatment 3 (Z plus) had 50% protein and 14% unsaturated fatty acid, but its final survival rate was very low (7%), and in comparison to other treatments it had little growth (3 mm), while in zoa III, the survival rate of this treatment was 88.8%. This decay maybe related to the situation of larvae in mysis stage.

The longest length of larvae was observed in treatment 4 (M.C.F), but its survival rate was 59.6%. According to the component of this diet, it had many different nutrients with high protein, such as meat powder, squid, egg, *Spirulina* ...), mineral, vitamins. It is an enriched food that is produced with industrial methods. In comparison to the produced *Spirulina* in this project, it had dark blue color, suitable granulation for larvae and instantly dissolved in water, while the previous produced *Spirulina* in treatment 1 had green color and difficult solvability in water. When the survival rate is decreased in mysis III, other larvae could take more food and then they have more growth.

Hagh Nejat et al. [2] evaluated *P. semisulcatus* larvae feeding in zoa stage with *Tetraselmis, Chaetoceros, Skeletonema* and *Chlorella* individually and syncretistic. The most survival rate in the individual feeding was 82% and the growth rate was 3.55 mm for *Skeletonema*, and in the syncretistic feeding, the best survival rate was 97% in combination with *Tetraselmis* and *Chaetoceros*. The general results showed that the growth rate of larvae in syncretistic feeding was larger than individual feeding.

In our research, the control (*Chaetoceros sp.*) had a fast growth rate with the highest survival rate after treatment 2 (68.5%). *Chaetoceros* had the best size among other microalgae that is used in larviculture for zoa feeding, as live food, which can provide lots of feeding necessities. In the other research, *Penaeus indicus* zoa I larvae reared successfully on the mixed algal feed of *Tetraselmis chuii* (25 cells/µl) and *Skeletonema costatum* (35 cells/µl), together with five *Artemia nauplii* per µl from mysis I onwards until metamorphosis in only 6-7 days [23]. While the survival percentage of *P. monodon* post larvae fed on *Arthrospira fusiformis* was significantly higher at inclusion levels of 10% and 20%. The results suggest that *Arthrospira fusiformis* cannot serve as the sole protein source in the diet of *P. monodon postlarvae*, but can be effectively used as a supplementary protein source [24]. This result was the same as our research. Generally, acombination of Z plus (consisting of 50 % protein) and powder *Spirulina* (dry weight protein 46.6%) was found as the best diet in this project.

Acknowledgments

We extend our appreciation to Dr. Neda Soltani and Dr. Shakori for identification of *Spirulina* species in Shahid Beheshti University. Thanks are also due to Mr Zendebudi, Dalirpour and Dr. Pazir for collaborating with us in Bushehr Bandargah Station. Furthermore, we like to thank Dr. Seif Abadi and Dr. Zohreh Amiri who kindly provided helpful advice for editing and statistical analysis.

References

1. Hagh Nejat M (2001) *Peneaus semisulcatus* zoa stage feeding, second educational seminar on *Peneaus semisulcatus* culture. Aquaculture development deputy.

2. Hagh Nejat M, Dalirpur GH, Ghaednia B, Mirbakhsh M, Alekhorshid M (2005) Comparison growth and survival of *Peneaus semisulcatus* zoa by 4 algae feeding, individual and combination. Iranian Fisheries Journal.

3. Jaime-Ceballos B, Hernandez-Llamas A, Garcia T, Perez-Jar L, Villareal H (2006) Substitution of *Chaetoceros mulleri* by *Spirulina platensis* meal in diets for *Litopenaeus schmitti* larvae. Aquaculture 260: 215-220.

4. Tri-panji, Suharyanto (2001) Optimization media low cost nutrient sources for growing *Spirulina platensis* and carotenoid production. Menara Perkebunan 69: 18-28.

5. Jimenez C, Cossio BR, Labella D, Xavier Niell F (2003) The feasibility of industrial production of spirulina *(Arthrospira)* in southern Spain. Aquaculture 217: 179-190.

6. Goksan T, Zekeriyaoglu A, Aysegul, Ilknur Ak (2007) The growth of *Spirulina platensis* in different culture systems under greenhouse condition. Turk J Biol 31: 47-52.

7. Volkmann H, Imianovsky U, Oliveira JLB, Sant Anna ES (2008) Cultivation of *Arthrospira platensis* in desalinator wastewater and salinated synthetic medium: protein content and amino acid profile. Brazilian Journal of Microbiology 39: 1-4.

8. Sandbank E, Hepher B (1978) The utilization of microalgae as a feed for fish additive for fish. Arch Hydrobiology Suppl Ergeb Limnol 11: 108-20.

9. Nakagawa H, Gomez-Diaz G (1995) Usefulness of *Spirulina sp.* meal as feed additive for giant freshwater prawn, *Macrobrachium rosenbergii*. Suisanzoshoku 43: 521-526.

10. Hanel H, Broekman D, de Graaf S, Schnack D (2007) Partial replacement of fishmeal by lyophilized powder of the microalgae *Spirulina platensis* in pacific white shrimp diets. Open Mar Biol J 1: 1-5.

11. Ronquillo JD, SiashoT, McKinley RS (2006) Early developmental stages of green tiger prawn, *Peneaus semisulcatus* de Haan. Hydrobiologia 560: 175-196.

12. Al-Ameeri AA, Cruz EM (1998) Effect of sand substrate on growth and survival of *Peneaus semisulcatus* de Haan juveniles. J Aquaculture Trop 13: 239-244.

13. Sadek S, Moreau M (2000) Performance of *Macrobrachium rosenbergii* and *Peneaus semisulcatus* under mono and mixed culture systems, and their suitability for polyculture with Florida red tilapia, in Egypt. J Aquaculture Trop 15: 97-107.

14. Aktas M, Kumlu M, Eroldogan OT (2003) Off- season maturation and spawning of *Peneaus semisulcatus* by eyestalk ablation and/or temperature-photoperiod regimes. Aquaculture 228: 361-370.

15. Shiau SY (1998) Nutrient requirements of penaeid shrimps. Aquaculture 164: 77-93.

16. Al-Ameeri AA, Cruz EM, Al-Sharrah T (2006) Growth and performance of *Peneaus semisulcatus* de Haan fed with two commercial shrimp feeds. Aquaculture Res 37: 1507-1515.

17. Jourdan P (2001) Manual of small scale *Spirulina* culture. Antenna Technologies.

18. Jaime-Ceballos B, Villareal H, Garcia T, Perez-Jar L, Alfonso E (2005) Effect of Spirulina platensis meal as feed additive on growth, survival and development in Litopenaeus schmitti shrimp larvae. Revista de Investigaciones Marinas 26: 235-241.

19. Ingthamjitr S (1989) Use of *Spirulina* in the culture of *Penaeus monodon* larvae. Asian Inst of Technology, Bangkok, Thailand.

20. Hemtanon P, Direkbusarakom S, Bunyaviwat V (2008) Application of *Spirulina platensis* for prevention of white spot syndrome virus in post larvae and juvenile black tiger shrimp *(Penaeus monodon)*. Songklanakavin J Sci Technol, Lund University, Sweden 27: 253.

21. Ronquillo JD, Matias JR, Siasho T, Yamasaki S (1997) Culture of *Tetraselmis tetrathele* and its utilization in the hatchery production of different penaeid shrimps in Asia. Hydrobiologia 358: 237-244.

22. Shuli C, Baoqing X (2001) Experimental studies on *Spirulina platensis* as food for mysis of *Penaeus orientalis*, Mar Sci Bull.

23. Kumlu M (1998) Larval growth and survival of *Penaeus indicus* (Decapoda: Penaeidae) on live feeds. Tr J Biol 22: 235-245.

24. Ketan VT, Thresiamma j, Mohanakumaran N, Bajaniya VC (2010) Evaluation of *Arthrospira fusiformis* as a protein source in the diet of *Penaeus monodon Fabricius*. J Appl Aquaculture 22: 109-116.

2

Effect of Physico-Chemical Parameters on Crabs Biodiversity

Varadharajan D*, Soundarapandian P and Pushparajan N

Faculty of Marine Sciences, Centre of Advanced Study in Marine Biology, Annamalai University, Parangipettai-608502, Tamil Nadu, India

Abstract

If physico-chemical parameters are not suitable, it will affect the distribution of crabs. So, it is designed to know the effect of physico-chemical parameters on the biodiversity of crabs from Arukkattuthurai to Pasipattinam. Minimum atmospheric temperature was recorded in the months of November and December 26.4°C (stations 2 and 8) and maximum 32.3°C was recorded in the months of April and May (stations 2 and 5). Minimum water temperature was recorded in the month of November 26.3°C (stations 2 and 8) and maximum 31.5°C was recorded in the months of April and May (stations 2 and 5). The salinity was minimum in month of December 25.5% (stations 2, 5 and 9) and maximum was recorded in the month of May 35.5 (stations 2, 5 and 9). The minimum pH 7.6 was recorded in the month November (stations 2, 5 and 8) and maximum was 8.3 in the month of May (stations 2, 5 and 8). Minimum dissolved oxygen 2.58 mg/L was recorded in the month of May (stations 2, 5 and 7) and maximum was 5.83 mg/L recorded in the month of November (stations 2, 5 and 8). Light extension co-efficient was recorded minimum, 1.43 Mcm^{-1} during the summer season in April (stations 5, 6 and 8), and maximum 7.95 Mcm-1, during monsoon season in November (stations 5, 6 and 8). The minimum turbidity 24.0 NTU was recorded during the monsoon in April (stations 2, 3 and 5) and maximum 35. 0 NTU was recorded during summer in December (stations 2, 3 and 5).

Keywords: Physico-chemical; Crab; Biodiversity; Distribution; Affect

Introduction

The coastal ecosystems provide food and other resources, waste disposal, recreation and inspiration [1]. The crustaceans are highly sensitive to pollution [2-4], and their distributions are strongly influenced by the physico-chemical parameters [5]. Nutrients in coastal waters have caused a number of environmental problems, such as death of benthic fauna, decapod crabs occurrence of nuisance algal blooms [6,7], and the disappearance of seagrass and mangroves [7]. Physico-chemical variations induce changes in immune status of crustaceans. These physico-chemical variations are often stressing crustacea, resulting in a reduction of immune vigour. Water quality monitoring has one of the highest priorities in environmental protection policy. The main objective is to control and minimise the incidence of pollutant oriented problems, and to provide water of appropriate quality to serve various environmental purposes [8,9]. The environmental parameters of coastal areas are very important, because the variations in the physico-chemical properties, such as temperature, salinity, pH, dissolved oxygen and nutrients influence on the crustaceans abundance and life cycles. So, in the present study, effect of physico-chemical parameters on the biodiversity of crabs was studied all along Arukkattuthurai to Pasipattinam coast.

Materials and Methods

The physico-chemical parameters were selected for ten different stations *viz.,* Arukkattuthurai (Station-1), Pointcalimere (or) Kodikkarai (Station-2), Muthupettai (Station-3), Adirampattinam (Station-4), Mallipattinam (Station-5), Sethubavachatram (Station-6), Kattumavadi (Station-7), Manamelkudi (Station-8), Jegathapattinam (Station-9) and Pasipattinam (Station-10). The rainfall data were collected from the meteorological unit (Govt. of India). The surface water samples were collected monthly from all stations (stations 1 to 10) during the study period, from January 2010 to December 2010. The temperature (Atmospheric and Water) and pH were recorded immediately after collection, using a standard centigrade thermometer of 0.01°C accuracy and a field pH ELICO Grip pH meter (Model LC-120), respectively. Light penetration in the water column was measured with the help of a Secchi disc, and the Light Extinction Co-efficient (LEC) was calculated using Poole and Atkins [10] formula. Salinity was determined by a Refrectometer (Model-2, ERMA, Japan), while Winkler' method [11] was adopted for oxygen determination, and turbidity was determined by using the standard Turbidity meter (LP-2000).

For the analysis of nutrients, surface water samples were collected in clean polythene bottles and kept immediately in an icebox, and transported to the laboratory. The water samples were than filtered using a Millipore filtering system and analyzed for dissolved inorganic phosphate, nitrate, nitrite, reactive silicate and ammonia, by adopting standard procedure of Strickland and Parsons [11]. All the statistical analysis was performed by using SPSS 16 software, particularly Box Plot Method.

Results

Rainfall

Total rainfall recorded from the study area was on average of 788 mm. Minimum rainfall was recorded during the month of March (27.5 mm), and maximum rainfall was recorded during the month of November (289.1 mm) in all stations. There was no rainfall during January, February, April and May.

***Corresponding author:** Varadharajan D, Faculty of Marine Sciences, Centre of Advanced Study in Marine Biology, Annamalai University, Parangipettai-608502, Tamil Nadu, India, E-mail: heartvaradhan@gmail.com

Atmospheric temperature

In the present study, minimum atmospheric temperature was recorded in the months of November and December 26.4°C (stations 2 and 8), and maximum 32.3°C was recorded in the months of April and May (stations 2 and 5) (Figure 1).

Water temperature

In the present study, minimum water temperature was recorded in the month of November 26.3°C (stations 2 and 8), and maximum 31.5°C was recorded in the months of April and May (stations 2 and 5) (Figure 2).

Salinity (%)

The salinity was minimum in the month of December 25.5% (stations 2, 5 and 9) and maximum was recorded in the month of May 35.5% (stations 2, 5 and 9) (Figure 3).

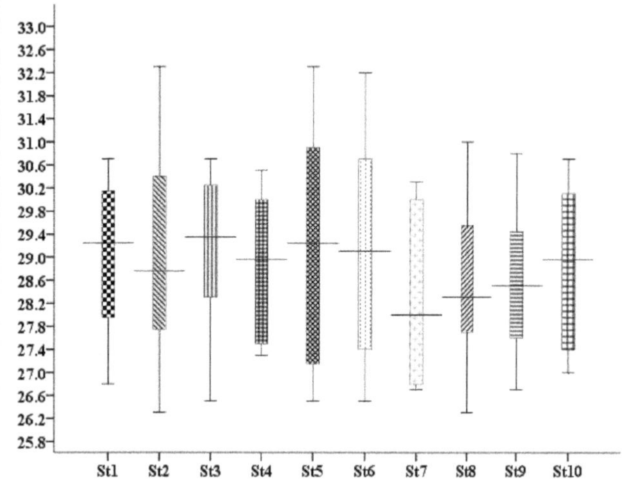

Figure1: Atmospheric temperature (°C) in different stations from January-2010 toDecember-2010.

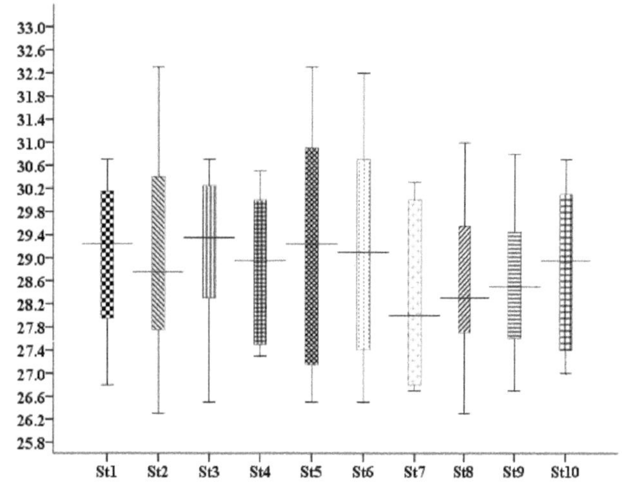

Figure 2: Water temperature (˚C) in different stations from January-2010 to December-2010.

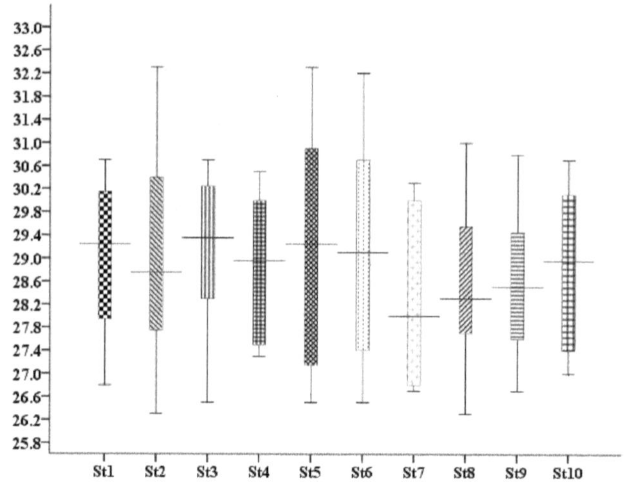

Figure 3: Salinity (%) in different stations from January-2010 to December-2010.

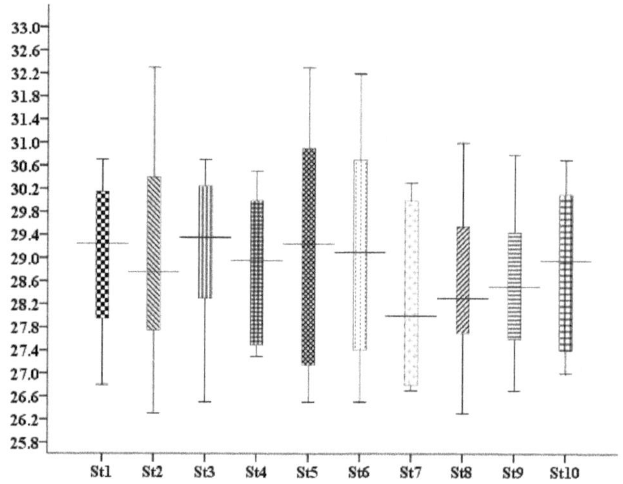

Figure 4: pH in different stations from January-2010 to December-2010.

pH (Hydrogen ion concentration)

The minimum pH 7.6 was recorded in the month November (stations 2, 5 and 8), and maximum was 8.3, recorded in the month of May (stations 2, 5 and 8) (Figure 4).

Dissolved Oxygen (mg/L)

The minimum dissolved oxygen 2.58 mg/L was recorded in the month of May (stations 2, 5 and 7), and maximum was 5.83 mg/L, recorded in the month of November (stations 2, 5 and 8) (Figure 5).

Light extinction co-efficient (Mcm^{-1})

Light extinction co-efficient was recorded minimum 1.43 Mcm^{-1} during the summer season in April (stations 5, 6 and 8), and maximum

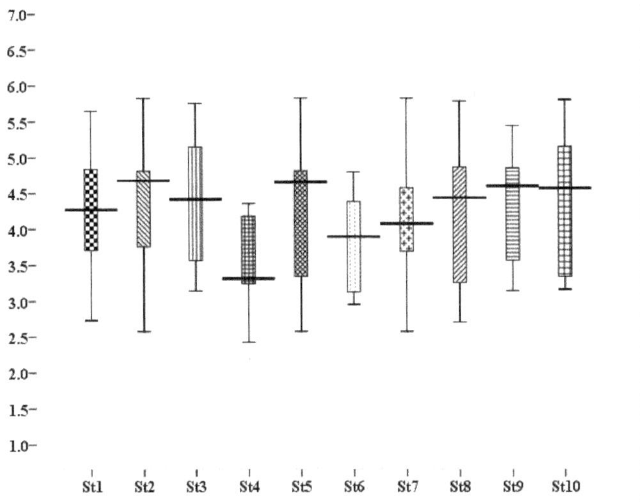

Figure 5: Dissolved Oxygen (mg/L) in different stations from January-2010 to December-2010.

7.95 Mcm^{-1} during monsoon season in November (stations 5, 6 and 8) (Figure 6).

Turbidity (NTU)

The minimum turbidity 24.0 NTU was recorded during the monsoon in April (stations 2, 3 and 5), and maximum 35.0 NTU was recorded during summer in December (stations 2, 3 and 5) (Figure 7). Outliers show to the extreme levels of the values.

Total suspended solids (mg/L)

The total suspended solids were varied from 31.75 to 68.4 mg/L. Minimum was recorded during the post monsoon season in January (stations 2, 3 and 5), and maximum during the monsoon season in December (stations 2, 3 and 5) (Figure 8). Outliers show to the extreme levels of the values.

Biological oxygen demand (mg/L)

The BOD values were varied from 1.137 to 3.214 mg/L. Minimum was recorded during monsoon season in November (stations 2, 3 and 5), and maximum during April (stations 2, 3 and 5) (Figure 9). Outliers show to the extreme levels of the values.

Chemical oxygen demand (mg/L)

The COD values were varied from 0.248 to 0.572 mg/L. Minimum was recorded during monsoon in December (stations 2, 3 and 5), and maximum during the summer season in June (stations 2, 3 & 5) (Figure 10). Outliers show to the extreme levels of the values.

Nutrients

Nitrate ($\mu M/1^{-1}$): Nitrate concentration was varied from 0.317 to 5.214 $\mu M/1^{-1}$. Minimum was recorded during summer season in April (stations 2, 3 and 5), and maximum during monsoon season in November (stations 2, 3 and 5) (Figure 11). Outliers show to the extreme levels of the values.

Nitrite ($\mu M/1^{-1}$)

Nitrite concentration was varied from 0.281 to 3.572 $\mu M/1^{-1}$. Minimum was recorded during May (stations 2, 3 and 5) and maximum during November (stations 2, 3 and 5) (Figure 12). Outliers show to the extreme levels of the values.

Ammonia ($\mu M/1^{-1}$)

The ammonia concentration was varied from 0.845 to 3.423 $\mu M/1^{-1}$. Minimum was recorded during postmonsoon in March (stations 2, 3 and 5), and maximum during November (stations 2, 3 and 5) (Figure 13). Outliers show to the extreme levels of the values.

Total Nitrogen ($\mu M/1^{-1}$)

The total nitrogen level was ranged from 5.217 to 25.32 $\mu M/1^{-1}$. Minimum was recorded in August (stations 2, 3 and 5), and maximum in November (stations 2, 3 and 5) (Figure 14). Outliers show to the extreme levels of the values.

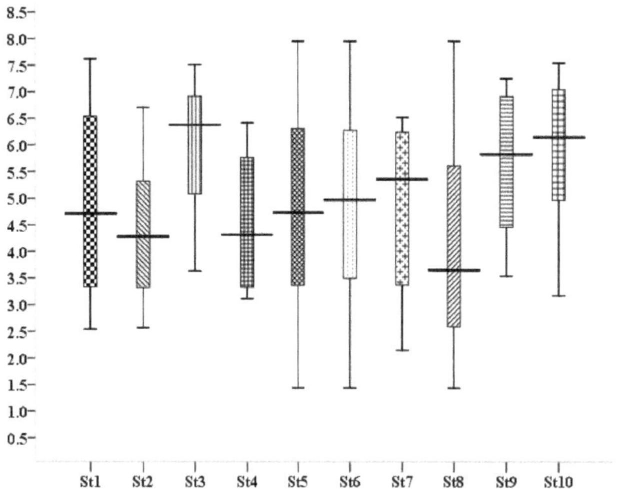

Figure 6: Light Extinction Co-Efficient **(**Mcm-1**)** in different stations from January- 2010 to December-2010.

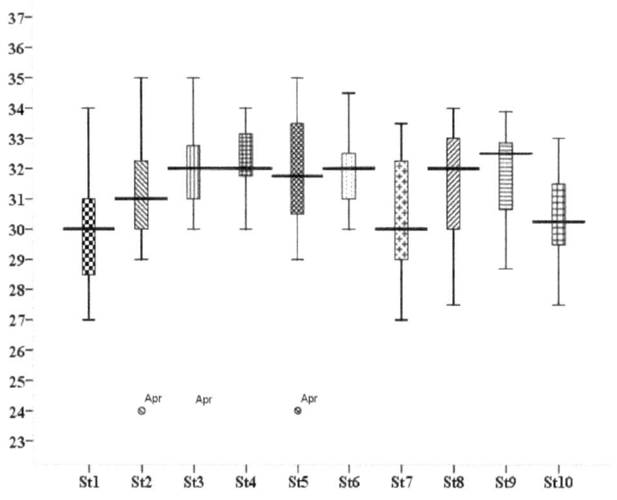

Figure 7: Turbidity (NTU) in different stations from January-2010 to December-2010.

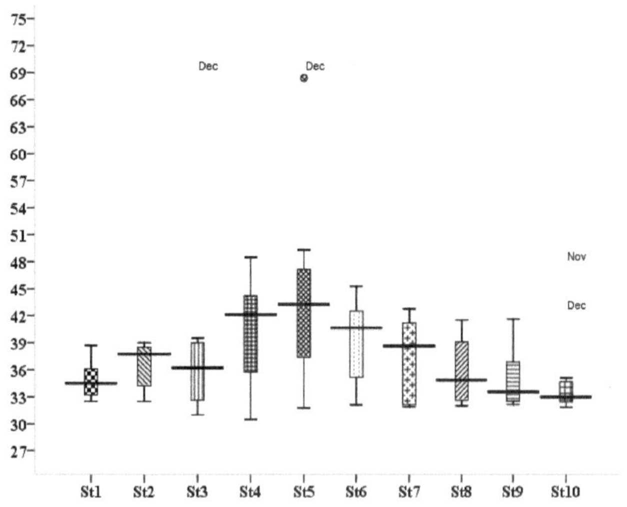

Figure 8: Total Suspended Solids (mgl/1^{-1}) in different stations from January-2010 to December-2010.

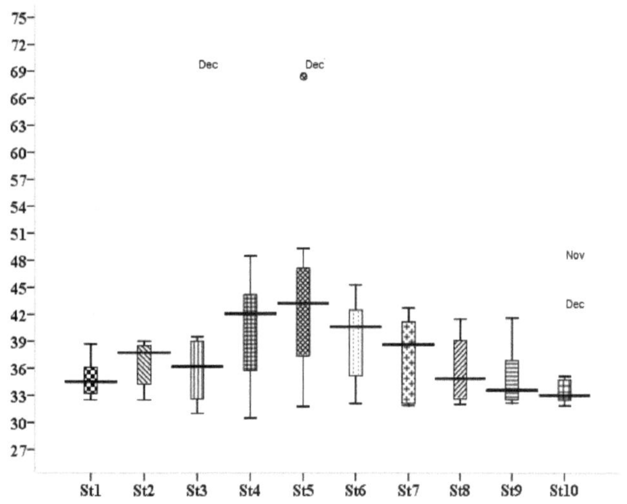

Figure 9: Biological Oxygen Demand (mgl/1^{-1}) in different stations from January-2010 to December-2010.

Inorganic phosphate (µM/1^{-1})

The phosphate concentration was recorded minimum 0.259 µM/1^{-1} during the post monsoon season in January (stations 4, 7 and 10), and maximum 4.96 µM/1^{-1} during monsoon season in November (stations 2, 3 and 6) (Figure 15). Outliers show to the extreme levels of the values.

Total Phosphate (µM/1^{-1})

The total phosphate level was ranged from 0.312 to 2.69 µM/1^{-1}. Minimum was recorded in May (stations 2, 3, 5 and 9), and maximum in November (stations 2, 3, 5 and 9) (Figure 16). Outliers show to the extreme levels of the values.

Reactive silicate (µM/1^{-1})

Reactive silicate concentration was varied from 1.173 to 3. 021 µM/1^{-1}. Minimum was recorded during post monsoon in March (stations 1, 3 and 6), and maximum during monsoon season in December (stations 1, 3 and 6) (Figure 17). Outliers show to the extreme levels of the values.

Discussion

Climate change is an important factor in determining the past and future distributions of crab biodiversity [12-15]. Crustaceans life patterns are strongly related to environmental factors [16,17]. The eutrophication is also spoiling the biodiversity [18]. Physical and chemical factors are highly affected by crustacean's inhabitants. This study is agreement with earlier studies [19,20]. In general, the organic waste dump caused environmental stress in coastal waters, which resulted to low landing of important crustacean fisheries, and affect the diversity of decapod crabs [21-23]. The physico-chemical parameters

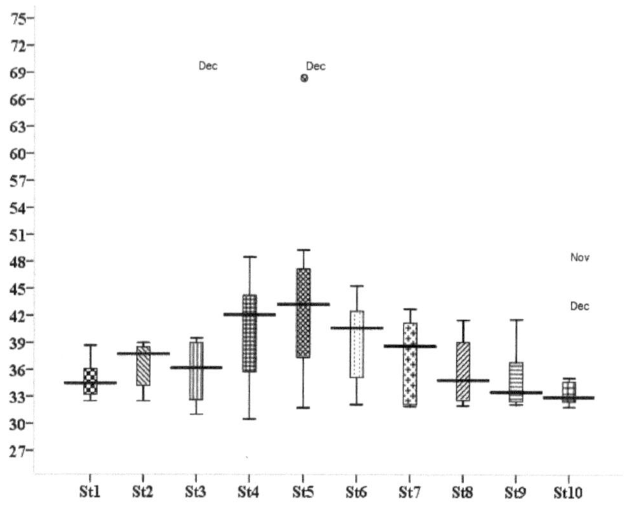

Figure 10: Chemical Oxygen Demand (mgl/1^{-1}) in different stations from January-2010 to December-2010.

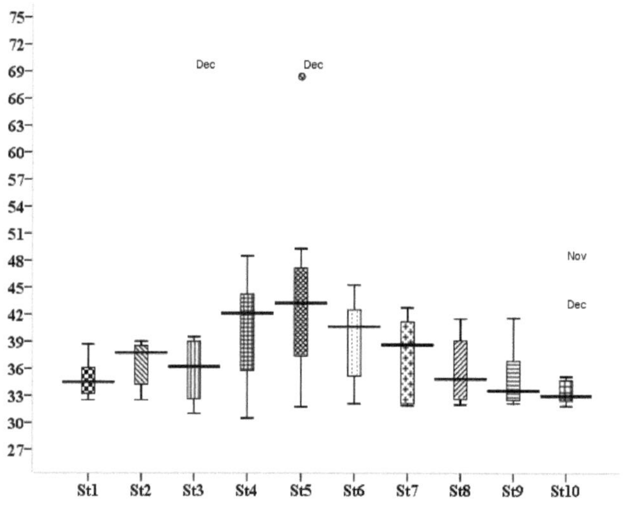

Figure 11: Nitrate (µM/1^{-1}) in different stations from January-2010 to December-2010.

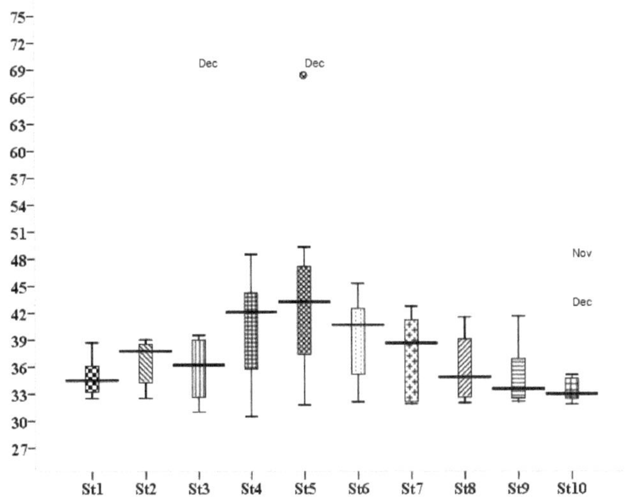

Figure 12: Nitrite (μM/1^{-1}) in different stations from January-2010 to December-2010.

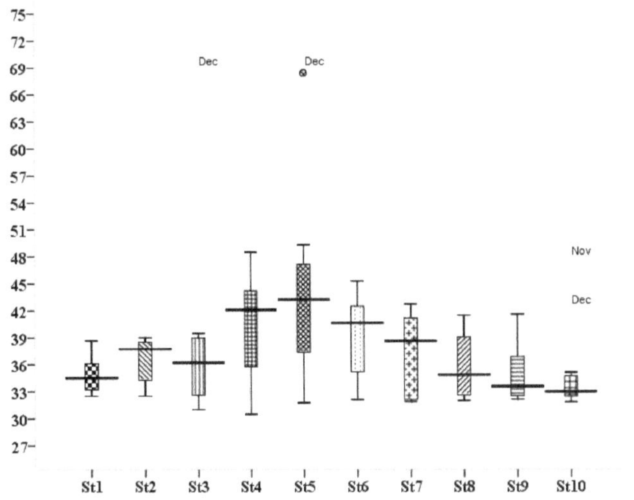

Figure 13: Ammonia (μM/1^{-1}) in different stations from January-2010 to December-2010.

are vital ecological factors, as it directly affects oxygen consumption, metabolism, growth, moulting, hormones and survival of crustaceans [24,25].

Rainfall is one of the important factors, which affect the distribution of aquatic animals. The substances present in the air are affected, and also toxic gases and chemicals dissolved so far, entrapped in rain activities. The mangroves, seagrass, salt marshes and coral reef also contaminated due to raining activities [26,27]. Especially where the large freshwater rivers enter the sea, turbid waters are thick with sediment during the annual monsoon season. Huge volumes of water carry huge volumes of phosphate loaded mud, especially off agricultural land, by increasing the concentrations of such substance as nutrients. This could be a major problem for the seagrass and coral reef associated decapods. It makes a great difference of temperature, salinity and turbidity as important factors associated with crustacean's abundance [28].

In the present study, summer peaks and monsoonal troughs in air and surface water temperatures were observed by several workers from southeast coast of India [29,30]. Tides controlled changes in temperature are distinct during monsoon months, but during other months, heating of the surface water by solar radiation greatly over shadowed the small changes in temperature, such as those induced by tides. The temperature spread of the spring may be the result of the influx of heat from the sun and atmosphere. It is a limiting factor in the aquatic environment. It will affect the metabolic activities, growth, oxygen consumption, reproduction, moulting, survival, distribution and migratory behaviours of crustaceans [31]. The temperature increased blood cell numbers [32], affects blood clotting times [33] and also the phagocytosis [34], affects the antibacterial activity by haemocytes [35], reduction of metabolic rate [36], and change in osmotic pressure of the haemolymph [37]. It has

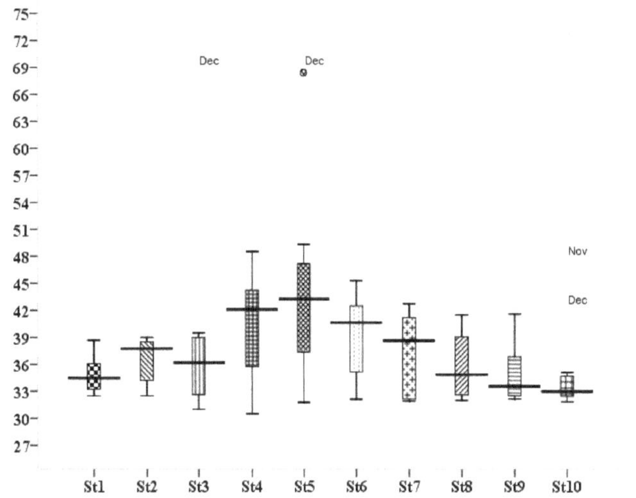

Figure 14: Total Nitrogen (μM/1^{-1}) in different stations from January-2010 to December-2010.

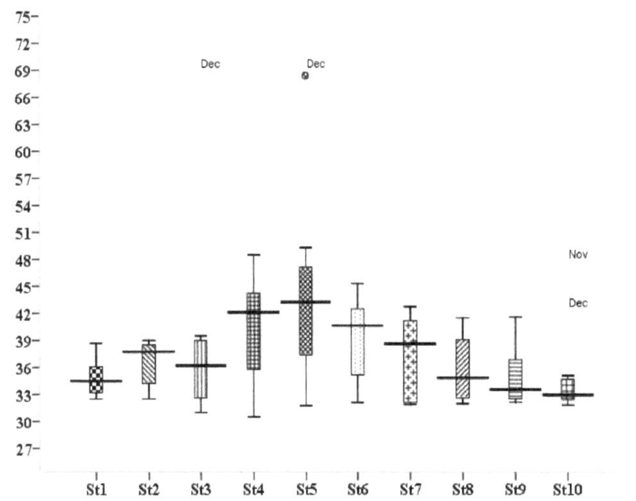

Figure 15: Inorganic Phosphate (μM/1^{-1}) in different stations from January-2010 to December-2010.

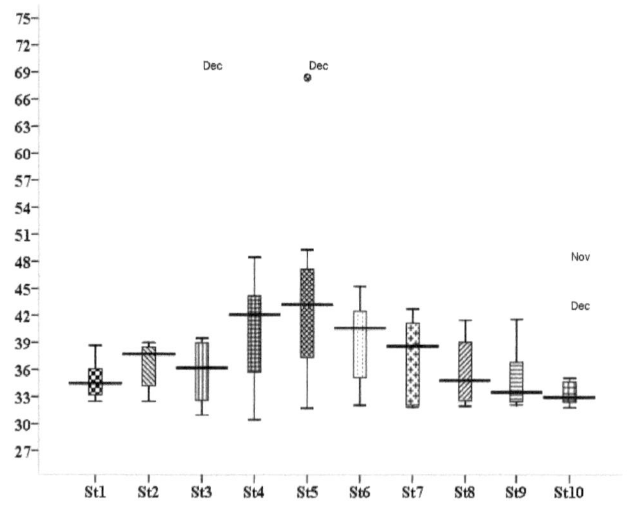

Figure 16: Total Phosphate (µM/1⁻¹) in different stations from January-2010 to December-2010.

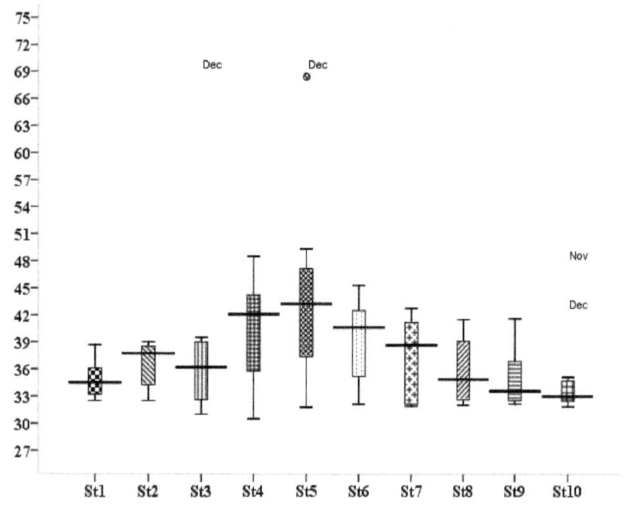

Figure 17: Reactive Silicate (µM/1⁻¹) in different stations from January-2010 to December-2010.

been direct effect on other environmental parameters, such as salinity and oxygenation of the water. The effects of rain on the physiology of aquatic organisms are discussed [38,39] in *Penaeus notalis* [40], *P. indicus* [41], *Macrobrachium olfersii* [42], and in gammarid, *Echinogammarus berrilloni* [43].

The salinity is an ecological master factor in the distribution of living organisms, likely to influence decapod crabs distribution and production of the coastal ecosystems [44]. The salinity was found to be high during summer season and low during the monsoon season at the stations [45]. The spectral distribution clearly shows that the premonsoon and summer as the season of abundance and minimum catch, and also no catch during monsoon and spring. As a direct relationship between the trawl catches and salinity, temperature as well as plankters, and an inverse relationship between the rainfall and trawl catches. This is agreement with earlier studies by Sudarsan [46], John Samuel et al. [47] and Varadharajan et al. [48]. The monsoon is the period of spawning of decapod crabs, and the spawning activity is related to the intensity of rainfall. The lower temperature and salinity conditions prevailing throughout the period may be the causative factors of spawning [49-51]. Even in the same position, the size at maturity of individual species of the decapod crabs may vary because the moult occurs over a wide range of size. The fast moulting may possible be fast for a considerable amount of time, when the breeding activity is high [52,53].

The pH is another important parameter affecting crab diversity and distribution in an ecosystem. The uptake of CO_2 by the photosynthesizing organisms, especially phytoplankton from the sea, could have increased the pH levels. It can be due to the influence of seawater penetration and high biological activity [54], and photosynthetic activity [55]. It was also quite low during floods in the peak monsoon season, reduction of salinity and temperature, and decomposition of organic matter [56]. The alkaline pH was found to be associated with more number of crab species. However, with increasing pH, the number of species has been reported to decrease [54]. In decapods, pH influences the metabolism, physiology and maturation process [57].

One of the most important abiotic factors influencing young stages, such as zoea and megalopa in aquatic ecosystem, is the dissolved oxygen [58]. This parameter is more critical because it shows whether there is sufficient oxygen in the water for marine life to survive. Pollution is manifested in the low dissolved oxygen levels and in high nutrient levels in these waters, which can lead to an imbalance of crustacean's communities through the food web. DO concentrations clearly influence the behaviour of decapods [59,60], and life strategies on the basis of oxygen consumption and energy content [61]. Many previous studies reported that oxygen deficiency and stress condition of crustaceans [2,62]. In summer season, dissolved oxygen decreased due to increased temperature and salinity of water [63], this is attributed to the lesser input of freshwater into the study areas [64]. The current and its associated transport of oxygen and food materials strongly influence the growth of decapod species. Oxygen consumption has been reported in several authors in many species of crustaceans [65,66]. The respiration rates influence the metabolism in crustaceans [67-69], and anaerobic metabolism [67,70], will reduce growth and moulting frequency [71], and cause mortality [72], reduction of metabolic rate [36], and change in osmotic pressure of the haemolymph [37].

Light extinction co-efficient at sampling area was high during the monsoon season due to the monsoonal rains, coupled with low intensity of solar radiation and higher concentration of dissolved organic matter and suspended sediments, which resulted in increased turbulence in the water. The post monsoon season at the sampling area due to the removal of suspended materials from the water column and cessation of freshwater flow, reduces the abiogenic turbidity [73]. The marine invertebrates, light from a significant external signal [74] to control various reproductive activities [75]. The initiation and synchronization of spawning, gonad maturation and sex determination are light-dependent reproductive activities and appear to be mediated through photo-neuroendocrine pathways [76]. The extension of photoperiod lengthens the period of reproduction and delays the moult [77]. In decapod crustaceans such as *Pachygrapsus marmoratus* [78] and *Palaemonetes pugio* [79], the administration of longer photoperiod accelerates reproduction. The above mentioned examples clearly suggest the role of day length on the initiation of reproductive processes in lower, as well as higher crustaceans.

Surface water temperatures increased and decreased generally day by day in the study periods. The temperatures in the low tide were generally higher than those in high tide. The variation of the turbidity in the sea may be fairly complicated in the near shore region. Since, the high turbidity reduces the growth of phytoplankton, seaweeds and zooplankton; so naturally, numerical density and biomass of decapods species also reduced [80]. The meteorological conditions, especially the strong wind, affect changes of the turbidity. High values were associated with heavy freshwater flow and turbidity during the northeast season. Besides, this visibility as well was poor due to the cloudy weather, towards the end of December. Soil conditions, organic matter, plankters and other minute organisms cause turbidity in water, recognized as a valuable limiting factor in the crustacean's productivity of water bodies [81]. So, the provisional changes are mostly observed and sensitive sites almost immediately, leaving the polluted area [3]. The relation between the fishing conditions and oceanographical conditions, especially the turbidity was little understood, because of the unfavorable meteorological conditions.

The total suspended solids degrade optical water quality by reducing water clarity and decreasing light available to support photosynthesis [81]. TSS levels and fluctuations influence aquatic life, from phytoplankton to crustaceans. It is especially when the individual particles are small; carry many substances that are harmful or toxic. As a result, suspended particles are often the primary carrier of these pollutants to coastal zones of oceans where they settle. In coastal zones, these fine particles are food source for filter feeders, which are part of the food chain, leading to biomagnification of chemical pollutants in crustaceans and ultimately, in man. In intertidal area, however, deposition of fine particles effectively removes pollutants from the overlying water by burying them in the bottom sediments of the coastal zones. In coastal areas where erosion is a serious problem, suspended solids can blanket the coastal bed, thereby, destroying crustacean's habitat [82].

The low BOD values recorded during the monsoon season in the sampling areas indicated effective assimilation of organic load. Biochemical oxygen demand is the amount of oxygen utilized by microorganisms to stabilize the organic matter [64]. This process can severely reduce the available oxygen, so that crustaceans and other aquatic life are suffocated. In crustaceans, the oxygen uptake of *Pachygrapus crassipes* and *Gecarcinus lateralis* was found to rise slowly from the intermoult, until few days after moult. The oxygen consumption of *C. sapidus* from proecdysis to postecdysis was studied by Elizabeth et al. [83]. Taylor [84] reported the respiratory responses of *C. meanus* to changes in environmental salinity. Effect of temperature, salinity on oxygen consumption in a freshwater population of *Paleamonetes antennarius* was revealed by Giuseppe [85]. Davenport and Wong [86] have studied the behaviroural responses of *S. serrata* to salinity and low oxygen tension.

The COD is a measure of the oxygen equivalent of the organic matter content of water that is susceptible to oxidation by a strong chemical oxidant. Maximum COD values were observed during summer seasons because of decrease in land wastes inflow, so far increases in temperature, salinity and plankters productivity and oxygen utilization of microbes involved in decomposition activities, and also due to the cumulative effect of higher wind velocity [54,87]. Minimum values of COD was observed during the monsoon season because of the presence of heavy river run off, increased mixing of land wastes into the coast and decreased biological activity because of decreased temperature and salinity. Thus, COD is a reliable parameter for judging the extent of pollution in water [82]. The COD of water increases with increasing concentration of organic matter [54].

High concentration of nitrate observed during the monsoon season might be due to the heavy rainfall and wastes [88]. The minimum concentration of nitrate observed during the summer season could be due to its utilization by phytoplankton, as evidenced by their high population density and also decreased freshwater flow during this period [30]. The marine systems are generally nitrogen limited, excessive nitrogen inputs can result in water quality degradation due to toxic algal blooms, oxygen deficiency, habitat loss, decreases in biodiversity and fishery losses.

The higher nitrite values recorded during monsoon season in the month of November could be due to the increased planktonic excretion, oxidation of ammonia and reduction of nitrate, by recycling of nitrogen, and also due to bacterial decomposition of plankton detritus present in the environment [89]. Further, the denitrification and air sea interaction exchange of chemicals are also responsible for this increased value [90]. The low nitrite value recorded during summer season may be due to less freshwater inflow and high salinity [91]. Maximum nitrite values recorded in monsoon season could be mainly due to the organic materials received from the catchments areas during the receding or outgoing tide [54,91]. Further, excretion of phytoplankton, reduction of nitrate and oxidation of ammonia could all contribute together or individually to increase the concentration of nitrite in the environment [92]. The nitrites are probably the most important environmental variable, because it directly affects the behavior, influencing growth, reproduction, metabolic activities, and abnormal conditions of crustaceans inhabitants.

Algae blooms cloud the water, blocking the light from reaching important underwater grasses that provide food and habitat for zoea to adult crustaceans. Changing the type of nitrogen entering the water from nitrate to ammonia may also cause a fundamental change in the algae community, the building block for the aquatic food web. Species that best accumulate ammonia may be different from those that have thrived on nitrate [30]. Algae fueled by too many nutrients are the largest water quality problem in the environment. The blooms provide food and shade, prevent the growth of undesirable benthic algae, reduce toxic ammonia concentrations and increase oxygen levels. Ammonia is toxic to aquatic life, which affects crustaceans in various ways, such as food into the energy, nutrients and proteins they use for survival and growth, and also cause impairment in numerous organs [93]. Lethal and sub-lethal effects of ammonia on crustaceans, with respect to metabolism and osmoregulation were studied by Chen and Lei [94]. Higher concentration of ammonia was observed during the monsoon season in all sampling stations, but reported in higher values in few sites (2, 3 and 5), particularly in the month of November. Lower concentration of ammonia was observed during the postmonsoon season in all sampling stations, at the same time reported in lower values in few sites (2, 3 and 5), mainly in the month of March. The higher concentration could be partially due to the death and subsequent decomposition of phytoplankton, and also due to the excretion of ammonia by plankton as opined by Segar and Hariharan [95].

The main cause of eutrophication involves the enrichment of water by excess nutrients. It can cause serious problems in the coastal zone through disturbance of ecological balances and fisheries, ultimately interfering with recreational activities, and also quality of marine life [96]. Due to combination of processes affecting the nitrogen cycle in water, the reduction of nitrogen from agricultural lands to coastal ecosystem is observed. It has important indirect effects, such as sea grasses provide food and shelter for crustaceans cause many diseases. In the present study, total nitrogen in water was high during the monsoon season, especially in the month of November. Lower concentration was observed in the month of August. The higher concentration of nitrogen is due to release from the decay of a large number of phytoplankton [97].

High concentration of phosphate was recorded in the present study during monsoon season, might have resulted from the regeneration of phosphate from the bottom muds, and subsequent release of the same in water column by turbulence and mixing caused by heavy winds prevailing during rainy season. Low concentration of phosphate was observed during the post monsoonal season due to the decreased land drainage, sewage and fertilizer disposal from the agricultural lands. As the water masses are without stratification throughout the year, and are in constant contact with the bottom, the regeneration of phosphate taking place at the bottom is constantly utilised by the phytoplankton. The seasonal variations in the phosphate content of the coastal waters have been conducted by Marichamy and Siraimeetan [98] at Madras, Jayaraman [99], in Gulf of Mannar and Palk Bay, George [100] and Subrahmanyan [101] at Calicut by Reddy et al. [102] for the shelf waters of the Arabian Sea, Seshappa and Jayaraman [103] have studied the phosphates in the mud banks at Calicut and Qasim [104] in the cochin backwaters.

The reactive silicate was maximum in the monsoon season, mainly in the month of December, may be due to heavy inflow of monsoonal freshwater derived from land drainage carrying reactive silicate, leached out from rocks. Further, due to the turbulent nature of water, the reactive silicate from the bottom sediment might have been exchanged with overlying water [105]. In addition to phytoplankton uptake, some other processes like absorption and co-precipitation of soluble silicon might also govern the distribution of dissolved reactive silicate in the environment [90]. The low postmonsoon values could also be attributed to uptake of reactive silicates by phytoplankton for their biological activity [106].

The pollution is due to many pathogens, as well as toxic chemicals can exert harmful effect on crustaceans, and also the health of consumers [107]. The contaminated food, sediment, suspended particles as well as water [108,109], contribute to the bioaccumulation of these compounds in crustaceans. Crustaceans have been unlocking vascular systems, in which many haemocytes generously flow in haemolymph [110,111]. Circulating haemocytes of the immune, functions against parasites and microbes [112]. Since, the moult cycle influences the status of a number of physiological processes, and since, an animal's physiology influences its behavior and its interaction with its environment also the affected haemocyte antibacterial activity of decreased immune mechanism of decapods [113,114].

Crabs are opportunistic omnivores, eating on a variety of foods, with a preference for plants and animal food, in general. Many crabs are not bottom dwellers. These are the active swimmers, living near the surface, feeding on plankton. The abundance of invertebrate that are known as major copepod grazers, such as chaetognaths, ctenophores and medusae, as suggested by Roff et al. [115] and Mills [116]. Availability plants materials are one of the important factors determining the special composition and diversity of crustaceans [117,118]. The mangrove, seagrass beds and salt marshes are acting as a nursery ground, which provide feed for zoea, megalopa, adult and all stages of the decapod crabs. Zooplankton communities in coastal area such as periphyton, along with protozoans, bacteria and detritus, can serve as an indirect food source for crustacean's young stages to adult. Utilising the high level of nutrients available in the surface waters, the chlorophyll concentration in diatoms proliferate and occur in high abundance [119,120]. This study is agreement with earlier studies [13,121].

In a coastal ecosystem, zooplankton forms an important link in the food chain, from primary to tertiary level leading to the production of crustacean's fishery. It has been well established that potentials of pelagic crustaceans, either directly or indirectly, depend on zooplankton [122]. By asset of sheer profusion and intermediatary role between phytoplankton and crustaceans, they are considered as the chief index of utilization of aquatic biotope at the secondary trophic level. The herbivorous zooplanktons are efficient grazers of the phytoplankton, and have been referred to as living machines transforming plant material into animal tissue. Hence, they play an important role as the intermediaries for nutrients and energy transfer between primary and tertiary trophic large density, shorter life span, drifting nature, high group and species diversity levels [123]. Due to their and different tolerance to the stress, they are being used as the indicator organisms for the physical, chemical and biological processes in the aquatic ecosystem [124]. Although most of the species of plankters are helpful in fisheries, a few are detrimental to the fishery potential. The copepods have been considered as being primarily major grazers on phytoplankton in most marine ecosystems [125]. It has been grazers of phytoplankton, but also as being an important copepod prey [126]. Certain zooplankton predators, such as decapod larvae, arrow worms and the copepods may voraciously feed on megalopa, decimating the adult crab's population. The dominant grazers of phytoplankton, copepods, in turn, function as prey for crustaceans also provided in energy and support developmental stages [117], and are natural feed on crustaceans [127]. This study is in agreement with earlier studies [128,129]. In the present study, physico-chemical parameters showed exceptional values in some locations, and also these could be compared to the effect of decapod crab diversity for different manner in this coastal environment. Marine crustaceans are under the influence of numerous environmental factors. The awareness needed in belongings of over exploitation and pollution is increasingly obvious and serious, as for loss of food species, microbial diseases of marine decapods, contamination of edible crustaceans. However, in marine crustaceans, there is a scarcity of data to support the assumption that environmental changes induce a modification of the pollution, leading to an enhanced susceptibility to infectious disease agents.

References

1. Uthoff D (1996) From traditional use to total destruction forms and extent of economic utilization in the Southeast Asian mangroves. Nat Res Dev 43: 58-97.
2. DelValls TA, Conradi M, Garcia-Adiego E, Forja JM, Gomez-Parra A (1998) Analysis of macrobenthic community structure in relation to different environmental sources of contamination in two littoral ecosystems from the Gulf of Cadiz (SW Spain). Hydrobiologia 385: 59-70.
3. Bat L, Akbulut M, Sezgin M, Culha M (2001) Effects of Sewage Pollution the Structure of the Community of *Ulva lactuca*, *Enteremorpha linza* and Rocky Macrofauna in Disliman of Sinop. Turk J Biol 25: 93-102.
4. Guerra-Garcia JM, Garcia-Gomez JC (2004) Crustacean assemblages and sediment pollution in an exceptional case study: A harbour with two opposing entrances. Crustaceana 77: 353-370.
5. Sindermann CJ (1979) Pollution-associated diseases and abnormalities of fish and shellfish: a review. Fish Bull 76: 717-749.
6. Lapointe BE, Barile PJ, Matzie WR (2004) Anthropogenic nutrient enrichment of seagrass and coral reef communities in the lower Florida Keys: Discrimination of local versus regional nitrogen sources. J Exp Mar Biol Ecol 308: 23-58
7. IMA (2006) Development of a National Programme of Action (NPA) for the protection of the marine environment from land-based sources and activities. Background paper by The Institute of Marine Affairs of Trinidad and Tobago.

8. Bockstael NE, Hanemann WM, Kling CL (1987) Estimating the value of water quality improvements in a recreational demand framework. Water Resour Res 23: 951-960.

9. Sargaonkar A, Deshpande V (2003) Development of an overall index of pollution for surface water based on a general classification scheme in Indian context. Environ Monit Assess 89: 43-67.

10. Poole HH, Atkins WRG (1929) Photoelectric measurements of submarine illumination throughout the year. J Mar Biolog Assoc UK 16: 297-324.

11. Strickland JDH, Parsons TR (1972) A practical hand book of seawater analysis. (2ndedn), Bulletin (Fisheries Research Board of Canada), Ottawa, Fisheries Research Board of Canada.

12. Vernberg FJ, Vernberg WB (1970) Lethal limits and the zoogeography of the faunal assemblages of coastal Carolina waters. Mar Biol 6: 26-32.

13. Peterson BJ, Howarth RW (1987) Sulfur, carbon, and nitrogen isotopes used to trace organic matter flow in the salt-marsh estuaries of Sapelo Island, Georgia. Limnol Oceanogr 32: 1195-1213.

14. Root TL, Price JT, Hall KR, Schneider SH, Rosenzweig C, et al. (2003) Fingerprints of global warming on wild animals and plants. Nature 421: 57-60.

15. Rosenzweig C, Karoly D, Vicarelli M, Neofotis P, Wu Q, et al. (2008) Attributing Physical and Biological Impacts to Anthropogenic Climate Change. Nature 453: 353-357.

16. Macpherson E (2002) Large-scale species-richness gradients in the Atlantic Ocean. Proc Biol Sci 269: 1715-1720.

17. Hiddink JG, Ter Hofstede R (2008) Climate induced increases in species richness of marine fishes. Glob Chang Biol 14: 453-460.

18. Elmanama AA, Afifi S, Bahr S (2006) Seasonal and spatial variation in the monitoring parameters of Gaza Beach during 2002-2003. Environ Res 101: 25-33.

19. Collins VG (1983) The distribution and Ecology of Bacteria in freshwater. Proceedings Soil Water Treatment and Examination 12: 40-73.

20. Mukhtar MD, Deeni YY (1998) Microbiological and Physicochemical. Studies on Salata river, South-Central Kano Metropolis, 9th/10th Annual conferences, Proceedings: Nigerian Ass Aquat Sci 241-250.

21. Wyatt T, Yolarda P (1992) Harmful algal bloom: UNESCO, An IOC Newsletter on topic Algae and Algal bloom 62: 1-8.

22. Appasamy P, Lundqvist J (1993) Water supply and wastes disposal strategies for Madras. Ambio 22: 442-448.

23. Lemly AD (1996) Wastewater discharges may be most hazardous to fish during winter. Environ Pollut 93: 169-174.

24. Chen S, Wu J, Huner JV, Malone RF (1995) Effects of temperature upon ablation-to-molt interval and mortality of red swamp crawfish (*Procambarus clarkii*) subjected to bilateral eyestalk ablation. Aquaculture 138: 191-204.

25. Medesani DA, López Greco LS, Rodríguez EM (2001) Effects of cadmium and copper on hormonal regulation of glycemia by the eyestalks in the crab *Chasmagnathus granulata*. Bull Environ Contam Toxicol 66: 71-76.

26. Lipp EK, Farrah SA, Rose JB (2001) Assessment and impact of microbial fecal pollution and human enteric pathogens in a Coastal Community. Mar Pollut Bull 42: 286-293.

27. Boehm AB, Grant SB, Kim JH, Mowbray SL, McGee CD, et al. (2002) Decadal and shorter period variability of surf zone water quality at Huntington Beach, California. Environ Sci Technol 36: 3885-3892.

28. Ryther JH, Hall JR, Pease AK, Bakun A, Jones MM (1966) Primary organic production in relation to the chemistry and hydrography of the western Indian Ocean. Limnol Oceanogr 11: 371-380.

29. Prabha Devi L (1986) Environmental inventory of tidal and gradient zones of Coleroon estuary, southeast coast of India, Ph.D. Thesis, Annamalai University 241.

30. Vasantha K (1989) Studies on hydrobiology and decomposition of macrophytes in Portonovo marine environment Southeast coast of India. Ph.D. Thesis, Annamalai University, India 252.

31. Le Moullac G, Haffner P (2000) Environmental factors affecting immune responses in Crustacea. Aquaculture 191: 121-131.

32. Truscott R, White KN (1990) The influence of metal and temperature stress on the immune system of crabs. Funct Ecol 4: 455-461.

33. Dean JM, Vernberg FJ (1966) Hypothermia and blood of crabs. Comp Biochem Physiol 17: 19-22.

34. Paterson WD, Stewart JE (1974) In vitro phagocytosis by hemocytes of the American lobster *(Homarus americanus)*. Journal of the Fisheries Research Board of Canada 31: 1051-1056.

35. Chisholm JRS, Smith VJ (1994) Variation of antibacterial activity in the haemocytes of the shore crab, Carcinus maenas, with temperature. J Mar Biolog Assoc U.K. 74: 979-982.

36. Hill AD, Taylor AC, Strang RHC (1991) Physiological and metabolic responses of the shore crab *Carcinus maenas* (L.) during environmental anoxia and subsequent recovery. J Exp Mar Biol Ecol 150: 31-50.

37. Charmantier G, Soyez C, Aquacop (1994) Effect of molt stage and hypoxia on osmoregulatory capacity in the peneid shrimp *Penaeus vannamei*. J Exp Mar Biol Ecol 178: 233-246.

38. Prosser CL, Brown FA (1961) Comp Ani Physiol. (2nd Ed.), W.B. Saunders Co., Philadelphia, USA.

39. Kinne O (1970) Environmental factors. Part I (Ed), Wiley Interscience, London.

40. Ramos Trujillo L, Oliva Suarez M (1984) Energetic metabolism of juveniles of pink shrimp *Penaeus notialis*. Rev Invest Mar 5: 35-36.

41. Kutty MN, Murugapoopathy G, Krishnan TS (1971) Influence of salinity and temperature on the oxygen consumption in young juveniles of Indian prawn *Penaeus indicus*. Mar Biol 11: 125-131.

42. McNamara JC, Moreira GS, Moreira PS (1983) The effect of salinity on respiratory metabolism, survival and moulting in the first zoea of *Macrobrachium amazonicum* (Heller) (Crustacea, Palaemonidae). Hydrobiologia 101: 239-242.

43. Ciavatti G, Vincent M (1982) Comparative study of the respiratory metabolism of two populations of Gammaridae: *Gammarus pulex* and *Echinogammarus herilloni*. Ann Stu Biol Bessee-en-chandesse 16: 158-170.

44. Chandra Mohan P, Sreenivas N (1998) Diel variations in zooplankton populations in mangrove ecosystem at Gaderu canal, Southeast coast of India. Indian J Mar Sci 27: 486-488.

45. Gowda G, Gupta TRC, Rajesh KM, Gowda H, Lingadhal C, et al. (2001) Distribution of phytoplankton pigment in the Nethravathi estuary. Indian J Fish 49: 267-273.

46. Sudarsan D (1985) Results of exploratory survey around the Andaman Islands. Bulletin of the Exploratory Fisheries Project 7: 1-43.

47. John Samuel N, Thirunavukkarasu N, Soundarapandian P, Shanmugam A, Kannupandi T (2004) Fishery potential of commercially important portunid crabs along Parangipettai coast. In: Proceedings of Ocean Life Food & Medicine Expo 165-173.

48. Varadharajan D, Soundarapandian P, Dinakaran GK, Vijakumar G (2009) Crab Fishery Resources from Arukkattuthurai to Aiyammpattinam, South East Coast of India. Curr Res J Biol Sci 1: 118-122.

49. Menon MK (1953) A note on the bionomics and fishery of the swimming crab *Neptunus sanguinolenlus* (Herbst) on the Malabar Coast. Journal of the Zoological Society of India 4: 177-184.

50. Prasad RR, Thampi PRS (1953) Contribution to the biology of the blue swimming crab, *Neptunus pelagicus* (Linnaeus), with a note on the zoea of *Thalamita crenata Latreille*. Journal of the Bombay Natural History Society 51: 674-689.

51. Chhapgar BF (1956) On the breeding habits and larval stages of some crabs. Rec Indian Mus 54: 33-52.

52. Okafor FC (1988) Rural development and the Environment: Degradation Versus protection. In: *Environmental Issues and Management in Nigeria Development,* P. O. Sada and F.O. Oemerho (Eds.), Evans Brothers (Nigeria Publishers) Limited, Ibadan 150-163.

53. Lawal-Are AO, Kusemiju K (2000) Size composition, growth pattern and feeding habits of the blue crab, *Callinectes amnicola* (De Rocheburne) in the Badagry Lagoon, Nigeria. J Sci Res Dev 5: 169-176.

54. Das J, Das SN, Sahoo RK (1997) Semidiurnal variation of some physico-chemical parameters in the Mahanadi estuary, East coast of India. Ind J Mar Sci 26: 323-326.

55. Subramanian B, Mahadevan A (1999) Seasonal and diurnal variation of hydrobiological characters of coastal water of Chennai (Madras), Bay of Bengal. Indian J Mar Sci 28: 345-415.

56. Zhingde MD, Nair RV, Govindan K, Sabnis MM (1988) Marine environmental impact assessment of proposed tidal power development in the Gulf of Kutch. Proc Symp on tidal power plant development in the Gulf of Kutch, Central Electricity Authority, New Delhi 205-208.

57. Muthu MS, Laxminarayana A (1977) Induced maturation and spawning of Indian penaeid prawns. Indian Journal of Fisheries 24.

58. Das AK (2000) Limno-Chemistry of some andhra pradesh reservoirs. Journal of the Inland Fisheries Society of India 32.

59. Riedel B, Zuschin M, Haselmair A, Stachowitsch M (2008) Oxygen depletion under glass: Behavioural responses of benthic macrofauna to induced anoxia in the Northern Adriatic. J Exp Mar Biol Ecol 367: 17-27.

60. Haselmair A, Stachowitsch M, Zuschin M, Riedel B (2010) Behaviour and mortality of benthic crustaceans in response to experimentally induced hypoxia and anoxia *in situ*. Mar Ecol Prog Ser 414: 195-208.

61. Company JB, Sarda F (1998) Metabolic rates and energy content of deep-sea benthic decapod crustaceans in the western Mediterranean Sea. Deep Sea Res Part I Oceanogr Res Pap 45: 1861-1880.

62. Pearson TH, Rosenberg R (1978) Macro benthic succession in relation to organic enrichment and pollution of the marine environment. Oceanogr Mar Biol Ann Rev 16: 229-311.

63. Naz M, Turkmen M (2004) Phytoplankton Biomass and Species Composition of Lake Golbasi (Hatay-Turkey). Turk J Biol 29: 49-56.

64. Bijoy Nandan S, Abdul Azis PK (1990) Studies on BOD5 and dissolved oxygen in the Kadinamkulam Kayal, Southern Kerala. Mahasagar 23: 95-101.

65. Sumpton WD, Smith GS (1990) Effect of temperature on the emergence, activity and feeding of male and female sand crabs *(Portunus pelagicus)*. Australian Journal of Marine and Freshwater Research 41: 545-550.

66. Kunzmann A, Schmid M, Yuwono E (2007) Routine respiration and activity of the Indonesian mangrove crab *Scylla serrata* (Forskal, 1775), (Decapoda, Brachyura). Crustaceana 80: 77-95.

67. Bridges CR, Brand AR (1980) The effect of hypoxia on oxygen consumption and blood lactate levels of some marine crustacea. Comp Biochem Physiol A Physiol 65: 399-409.

68. Childress JJ, Cowles DL, Favuzzi JA, Mickel TJ (1990) Metabolic rates of benthic deep-sea decapod crustaceans decline with increasing depth primarily due to the decline in temperature. Deep Sea Res A 37: 929-949.

69. Henry RP (1994) Morphological, behavioral and physiological characterization of bimodal breathing crustaceans. Amer Zool 34: 205-215.

70. Johnson IT (1985) Studies on some responses of *Carcinus maenas* (L.) and other brachyurans to hypoxia and aerial exposure.

71. Allan GL, Maguire GB (1991) Lethal levels of low dissolved oxygen and effects of short-term oxygen stress on subsequent growth of juvenile *Penaeus monodon*. Aquaculture 94: 27-37.

72. Madenjian CP, Rogers GL, Fast AW (1987) Predicting night time dissolved oxygen loss in prawn ponds of Hawaii: Part I. Evaluation of traditional methods. Aquac Eng 6: 191-208.

73. Jegadeesan P (1986) Studies on environmental inventory of the marine zone of Coleroon estuary and inshore waters of Pazhaiyaru, southeast coast of India, Ph. D Thesis Annamalai University, India 277.

74. Pittendrigh CS (1981) Circadian organization and the photoperiodic phenomena. In: B K Follett (Ed.) Biological Clocks in Reproductive Cycles, Bristol: John Wright 1-35.

75. Segal E (1970) Light invertebrates. In: Kinne O (ed.), Marine ecology, Environmental factors, Part I Wiley London I: 59-211.

76. Scharrer E (1964) Photo-neuroendocrine systems general concepts. Ann N Y Acad Sci 117: 13-22.

77. Mocquard JP, Besse G, Juchault P, Legrand JJ, Martin G, et al. (1976) Dur6e de la pgriode de reproduction chez les femelles de l'oniscoide *Porcellio dilatatus* Brandt, suivant les conditions delevage: Temperature, photoperibde et groupement. Vie Milieu 26: 51-76.

78. Pradeille-Rouquette M (1976) Étude de la fonction de reproduction chez les femelles du crabe Pachygrapsus marmora (F.) et de différents facteurs qui lui sont liés (1) = Study of the reproduction of female Pachygrapsus marmoratus (F.) and of different linked factors (1). Cah Biol Mar 17: 387.

79. Little G (1968) Induced winter breeding and larval development in the shrimp, *Palaemonetes Pugio Holthuis* (Caridea, Palaemonidae). Crustaceana 2: 19-26.

80. Miner JG, Stein RA (1996) Detection of predators and habitat choice by small bluegills: effects of turbidity and alternative prey. Trans Am Fish Soc 125: 97-103.

81. Kishore K, Joshi BD, Deepali K (2005) Physicochemical characteristics of pond water at Kanpur village in Bareilly district (UP). Him J Environ Zool 19: 89-92.

82. Karlsson PE, Uddling J, Braun S, Broadmeadow M, Elvirae S, et al. (2004) New critical levels for ozone effects on young trees based on AOT40 and simulated cumulative leaf uptake of ozone. Atmos Environ 38: 2283-2294.

83. Elizabeth GL, Paul AH (1976) Oxygen consumption of the blue crab, *Callinectes sapidus* rathbun, from proecdysis to postecdysis. Comp Biochem Physiol A Physiol 54: 55–60.

84. Taylor AC (1977) The respiratory responses of *Carcinus maenas* (L.) to changes in environmental salinity. J Exp Mar Biol Ecol 29: 197-210.

85. Giuseppe JDV (1987) Salinity responses in brackish water populations of the freshwater shrimp *Palaemonetes antennarius*—I. Oxygen consumption. Comp Biochem Physiol A Physiol 87: 471-478.

86. Davenport J, Wong TM (1987) Responses of adult mud crabs (*Scylla serrata)* (Forskal) to salinity and low oxygen tension. Comp Biochem Physiol A Physiol 86: 43-47.

87. Mitra A, Patra KC, Panigrahy RC (1990) Ecology of Planktonic copepods in the Madarmani creek of West Bengal, India. Indian J Mar Sci 19: 278-281.

88. Anbazhagan P (1988) Hydrobiology and benthic ecology of Kodiakkarai coastal sanctuary (Southeast coast of India). Ph.D. Thesis, Annamalai University, India 208.

89. Jayaprakash M, Srinivasalu S, Jonathan MP, Mohan VR (2005) A baseline study of physico-chemical parameters and trace metals in waters of Ennore Creek, Chennai, India. Mar Pollut Bull 50: 583-589.

90. Choudhury SB, Panigrahy RC (1991) Seasonal distribution and behavior of nutrients in the creek and coastal waters of Gopalpur, East coast of India: Mahasagar 24: 81-88.

91. Murugan A, Ayyakkannu K (1991) Ecology of Uppanar backwater, Cuddalore: 1. Physico-chemical parameters. Mahasagar 24: 31-38.

92. Dahanayaka DDGL, Jayamanne SC, Jinadasa SUP (2007) Hydrobiological aspects of Palk Bay and Palk Strait area. J Nat Aquat Resour Res Dev Agency 38: 33-44.

93. Colt JE, Armstrong DA (1979) Nitrogen toxicity to fish, crustaceans and molluscs. Bio-Engineering Symposium, Fish Culture Section of the American Fisheries Society, Traverse City, Michigan.

94. Chen JC, Lei SC (1990) Toxicities of ammonia and nitrite to *Penaeus monodon* Juveniles. J World Aquac Soc 21: 300-306.

95. Segar K, Hariharan V (1989) Seasonal distribution of nitrate, nitrite, ammonia and plankton in effluent discharge area of Mangalore, west coast of India. Indian J Mar Sci 18: 170-173.

96. Vinayachandran PN, Mathew S (2003) Phytoplankton bloom in the Bay of Bengal during the northeast monsoon and its intensification by cyclones. Geophys Res Lett 30.

97. Edward JKP, Ayyakkannu K (1991) Studies on the ecology of plankton community of Kollidam estuary, Southeast coast of India: 1. Phytoplankton,. Mahasagar 24: 89-97.

98. Marichamy R, Siraimeetan P (1979) Hydrobiological Studies in the Madras coastal waters of Tuticorin, Gulf of Mannar. Journal of the Marine Biological Association of India 21: 67-76.

99. Jayaraman R (1954) Seasonal variations in salinity, dissolved oxygen and nutrient salts in the inshore waters of the Gulf of Mannar and Palk Bay near Mandapam (S. India). Indian Journal of Fisheries 1: 345-364.

100. George PC (1953) The marine plankton of the coastal waters of Calicut with observations on the hydrological conditions. Journal of the Zoological Society of India 5: 76-107.

101. Subrahmanyan R (1959) Studies on the phytoplankton of the West Coast of India. Proceedings of the Indian Academy of Sciences. Section B 50: 189-252.

102. Reddy C, Gangadhara V, Sankaranarayanan VN (1968) Distribution of nutrients in the shelf waters of the Arabian Sea along the west coast of India. Bull Natn Inst Sci India 38: 206-220.

103. Seshappa O, Jayaraman R (1956) Observations on the composition of bottom muds in relation to the phosphate cycle in the inshore waters of the Malabar Coast. Proceedings of the Indian Academy of Sciences 43: 288-301.

104. Qasim SZ (1969) Some problems related to the food chain in a tropical estuary. ed J H Steele (Edinburgh: Oliver & Boyd), 45-51.

105. Rajasegar M (2003) Physico-chemical characteristic of the Vellar estuary in relation to shrimp farming. J Environ Biol 24: 95-101.

106. Qasim SZ (1980) Adaptations in phytoplankton to changing conditions in tropical estuaries. Mahasagar 13: 117-124.

107. Philips S, Laanbroek HJ, Verstraete W (2002) Origin causes and effects of increased nitrite concentrations in aquatic environments. Rev. Environ. Sci. Biotechnol 1: 115-141.

108. Abrams PA, Roth J (1994) The responses of unstable food chains to enrichment. Evol Ecol 8: 150-171.

109. Berard A, Pelte T, Druart JC (1999) Seasonal variations in the sensitivity of Lake Geneva phytoplankton community structure to atrazine. Arch Hydrobiol 145: 277-295.

110. Soderhall K, Smith VJ (1983) Separation of the haemocyte populations of *Carcinus maenas* and other marine decapods, and prophenoloxidase distribution. Dev Comp Immunol 7: 229-239.

111. Destoumieux D, Bulet P, Loew D, Dorsselaer AV, Rodriguez J, et al. (1997) Penaeidins, a new family of antimicrobial peptides isolated from the shrimp *Penaeus vannamei* (Decapoda). J Biol Chem 272: 28398-28406.

112. Bauchau AG (1981) Crustaceans, Invertebrate Blood Cell vol 2. In: N.A. Ratcliffe, A.F. Rowley, Editors Academic Press, New York 386-420.

113. Chisholm JRS, Smith VJ (1992) Antibacterial activity in the haemocytes of the shore crab, *Carcinus maenas*. J Mar Biol Assoc UK 72: 529-542.

114. Haug T, Kjuul AK, Stensvåg K, Sandsdalen E, Styrvold OB (2002) Antibacterial activity in four marine crustacean decapods. Fish Shell Immunol 12: 371-385.

115. Roff JC, Turner JT, Webber MK, Hopcroft RR (1995) Bacterivory by tropical copepod nauplii: extent and possible significance. Aquat Microb Ecol 9: 165-175.

116. Mills CE (1995) Medusae, siphonophores and ctenophores as planktivorous predators in changing global ecosystems. ICES J Mar Sci 52: 575-581.

117. Epifanio CE (1995) Transport of blue crab (*Callinectes sapidus*) larvae in the waters off Mid-Atlantic States. Bull Mar Sci 57: 713-725.

118. Dittel AI, Epifanio CE, Cifuentes LA, Kirchman DL (1997) Carbon and nitrogen sources for shrimp postlarvae fed natural diets from a tropical mangrove system. Estuar Coast Shelf Sci 45: 629-637.

119. Madhupratap M, Nair SRS, Haridas P, Padmavati G (1990) Response of zooplankton to physical changes in the environment: coastal upwelling along the central west coast of India. Journal of Coastal Research 6: 413-426.

120. Sawant SS, Madhupratap M (1996) Seasonality and composition of phytoplanlton in the Arabian Sea. Curr Sci 71: 869-873.

121. Stribling JM, Cornwell JC (1997) Identification of important primary producers in a Chesapeake Bay tidal creek system using stable isotopes of carbon and sulfur. Estuaries 20: 77-85.

122. Butterfield NJ (2001) Ecology and evolution of Cambrian plankton. In the ecology of the Cambrian radiation. (eds A. Y. Zhuravlev & R. Riding), 200–216

123. Robertson AI, Alongi DM, Boto KG (1992) Food chains and carbon fluxes. Coastal and Estuarine Studies 41: 293-326.

124. Gajbhiye SN (2002) Zooplankton–Study methods, importance and significant observations. Proceedings of the National Seminar on Creeks, Estuaries and Mangroves-Pollution and Conservation 21-27.

125. Shanks AL, McCulloch A, Miller J (2003) Topographically generated fronts, very nearshore oceanography and the distribution of larval invertebrates and holoplankters. J Plankton Res 25: 1251-1277.

126. Calvert M, Adsett K, Simpson J (2000) Land Use Mapping of the Fitzroy Catchment: Fitzroy Catchment Implementation Project: Final Report. Department of Natural Resources.

127. McVey JP (1993) CRC Hand Book of Mariculture. Vol 1 Crustacean Aquaculture. (2ndedn), CRC Press.

128. Steidinger KA, Walker LM (1984) Marine Plankton Life Cycle Strategies. Boca Raton, Fla, CRC Press.

129. Trussell GC, Smith LD (2000) Induced defenses in response to an invading crab predator: an explanation for historical and geographic phenotypic change. Proc Nat Acad Sci 97: 2123-2127.

3

Biochemical Composition of Edible Crab, *Podophthalmus Vigil* (Fabricius)

Soundarapandian P[1]*, Ravichandran S[2] and Varadharajan D[1]

[1]*Faculty of Marine Sciences, Centre of Advanced Study in Marine Biology, Annamalai University, Parangipettai - 608 502, Tamil Nadu, India*
[2]*Department of Zoology, Government Arts College, Kumbakonam, Tamil Nadu, India*

Abstract

Study of proximate composition in different sexes is important to know the nutritional status of each sex. In the present study, biochemical composition (protein, carbohydrate and lipid) were studied male, female and berried crabs of *Podophthalmus vigil*. The protein content was maximum in females (23.47%) when compared to males (21.53%) and berried females (20.93%). The carbohydrate content was significantly higher in berried females (2.76%) and lesser in males (2.09 %) and females (2.06%). The lipid contents were significantly higher in females (1.09%) and berried females (1.05%) than males (0.32%). Ash content was maximum in females (0.99%) and berried female (0.98%) than males (0.31%). The moisture content was maximum in berried females (79%) followed by males (75%) and females (74%). From the results, it is confirmed that organic constituents of the females are nutritionally superior over male crabs.

Keywords: Long eyed swimming crab; Protein; Carbohydrate; Lipid; Biochemical compositions

Introduction

Seafood is easily digestible because it has very little connective tissue. For that reason, fish is recommended in many special diets. The knowledge of the chemical composition of any edible organisms is extremely important since the nutritive value is reflected in its biochemical contents. A newer species should be recommended for human consumption only after assessing the nutritive value of the species with regards to its nutritional merits. Even though large numbers of marine crabs are suitable for human consumption, our knowledge on its nutritive value is fragmentary. A lot of studies examined the proximate composition of different crab species in different part of the world viz., *C. maenas* [1], *P. pelagicus* [2], queen crab, *Chionoecetes opilio* [3], *P. pelagicus* and *P. sanguinolentus* [4], *S. serrata* [5,6], *C. longimanus* [7], deep water crabs, *M. simithi* and *Chaceon affinis* [8], *Mesopodopsis slabberi* [9], *S. tranquebarica* [10,11], *C. lucifera* [12] and *P. sanguinolentus* [13]. However, available data on the proximate composition of crab *Podopthalmus vigil* are limited. So in the present study, an attempt has been made to evaluate the nutritional status of the edible portunid crab *P. vigil.*

Materials and Methods

The male, female and berried females of *Podophthalmus vigil* were procured from Parangipettai (Lat. 11°21' N; Long. 79° 46' E) landing centers. The carapace of the crabs was opened and the edible parts of muscle tissues were removed with sharp forceps (Plates 1-3). The removed muscle tissues were homogenized with pestle and mortar. The grounded muscles were then freeze dried and powdered and eventually stored in refrigerator for further analysis. The protein [14], carbohydrate [15], lipid [16], ash and moisture contents were estimated by adopting the standard method.

Statistical Analysis

The data were subjected to One-way analysis of variance (ANOVA) and difference between means were determined by Duncan's multiple range tests ($p<0.05$) using SPSS version 17.0.

Results

Protein

The protein content was maximum in females (23.47 ± 0.36) when compared to males (21.53 ± 0.16) and berried females (20.93 ± 0.14). The protein values were significantly different between different sexes (Table 1 and Figure 1).

Carbohydrate

The carbohydrate content was significantly higher in berried females (2.76 ± 0.24) and lesser in males (2.09 ± 0.22) and females (2.06 ± 0.10). However, according to Duncan's multiple range tests, the carbohydrate values of malse and females were not significantly different (Table 1 and Figure 2).

Lipid

The lipid contents were significantly higher in females (1.09 ± 0.20) and berried females (1.05 ± 0.22) than males (0.32 ± 0.10). But the lipid

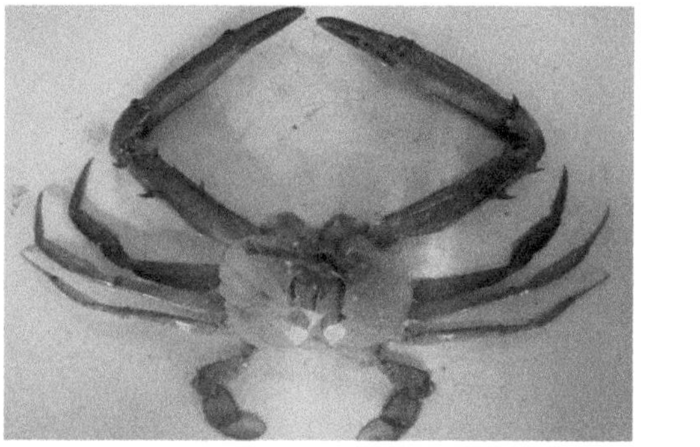

Plate 1: Edible part of the muscle tissue in *P. vigil.*

***Corresponding author:** P Soundarapandian, Faculty of Marine Sciences, Centre of Advanced Study in Marine Biology, Annamalai University, Parangipettai-608 502, Tamil Nadu, India, E-mail: soundsuma@gmail.com

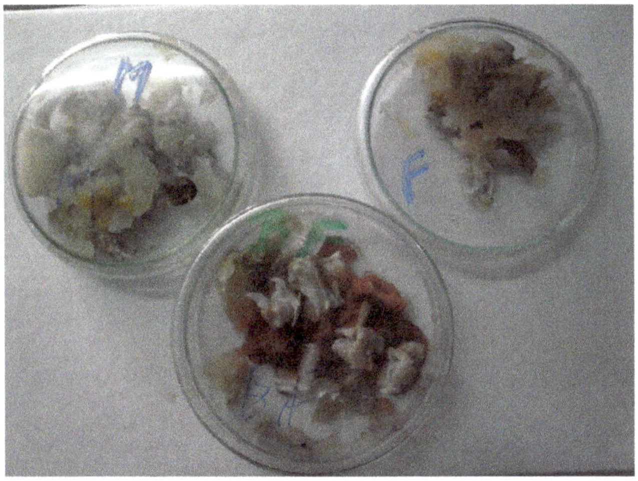

Plate 2: Fresh muscle tissue from male, female and berried females of *P. vigil.*

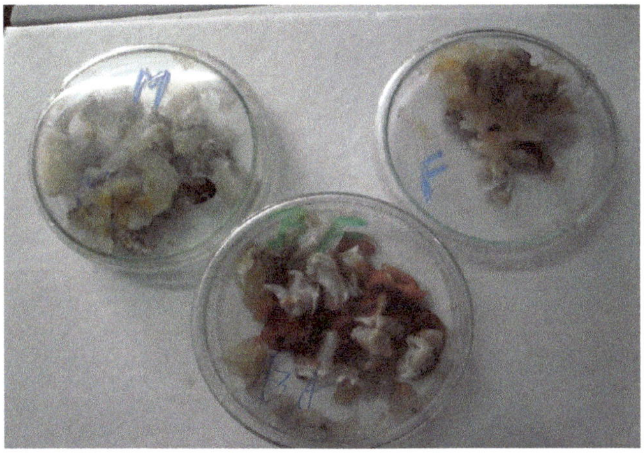

Plate 3: Dried muscle tissue from male, female and berried females of *P. vigil.*

Constituents	Male	Female	Berried
Protein	21.53^{b} ± 0.16	23.47^{a} ± 0.36	20.93^{c} ± 0.14
Carbohydrate	2.09^{b} ± 0.22	2.06^{b} ± 0.10	2.76^{a} ± 0.24
Lipid	0.32^{b} ± 0.10	1.09^{a} ± 0.20	1.05^{a} ± 0.22
Ash	0.31^{b} ± 0.12	0.99^{a} ± 0.11	0.98^{a} ± 0.28
Moisture	75.0^{b} ± 0.18	74.0^{c} ± 0.23	79.0^{a} ± 0.23

Different superscripts in a row are significantly different (P<0.05).

Table 1: Proximate composition of the muscle tissues of male, female and berried females (*P. vigil)*. Values are expressed in % on dry weight basis except moisture. (Values are mean of three values ± SE).

contents were very similar both berried females and females (Table 1 and Figure 3).

Ash

Ash content was maximum in females (0.99 ± 0.11) and berried females (0.98 ± 0.28) than males (0.31 ± 0.12). But the values were not significantly varied between females and berried females (Table 1 and Figure 4)

Moisture

The moisture content was maximum in berried females (79 ± 0.23) followed by males (75 ± 0.18) and females (74 ± 0.23). The differences were statistically significant (Table 1 and Figure 5).

Conclusion and Discussion

The protein performs a wide range of functions and provides energy of 5.65 Kcal/g. The recommended dietary allowance (RDA) for protein is 1 g/kg body weight (adults). The protein requirements vary with age, physiological status and stress. More proteins are utilized by growing infant's children, pregnant women, lactating women and individuals during infections and illness or stress. In the present study, protein

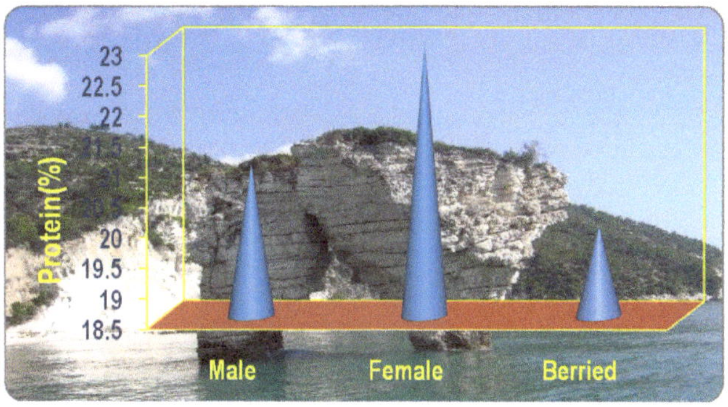

Figure 1: Protein content in the muscle tissues of male, female and berried females of *P. vigil.*

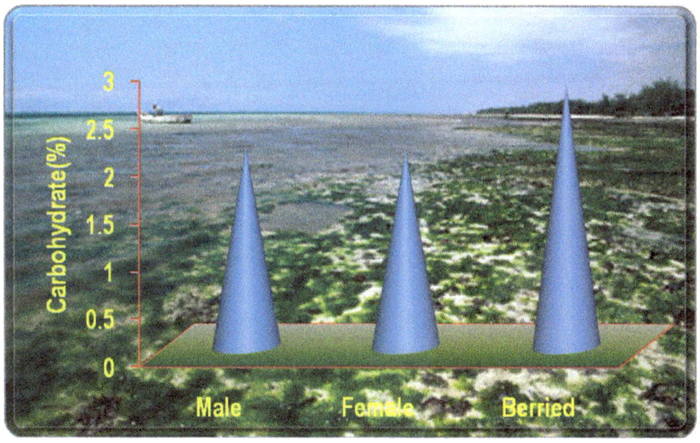

Figure 2: Carbohydrate content in the muscle tissues of male, female and berried females of *P. vigil.*

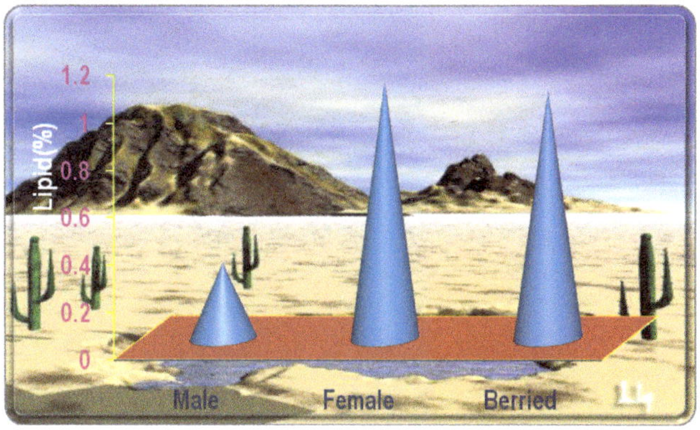

Figure 3: Lipid content in the muscle tissues of male, female and berried females of *P. vigil.*

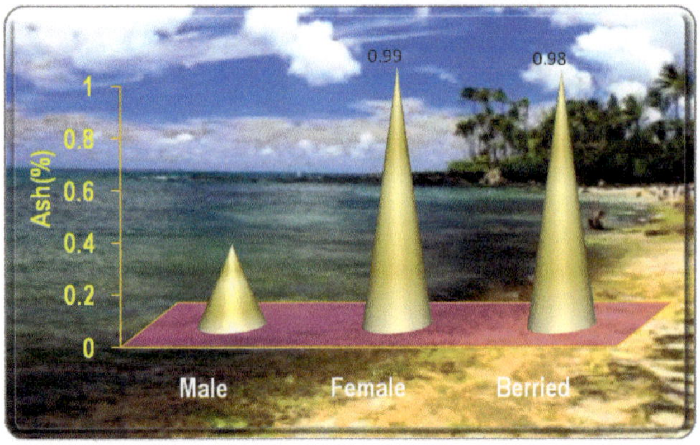

Figure 4: Ash content in the muscle tissues of male, female and berried females of *P. vigil*.

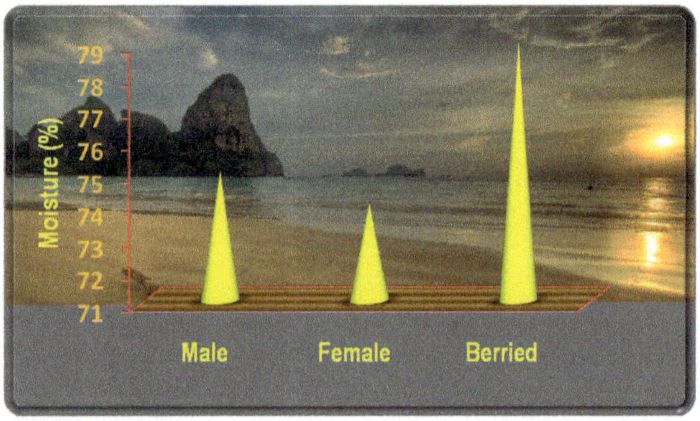

Figure 5: Moisture content in the muscle tissues of male, female and berried females of *P. vigil*.

content was higher in females (23.47%) than that of males (21.53%) and berried females (20.93%). Similar trend was already reported in freshwater prawns, *Macrobrachium idella idella* [17], *M. idae* [18], *M. scabriculam* [19] and in crab, *S. serrata* [20]. The fall in protein content, which is very well pronounced in berried females suggested that the protein in the muscle may be mobilized for the gonadal development. Similar trend was observed by Sriraman [21] in shrimp, *Penaeus merguiensis* and in freshwater prawn, *M. idae*. The females of *P. indicus* and *Metapenaeus monoceros* were increasingly proteinacious than males. Almost similar values of protein were already reported in *P. vigil* (15.75 to 20.16%) by Radhakrishnan and Natarajan [22] and in *C. affinis* was 17.8% [8]. Observed the protein content in *S. serrata* with egg (19.16%), without egg (20.92%), body meat (16.8%) and claw meat (16.28%) [23-25]. In *S. serrata*, the protein content of the body meat and claw meat was 20.11% and 18.54% respectively [5].

The protein content of various crabs was reported by different authors. The protein content of *P. pelagicus* and *P. sanguinolentus* was 0.47 to 15.91% and 12.81 to 13.6% respectively [24]. It was noticed that the protein values in cooked crab of S. *serrata* were ranged from 14.43 to 18.96% [25]. It was reported that the protein value in blue crab was 17.17% [26]. The assessed protein values in *C. smithii* were 59.8 to 71% in dry matter basis [27]. The protein values in *S. tranquebarica* were recorded from different parts *viz*., body meat (65.48 to 72.24%), claw meat (69.5 to 80.29%) and leg meat (69.47 to 74.7%) [10]. The protein content in *S. tranquebarica* varied between 67.62-72.47% [11]. The protein content in *P. sanguinolentus* was 32.6% [13] and 41.64% protein content was recorded in *C. lucifera* [12].

Carbohydrates are chemically defined as aldehyde or ketone derivatives of higher polyhydric alcohols. In the present study, carbohydrate content was higher in berried females than in males and females. Similar observations were made in prawns *M. idae* [18] and *M. scabriculam* [19]. The previous studies were suggested that the carbohydrate in the muscle tissues were varied from 0.3 to 0.63% in *P. vigil* [22], 2.4 to 3.4% in *C. smithii* [27], 0.17% in body meat, 0.24% in claw meat of *S. serrata* [25], 0.16 to 0.55% in *P. pelagicus* and 0.44 to 0.73% in *P. sanguinolentus* [24]. In *S. tranquebarica*, the carbohydrate values of body meat, claw meat and the leg meat was 0.59 to 2.23%, 0.68 to 2.87% and 0.76 to 2.76% respectively [10]. The carbohydrate content of *P. sanguinolentus* [13] and *C. lucifer* [12] was 1.42% and 2.89% respectively.

In the present study, lipid content of the females (1.09%) was maximum followed by berried females (1.05%) and males (0.32%). The previous studies were suggested that the lipids in the muscle tissues were varied from 5.13 to 9.73% in *P. vigil* by [22]. 6.2 to 7.6% in *C. smithii* [27], 0.7% in *Chaceon affinis* [8], 1.5 in blue crab [26]. 1.65% in body meat, 2.01% in claw meat of *S. serrata* [25], 0.43% in egg, 0.7% in without egg, 1.07% in body meat, 1.0% in claw meat of *S. serrata* [26], 3.3 to 5.6% in *P. pelagicus* and 3.8 to 5.5% in *P. sanguinolentus* [24], 1.6 to 0.9% in body meat, 1.83 to 2.06% in claw meat, 1.58 to 2.08% in leg meat of *S. tranquebarica* [10], 1.85% in *C. lucifera* [12], 1.81% in *P. sanguinolentus* [13], respectively.

The ash composition is maximum both in females and berried females than males. The ash content does not match any crab species that was already studied. The ash content of *S. tranguebarica* was reported to be 17-19% [11]. However in *P. sanguinolentus,* it was ranged from 7.62-7.83% [13]. The body moisture content of the present investigation was 79%, 74% and 75% in the berried, females and males, respectively. Similar trend was already observed in *M. scabriculum* [17] and *M. idae* [18]. Most of the previous studies in crabs show that moisture contents were similar to the present study. A moisture level of 69.52 to 80.51% and 67.44 to 82.04% in *P. pelagicus* and *P. sanguinolentus* [24] was reported. More or less similar moisture values of 69.54 to 74.46% were reported in *P. vigil* [22]. The moisture value was observed in *S. serrata* with egg, without egg, body meat and claw meat as 78.02%, 77.2%, 80.19% and 82.94%, respectively [25]. The moisture values in cooked crab were observed ranging from 78.4 to 83.5% [5]. The moisture content ranged from 73.5 to 81.8% in body meat, 73.5 to 80.16% in claw meat and 73.23 to 79.6% in leg meat of *S. tranquebarica* [10]. Moisture content ranged between 74.3 to 75.44% in *S. tranquebarica* [11] and it varied from 78.24-79.15% in *P. sanguinolentus* [13]. From the results, it is confirmed that organic constituents of berried and non berried females are nutritionally superior over male crabs.

References

1. Heath JR, Barnes H (1970) Some changes in biochemical composition with season and during the moulting cycle of the common shore crab, *Carcinus maenas* (L). J Exp Mar Biol Ecol 5: 199-233.
2. Badawi HK (1971) On the chemical composition of the red crab *Portunus pelagicus*. Mar Biol 11: 198-200.
3. Addission RF, Ackman RG, Hingley J (1972) Lipid composition of the queen crab *Chinoecetes opilio*. Journal of the Fisheries Research Board of Canada 29: 407-411.
4. Pillai KK, Nair NB (1973) Observation on the breeding biology of some crabs from south west coast of India.
5. Prasad PN, Neelakandan B (1989) Proximate and essential amino acid composition in edible crab *Scylla serrata*. Comp Physiol Ecol 14: 34-37.

6. Zafar M, Siddiqui MZH, Hoque MA (2004) Biochemical composition in *Scylla serrata* (Forskal) of Chakaria, Sundarban area, Bangladesh. Pakistan J Biol Sci 7: 2182-2186.

7. Anil MK, Suseelan C (2001) Experimental fattening of the green mud crab *Scylla oceanica*. In: Perspectives in Mariculture. Menon NG and Pillai PP, (eds.). The Marine Biological Association of India 95-110.

8. Vasconcelos P, Braz NR (2001) Proximate composition of the deep-sea crab, *Chaceon affinis* from an exploratory fishery off Maderi island (Portugal - Eastern Central Atlantic). Northwest Atlantic Fisheries Organization 1-6.

9. Azeiteiro UM, Fonseca J, Marques JLS (2001) Biometry, estimates of production and seasonal variation in the biochemical composition of *Mesopodopsis slabberi* (Van Beneden, 1861) (*Crustacea: Mysidacea*). Bol Inst Esp Oceanogr 17: 15-25.

10. Thirunavukkarasu N (2005) Biology, nutritional evaluation and utilization of mud crab *Scylla tranquebarica* (Fabricius, 1798). Ph.D. Thesis, Annamalai University, India 126.

11. Manivannan K, Sudhakar M, Murugesan R, Soundrapandian P (2010) Effect of feed on the Biochemical composition of commercially important mud crab *Scylla tranquebarica* (Fabricius, 1798). Int J Anim Vet Adv 2: 16-20.

12. Murugesan R, Soundarapandian P, Manivannan K (2008) Effect of Unilateral Eyestalk Ablation on the Biochemical changes of edible portunid crab *Charybdis lucifera* (Fabricus). J Fish Aquat Sci 3: 82-86.

13. Sudhakar M, Manivannan K, Soundarapandian P (2009) Nutritive value of hard and soft shell crabs of *Portunus sanguinolentus* (Herbst). Int J Ani Vet Adv 1: 44-48.

14. Raymont JEG, Austin J, Linford E (1964) Biochemical studies on marine zooplankton. I. The biochemical composition of *Neomysis integer*. J Cons Int Explor Mer 28: 354-363.

15. Dubois M, Giles KA, Hamilton JK, Rebors PA, Smith F (1956) Calorimetric method for determination of sugars and related substances. Analytical Chemistry 28: 350-356.

16. Folch J, Lees M, Sloane-Stanley GH (1957) A simple method for the isolation and purification of total lipids from animal tissues. J Biol Chem 226: 497-509.

17. Dinakaran GK, Soundarapandian P (2009) Biochemical Status of Edible Palaemonid Prawn *Macrobrachium idella idella* (Hilgendorf, 1898). Adv J Food Sci Tech 1: 19-26.

18. Dinakaran GK, Soundarapandian P, Chandra SK (2009) Proximate composition of edible palaemonid prawn *Macrobrachium idae*. Current Research Journal of Biological Sciences 1: 78-82.

19. Dinakaran GK, Soundarapandian P, Tiwary AK (2010) Nutritional status of edible palaemonid prawn *Macrobrachium scabriculum* (Heller, 1862). European Journal of Applied Sciences 2: 30-36.

20. Khan PA (1992) Biochemical composition, minerals (calcium and iron) and chitin content of two portunid crabs *Scylla serrata* and *Portunus pelagicus* M.Sc. thesis, Institute of Marine Sciences, Chittagong University 11-112.

21. Sriraman K (1998) Biological and biochemical studies on the prawns of Portonova coast (Crustacea: Decqpoda: Macrura). Ph.D. Thesis, Annamalai University, India 1-130.

22. Radhakrishnan CK, Natarajan R (1979) Nutritive value of the crab *Podophthalamus vigil* (Fabricius). Fishery Technology 16: 37-38.

23. George C, Gopakumar K (1987) Biochemical studies on crab *Scylla serrata*. Fish Tech 24: 57-61.

24. Radhakrishnan CK (1979) Studies on Portunid crabs of Porto Novo (Crustacea: Decapoda: Brachyura). Ph.D. Thesis, Annamalai University, India 1-215.

25. George C, Gopakumar K, Perigreen PA (1990) Frozen storage characteristics of raw and cooked crab (*Scylla serrata*) segments, body meat and shell on claws. J Mar Biol Ass India 32: 193-197.

26. (2011) Welcome to the USDA national nutrient database for standard reference. USDA.

27. Balasubramanian CP, Suseelan C (2001) Biochemical composition of the deep water crab *Charybdis smithii*. Indian Journal of Fisheries 48: 333-335.

4

Contribution of Polychaetes in Feeding Capability of Commercially Important Crabs, South East Coast of India

D Varadharajan* and P Soundarapandian

Faculty of Marine Sciences, Centre of Advanced Study in Marine Biology, Annamalai University, Tamil Nadu, India

Abstract

The food is very important for growth, metabolism, parasitic, migratory movement, maturity, reproduction, breeding and survival of all life stages. A commercially important crab has a great demand in the market because of its high nutritional value and medicinal use. The study of food and feeding habits of crab has a variety of importance in fishery biology. For well grow farming a thorough knowledge about the food and feeding habit is important. The studies of interactions between the links in complex food chains can lead to an understanding of crab gain their food sources. The examination of gut contents of crab provides information about particular crab in the ecosystem.

Keywords: Food; Growth; Survival; Farming; Commercial crabs

Introduction

Portunidae of the swimming crabs are common in tropical and subtropical estuarine and near shore habitats. Crabs are filter feeders, detritus feeders, plant and carrion feeders, scavengers and predators and many of them use more than one method of feeding and various sources of food [1]. The polychaete from a number of habitats both intertidal, subtidal and the plant substrata of mangrove, seagrass and seaweed provide feed, spawning and nursery grounds for many organisms and a vast range of direct and indirect products and benefit their role in the energy flow and the diet of crabs [2-5]. The number of species of marine crabs is edible, the large ones and highly esteemed species suitable for coastal aquaculture. The crab farmers as well require knowing how much feed their farm animals have actually consumed but this can be difficult to judge. On the basis of the nature of the food taken, this is often overlapping different feeding categories in decapods. The knowledge of the food and feeding of crab is very important ones since to start hatchery or farming. The distribution of marine benthos varies both temporally and spatially due to the roughness of their occurrence [5,6]. The main difficulty when studying the diet of crab is the identification of gut contents to a specific level. So far the regarding of polychaetes the presence of certain structures which are more resistant to digestion, such as setae, uncini, hooks, mouthpieces and opercula, allows the determination to family level. In the present study, gut contents of commercially important crabs were analysing, there are contain a number of carnovore particle, however, specically were identified the main feed of crabs in annelids polychaetes particle. Since, the study was indicating the loss of species diversity, due to effects of predation.

Materials and Methods

The study was conducted from Sethubavachatram it is situated along the Palk Strait in the southeast part of India. Seasonal sampling at different depths 0-5, 5-10, 10-15 and 15-30 m was carried out by means of an otter trawl in collecting crab species in different marine sites. For this study, 12 species of commercial brachyuran crabs were selected. After collection, the crabs were stored in ice boxes and the gut contents were removed and fixed 10% buffered formalin. The gut contents were later analyzed in the laboratory the presumed polychaete material was sorted and preserved in 70% alcohol for further identification, based on the presence of structures as body segment of antennae, prostomium, peristomium, parapodium, setae, uncini, hooks, mouthpieces and opercula. An attempt was made to identify Polychaeta in the gut contents to species level by investigating the fragments with a microscope. However, this method tends to overestimate substance with hard parts and small substance such as annelids polychates which contribute little to the relative volume of the gut contents [7]. The identified preys of polychaetes were using standard references.

Results

Commercial crab gut contents identified different family Polychaetes in different structures

The Aphroditidae family species parts were identified in different crab guts in the following order; *S. serrata>P. pelagicus>S. tranquebarica>C. feriata>C. lucifera>C. variegata>C. granulata>P. sanguinolentus>C. gladiator>P. vigil>C. natator>C.truncata* (Figure 1). The Ampharetidae family species parts were identified in different crab guts in the following order; *S. serrata>P. pelagicus>C. gladiator>C. granulata>S. tranquebarica>C. natator>C. variegata>P. sanguinolentus>P. vigil>C. lucifera>C. feriata>C. truncata* (Figure 2). The Capitellidae family species parts were identified in different crab guts in the following order; *S. serrata>P. vigil>S. tranquebarica>P. sanguinolentus>C. granulata>P. pelagicus>C.truncata>C. natator>C. variegata>P. gladiator>C. feriata>C. lucifera* (Figure 3). The Cirratulidae family species parts were identified in different crab guts in the following order; *S. serrata>C. natator>S. tranquebarica>P. sanguinolentus>P. gladiator>C. lucifera>C. granulata>P. pelagicus>P. vigil>C. feriata>C. truncata>C. variegata* (Figure 4). The Cossuridae family species segments were identified in different crab guts in the following order; S. *serrata>C. feriata>S. tranquebarica>P. gladiator>P. vigil>C. lucifera>C. granulata>C. truncata>P. pelagicus>P. sanguinolentus>C. natator>C. variegata* (Figure 5). The Dorvillidae family species parts

***Corresponding author:** D Varadharajan, Faculty of Marine Sciences, Centre of Advanced Study in Marine Biology, Annamalai University, Parangipettai-608 502, Tamil Nadu, India, E-mail: heartvaradhan@gmail.com

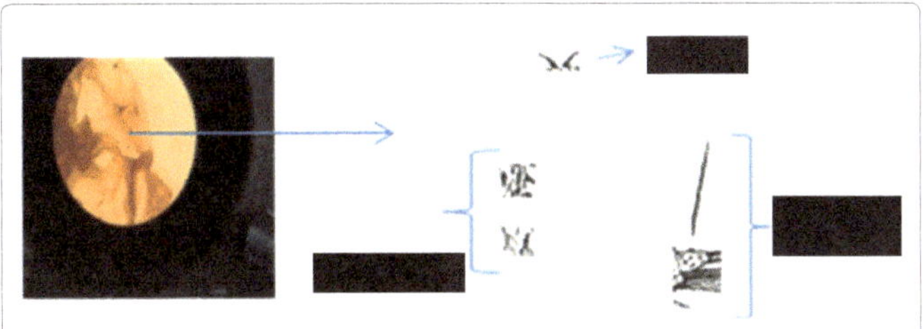

Figure 1: Family Aphroditidae.

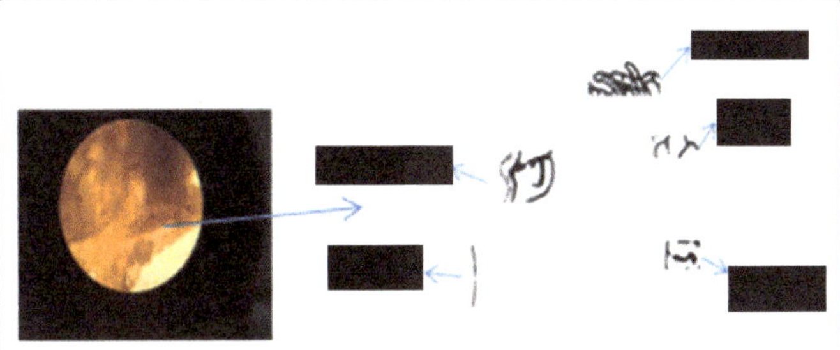

Figure 2: Family Ampharetidae.

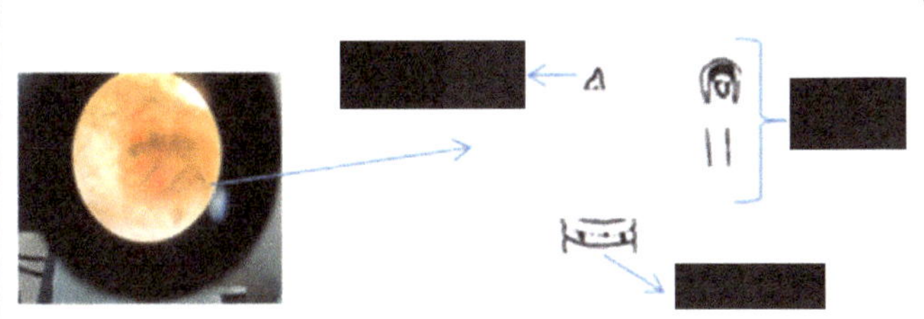

Figure 3: Family Capitellidae.

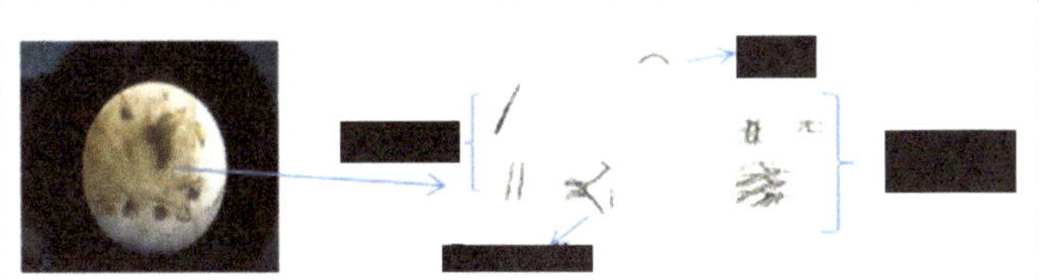

Figure 4: Family Cirratulidae.

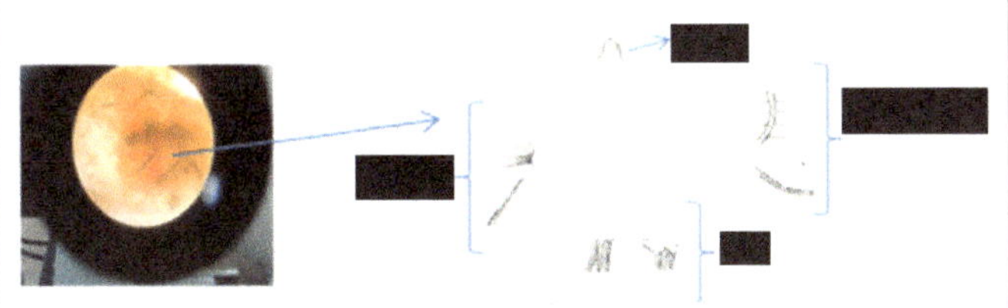

Figure 5: Family Cossuridae.

were identified in different crab guts in the following order; *S. serrata>S. tranquebarica>P. pelagicus>P. sanguinolentus>P. vigil>C. natator>C. variegata>C. gladiator>C. granulata>C.truncata>C. feriata> C. lucifera* (Figure 6). The Eunicidae family species segments were identified in different crab guts in the following order; *S. tranquebarica>S. serrata>P. sanguinolentus>C. gladiator>P. pelagicus>P. vigil>C. feriata>C. natator>C.variegata>C. granulata>C. lucifera>C.truncata* (Figure 7).

The Goniadidae family species parts were identified in different crab guts in the following order; *S. serrata>S. tranquebarica>P. pelagicus>P. sanguinolentus>P. gladiator>C. natator>C. granulata>C. variegata>C.truncata>P. vigil>C. lucifera>C. feriata* (Figure 8). The Glyceridae family species parts were identified in different crab guts in the following order; *S. serrata>S. tranquebarica>P. pelagicus>P. sanguinolentus>C. gladiator>C. feriata>C. variegata>C. truncata>P. vigil>C. lucifera>C. natator>C. granulata* (Figure 9). The Hesionidae family species segments were identified in different crab guts in the following order; *S. serrata>S. tranquebarica>P. vigil>P. gladiator>C. natator>C. granulata>P. pelagicus>C. lucifera>P. sanguinolentus>C. variegata>C. feriata>C. truncata* (Figure 10). The Maldanidae family species parts were identified in different crab guts in the following order; *P. vigil>C. granulata>S. tranquebarica>P. pelagicus>C.*

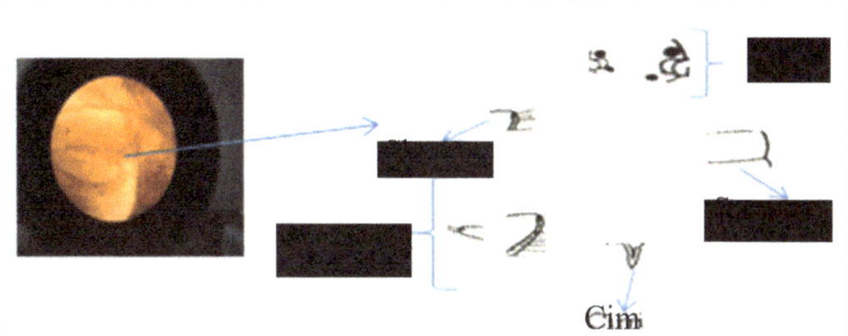

Figure 6: Family: Dorvillidae.

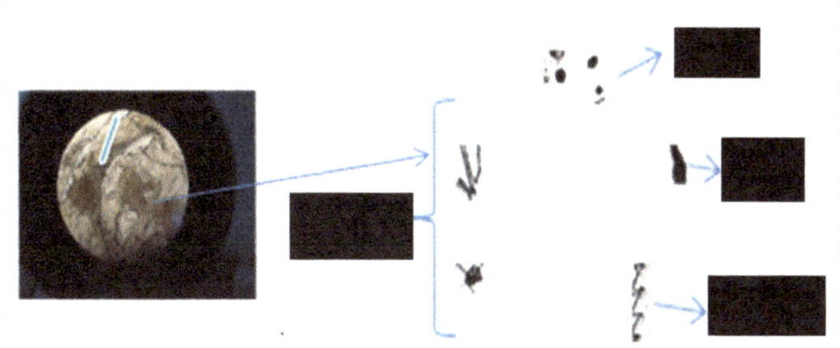

Figure 7: Family: Eunicidae.

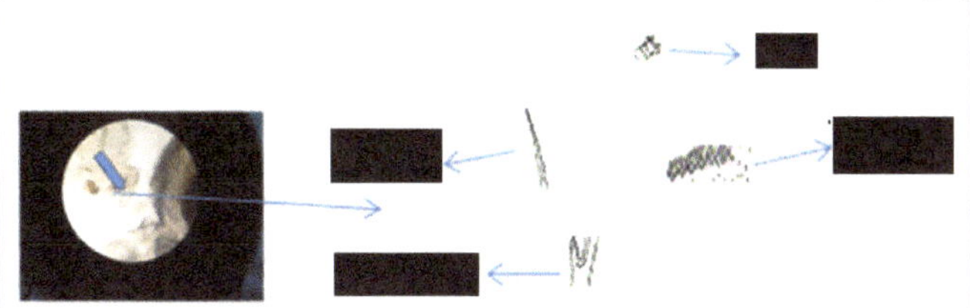

Fiugure 8: Goniadidae.

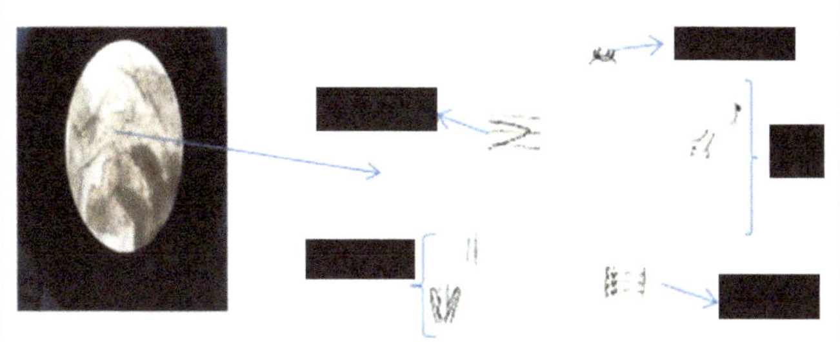

Figure 9: Family Glyceridae.

lucifera>*S. serrata*>*P. sanguinolentus*>*P. gladiator*>*C. feriata*>*C. natator*>*C. variegata*>*C. truncata* (Figure 11). The Nephtyidae family species segments were identified in different crab guts in the following order; *S. serrata*>*S. tranquebarica*>*P. sanguinolentus*>*C. feriata*>*C. natator*>*C. variegata*>*C. granulata*>*P. pelagicus*>*P. gladiator*>*C. truncata*>*C. lucifera*>*P. vigil* (Figure 12). The Nereididae family species parts were identified in different crab guts in the following order; *S. serrata*>*S.tranquebarica*>*P. pelagicus*>*P. sanguinolentus*>*C. variegata*>*P. gladiator*>*P. vigil*>*C. natator*>*C. lucifera*>*C. granulate*>*C. feriata*>*C. truncata* (Figure 13). The Onuphidae family species segments were identified in different crab guts in the following order; *S. serrata*>*C. gladiator*>*C. granulata*>*P. pelagicus*>*P. sanguinolentus*>*S. tranquebarica*>*P. vigil*>*C. feriata*>*C. natator*>*C. lucifera*>*C. truncata*>*C. variegata* (Figure 14). The Ophelidae family species parts were identified in different crab guts in the following order; *S. serrata*>*S. tranquebarica*>*P. pelagicus*>*C. lucifera*>*P. sanguinolentus*>*P. vigil*>*C. feriata*>*C. granulata*>*C. natator*>*C. gladiator*>*C. variegata*>*C. truncata* (Figure 15).

The Pectinariidae family species segments were identified in different crab guts in the following order; *S. serrata*>*P. sanguinolentus*>*C. feriata*>*C. truncata*>*S. tranquebarica*>*P. vigil*>*C. natator*>*C. lucifera*>*C. variegata*>*C. granulata*>*P. pelagicus*>*C. gladiator* (Figure 16). The Pilargidiidae family species segments were identified in different crab guts in the following order; *S. serrata*>*P. pelagicus*>*C. gladiator*>*P. vigil*>*C. feriata*>*C. granulate*>*C. lucifera*>*C. natator*>*S. tranquebarica*>*C. variegata*>*C. truncata*>*P. sanguinolentus* (Figure 17). The Pisionidae family species parts were identified in

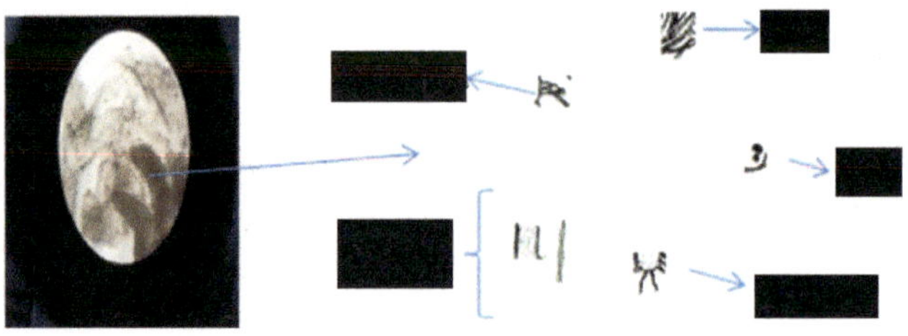

Figure 10: Family Hesionidae.

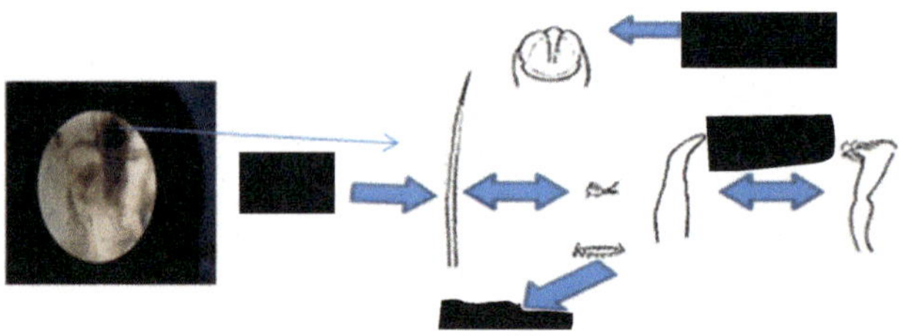

Figure 11: Family Maldanidae.

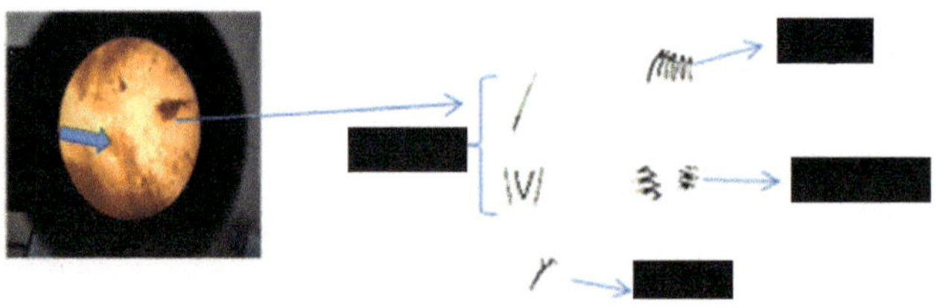

Figure 12: Family Nephtyidae.

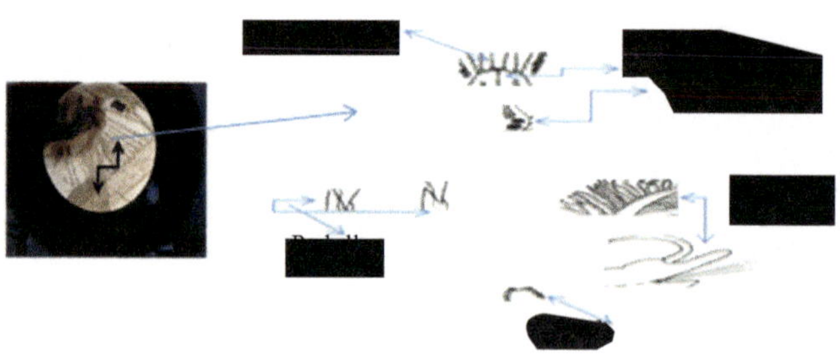

Figure 13: Family: Nereididae.

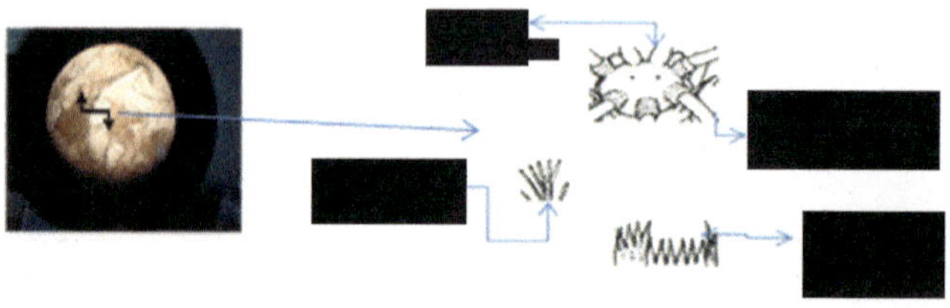

Figure 14: Family Onuphidae.

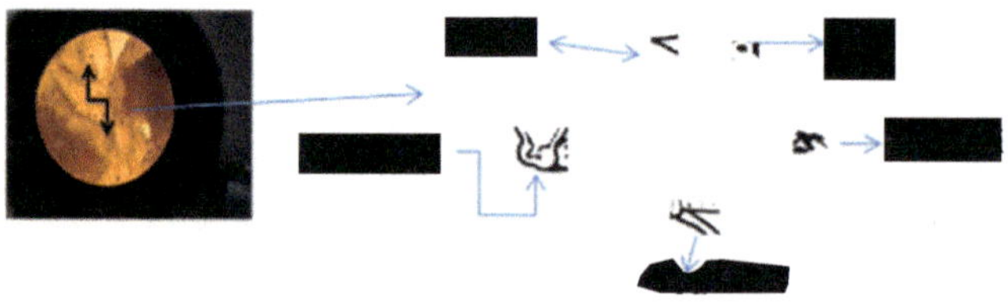

Figure 15: Family Ophelidae.

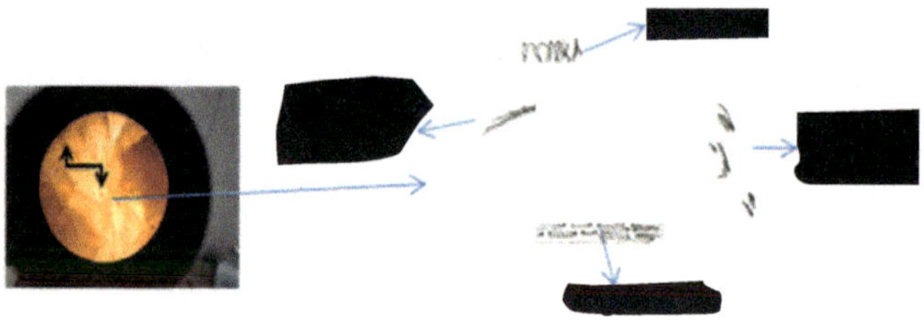

Figure 16: Family Pectinariidae.

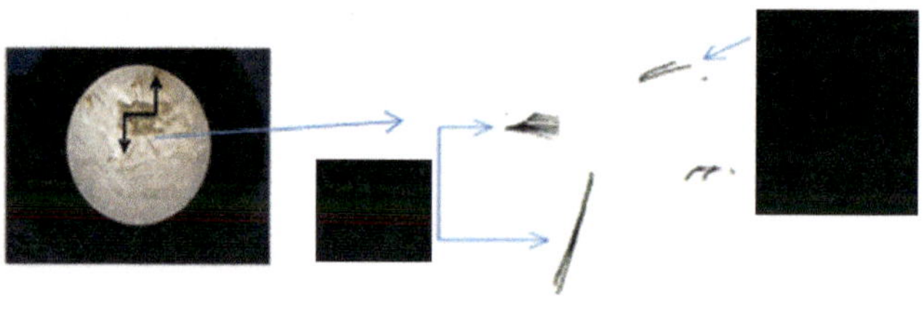

Figure 17: Family Pilargidiidae.

different crab guts following order; *S. tranquebarica*>*S. serrata*>*P. gladiator*>*P. sanguinolentus*>*C. natator*>*C. granulata*>*P. vigil*>*C. feriata*>*C. lucifera*>*P. pelagicus*>*C. variegata*>*C. truncata* (Figure 18). The Sabellariidae family characters were identified in crab guts from; *S. serrata*>*S. tranquebarica*>*P. sanguinolentus*>*C. feriata*>*P.*

pelagicus>*C. lucifera*>*C. variegata*>*C. truncata*>*P. gladiator*>*P. vigil*>*C. granulate*>*C. natator* (Figure 19). The Sabellidae family characters were identified in crab guts from; *S. serrata*> *S. tranquebarica*>*C. lucifera*>*C. natator*>*P. gladiator*>*P. vigil*>*P. pelagicus*> *P. sanguinolentus*>*C. granulate*>*C. truncata*>*C. variegata*>*C. feriata* (Figure 20). The Spionida family characters were identified in crab guts from; *S. serrata*>*P. gladiator*>*P. pelagicus*>*C. feriata*>*C. variegata*>*S. tranquebarica*>*P. sanguinolentus*>*C. lucifera*>*C. natator*>*C. granulata*>*P. vigil*>*C. truncata* (Figure 21). The Syllidae family characters were identified in crab guts from; *S. serrata*>*P. vigil*>*C.variegata*>*P. sanguinolentus*>*C. natator*>*C. granulata*>*S. tranquebarica*>*P. pelagicus*>*C. lucifera*>*C. truncata*>*P. gladiator*>*C. feriata* (Figure 22). The Terebellidae family characters were identified in crab guts from; *S. serrata*>*S. tranquebarica*>*P. vigil*>*C. feriata*>*P. sanguinolentus*>*P. gladiator*>*P. pelagicus*>*C. lucifera*>*C. truncata*>*C. natator*>*C. variegata*> *C. granulate* (Figure 23). Unidentified worms: The polychaetes worms characters were identified in crab guts from; *S. serrata*>*P. pelagicus*>*C. feriata*>*P. gladiator*>*P. sanguinolentus*>*P. vigil*>*C. natator*>*S. tranquebarica*>*C. variegata*>*C. granulata*>*C. truncata*>*C. lucifera* (Figure 24).

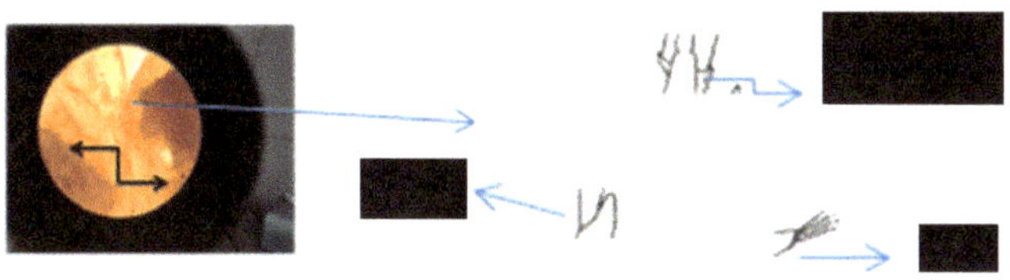

Figure 18: Family Pisionidae.

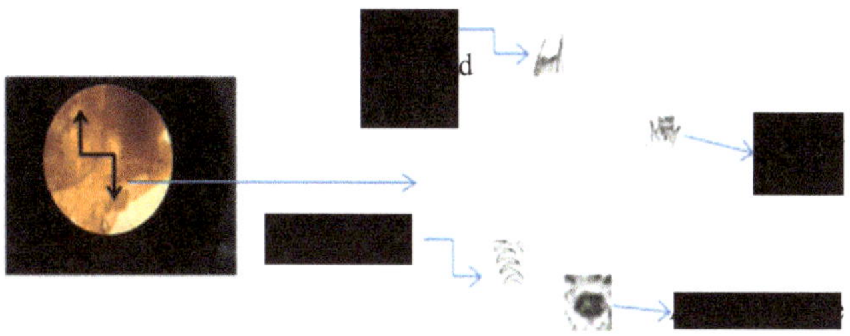

Figure 19: Family Sabellariidae.

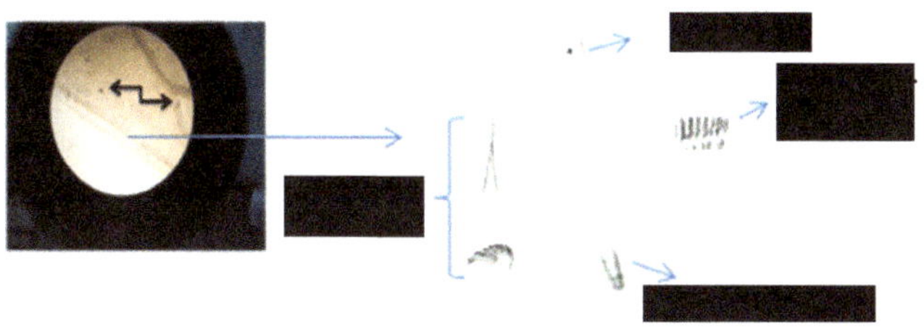

Figure 20: Family Sabellidae.

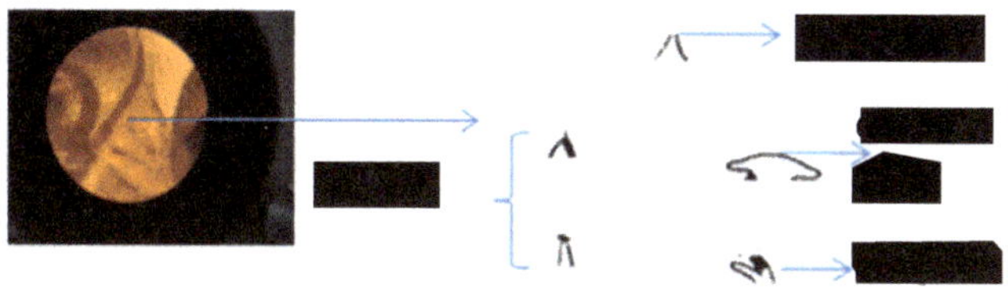

Figure 21: Family Spionida.

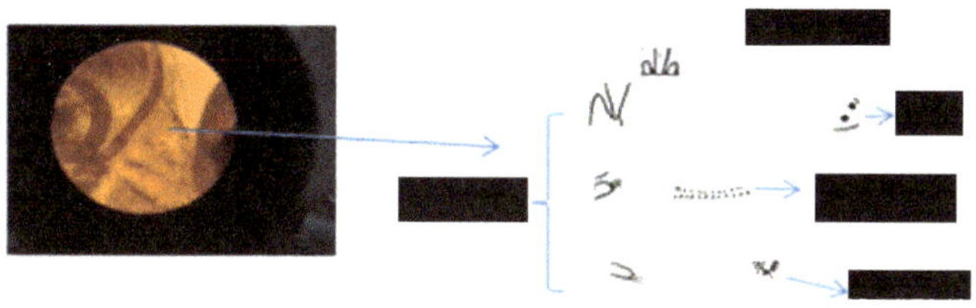

Figure 22: Family Syllidae.

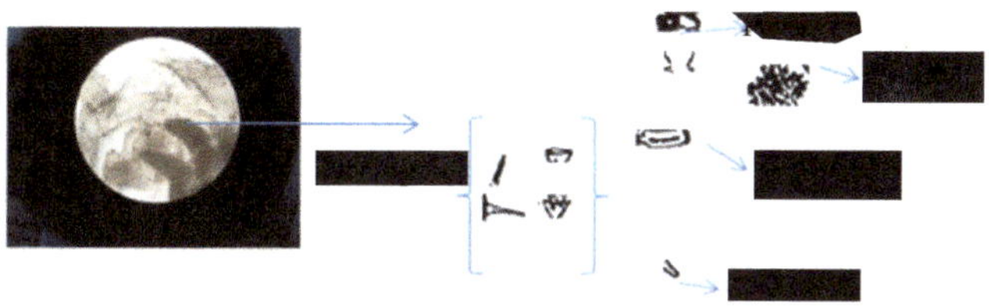

Figure 23: Family Terebellidae.

Figure 24: Polychaetes Worms.

Discussion

In the present study were examined twelve commercial crabs were collected at a sampling station in different depths and then all crab gut contents analyzed the annelids polychaetes different structure their presence. Most upon the polychaete species under the families like Nereidae, Nephtheidae, Onuphidae, Glyceridae, Spionidae Cirratulidae, Capitellidae, Dorvilleidae, Eunicidae, Spionidae, Maldanidae, Sabellariidae and Syllidae are the major biomass producers among annelids and these form an important food for different species of crabs. The diet variations indicated in the present study are also supported by the observations of that the length of the alimentary canal is indicative of the food preferences of the crabs with carnivores having the largest gut length and detritivores the smallest present the reason of the diversity of annelid polychaetes suitable substratum probably occurs [5,6,8,9]. Differences in the relative length of the alimentary tract of sub adult and adult crabs and attributed it to differences in their feeding habits. This observation is supported by the fact that larger crabs have the capability to digest even low quality food and therefore often meet their energy demands by consuming annelid polychaetes. It is reported that crab metabolism an influence on feeding behavior and feed intake by brachyuran crabs is such that it meets their energy requirements [10]. Thus if a diet has a low energy value, crab will compensate by eating more within the limits of its gut contents capacity [11]. For example *S. tranquebarica* and *S. serrata* have reported with comparatively higher fecundity [12]. The portunid crabs breeding activity mostly throughout the year probable, this term agrees with previous studies [13-15]. The changes in the seasons of berried crab availability in the coasts might be due to different monsoonal periods, current patterns and environmental parameters [16,17]. In the present investigations the peak breeding seasons also depend on the available feed and also the surrounding water conditions.

A large proportion of the crab had empty gut contents shortly after high tide. Since the polychaetes are commonly a major component of crab feeds. Such complex pathways of energy flow intensely affect web dynamics, spreading the effects of consumption and efficiency throughout the web [18]. The different food items indicated seasonal variations in diet composition with polychaetes than ranked first followed by plankton during the premonsoon season. The crab gut contents mostly polychaetes and digested matter dominated during the monsoon and post monsoon period. The summer season presents in the crabs gut contents there polychaetes and mud and plant materials. This observation agrees with results, previously reported by Menon and Patel et al. [19,20]. The many crabs are opportunistic feeders eating what is available within a more otherwise less restricted range of items and changes in number of prey types reflect only this plasticity as well as the variability of prey in the particular biotope where the crab is feeding [21,22]. The observations are supported by several reports which indicate polychaetes parts is not easily digest in crab gut contents reason of the segregation of digest enzymes is low. Other food particles the digestion of detritus can supply the consumer with energy and therefore detrital aggregates which are not a normally acceptable food resource may serve as a short term food resource if the consumer can tolerate a temporary nutritional deficiency [16,20,21]. The rapid increase in the feeding intensity in the berrieds and progressively thereafter was coinciding with the length at which maturation is initiated and spawning activities predominate in this species [13,14,17,23]. Typically, when crabs approach maturity they are found to increase the energy devoted to gonad development by increased feeding intensity which supports the observations in this study [24]. Although the strength of feeding in brachyuran crab has been reported to vary with maturity and spawning conditions, being minimal during spawning, high in the maturing group and maximum during post spawning period no such decline in feeding activity of sub adult, adult and berried crabs was observed in the present study [25]. This study agrees with the results previously observed by Rosas et al. [26]. The observation that egg production in many crabs depends more on the energy intake during the spawning season than on the energy reserves accumulated earlier and increased feeding during maturation which enables development of gonadal growth without slowing of somatic growth [27,28]. The situation for the crabs is worse because the temporary habitats occupied by these species are difficult to sample quantities [29]. Even if reliable estimates of abundance were available, estimating the strength of links would still be problematic because determining exactly how many prey predators consume is a difficult task, to the extent that such information is rarely available [30]. The lack of oxygen in bottom waters changes the different types of organisms that exist on the bottom. It is living species change from long living, deep burrowing and slow growing animals to species that live on the sediment surface. They are little and fast growing and can tolerate low concentrations of oxygen otherwise can be difficult to grow. Since, the predators can solve for this kind of problems such as food and feeding activities. The crabs are remarkably adaptable crustaceans that will feed on a wide variety of foods, although some foods are favored over others and individual crabs may have individual preferences. A particular diet may be chosen to seek weight loss or weight gain. In the present investigations clearly indicate that's the crabs prey mostly bottom living organisms it's highly like in annelids polychates. It can play an important role in the regeneration and recycling of nutrients, as their appearance and vital source of food item for the crabs. The culturing time mostly the feeding problems arise due to overfeeding, if given excess food also means more waste. The more food will putrefy and degrade the water quality. In the present study indicates that the interactions between the links in complex food chains can lead to an understanding of how specific organisms gain their food sources. To advance the knowledge of the feeding preferences of brachyuran crabs an understanding of the broader aquatic community is a requisite. However, the scientific knowledge on the crab's food and feeding habits is limited. Hence, this information provides baseline information that's useful for sustainable exploitation and management of fisheries activities.

References

1. Sastry AN (1983) Ecological aspects of reproduction, p179-270. In: W.B. Vernberg (Ed.). The Biology of Crustacea: Environment adaptations. New York, Academic Press 8: 383.
2. Pearson TH, Rosenberg R (1978) Macrobenthic succession in relation to organic enrichment and pollution of the marine environment. Oceanogr Mar Biol Ann Rev 16: 229-311.
3. Schembri PJ (1981) Feeding in Ebalia tuberosa (Pennant) (Crustacea: Decapoda: Leucosiidae). J Exper Mar Biol Ecol 55: 1-10.
4. Hall SJ, Raffaelli D, Turrell WR (1990) Predator-caging experiments in marine systems: A reexamination of their value. Am Nat 136: 657-672.
5. Ansari ZA, Ingole B, Baneerjee S, Parulekar AH (1986) Spatial and temporal changes in benthic macrofauna from Mandovi and Zuari estuaries of Goa, west coast of India. Indian J Mar Sci 15: 223-229.
6. Varadharajan D, Soundarapandian P, Gunalan B, Babu R (2010) Seasonal Abundance of Macro Benthic Composition and Diversity along the South East Coast of India. European J Appl Sci 2: 1-5.
7. Williams MJ (1981) Methods for analysis of natural diet in portunid crabs (Crustacea; Decapoda: Portunidae). J Exp Mar Biol Ecol 52: 103-111.
8. Hill BJ (1976) Natural food, foregut clearance-rate and activity of the crab *Scylla serrata*. Mar Biol 34: 109-116.
9. Haefner, PA Jr (1990) Natural diet of Callinectes ornatus (Brachyura : Portunidae) in Bermuda. J Crust Biol 10: 236-246.
10. Stehlik LL (1993) Diets of the brachyuran crabs Cancer irroratus, C. borealis and Ovalipes ocellatus in the New York Bight. J Crust Biol 13: 723-735.
11. Williams MJ (1982) Natural food and feeding in the commercial sand crab Portunus pelagjcus Linnaeus, 1766 (Crustacea: Decapoda: Portunidae) in Moreton Bay, Queensland. J Exp Mar Biol Ecol 59: 165-176.
12. Pillai KK, Nair NB (1971) The annual reproductive cycles of *Uca annulipes, Portunus pelagicus* and *Metapenaeus affinis* (Decapoda: Crustacea) from the South-west coast of India. Mar Biol 11: 152-166.
13. Chhapgar BF (1956) On the breeding habits and larval stages of some crabs. Rec Indian Mus 54: 33-52.
14. Knudsen IW (1964) Observations on the reproductive cycles and ecology of the common Brachyura and crab-like Anomura of the Puget South, Washington. Pac Science 18: 3-33.
15. Ryan EP (1967) Structure and function of the reproductive system of the crab Portunus sanguinolentus (Herbst) (Brachyura : Portunidae). II. Female system. Proc Symp Crust Mar Biol Ass India 522-544.
16. Prasad RR, Tampi PRS (1953) A contribution to the biology of the blue swimming crab, Neptunus pelagicus (Linnaeus), with a note on the zoea of Thalamita crenata Latrelle. J Bombay Nat Hist Soc 51: 674-689.
17. Varadharajan, D, Soundarapandian P, Dinakaran GK, Vijakumar G (2009) Crab Fishery Resources from Arukkattuthurai to Aiyammpattinam, South East Coast of India. Curr Res J Biol Sci 1:118-122.
18. Polis GA, Strong DR (1996) Food web complexity and community dynamics. Am Nat 147: 813-846.

19. Menon MK (1952) A Note on the Bionomics and Fishery of the Swimming Crab Neptunus sanguinolenlus (Herbst) on the Malabar Coast. J Zool Soc India 4: 177-184.

20. Patel NM, Chhaya ND, Bhaskaran (1979) Stomach contents of Portunus pelagicus (Linn.) from AD net catches. Indian J Mar Sci 8: 48-49.

21. Hill BJ (1980) Effect of temperature on the feeding activity in the mud crab Scylla serrata. Mar Biol 59: 189-192.

22. Sumpton WD, Smith GS (1990) Effect of temperature on the emergence activity and feeding of male and female sand crabs (Portunus pelagicus). Australian J Mar Freshwater Res 41: 545-550.

23. Hines AH (1982) Allometric constraints and variables of reproductive effort in brachyuran crabs. Mar Biol 69: 309-320.

24. Hartnoll R (1978) The determination of relative growth in crustacea. Crust 34: 281-293.

25. Pillai KK, Nair NB (1968) Observations on the reproductive cycles of some crabs from south-west coast of India. J Mar Biol Ass India 10: 384-385.

26. Rosas C, Lazaro-Chavez E, Buckle-Ramirez F (1994) Feeding habits and food niche segregation of Callinectes sapidus, C. rathbunae and C. similis in a sub-tropicalcoastallagoon of the Gulf of Mexico. J Crust Biol 14: 371-382.

27. Wieczorek SK, Hooper RG (1995) Relationship between diet and food availability in the snow crab Chinoecetes opilio in Bonne Bay, Newfoundland. J Crus Biol 15: 236:247.

28. Wassenberg TJ, Hill BJ (1987) Feeding by the sand crab *Portunus pelagicus* on material discarded from prawn trawlers in Moreton Bay, Australia. Mar Biol 95: 387-393.

29. Sheaves MJ, Baker R, Johnston RW (2006) Marine nurseries and effective juvenile habitats: an alternative view. Mar Ecol Prog Ser 318: 303- 306.

30. Hall SJ, Raffaelli DG (1993) Food webs: theory and reality. Adv Ecol Res 24: 187-239.

5

Distribution and Abundance of Amphioxus With Relation To Sediment Characteristics from South East Coast of India

A Babu*, P Sampathkumar, T Balasubramanian, D Varadharajan and T Manikandarajan

Faculty of Marine Sciences, Annamalai University, Parangipettai – 608 502, Tamil Nadu, India

Abstract

Biodiversity, the variety of life and understood of species survive. Lancelet plays an important role in marine benthic food chain. A study was carried out to observe the abundance and diversity of marine bottom lancelet and observations on sediment characteristics. The lancelet were collected station 1 at Cooum the maximum of 98.12 No/m^2 was observed in site IV and minimum of 4.22 No/m^2 was observed in site II. Station 2 at Muttukadu the maximum of 165.12 No/m^2 was observed in site I and minimum of 1.10 No/m^2 was observed in site IV. Station 3 at Pondicherry the maximum of 98.8 No/m^2 was recorded in site II and minimum of 2.35 No/m^2 was recorded in site IV. At all station when compared the maximum species was observed in Muttukadu and minimum in Cooum during the study period.

Keywords: Lancelet; Diversity Record; *Branchiostoma* Sp; Sediment Characteristics

Introduction

A study on the abundance of amphioxus and larvae was to estimate the marine resources or biodiversity. Most of the amphioxus are marine habitat, particularly in larvae are pelagic and adults both in pelagic and benthic nature. It produces young ones in open sea at both pelagic and benthic [1,2]. Amphioxus is one among the macrobenthic organisms which are retained on 0.5 mm screen [3]. Benthic communities represented by diversified population within the sediments, it play an important role in enhancing fish production through a food chain [4,5]. Benthic fauna have been found to vary greatly in sensitivity to various types of pollution. Benthic communities often have been regarded as good indicators of organic pollution and also free from organic pollution because of their constant presence. The present study focused with the density and distribution of amphioxus with respect to sediments characters in the study area.

Materials and Methods

The study was conducted in three different stations, Cooum (station1), Muttukadu (station 2), Pondicherry (station 3) and each station has 4 sampling sites depend on the distances and depth during the year of January 2008-December 2008. Sample collection was using a Peterson grab (0.0256 m^2). Two samples were collected at each site, one for benthic faunal analysis and one for sediment analysis. Specimens of amphioxus were collected, enumerated and preserved with 5% buffered formalin. After collecting samples, they were emptied into a plastic tray. The larger organisms were picked out immediately from the sediment and then sieved through 0.5 mm mesh screen. The procedure adopted for sampling was followed the standard method proposed by Mudroch and MacKnight [6]. This procedure was repeated number of times, gradually breaking up the sediment until the majority of the mud and suspended specimens were removed. The organisms retained by the sieve were placed in a labelled container and fixed in 5% formalin.

Subsequently, the organisms were stained with Rose Bengal solution (0.1 g in 100 ml of distilled water) for enhanced visibility during identification, once the fixative was added, each sealed sample container was gently upturned and rotated to distribute the formalin evenly throughout the sieved sample. At all stages of the sieving, care was taken to individually remove noticed fragile animals. Once back at the laboratory, the sieved samples were gently but thoroughly washed in freshwater. The samples were preserved with 75% alcohol for further identification. The density of macro fauna amphioxus is expressed as number of individuals/m^2).

Sediment analysis

For sediment analysis, the sediment dried by sunshade and then dried samples used for the estimation of total organic carbon by conventional methods; they were dry-sieved with the help of a mechanical shaker. The sand, silt and clay fraction was determined as the amount of sediment in different sieves.

Results

Total organic carbon (mg/g)

In site wise, the concentration of total organic carbons varied at Cooum, the minimum of 1.794 mg/g was observed at site II and the maximum of 7.632 mg/g at site I (Figure 1a). At Muttukadu, the minimum of 2.345 mg/g was observed at site I and the maximum of 5.840 mg/g at site IV (Figure 1b). At Pondicherry, the minimum of 0.760 mg/g was observed at site I and the maximum of 6.880 mg/g at site IV (Figure 1c). The station wise the minimum was observed at Pondicherry and the maximum at Cooum during the study period (Figure 1d).

Sand (%)

In site wise, sand varied at Cooum minimum of 62.42% was observed at site I and the maximum of 93.67% at site IV (Figure 2a). At Muttukadu, the minimum of 2.345 % at site I and the maximum of 5.840 % at site IV (Figure 2b). At Pondicherry the minimum of 0.760 % at site I and the maximum of 6.880 % at site IV (Figure 2c). The

***Corresponding author:** A Babu, Faculty of Marine Sciences, Centre of Advanced Study in Marine Biology, Annamalai University, Parangipettai-608 502, Tamil Nadu, India, E-mail: babumarine@gmail.com

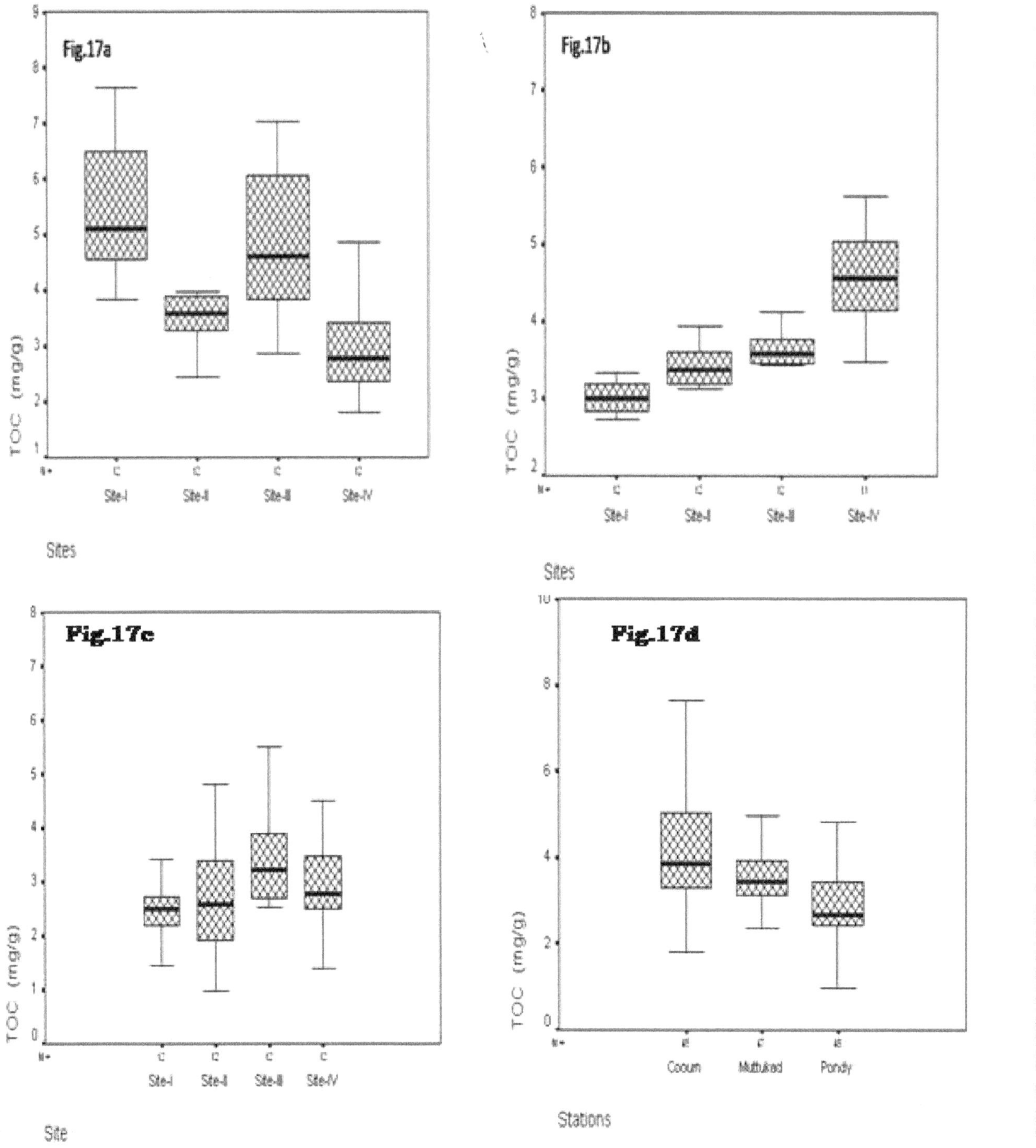

Figure 1: a,b,c Total organic carbon changes at different sampling sites in Cooum, Muttukadu, Pondicherry and 17d, changes at different sampling stations respectively.

station wise, minimum was observed at Cooum, and the maximum at Pondicherry (Figure 2d).

Silt (%)

In site wise, silt varied at Cooum, the minimum of 5.34 % at site I and the maximum of 22.13% at site III (Figure 3a). At Muttukadu, the minimum of 4.82 % at site II and the maximum of 18.020 % at site IV (Figure 3b). At Pondicherry the minimum of 1.860 % at site III and the maximum of 17.781 % at site IV (Figure 3c). The station wise minimum was observed at Pondicherry and the maximum at Cooum (Figure 3d).

Clay (%)

In site wise, clay varied at Cooum, the minimum of 0.15% in site IV and the maximum of 17.86 % in site I (Figure 4a). At Muttukadu, the minimum of 0.235% at site I and the maximum of 14.51 % at site IV (Figure 4b). At Pondicherry, the minimum of 0.121% at site IV and the

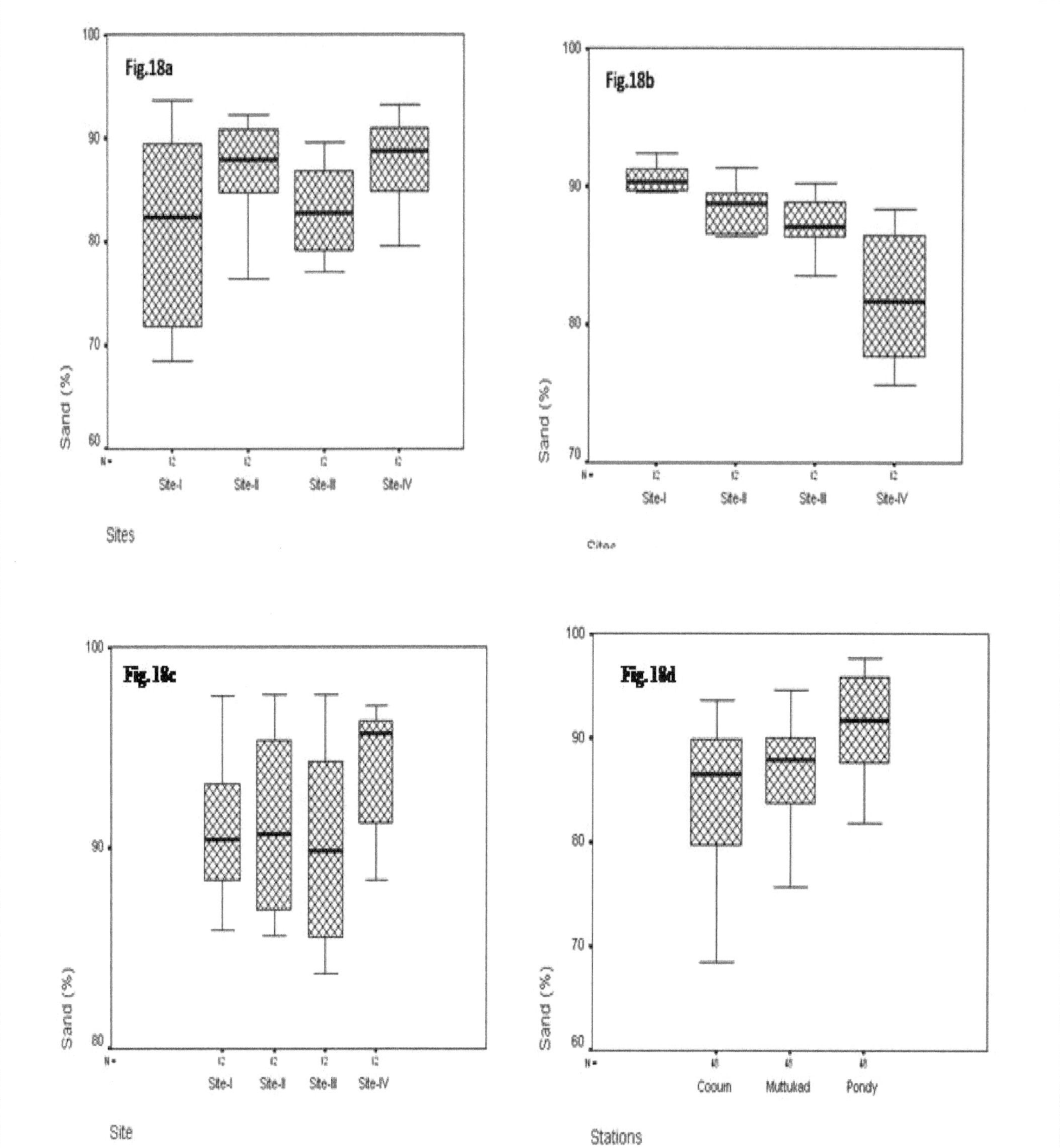

Figure 2: a,b,c Sand changes at different sampling sites in Cooum, Muttukadu, Pondicherry and 18d, changes at different sampling stations respectively.

maximum of 1.828 % at site III (Figure 4c). The station wise minimum was observed at Cooum and the maximum at Pondicherry during the study period (Figure 4d).

Branchiostoma lanceolatum (No/m^2)

In site wise *B. lanceolatum* was observed during the study period, At Cooum the minimum of 0.22 No/m^2 was observed at site III and the maximum of 23.12 No/m^2 in site IV (Figure 5a). At Muttukadu minimum of 0.10 No/m^2 was observed at site IV and the maximum of 98.12 No/m^2 at site I (Figure 5b). At Pondicherry the minimum of 0.35 No/m^2 was observed at site IV and the maximum of 32.8 No/m^2 at site III (Figure 5c). The station wise maximum was observed at Muttukadu and minimum in Cooum during the study period (Figure 5d).

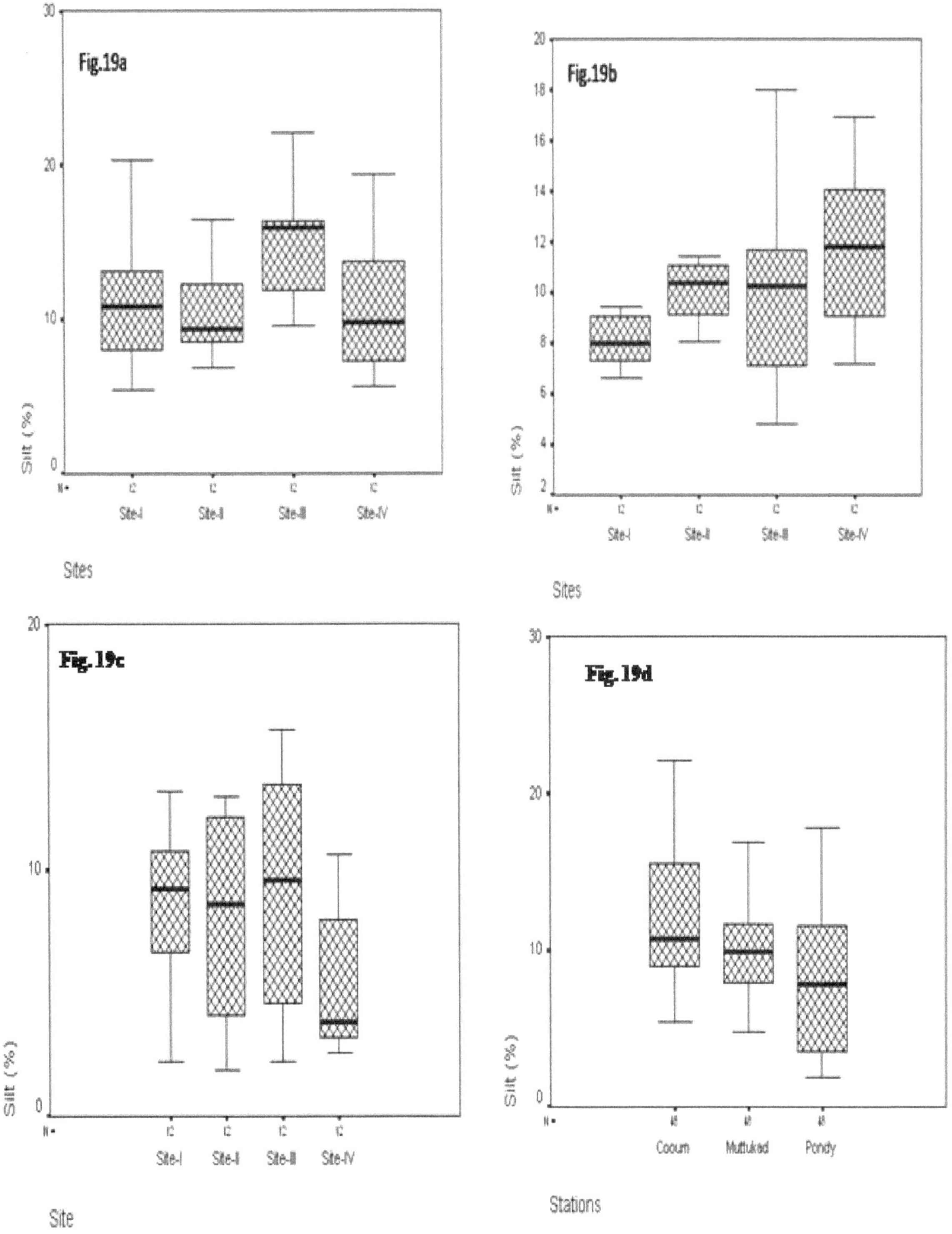

Figure 3: a,b,c Silt changes at different sampling sites in Cooum, Muttukadu, Pondicherry and 19d, changes at different sampling stations respectively.

Branchiostoma belcheri (No/m^2)

In site wise, *B. belcheri* was observed during the study period, At Cooum, the minimum of 0.22 No/m^2 were observed at site III and the maximum of 33.12 No/m^2 at site IV (Figure 6a). At Muttukadu, the minimum of 0.10 No/m^2 was observed at site IV and the maximum of 99.12 No/m^2 in site I (Figure 6b). At Pondicherry, the minimum of 0.35 No/m^2 in site IV and the maximum of 72.8 No/m^2 at site III (Figure 6c). The station wise maximum was recorded in Muttukadu and minimum

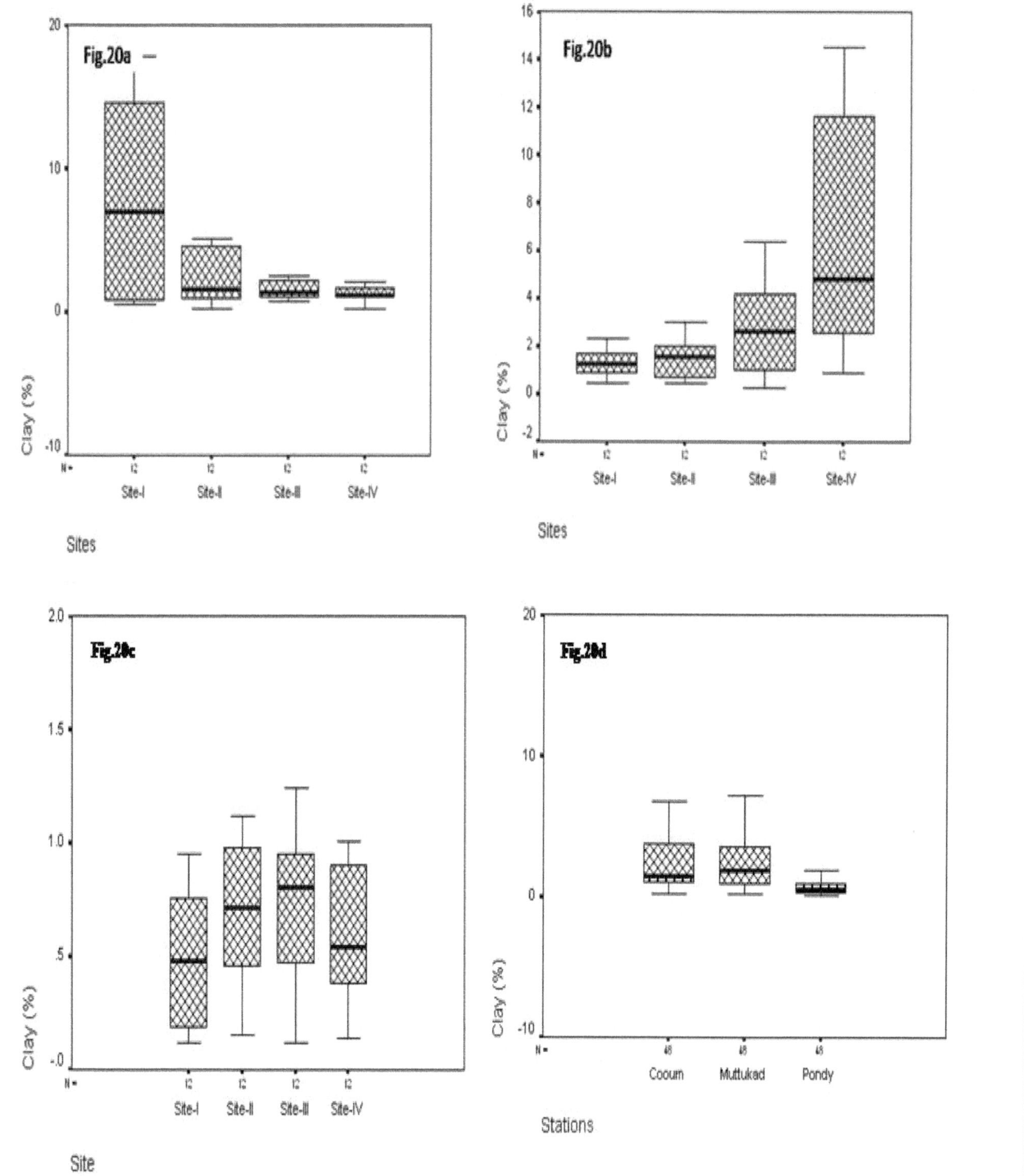

Figure 4: a,b,c Clay changes at different sampling sites in Cooum, Muttukadu, Pondicherry and 20d, changes at different sampling stations respectively.

in Cooum during the study period (Figure 6d). Statistical significant changes were occurred in sampling stations and site ($P \leq 0.05$).

Branchiostoma indicum **(No/m^2)**

In site wise, *B. indicum* was observed during the study period, at Cooum the minimum of 0.12 No/m^2 was observed at site IV and the maximum of 0.22 No/m^2 at site IV (Figure 7a). At Muttukadu the minimum of 0.10 No/m^2 was observed at site IV and the maximum of 0.12 No/m^2 in site I (Figure 7b). At Pondicherry the minimum of 0.35 No/m^2 was observed at site IV and the maximum of 0.8 No/m^2 in site I (Figure 7c). The station wise maximum was observed at Muttukadu and minimum in Cooum during the study period. (Figure

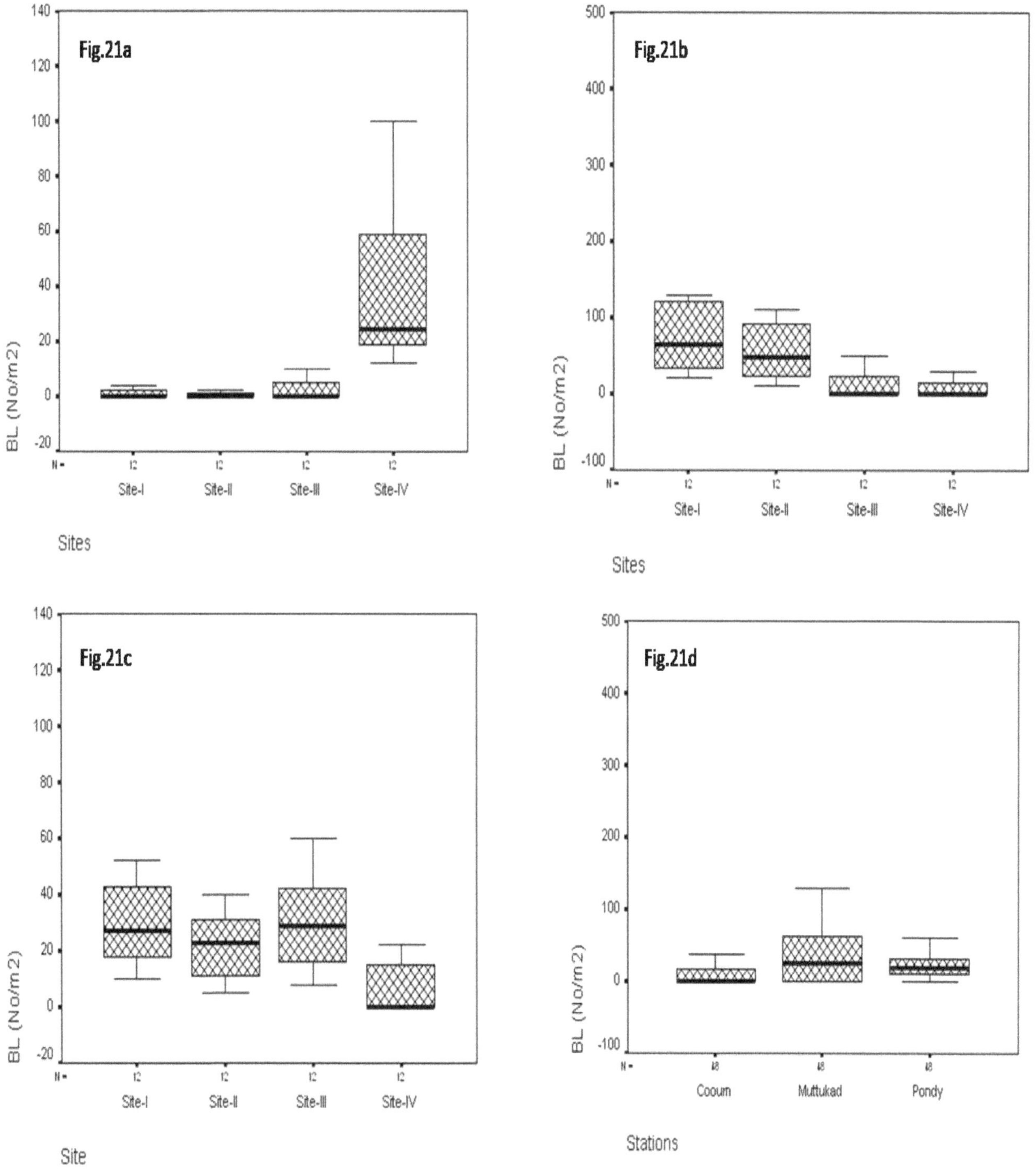

Figure 5: a,b,c *Branchiostoma lanceolatum* species at different sampling sites in Cooum, Muttukadu, Pondicherry and 21d, species at different sampling stations respectively.

7d). Statistical significant changes were occurred in sampling stations and site ($P \leq 0.05$).

Branchiostoma malayanum (No/m^2)

The *Branchiostoma malayanum* were recorded during the study period at Cooum the minimum of 0.22 No/m^2 at site IV and the maximum of 99.12 No/m^2 at site IV (Figure 8a). At Muttukadu the minimum of 0.10 No/m^2 was observed at site IV and the maximum of 152.12 No/m^2 in site I (Figure 8b). At Pondicherry, minimum of 0.35 No/m^2 was observed at site IV and the maximum of 98.8 No/m^2 were observed at site II (Figure 8c). The station wise, maximum was recorded in Muttukadu and minimum in Cooum during the study

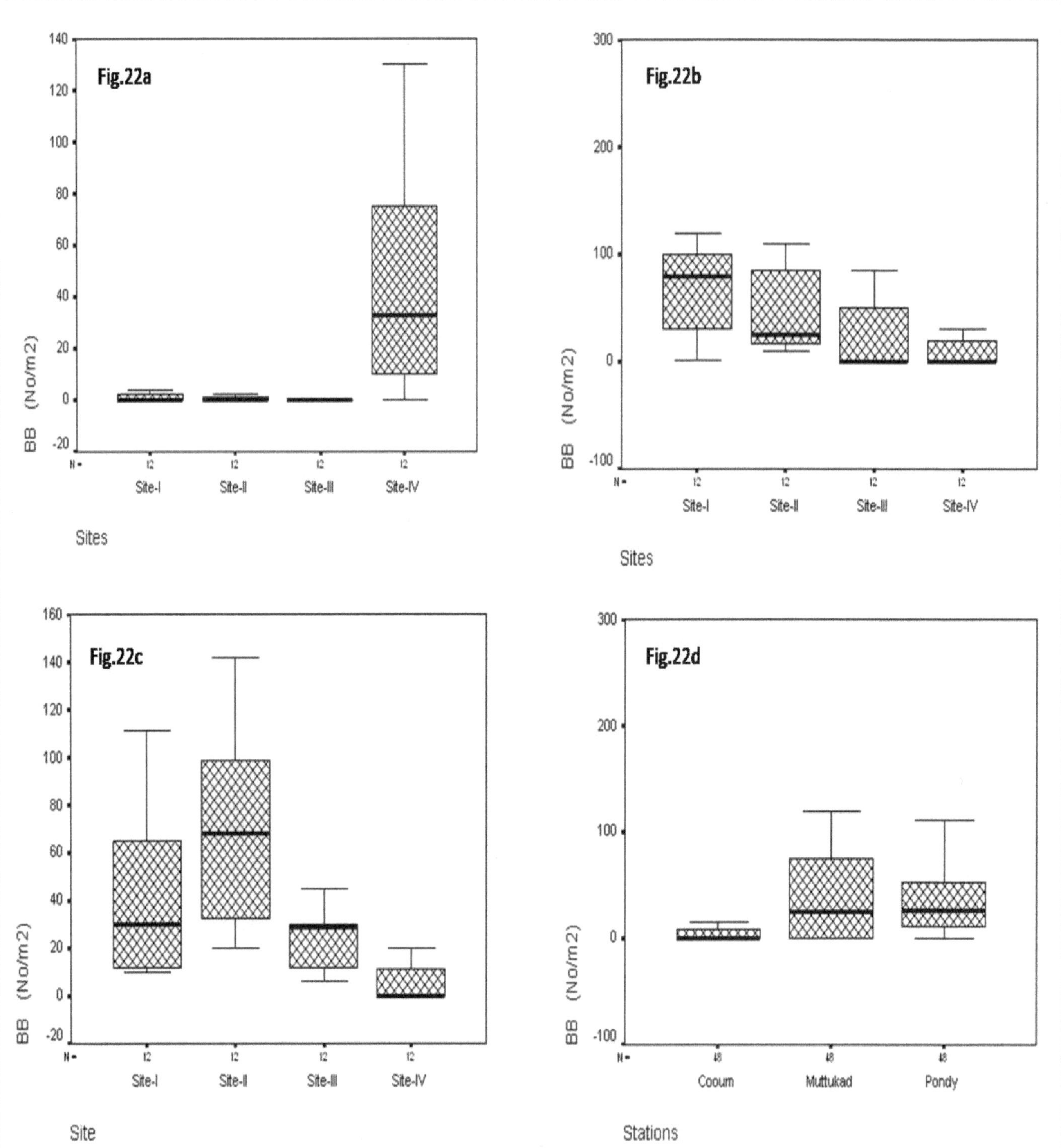

Figure 6: a,b,c *Branchiostoma belcheri* species at different sampling sites in Cooum, Muttukadu, Pondicherry and 22d, species at different sampling stations respectively.

period (Figure 8d). Statistical significant changes were occurred in sampling stations and site ($P \leq 0.05$).

Amphioxus population density (No/m^2)

In site wise amphioxus was observed during the study period at Cooum the minimum of 4.22 No/m^2 at site II and the maximum of 98.12 No/m^2 at site IV (Figure 9a). At Muttukadu, the minimum of 1.10 No/m^2 was observed at site IV and the maximum of 165.12 No/m^2 at site I (Figure 9b). At Pondicherry, the minimum of 2.35 No/m^2 was observed at site IV and the maximum of 98.8 No/m^2 at site II (Figure 9c). The station wise maximum was observed at Muttukadu and minimum in Cooum during the study period (Figure 9d).

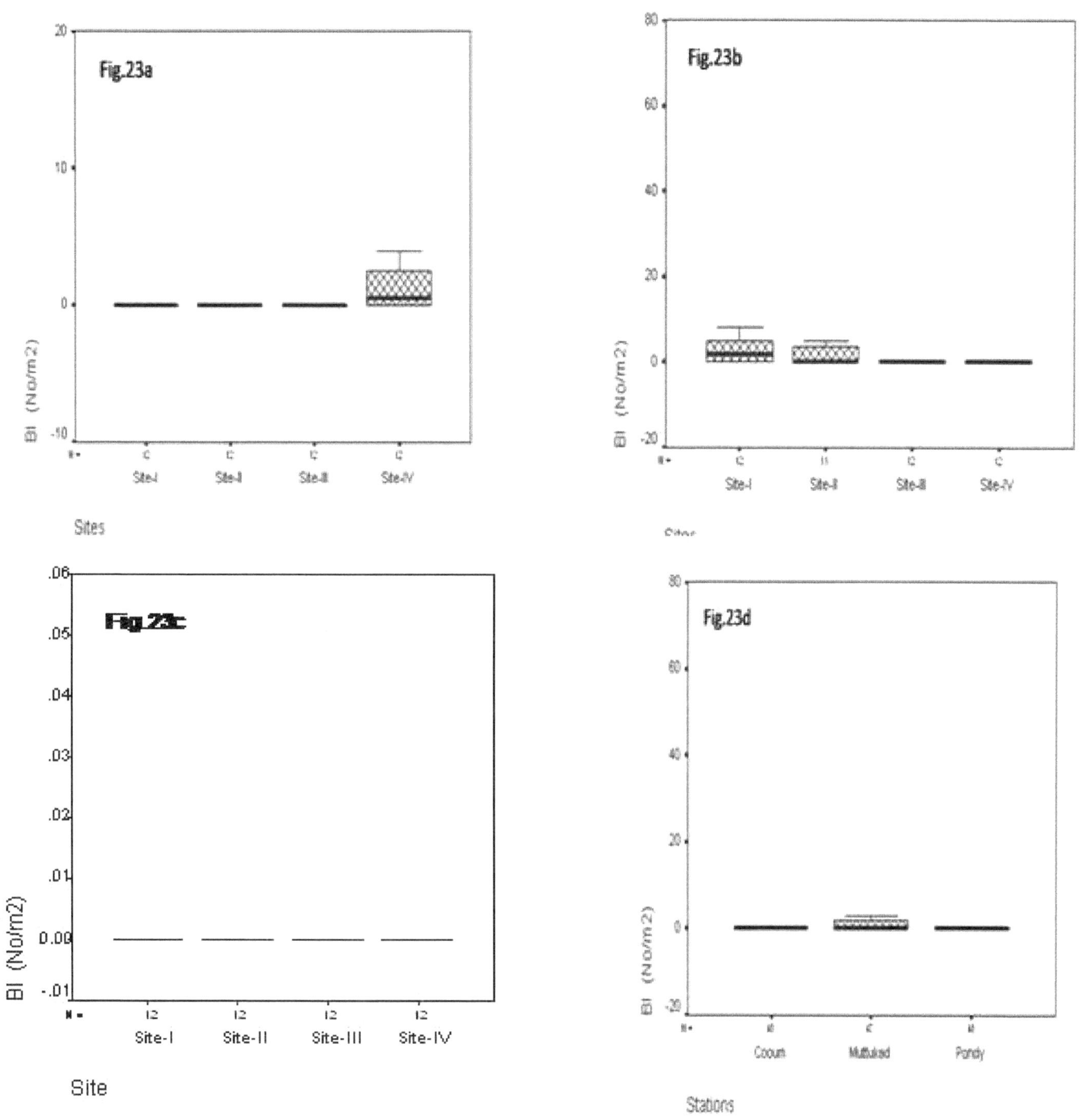

Figure 7: a,b,c *Branchiostoma indium* species at different sampling sites in Cooum, Muttukadu, Pondicherry and 23d, species at different sampling stations respectively.

Statistical interpretation

Analysis of variance test results (ANOVA) for comparison of sediment characteristics between stations 2, 3 and 4 are given in table 1. Cluster analysis is technique in which entities are sequentially linked together according to their similarity or dissimilarity producing a two dimensional hierarchical structure (dendrogram). In the present study, the cluster analysis was employed to know the similarity between samples and the results are shown in figure 10. Dendrogram (Figure 10) and MDS ordination (Figure 11) were showed that the distribution and abundance of *Branchiostoma* spp. between three stations, among them Muttukadu and Cooum showed similar distribution and abundance whereas Pondicherry showed different distribution and abundance pattern (Table 2 and Figure 12). The dendrogram clearly revealed that (Figure 13) Muttukadu and Cooum were grouped at the highest level of similarity (85%), followed by Pondicherry group at the next level of similarity. It is indicates that the excellent ordination pattern between the samples and stations (Figure 13). In order to ascertain the

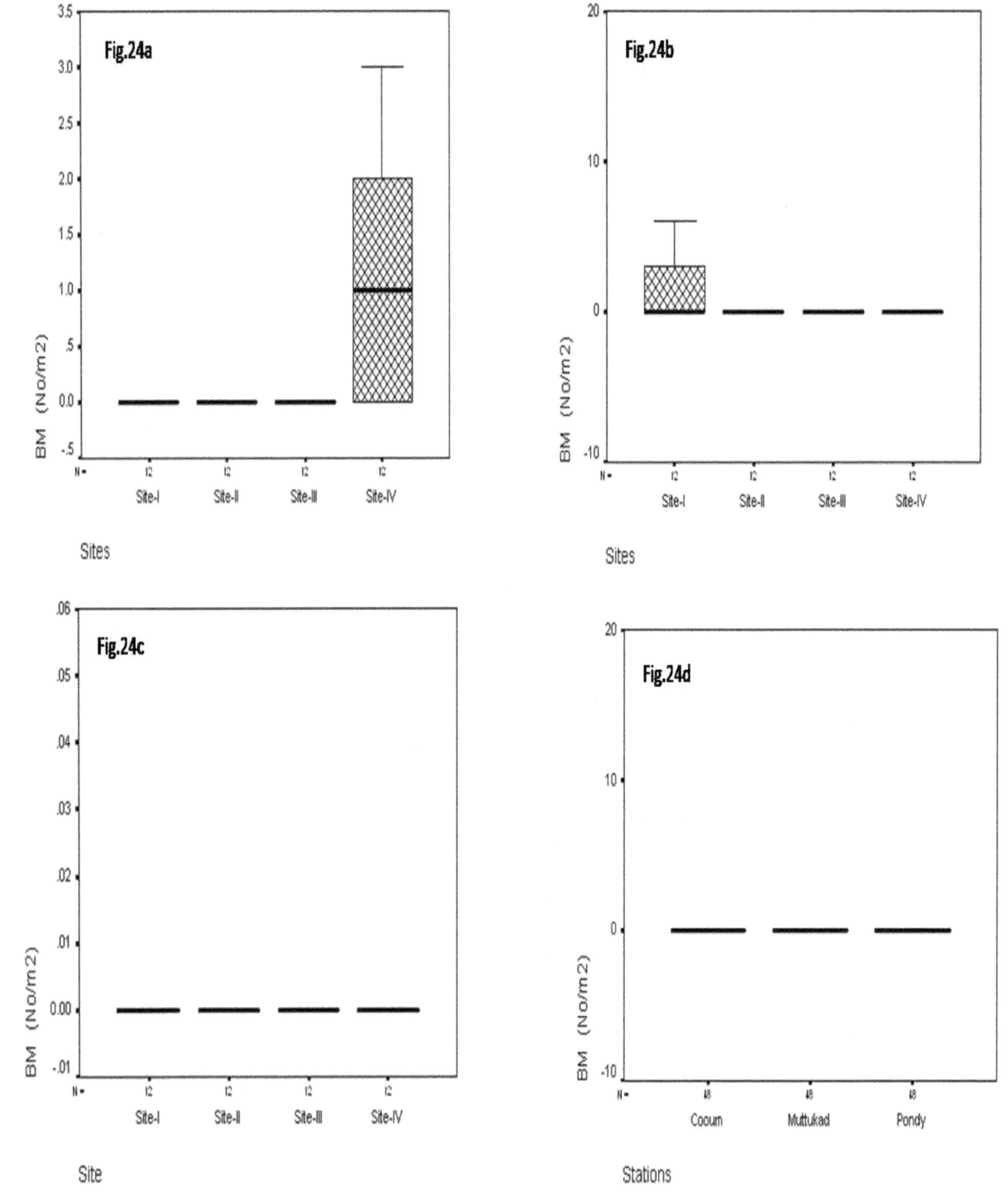

Figure 8: a,b,c *Branchiostoma malayanum* species at different sampling sites in Cooum, Muttukadu, Pondicherry and 24d, species at different sampling stations respectively.

diversity pattern graphically, K- dominance plot was drawn combine for 3 stations (Figure 14) and shown the pronounced variation between these 3 stations. The trend observed in diversity index was quite evident here as well. The plots are drawn for 3 stations blue colour indicating the presence of more number of species at Muttukadu. As the percentage contribution of each species was added, the curve extended horizontally giving `S` shape, before reaching the cumulative 100%.

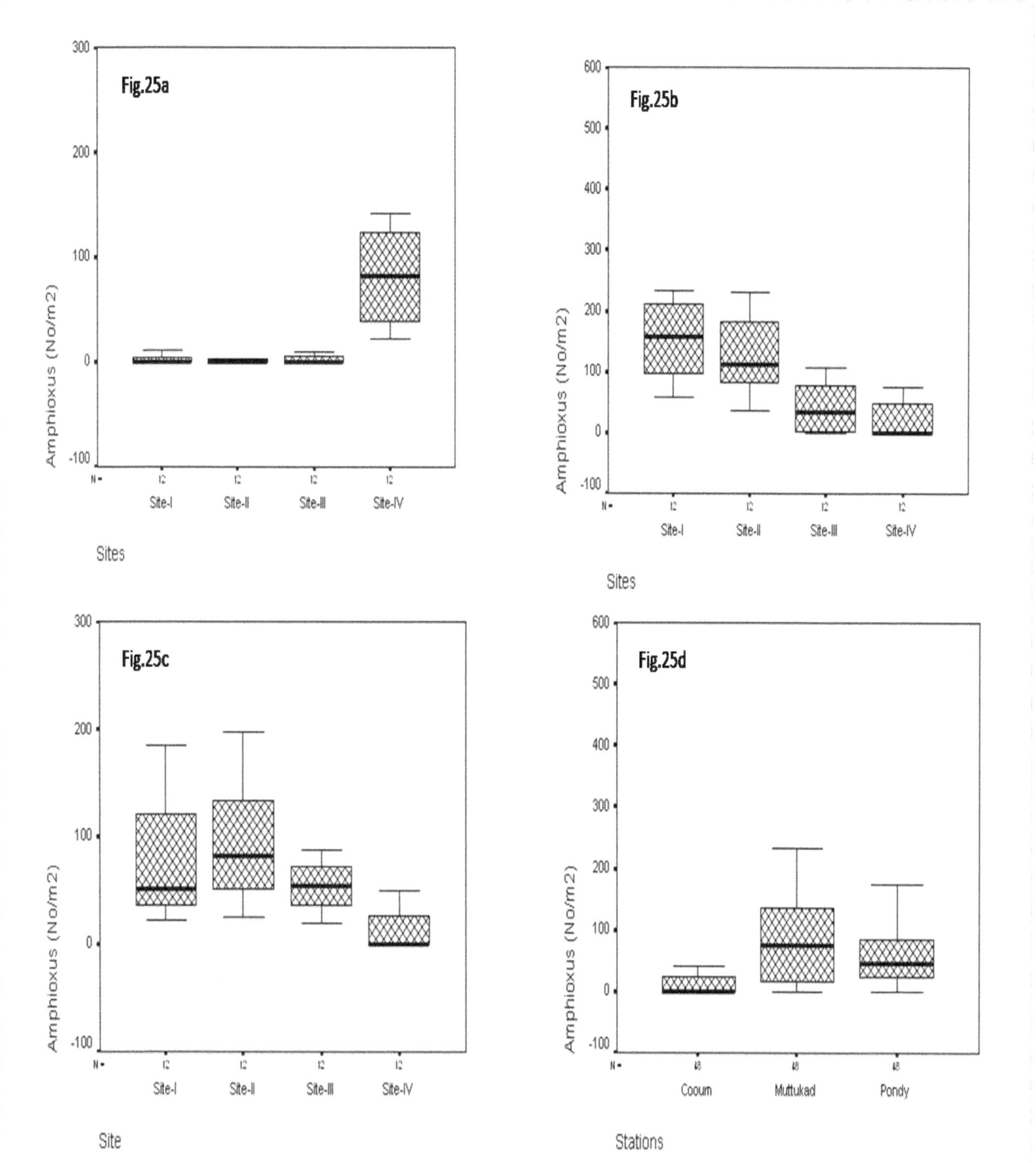

Figure 9: a,b,c Amphioxus Nos. at different sampling sites in Cooum, Muttukadu, Pondicherry and 25d, nos. at different sampling stations respectively.

Discussion

Order Amphioxiformes, Family Branchiostomidae, amphioxus is a lancelet in the subphylum Cephalochordata. The cephalochordate amphioxus the closest invertebrate relative to the vertebrates, possessing a vertebrate-like body plan, with nothocord, hollow dorsal nerve cord, segmented muscle are arranged in blocks called myomeres, perforated pharyngeal region slits and post anal tail and tail that runs past the anus [3,6,7]. However, amphioxus is devoid of most of the complex features of vertebrates, as an elaborated brain or paired

Source	Type III Sum of Squares	df	Mean Square	F	Sig.
STATIONS	98788.722(a)	2	49394.361	10.934	0.000
SITE	65573.833(a)	3	21857.944	4.566	0.004
MONTH	68961.222(a)	11	6269.202	1.241	0.266
TOTAL ORGANIC CARBON (mg/g)					
Corrected Model	623569.012(a)	99	6298.677	2.415	0.001
Intercept	395412.178	1	395412.178	151.577	0.000
TOC (mg/g)	623569.012	99	6298.677	2.415	0.001
Error	112171.967	43	2608.650		
Total	1226352.000	143			
Corrected Total	735740.979	142			
SAND (%)					
Corrected Model	646839.889(a)	124	5216.451	1.114	0.413
Intercept	467069.923	1	467069.923	99.774	0.000
SAND (%)	646839.889	124	5216.451	1.114	0.413
Error	88944.000	19	4681.263		
Total	1229056.000	144			
Corrected Total	735783.889	143			
SILT (%)					
Corrected Model	575222.722(a)	131	4391.013	0.328	0.999
Intercept	435261.316	1	435261.316	32.531	0.000
SILT (%)	575222.722	131	4391.013	0.328	0.999
Error	160561.167	12	13380.097		
Total	1229056.000	144			
Corrected Total	735783.889	143			
CLAY (%)					
Corrected Model	557591.472(a)	124	4496.705	0.479	0.991
Intercept	413958.855	1	413958.855	44.139	0.000
CLAY (%)	557591.472	124	4496.705	0.479	0.991
Error	178192.417	19	9378.548		
Total	1229056.000	144			
Corrected Total	735783.889	143			
DEPTH (m)					
Corrected Model	298368.438(a)	68	4387.771	0.752	0.883
Intercept	337316.631	1	337316.631	57.837	0.000
DEPTH (m)	298368.438	68	4387.771	0.752	0.883
Error	437415.451	75	5832.206		
Total	1229056.000	144			
Corrected Total	735783.889	143			
Branchiostoma lanceolatum **(No/m²)**					
Corrected Model	640283.108(a)	43	14890.305	15.592	0.000
Intercept	644378.171	1	644378.171	674.736	0.000
BL (No/m²)	640283.108	43	14890.305	15.592	0.000
Error	95500.781	100	955.008		
Total	1229056.000	144			
Corrected Total	735783.889	143			
Branchiostoma belcheri **(No/m²)**					
Corrected Model	595253.657(a)	43	13843.108	9.851	0.000
Intercept	757434.095	1	757434.095	538.983	0.000
BB (No/m²)	595253.657	43	13843.108	9.851	0.000
Error	140530.232	100	1405.302		
Total	1229056.000	144			
Corrected Total	735783.889	143			
Branchiostoma malayanum **(No/m²)**					
Corrected Model	36035.114(a)	7	5147.873	1.001	0.434
Intercept	89169.516	1	89169.516	17.331	0.000
BM (No/m²)	36035.114	7	5147.873	1.001	0.434
Error	699748.775	136	5145.212		
Total	1229056.000	144			
Corrected Model	36035.114(a)	7	5147.873	1.001	0.434

Table 1: Analysis of variance (Two way) dependent variable Amphioxus No/m² were expressed in different stations Cooum, Muttukadu and Pondicherry.

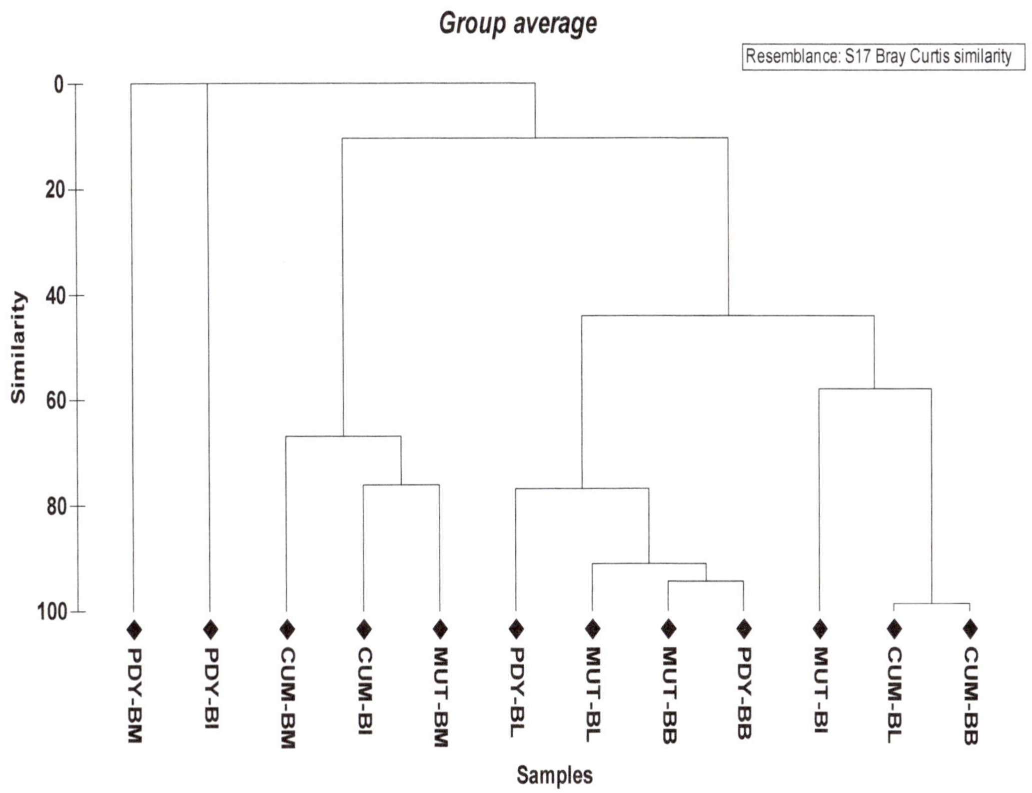

Figure 10: Dendrogram for hierarchical clustering (group-average linking) of the amphioxus recorded in over all the station during the study period.

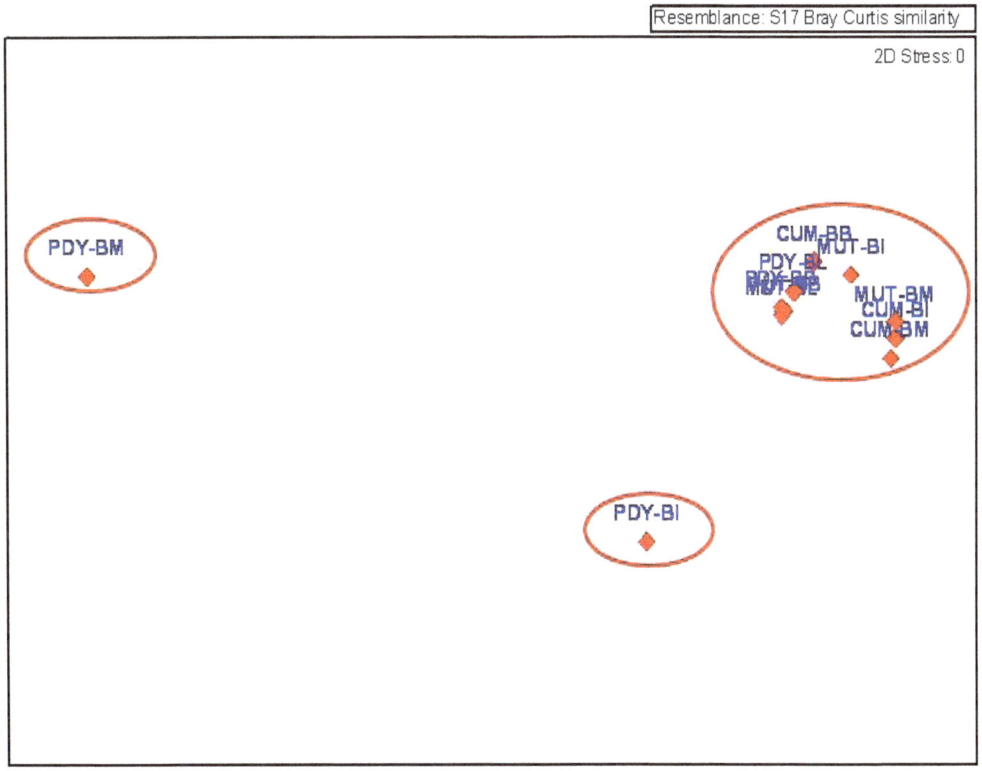

Figure 11: MDS ordination for the Amphioxus recorded in sampling station.

	S	N	d	J'	H'(log2)	1-Lambda'
CUM-BL	21	588	3.136405	0.845484	3.713634	0.898695
CUM-BB	17	600	2.5012	0.795957	3.253444	0.864307
CUM-BI	6	27	1.517065	0.759718	1.963841	0.669516
CUM-BM	7	13	2.339227	0.968632	2.719295	0.910256
MUT-BL	33	2147	4.171106	0.87846	4.431298	0.927713
MUT-BB	33	1893	4.240703	0.917777	4.629627	0.951241
MUT-BI	13	234	2.199687	0.68583	2.537872	0.764352
MUT-BM	9	46	2.089514	0.899362	2.850909	0.846377
PDY-BL	41	1172	5.660537	0.940736	5.040041	0.963192
PDY-BB	40	1694	5.245568	0.922141	4.907566	0.959461
PDY-BI	0	0	0	0	0	0
PDY-BM	0	0	0	0	0	0

Table 2: Amphioxus diversity in station 1, 2 and 3 during the study period.

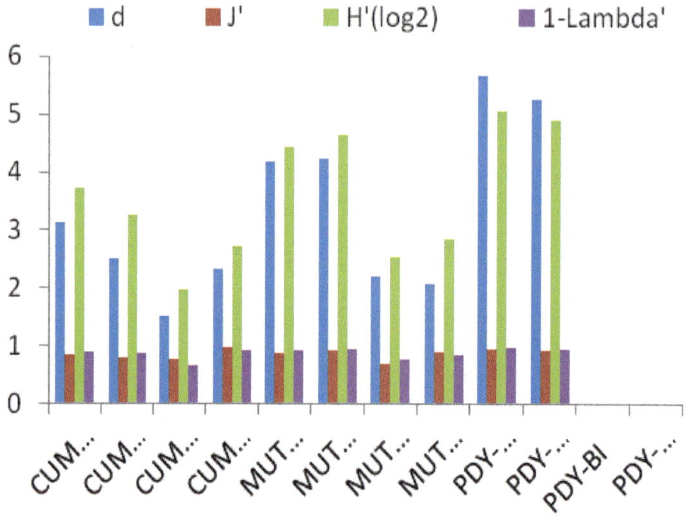

Figure 12: Amphioxus diversity in station 1, 2 & 3 during the study period.

fins. It is one of the marine invertebrate are inhabits in soft substrates and intertidal shallow regions. The evolutionary science it is used as a model organism to study the development of vertebrates. It has an assortment of cells and organs for sensing light and mechanical stimuli and also it has simple but plausible homologs of both the pineal and paired eyes of vertebrates. Male and female amphioxus is moreover similar and also due to differs only in the structure of gonads. The fish like structure of marine chordates are in the environment among their distribution the number of species usually found half buried in sand. Amphioxus is important for food in human and domesticated animals and evolution studies and they provide indications about the origin of the vertebrates [8,9]. The related modern studies are important especially biotechnology oriented research. Diversity of amphioxus is important because it provides own ecosystem and culture with unique and inspirational perspectives. It can create new ideas and changes that can be beneficial to an ecosystem.

In the present study the distributions and abundance of amphioxus at three stations and 12 sampling sites were examined by the basis of distance and depth, in which the results showed that amphioxus was present in all the three stations, mostly at near-shore sites expect Cooum (up to 1 km offshore). Where in the total of 7 out of the 12 sampling sites were recorded with the presence of amphioxus respectively, Cooum, Muttukadu and Pondicherry, the density from 3 to 478 ind/m^2, and 58.52 ind/m^2 on average. Muttukadu, amphioxus was present at two of the four sampling sites, with density from 40 to 230 ind/m^2, and 123.5 ind/m^2 on average. Cooum, amphioxus was present at all four sampling sites, with density ranging from 2 to 273ind/m^2, and 30.91 ind/m^2 on average. Pondicherry, amphioxus was also present at all the three sampling sites, with density ranging from 22 to 197 ind/m^2, and 62.51 ind/m^2 on average. Of all four of the 3 stations sampled, the highest abundance was located at Muttukadu site 1km, where amphioxus was present with density of 123.5 ind/m^2. The marine bottoms have a different benthic substratum nature of composition sand, silt and clay. The present study was considered the maximum amphioxus was observed in organically rich sediments [10-12]. It is one of the main reasons for the maximum species was abundance and diversity in these sampling stations during the study period. Species abundance and diversity maximum was observed in sand and sandy substrate and minimum of species abundance or poor was observed in clay substratum. The distributions and abundances of the amphioxus that produce sediments depend upon such environmental factors and waters in which the organisms live. This study were made in previous studies was represented by Ansari [12] has been reported maximum of fauna are associated with sandy substrate. Varadharajan et al. [13] has been reported the pollution is the major problem of the benthic species

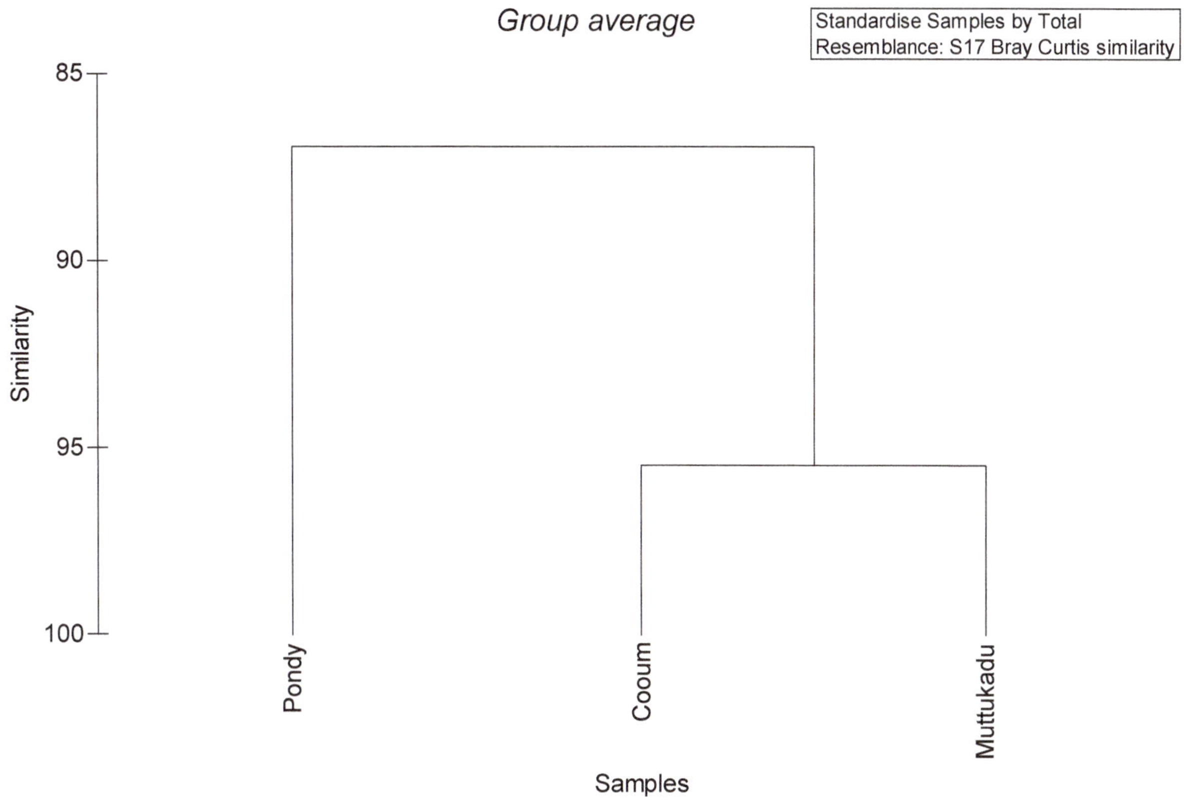

Figure 13: The dendrogram showing the similarity of sampling stations using by cluster analysis.

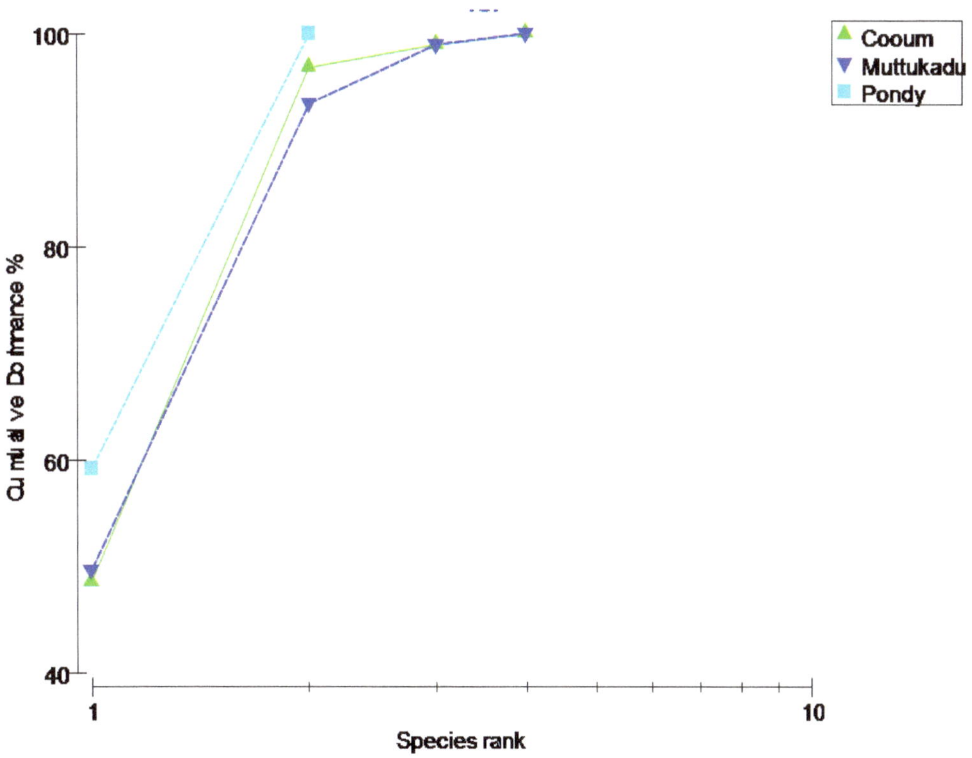

Figure 14: Diversity indices and species abundance of K- Dominance plot.

diversity, Manoharan et al. [14] has been observed substratum is important for benthic species survival and abundance. Sundaravarman et al. [15] has been reported physico-chemical parameters can affect the benthic species diversity. Fishing, climate change and pollution have left an indelible mark on virtually of the coastal ecosystem. Fishing are using with advances in fishing equipment, larger ships and new tracking technologies many species in these ecosystem have been reduced are significantly. The impact of marine species there are environmental problems that are linked to human lifestyles. However the amphioxus studies basically are very difficult ones, since the species identification is not easily. Due to the species the first time were reported along Indian coast it can be helpful for new species identification. The anthropogenic and manmade pollution should be avoiding these coastal ecosystems and monitoring is need for conservation and managements.

References

1. Baker CV (2008) The evolution and elaboration of vertebrate neural crest cells. Curr Opin Genet Dev 18: 536-543.
2. Da Silva LFB, Tavares M, Soares-Gomes A (2008) Population structure of the lancelet *Branchiostoma caribaeum* (Cephalochordata: Branchiostomidae) in the Baía de Guanabara, Rio de Janeiro, southeastern Brazil. Rev Bras Zool 25: 617-623.
3. Chin TG (1941) Studies on the biology of the Amoy amphioxus *Branchiostoma belcheri* Gray. Philipp J Sci 75: 369-421.
4. Coull BC (1973) Estuarine Meiofauna – A review, trophic relationships and microbial interactions 499-511.
5. Holland ND, Chen J (2001) Origin and early evolution of the vertebrates: new insights from advances in molecular biology, anatomy, and palaeontology. Bio Essays 23: 142-151.
6. Mudroch A, MacKnight SD (1994) CRC handbook of techniques for aquatic sediment sampling, (2nd Ed.), CRC Press, Boca Raton, FL, USA.
7. Kaneto S, Wada H (2011) Regeneration of amphioxus oral cirri and its skeletal rods: implications for the origin of the vertebrate skeleton. J Exp Zool B Mol Dev Evol 316: 409-417.
8. Poss SGaB, Boschung HT (1996) Lancelets (Cephalochordata: Branchiostomatidae): how many species are valid? Israel J Zool Suppl 42: 13-66.
9. Holland LZ, Laudet V, Schubert M (2004) The chordate amphioxus: an emerging model organism for developmental biology. Cell Mol Life Sci 61: 2290-2308.
10. Wang Y, Zhao PS, Wang J (2002) The comparison of the growth and development of the lancelets Branchiostoma belcheri tsingtaunese reared with natural and artificial seawater. Natural Science Journal of Hainan University 20: 62-66.
11. Zhang QJ, Zhong J, Fang SH, Wang YQ (2006) Branchiostoma japonicum and B. belcheri are distinct lancelets (Cephalochordata) in Xiamen waters in China. Zoolog Sci 23: 573-579.
12. Ansari ZA (1977) Macrobenthos of the Cochin backwater. Mahasagar- Bulletin of the National Institute of Oceanography 10: 169-171.
13. Varadharajan D, Soundarapandian P, Gunalan B, Babu R (2010) Seasonal Abundance of Macro Benthic Composition and Diversity along the South East Coast of India. Europ J Appl Sci 2: 1-5.
14. Manoharan J, Varadharajan D, Thilagavathi B, Priyadharsini S (2011) Biodiversity and abundance of benthos along the South East Coast of India. Advan Appl Sci Res 2: 554-562.
15. Sundaravarman K, Varadharajan D, Babu A, Saravanakumar A, Vijayalakshmi S, et al. (2012) A study of a marine benthic fauna with special reference to the environmental parameters, south east coast of India. Inter J Pharma Biol Arch 3: 1157-1169.

6

Assessment of Oxidative Stress and Histopathological Biomarkers in the Parablennius Incognitus Fish as Potential Contamination Indicators of the Bay of Sousse (Tunisia)

Tahar Gharred[1*], Azza Naija[1], Rawka Rayena Bouali[1], Zohra Haouas[2] and Benoît Chénais[3]

[1]*Research Unit of Genome, Immune Diagnosis and Valorization, Higher Institute of Biotechnology of Monastir, University of Monastir, Tunisia*

[2]*Laboratory of Histology Cytology and Genetics, Faculty of Medicine, Monastir 5019, Tunisia*

[3]*Laboratoire Mer Molécules Santé, EA2160, Université du Maine, Le Mans, France*

Abstract

Oxidative stress and histopathological biomarkers were investigated in the gills and liver of *Parablennius incognitus* fishes. Individuals sampled in the bay of Sousse were compared to fishes from a reference site, i.e. Ghdamsi Island (Tunisia). Severe biochemical and histological alterations were observed in fishes from Sousse bay and associated with urban discharges contamination. Several tissue alterations were also observed in gills, particularly lamellar epithelium detachment, disorganization of pillar cells, and hypertrophy of chloride cells. In the liver we noted particularly hypertrophy of hepatocytes, congestion and dilation of the central vein and sinusoid capillaries, pyknotic nucleus, and hepatic steatosis. The high reduction of superoxide dismutase and catalase activities in gills and of superoxide dismutase and glutathione peroxidase activities in the liver of contaminated fishes may explain the increase of lipoperoxidation in both organs of *P. incognitus*. The integrated biomarker response values found in individuals from the contaminated site were in good agreement with alteration of physico-chemical parameters and with the high level of Cd concentrations detected in water of the stressful place, i.e. the bay of Sousse. Oxidative stress and histopathological alterations were sensitive biomarkers to discriminate between fishes from the polluted Sousse bay site and those from the reference site, suggesting their potential utility in bio monitoring.

Keywords: *Parablennius incognitus*; Histology; Superoxide dismutase; Catalase; Glutathione peroxidase; Malondialdehyde; Oxidative stress; Pollution

Introduction

The marine littoral ecosystem and estuaries are threatened by increasing levels of various pollutants originating from human activities, urban, agricultural and industrial discharges. A considerable number of chemicals have already been released into the environment and persist in sediment, water and biota. Organic and inorganic chemicals are a major factor of aquatic organisms' poisoning [1]. In the recent years, environmental biomarkers have quickly evolved and have been extensively used for the assessment of the potential adverse effects of pollutants on the environment [2]. The assessment of biological effects of chemicals, from molecular to tissue levels, has been considered as an effective biological tool in the bio monitoring of marine ecosystem contamination [3]. Blenny fishes are interesting sentinel organisms for contaminants exposure in bio monitoring programs, since their diversity, their ubiquity, and their position in the trophic chain make them particularly exposed to sediment-associated contamination. In the Tunisian coast, fourteen Blenniidae species have been identified [4,5]. The use of Blennies for toxicity assays has been the subject of international recommendations [6-10]. However, only a few ecotoxicological studies have been conducted to date on Blenniidae species. For example, the works of Barhoumi et al. using *Salaria basilisca*, reported first the bioaccumulation of cadmium in the liver and gills of this Blennie fish, [11] and then the spatial and seasonal variability of some biomarkers, i.e. metallothioneins with their reduced and oxidized forms, labile zinc, 7-ethoxyresorufin-O-deethylase, DNA strand breaks and lipid peroxidation [12]. Beside, Messaoudi et al. [13] studied diverse antioxidant enzyme activities under laboratory controlled conditions. Thus, to the best of our knowledge, the present work is the first study that aims to assess the biochemical and histological biomarkers responses in the Blenniidae family in field conditions and particularly in the *P. incognitus* species from the Tunisian coast. Biochemical biomarkers are extensively used in bio monitoring of aquatic ecosystems using bivalves and fishes, as well as indicators of chemical complex pollution. They have been validated in crustaceans, bivalves and fishes from the Tunisian coast area [14-19]. The biomarker approach, coupled with a chemical analysis, was used to examine the environmental status of several sites in the Tunisian coast area, and to assess the relationships between pollutants and biological responses. The antioxidant enzymes such as catalase (CAT), superoxide dismutase (SOD) and glutathione peroxydase (GPx) are the primary enzymes of the antioxidant defence against oxidative damages generated by multiple environmental stressors and have been studied in fish of Tunisian marine environments [3,19]. Their induction reflects a specific response to variable pollutants including hydrocarbons and metals [20]. In addition, malonedialdehyde (MDA) is a biomarker of oxidative damage that reflects the state of lipid peroxidation of the membranes [21]. Finally, the integrated biomarker response (IBR) was computed from the biochemical biomarker measurements to assess the

***Corresponding author:** Tahar Gharred, Research Unit of Genome, Immune Diagnosis and Valorization, Higher Institute of Biotechnology of Monastir, University of Monastir, Tunisia, E-mail: gharred_tahar@yahoo.fr

ecological risk of the polluted area [in the Tunisian coast] and establish a possible correlation between metals, physico-chemical parameters and their biological effects. Altogether, when used with the Mediterranean fish *P.incognitus,* oxidative stress indicators and histopathological alterations were sensitive biomarkers to discriminate between fishes from the polluted Sousse bay site and those from the reference site, suggesting their potential utility in biomonitoring.

Materials and Methods

Studied area and fish sampling

P. incognitus was chosen because of its abundance on the Tunisian coast, which allowed spatial and seasonally comparisons. Specimens of *P. incognitus* (0.3-2.8 g weigh and 4-6 cm length) were manually collected in winter 2011, and immediately carried to the laboratory within 30 min in a 20 L tank of seawater. Twenty fishes per site were collected from Ghdamsi Island and Sousse bay (Tunisia). Thus, ten fishes per site were used for the biochemical analyses and the other ten served for the histopathological study. The two sites were selected in the east coast of Tunisia. The Sousse bay site, more precisely Sidi Abdelhamid (35°47' N, 10°40'E) is an industrial area located in the city of Sousse. This site is characterized by chronic discharges of urban and industrial pollutants by Hamdoun drain [22]. As a reference site, we selected Ghdamsi Island (Monastir, Tunisia) (36°02'N, 10°29' E), which is a relatively clean area, without industrial or urban influences (Figure 1). The reproductive season of *P.incognitus* is generally in summer (data under publication) and the sampling was carried out during the non-reproductive period to avoid interference of reproduction with the oxidative stress.

Physico-chemical analysis of water

Water quality was assessed at each fish sampling time. One surface water sample per site was taken out and analysed in situ for the following parameters: pH, dissolved oxygen (mg/L), turbidity (NTU: Nephelometric Turbidity Unit) and temperature (°C). All determinations were carried out in triplicate and the variation coefficient was usually less than 10%. For the Cadmium and Lead metal determinations, seawater samples were filtered at the time of collection through a 0.45 µm membrane and acidified for the determination of dissolved elements. All water samples were analysed by inductively coupled plasma/atomic emission spectrometry using OPTIMA 3300 RL apparatus and according to the standard NF EN ISO 11885.

Biochemical analysis

Tissues (liver and gills) of *P. incognitus* were homogenized in phosphate buffer (0.1 M; pH7.4) in a ratio of 1g of tissue per 3 ml buffer. This homogenate was centrifuged at 9500 rpm for 30 min at 4°C. Then, the recovered supernatant was conserved at-80°C until oxidative stress biomarkers measurements. The obtained supernatant was used for total protein determination, superoxide dismutase (SOD), catalase (CAT) and glutathione peroxidase (GPx) activities, lipid peroxidation (MDA) level, and protein determination.

Total protein determinations: Protein contents in the individual supernatant tissues were determined according to Bradford's method (1976) by using bovine serum albumin (BSA) as a standard. 50 µL of supernatant was added to 150 µL of distilled water and 2 mL of Bradford's reagent. After 5 min of incubation in the darkness, the absorbance was determined at λ=595 nm.

Superoxide dismutase (SOD) determination: SOD activity was determined using pyrogallol as a substrate by the method of Marklund and Marklund [23]. This method is based on pyrogallol oxidation by the superoxide anion and its dismutation by SOD. One unit (U) of total SOD is defined as the amount of enzyme required to inhibit the rate of pyrogallol autooxidation by 50%.

Catalase (CAT) determination: Catalase activity was determined by the method of Claiborne [24] measuring the rate of enzymatic decomposition of hydrogen peroxide determined as absorbance decrements at 240 nm. Reaction mixture (final volume of 1.78 mL) contained 1.56 mL of 100 mM phosphate buffer (pH 7.5) and 200 mL of 500 mM hydrogen peroxide After a 30 sec pre-incubation time, the reaction was started by the addition of 20 µL of cytosolic fraction. CAT activity was evaluated by kinetic measurement at 25°C. Results were expressed as µmol hydrogen peroxide consumed per min and per mg of protein.

Glutathione peroxidase (GPx) determination: GPx activity was performed according to the method described by Günzler et al. [25]. In

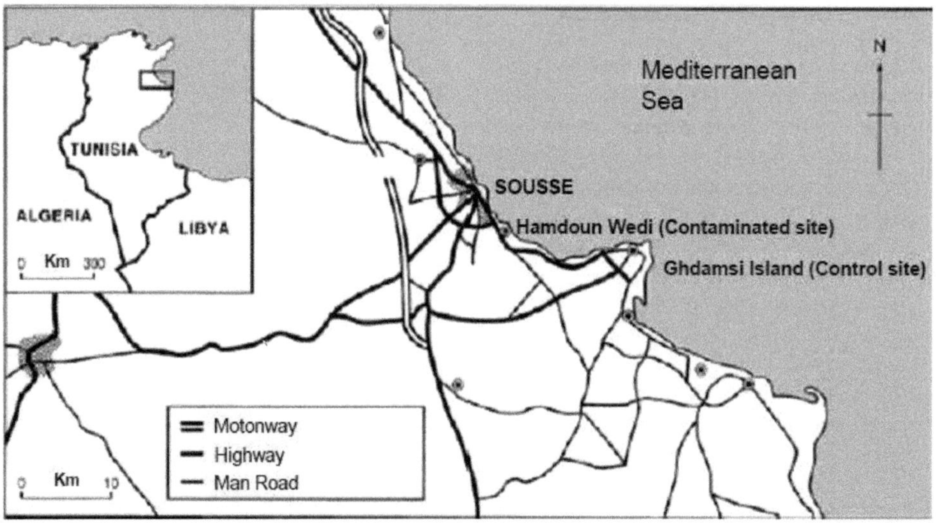

Figure 1: Map of Tunisia indicating the location of sampling sites in the bay of Sousse (Hamdoun Wadi) as a contaminated site and Ghdamsi Island (Monasti) as the reference site.

the typical assay, we used t-butylhydroxyde (t-BOOH) as oxidant and glutathionereductase (GR) allowing the reduction of GSSG to GSH by oxidizing NADPH to NADP. The activity was determined by measuring the decrement of NADPH at λ=340 nm. The activity is expressed as μmol of NADPH per min per mg of protein.

Lipid peroxidation (MDA) determination: Lipid peroxidation was estimated in terms of thiobarbituric acid reactive species production using 1,1,3,3-tetraethoxypropane as a standard. The reaction was determined at 532 nm using thiobarbituric acid reagent according to the method of Buege and Aust (1978). Malondialdehyde (MDA) content was expressed as nmol equivalent MDA per mg of protein.

Histopathological lesion determinations

Histopathological analysis was performed on the gills and liver of 10 specimens of *P. incognitus*. Sample tissues were fixed in Bouin's solution for 24 h, dehydrated with ethyl alcohol 95% and 100%, and then transferred into paraffin. The longitudinal sections of 4 to 5 μm were stained with Masson's trichrome color and observed with optical microscope.

Statistical analysis

The biochemical biomarkers data were expressed as mean ± SD (n=10). The biochemical data were first tested for normality and homogeneity of variance to meet statistical demands. The data were analyzed by the Student t-Test using the SPSS Statistical Program (Version 19.0); significance is noted as *, $p<0.05$ or **, $p< 0.01$.

Integrated Biomarkers Response determination (IBR)

The responses of the biochemical biomarkers were integrated based on the mathematic method developed by Beliaeff and Burgeot [26]. For each biomarker: (1) Calculation of mean and SD for each site. (2) Standardization of the data for each site, i.e. Xi'=(Xi-x)/S), where Xi' is the standardized value of the biomarker, Xi is the mean value of a biomarker from each site, x is the mean of the biomarker calculated for all sites, and S is the standard deviation calculated for the site-specific values of each biomarker. (3) Using standardized data, we add the value obtained for each site Xi' to the absolute (non-negative) value of the minimum value: B=Xi'+ | Xmin |, where B is the score of each biomarker and for each site. Then, we adjusted the lowest value in the set to zero for all the biomarkers. (4) Calculation of star plot areas by multiplication of the obtained value of each biomarker (Bi) with the value of the next biomarker, arranged as a set. (5) Summing-up of all values. The corresponding IBR value (average of different arrangements of biomarkers in the set) is: $\{[(B_1 \times B_2)/2]+[(B_2 \times B_3)/2] + \cdots + [(B_{n-1} \times B_n)/2]\}$ where B_{n-1} and B_n are two scores of two successive biomarkers.

	Reference site	Contaminated site	Tunisian norms
Temperature (°C)	27 ± 1[a]	32 ± 2	35
pH	7.29 ± 0.06	8.83 ± 0.08	6.6-8.5
Dissolved O_2 (mg/L)	9.6 ± 0.47	4.5 ± 0.45	>6
Turbidity (NTU)	1.08 ± 0.05	2.57 ± 0.08	N.A
Total nitrogen (mg/L)	10.5 ± 2	245 ± 11.53	30
Lead (mg/L)	0.277 ± 0.01	0.277 ± 0.02	0.5
Cadmium (mg/L)	0.0038 ± 0.001	0.0198 ± 0.003	0.005

[a]Data are reported as mean ± SD of three measurements of the same sample per site. N.A: not available.

Table 1: Physico-chemical analysis of water samples from the Sousse bay polluted site and from Ghdamsi Island (Monastir) reference site.

Results

Routine physico-chemical parameters differed between reference and contaminated sites. In the Sousse bay polluted site, temperature, turbidity, total nitrogen and pH were high; and the dissolved oxygen was reduced when compared to the reference site values (Table1). In addition, a high Cd level was also noted in the seawater of the contaminated site that exceeds the Tunisian authorized level.

Biochemical biomarkers responses

Reduced SOD and CAT activities and increased MDA level in gills: The activity of antioxidative enzymes SOD and CAT was significantly reduced ($p <0.05$) in fishes from the Sousse bay contaminated site, being 72% and 48% of the reference values, respectively (Table 2). However, no significant change ($p>0.05$) in GPx activity was observed. In contrast to the reduction of antioxidant enzymes, MDA level was significantly increased by 70% with respect to the reference value (Table2), indicating a high lipid peroxidation level in fishes from the polluted site compared to the reference individuals.

Reduced SOD and GPx activities and increased MDA level in liver: In the liver, the CAT activity, which is higher than in gills, showed no significant differences between fishes from the reference site and those from the contaminated site (Table 2). By contrast, the GPx activity, unchanged in gills, was reduced to 48% of the reference value in fishes from Sousse bay (Table 2).In addition, the SOD activity was decreased to 76% of the reference value. In the same way than in gills, the reduction of antioxidant enzyme activities (i.e. GPx and SOD in liver) was associated with a significant increase in lipid peroxidation, MDA level being greater of 27.5% relative to the reference value (Table 2).

Histopathological biomarkers

Gill histopathological lesions in contaminated fishes: Figure 2A showed the typical structure of the gills of *P. incognitus* from the reference site. The primary lamellae are lined by a thick stratified epithelium containing chloride cells. In addition a cartilaginous support composed of chondrocytes could be found in the centre of each primary lamella. The secondary lamellae, located on the coast side of the primary lamellae, contain erythrocytes and pillar cells (Figure 2A) as usual. When comparing histological sections of fishes of both contaminated and reference sites, an unusual tissue aspect appears in the gill of specimens from the contaminated site. The cartilage lost its chondrocytes in 40% of the samples. Limited hyperplasia was noted in individual fishes from the contaminated site, resulting in the fusion of secondary lamellae in 40% of the samples. These lamellae were narrowed with the development of telangiectasia in 60% of the samples. Similarly, the epithelium of secondary lamellae remained a single stratum and erythrocytes formed together an epithelioid line in 60% of the samples (Figure 2B).

Liver histopathology lesions in contaminated fishes: In the liver of fishes from the reference site, the tissue sections showed clearly the homogeneous structure of the parenchyma (Figure 2C). The central vein, the sinusoids and polygonal hepatocytes could be observed in the

Biomarker	Reference Site	Contaminated site	%
Gills			
Superoxide dismutase (SOD)	41.925 ± 8.985^{a,b}	32.526 ± 11.780^{a,b}	76%^{f}
Catalase (CAT)	3.106 ± 1.058^{a,c}	1.390 ± 0.512^{a,c*}	45%
Glutathione peroxidase (GPx)	1195.311 ± 520.825^{a,d}	1154.000 ± 173.30^{a,d}	N.A.
Lipid peroxidation (MDA)	0.040 ± 0.005^{a,e}	0.0510 ± 0.005^{a,e**}	127.50%
Integrated Biomarker Response (IBR)	0.812	1.364	
Liver			
Superoxide dismutase (SOD)	36.923 ± 11.845^{a,b}	26.937 ± 10.495^{a,b*}	72%
Catalase (CAT)	15.294 ± 11.888^{a,c}	79.745 ± 40.780^{a,c}	N.A.
Glutathione peroxidase (GPx)	973.373 ± 405.460^{a,d}	470.466 ± 282.908^{a,d*}	48%
Lipid peroxidation (MDA)	0.037 ± 0.007^{a,e}	0.063 ± 0.007^{a,e**}	170%
Integrated Biomarker Response (IBR)	0	7.025	

[a]Data are represented as the means ± SD; n = 10. Statistical differences between contaminated and reference groups were made at 0.05 confidence level. Statistical significance of the results is indicated as *, $p<0.05$; **, $p<0.01$. Enzymatic activity units are bμmol/min/mg protein for SOD, cμmol H2O2 consumed/min/mg protein for CAT, dμmol NADPH/min/mg protein for GPx, and e nmol TBARS/mg protein for MDA level.f For the oxidative stress biomarkers, the value measured in samples from the contaminated site is given as the % of the reference value when a significant difference was observed. N.A., not applicable.

Table2: Oxidative stress biomarkers and Integrated Biomarker Response (IBR) in gills and liver of P.incognitusfishes from reference and contaminated sites.

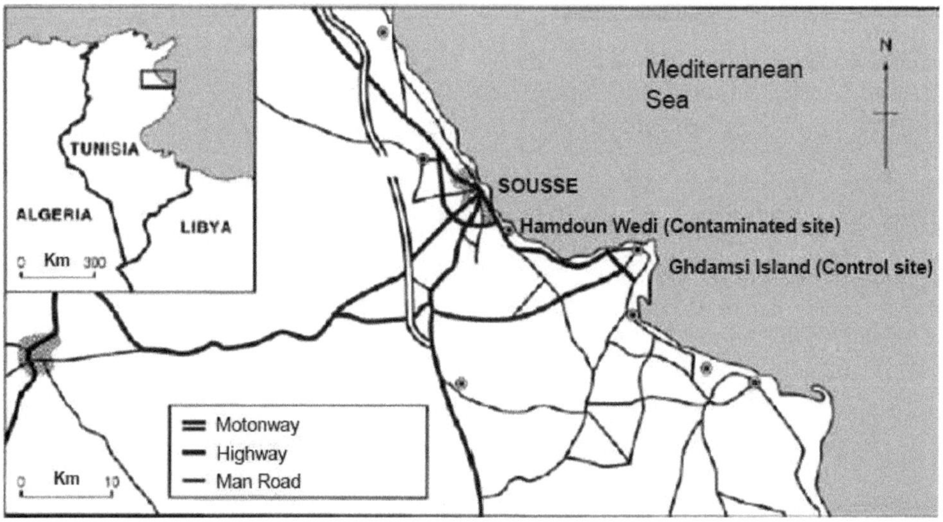

Figure 1: Map of Tunisia indicating the location of sampling sites in the bay of Sousse (Hamdoun Wadi) as a contaminated site and Ghdamsi Island (Monasti) as the reference site.

parenchyma (Figure 2C). Each hepatocyte contained a central spherical nucleus with concentrated chromatin near the nuclear membrane. In addition, the hepatocytes of reference fishes stained with Masson's trichrome showed a good storage of glycogene. The cytoplasm is homogeneous and composed of basophilic granulations. The nucleus is round with dense chromatin and individual nucleoli (Figure 2E). However, the microscopic observation of *P. incognitus* hepatic tissue of fishes from the contaminated site showed various tissue alterations. Hypertrophy of hepatocytes was seen in 40% of the samples, congestion and dilation of the central vein and sinusoid capillaries in 70% of them (Figure 2D, compared to Figure 2C). Associated to these pathological changes, a hepatic steatosis was also reported in 70% of the samples. Moreover, in 40% of the samples, the nucleolus has disappeared and the nucleus became pyknotic, displaying a dense chromatin, characteristic of necrotic cells (Figure 2F, compared to Figure 2E).

Discussion

Oxidative stress biomarkers responses

The present study investigated the oxidative stress and histopathological biomarkers responses in *P.incognitus* fishes from the bay of Sousse, an area of the Tunisian coast impacted by different anthropogenic activities. In the present work, we showed that the decrease in gills SOD, CAT, and hepatic SOD and GPx activities was highly associated with a significant increase of the lipid peroxidation in both organs of the *P. Incognitus* from Sousse bay. Various environmental pollutants may exert a negative effect on the antioxidative enzymes in particular metals, organic compounds such as polychlorobiphenyls (PCBs), and polyaromatics hydrocarbons (PAHs). We suspected that a mixture of chemicals might be present in the chronic discharges of Hamdoun Wadi site in the bay of Sousse. These environmental pollutants deeply affect the antioxidant/proxidant balance leading to the accumulation reactive oxygen species (ROS) and affecting the antioxidative enzymes [27,28]. Moreover, Vutukuru et al. [29] and Ates et al. [30] showed a decrease in SOD and CAT activity in the gills of *Oncorhynchus mykiss* and *Esomus danricus* treated with lead and copper. A recent ecotoxicological study on the *Solea solea* fishes from Sousse bay collected at the discharges point of Hamdoun Wadi, showed a high induction of hepatic and gills CAT activity [31]. The differences in CAT responses may be related to different fish species sensitivity

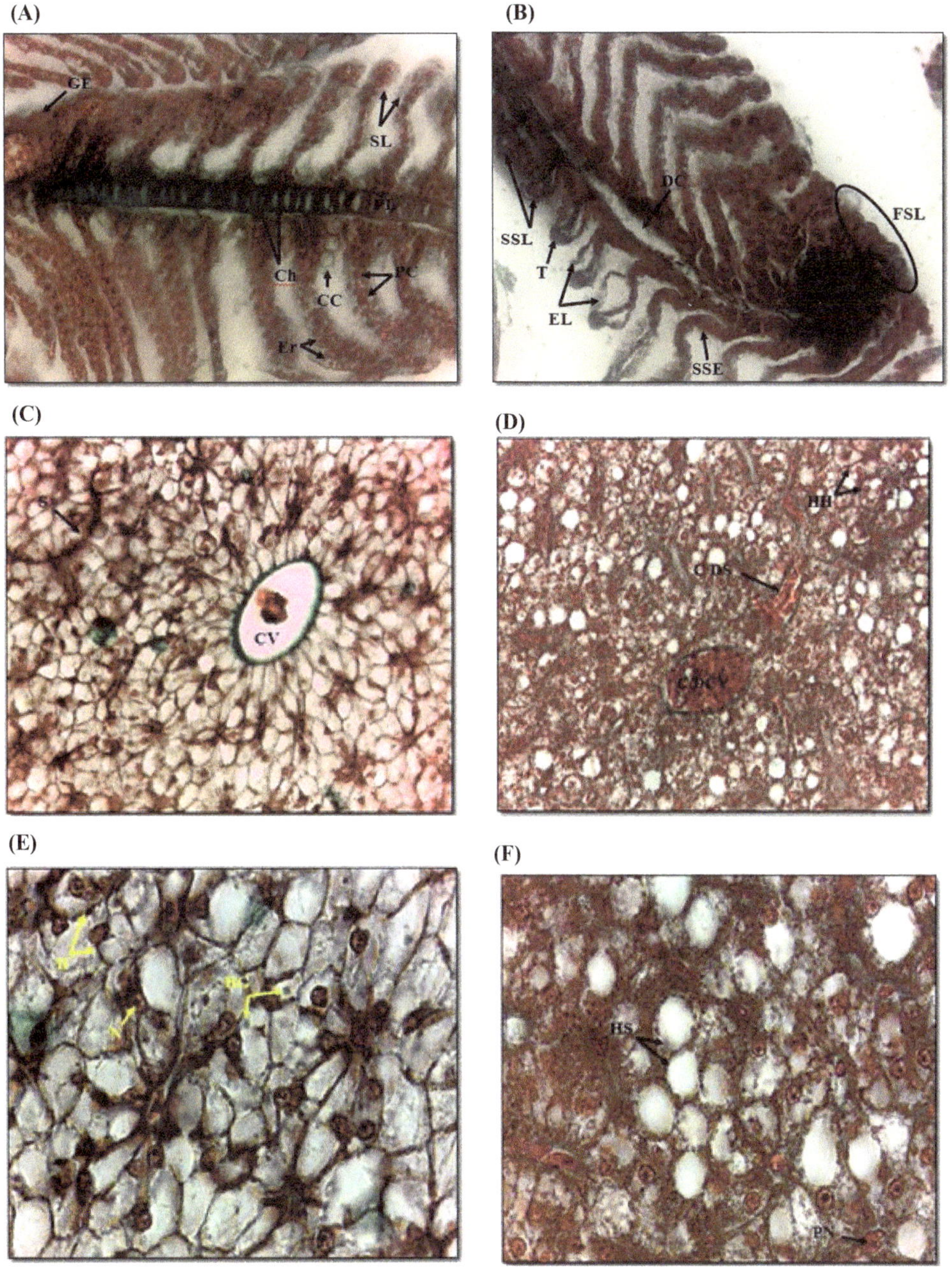

Figure 2: Histological analysis of *P. incognitus*gills (A,B) and liver (C-F) from reference site (A,C,E) and polluted Sousse bay site (B,D,E) colored by Masson trichrome. **(A)** Reference gills section showing the filament (GF), primary lamellae (PL), secondary lamellae (SL), pillar cell (PC), chondrocytes (Ch), gill epithelium (GE), chloride cells (CC) and erythrocytes (Er) (X40). **(B)** Gills section from fish of the contaminated sites howing a single stratum epithelium (SSE), fusion of secondary lamellae (FSL), telangiectasis (T), epithelium lifting (EL), shorting of secondary lamellae (SSL) and disappearance of chondrocytes DC (X40). **(C)** Liver section in reference *P. incognitus* showing sinusoids (S) and central vein (CV) (X40). **(D)** Liver section in *P. incognitus* from the contaminated site showing hypertrophy of hepatocytes (HH) congested and dilated sinusoids (C/DS) and congested and dilated central vein (C/DCV) (X40). **(E)** Liver section in reference *P. incognitus* showing hepatocytes (H) with basophilic. granulation (BG) and central and round nucleus (N) (X100). **(F)** Liver section in *P. incognitus* from the contaminated site showing pyknotic nucleus (PN) and hepatic steatosis (HS) (X40)..

to pollutants. Fishes under contamination stress are susceptible to the effects of ROS and have developed effective antioxidant defenses, including antioxidant substances (vitamin E, reduced glutathione and carotenoids) and enzymes (CAT, GR, GPx) [20]. Oxidative stress occurs when the rate of ROS generation exceeds the antioxidant defense system. Its deleterious effects include oxidation of proteins, DNA, and carbohydrate components, as well as peroxidation of unsaturated lipids in cell membranes [20]. CAT is known to protect the cell by reducing H_2O_2 to H_2O. CAT responds to a wide range of contaminants capable of ROS production such as PAHs, PCBs, heavy metals and pesticides by increase of enzyme activities. Thus, the anti-oxidative stress defenses were found altered in many field studies using crustaceans or fishes as bio indicators [31-34]. The results revealed that the activity of SOD and CAT in gills, and SOD and GPx in liver was highly repressed in the specimen collected from Sousse bay compared to those from the reference site. Such trends in antioxidant defense activities can be found in others native organisms at polluted sites according to the levels and duration of pollutant exposure. Such a repression antioxidant defense activities is similar to that reported by Ghedira et al. [34] in gills and digestive gland of the crab *Carcinus maenas* collected from the contaminated area of Bizerte lagoon (Tunisia). In this work, some differences in physico-chemical parameters (temperature, pH, dissolved oxygen, turbidity, total nitrogen and Cd) between reference and polluted sites were noted. These abiotic stresses highly altered the biochemical responses of fishes from the polluted site. The recent ecotoxicology study of Ben Khedher et al. [18] showed that the alteration of biochemical biomarkers of crabs from polluted area of Tunisian coast was associated with metals contamination and also with seawater physico-chemical parameters such as temperature and dissolved oxygen. The same authors suggested that the elevated temperature and the decrease of dissolved oxygen may increase in return the concentrations of dissolved metals and, indeed, higher concentrations of metals have been accumulated in the Mediterranean crab (*C. maenas*) tissues [35]. In the present work, IBR values found in the contaminated site is in good agreement with the alteration of physico-chemical parameters. Thus the high value of IBR found in the gills and in the liver was associated with the low level of the dissolved oxygen and the increase of temperature, pH, turbidity, and total nitrogen and Cd concentrations detected in water. The higher IBR values suggest that Sousse bay is a stressful place for fish life. Similarly, Damiens et al. [36] found that IBR values calculated from AChE, GST and CAT activities and TBARS concentrations in the three successive experiments were in good agreement with copper and PCB concentrations in transplanted mussels but not with PAH concentrations.

Histopathological biomarkers

Gills are considered an important route for uptake, bioconcentration and excretion of chemical compounds and a prime target to contaminants, due to the wide surface area in contact with the external medium and reduced distance between internal and external medium. In the present study, severe tissular lesions were noted in gills and liver of fishes from Sousse bay. We observed, particularly, the absence of chondrocytes in primary lamellae. The appearance of a single stratum epithelium, the fusion, the telangiectasia, the hyperplasia and the narrowing are also reported in the secondary lamellae. These results are consistent with the research of Koca et al. (2008) who investigated the tissue lesions of *Capito pectoralis* and *Chondrostoma nasus* after exposure to polluted water. The authors reported a passage from the double stratum to the single stratum epithelium in gills, causing a modification in capillary circulation due to the erythrocyte expansion. However, hyperplasia and fusion of secondary lamellae are defence mechanisms for species because they increase the distance through which the water pollutants are able to reach the blood circulation [37,38]. These effects slow free gas exchange [39,40], since they are probably related to the decrease in ATPase activity and/or the declining of sodium and chloride concentrations in the blood [41]. These lesions are increasing the barrier thickness between the water and the blood and therefore they decrease the oxygen uptake [42]. The liver plays key roles in the endogen molecule metabolism and detoxification of xenobiotics. Several molecular, biochemical and physiological processes may intercept the toxic effects of pollutants. The chronic exposure of fishes to contaminants may cause cellular process dysfunctions and tissue lesions. We have demonstrated a necrosis of hepatocytes and hepatic steatosis in fishes from the contaminated site. Others severe lesions such as congestion and dilation of the central vein and the sinusoid capillaries were also observed. Koca et al. [43] reported that congestion of the central vein, steatosis, nuclear pyknosis and hepatocyte necrosis, are the results of the exposure of *Barbus capito pectoralis* and *Chondrostoma nasus* to various types of pollutants. Otherwise, Arellano et al. [41] showed a hepatocellular necrosis, an increase in fat vacuolization and the congestion of central vein and sinusoid capillaries, in *Solea senegalensis* treated with copper. Moreover Braunbeck et al. [44] have demonstrated that alterations of hepatocytes are due to the increase of metabolic activity. The hepatocyte hypertrophy, the nuclear pyknosis and the dilatation of sinusoid capillaries has been shown to result from the inhibition of Na^+/H^+ transport by cadmium and from the depression of osmoregulation process. Hepatic steatisis is one of the detoxification reactions by Kupffer cells [43,45]. It results from the exhaustion of the rough endoplasmic reticulum and disruption of protein synthesis [46]. The observed vacuolization lesions may be due to the accumulation of glycogen particles in the hepatocytes [47].

Conclusions

Oxidative stress and histopathological biomarkers responses, representing different biological endpoints in the *P.incognitus* Blennie fishes, allowed to distinguish between sites receiving chronic domestic discharge and reference site in the Tunisian coastal area, and are thus useful for the assessment of environmental pollution effects. The multiple biomarker responses measured in two distinct tissues provided discrimination between sites with different levels of contamination. Therefore, it is important to note that the responses of histopathological biomarkers depend on the sampling site, and that they are sensitive tissue biomarkers appropriated to be integrated as a complementary biological tool to the biochemical biomarkers in the bio monitoring of marine ecosystems.

Acknowledgments

This work was supported by a fund from the Ministry of Scientific Research and Technology, University of Monastir, Tunisia (Research Unit of Genome, Immune Diagnosis and Valorization, Higher Institute of Biotechnology of Monastir, University of Monastir Tunisia).

References

1. Van der Oost R, Beyer J, Vermeulen NPE (2003) Fish bioaccumulation and biomarkers in environmental risk assessment: A review. Environ Toxicol Pharmacol 13: 57-149.
2. Viarengo A, Lowe D, Bolognesi C, Fabbri E, Koehler A (2007) The use of biomarkers in biomonitoring: A 2-tier approach assessing the level of pollutant-induced stress syndrome in sentinel organisms. Comparative Biochemistry and Physiology Part C 146: 281-300.
3. Jebali J, Banni M, Boussetta H (2012) Biochemical biomarkers in aquatic ecotoxicology: fundamental mechanisms, application and perspectives. 143-168 In Advance in environmental research, Daniel JA, Nova Science Publisher, New York 23: 323p.
4. GharredT, Ktari MH, Ben Salem M (1998) Systematic inventory of Bleaniidae of Tunisian coasts. Cybium 22: 99-105.
5. Gharred T, Ktari MH (2001) polymorphisme interspécifique des Blenniidea (poissons téléostéens) des côtes Tunisiennes. Bull Inst Natn Scien Tech Mer de Salammbo 28: 1-19.
6. Jacobsson A, Neuman E, Thoresson G (1986) The viviparous blenny as an indicator of environmental effects of harmful substances. Ambio 15: 236-238.
7. Jacobsson A, Neuman E, Olsson M (1993) The viviparous blenny as an indicator of effects of toxic substances. Fiskeriverket Kustlaboratoriet 538: 1-25.
8. Schladot JD, Backhaus F, Ostapczuk P, Emons H (1997) Eel-pout (Zoarces viviparus) as a marine bioindicator. Chemosphere 34: 2133-2142.
9. Noaksson E, Tjärnlund U, Ericson G, Balk L (1998) Biological effects on viviparous blenny exposed to Chrysene and held in synthetic as well as in natural brackish water. Mar Environ Res 46: 81-85.

10. Lyons BP, Bignell J, Stentiford GD, Feist SW (2004) The viviparous blenny (Zoarces viviparus) as a bioindicator of contaminant exposure: application of biomarkers of apoptosis and DNA damage. Mar Environ Res 58: 757-761.

11. Barhoumi S, Messaoudi I, Tmim D, Said K, Kerkeni A (2009) Cadmium bioaccumulation in three benthic fish species, *Salaria basilisca*, *Zosterisessor ophiocephalus* and *Solea vulgaris* collected from the Gulf of Gabes in Tunisia. Journal of Environmental Sciences 21: 980-984.

12. Barhoumi S, Messaoudi I, Gagné F, Kerkeni A (2012) Spatial and seasonal variability of some biomarkers in *Salaria basilisca* (Pisces:Blennidae): Implication for biomonitoring in Tunisian coasts. Ecological Indicators 14: 222-228.

13. Messaoudi I, Deli T, Kessabi K, Barhoumi S, Kerkeni A, et al. (2009) Association of spinal deformities with heavy metal bioaccumulation in natural populations of grass goby, Zosterisessor ophiocephalus Pallas, 1811 from the Gulf of Gabès (Tunisia). Environ Monit Assess 156: 551-560.

14. Banni M, Jebali J, Daubeze M, Clerandau C, Guerbej H, et al. (2005) Monitoring pollution in Tunisian coasts: application of a classification scale based on biochemical markers. Biomarkers 10: 105-116.

15. Messaoudi I, Barhoumi S, Said K, Kerkeni A (2009) Study on the sensitivity to cadmium of marine fish *Salaria basilisca* (Pisces: Blennidae). Journal of Environmental Sciences 21: 1620-1624.

16. Kessabi K, Kerkeni A, Saïd K, Messaoudi I (2009) Involvement of cd bioaccumulation in spinal deformities occurrence in natural populations of Mediterranean killifish. Biol Trace Elem Res 128: 72-81.

17. Bouraoui Z, Banni M, Chouba L, Ghedira J, Clerandeau C, et al. (2010) Monitoring pollution in Tunisian coasts using a scale of classification based on biochemical markers in worms Nereis (Hediste) diversicolor. Environ Monit Assess 164: 691-700.

18. Ben-Khedher S, Jebali J, Kamel N, Banni M, Rameh M, et al. (2013) Biochemical effects in crabs (Carcinus maenas) and contamination levels in the Bizerta Lagoon: an integrated approach in biomonitoring of marine complex pollution. Environ Sci Poll Res 20: 2616-2631.

19. Jebali J, Sabbagh M, Banni M, Kamel N, Ben-Khedher S, et al. (2013) Multiple biomarkers of pollution effects in Solea solea fish on the Tunisia coastline. Environ Sci Pollut Res 20: 3812-3821.

20. Ferreira M, Moradas-Ferreira P, Reis-Henriques MA (2005) Oxidative stress biomarkers in two resident species, mullet (Mugil cephalus) and flounder (Platichthys flesus), from a polluted site in River Douro Estuary, Portugal. Aquatic Toxicology 71: 39-48.

21. Almeida EA, Dias Bainy AC, Melo Loureiro AP, Martinez Regina R, Miyamoto S, et al. (2007) Oxidative stress in Perna perna and other bivalves as indicators of environmental stress in the Brazilian marine environment: Antioxidants, lipid peroxidation and DNA damage. Comparative Biochemistry and Physiology Part A 146: 588-600.

22. Annabi A, Kessabi K, Navarro A, Saïd K, Messaoudi I, et al. (2012) Assessment of reproductive stress in natural populations of the fish Aphanius fasciatus using quantitative mRNA markers. Aquatic Biology 17: 285-293.

23. Marklund S, Marklund G (1974) Involment of anion radical in the auto oxidation of pyrogallol and convenient assay for superoxide dismutase. Eur Biochem 47: 469-474.

24. Claiborne A (1985) Catalase activity. In: Greenwald R A (ed) CRC Handbook of Methods in Oxygen Radical Research CRC, Florida. 283-284.

25. Gunzler WA, Kremers H, Flohae L (1974) An improved coupled test procedure for glutathione peroxidase in blood. Z Klin Biochem 12: 444-448.

26. Beliaeff B, Burgeot T (2002) Integrated biomarker response (IBR): a useful graphical tool for ecological risk assessment. Environ Toxicol Chem 21: 1316-1322.

27. Padmini E, Rani MU (2009) Evaluation of oxidative stress biomarkers in hepatocytes of grey mullet inhabiting natural and polluted estuaries. Sci Total Environ 407: 4533-4541.

28. Toni C, Ferreira D, Kreutz LC, Loro VL, Barcellos LJG (2010) Assessment of oxidative stress and metabolic changes in common carp (Cyprinus carpio) acutely exposed to different concentrations of the fungicide tebuconazole. Chemosphere, in Press.

29. Vutukuru SS, Chintada S, Madhavi KR, Venkateswara Rao J, Anjaneyulu Y (2006) Acute effects of copper on superoxide dismutase, catalase and lipid peroxidation in the freshwater teleost fish, *Esomus danricus*. Fish Physiol Biochem 32: 221-229.

30. Ates B, Orun I, Talas ZS, Durmaz G, Yilmaz A (2008) Effects of sodium selenite on some biochemical and hematological parameters of rainbow trout (Oncorhynchus mykiss Walbaum, 1972) exposed to Pb2+and Cu2+Fish Physiol Biochem 34: 53-59.

31. Jebali J, Chicano-Galvez E, Banni M, Guerbej H, Boussetta H, et al. (2013) Biochemical responses in seabream (Sparus aurata) caged in-field or exposed to benzo(a)pyrene and paraquat. Characterization of glutathione S-transferases. Ecotoxicology and Environmental Safety 88: 169-177.

32. Fernandes D, Andreu-Sánchez O, Bebianno MJ, Porte C (2008) Assessment of pollution along the Northern Iberian shelf by the combined use of chemical and biochemical markers in two representative fish species. Environ Pollut 155: 327-335.

33. Pereira P, de Pablo H, Pacheco M, Vale C (2010) The relevance of temporal and organ specific factors on metals accumulation and biochemical effects in feral fish (Liza aurata) under a moderate contamination scenario. Ecotoxicol Environ Saf (73): 805–816.

34. Ghedira J, Jebali J, Banni M, Chouba L, Boussetta H, et al. (2011) Use of oxidative stress biomarkers in Carcinus maenas to assess littoral zone contamination in Tunisia. Aquat Biol 14:87-98.

35. Ben-Khedher S, Jebali J, Houas Z, Hmida N, Jrad A, et al. (2014) Metals bioaccumulation and histopathological biomarkers in Carcinus maenas crab from Bizerta lagoon, Tunisia. Environ Sci Pollut Res 21: 4343-4357.

36. Damiens G, Gnassia-Barelli M, Loquès F, Roméo M, Salbert V (2007) Integrated biomarker response index as a useful tool for environmental assessment evaluated using transplanted mussels. Chemosphere 66: 574-583.

37. Van Heerden D, Tiedt LR, Vosloo A (2004) Gill damage in Oreochromis mossambicus and Tilapia sparrmanii after short-term copper exposure. Int Congr Ser 1275: 195-200.

38. Figueiredo-Fernandes A, Ferreira-Cardoso JV, Garcia-Santos S, Monteiro SM, Carrola J, et al. (2007) Histopathological changes in liver and gill epithelium of Nile tilapia, *Oreochrmis niloticus*, exposed to waterborne copper. Pesq Vet Bras 27: 103-109.

39. Skidmore JF, Tovell PWA (1972) Toxic effect of zinc sulphate on the gills of rainbow trout. Water Res 6: 217-230.

40. Olson KR, Fromm PO (1973) A scanning electron microscopic study of secondary lamellae and chloride cells of rainbow trout (Salmo gairdneri). Z Zellforsch 143: 439-449.

41. Arellano JM, Storch V, Sarasquete C (1999) Histological changes and copper accumulation in liver and gills of the Senegales Sole, Solea senegalensis. Ecotoxicol Environ Saf 44: 62-72.

42. Nowak B (1992) Histological changes in gills induced by residues of endosulfan. Aquat Toxicol 23:65-83.

43. Koca S, Koca YB, Yildiz Ş Gürcü B (2008) Genotoxic and Histopathological Effects of Water Pollution on Two Fish Species, Barbus capito pectoralis and Chondrostoma nasus in the Büyük Menderes River, Turkey. Biol Trace Elem Res 122: 276-291.

44. Braunbeck T, Storch V, Hresch H (1990) Species-specific reaction of liver ultrastructure in zebrafish (Brachydarios rerio) and trout (Salmo gairdneri) after prolonged exposure to 4-chloroaniline. Arch Environ Contam Toxicol 19: 405-418.

45. Malik RN, Zeb N (2009) Assessment of environmental contamination using feathers of Bubulcus ibis L., as a biomonitor of heavy metal pollution, Pakistan. Ecotox 18: 522-536.

46. Oulmes Y, Negele RD, Braunbeck T (1995) Cytopathology of liver and kidney in rainbow trout (*Oncorhynchus mykiss*) after long-term exposure to sublethal concentrations of linuron. Dis Aquat Org 21: 35-52.

47. Wester PW, Canton JH (1986) Histopathological study of *Oryzias latipex* (medaka) after long-term β-hexachlorocyclohexane exposure. Aquat Toxicol 9: 21-45.

Effects of Cadmium Exposure on Reproduction and Survival of the Planktonic Copepod *Centropages ponticus*

Ensibi Cherif[1]*, Olivier Pringault[2], Wifek Hannaoui[1] and Mohamed Nejib Daly Yahia[2]

[1]University of Carthage, Faculty of Sciences of Bizerte, Laboratory "Biodiversity and functioning of aquatic systems", 7021 Zarzouna, Bizerte, Tunisia
[2]Laboratoire Ecosystèmes Lagunaires, UMR 5119 CNRS-UM2-IFREMER- IRD ECOLAG, Station Méditerranéenne de l'environnement Littoral, Sète. Université de Montpellier II, France

Abstract

Heavy metal pollution is a growing environmental problem in marine coastal waters; for this reason the monitoring of water pollution needs the availability of test-species able to furnish reliable and cheap results. From this point of view, copepods are important test organisms owing to their wide distribution and for their key-position within the food web. As *Centropages ponticus* seems to be a promising target-species in ecotoxicology. The present study assesses the toxic effects of cadmium (Cd) on survival and reproductive performance of *C. ponticus*. A microcosm experiment was carried out for this study. Copepods were exposed to cadmium that ranged in concentration from 0.1 to 200 µg Cd.L^{-1} and effects were examined after 24 hours. The results show that treatments with cadmium significantly affect survival in copepods at concentration equal or above 0.2 µg.L^{-1}. In order to assess sublethal effects of cadmium on reproduction; the test organisms were exposed to four nominal Cd concentrations 0, 0.2, 0.4, 1 µg.L^{-1}. The results indicated that egg production decreased with the increase of metal concentration. An increase of the Cd concentration reduced the number of hatched nauplii. Thus, survival and reproduction in *C. ponticus* as a model test species could be effective physiological markers to monitor marine metal pollution.

Keywords: Heavy metals; Egg production; Egg hatching success; Mortality; Ecotoxicology

Introduction

Coastal and marine ecosystems are usually contaminated with high concentrations of anthropogenically introduced heavy metals [1]. In these environments, heavy metals can cause risks to organism's survival and general biodiversity because of their toxicity, persistence and bioaccumulation characteristics [2-4] Metals are dangerous because they tend to bioaccumulate. The extent of bioaccumulation of metals is dependent on the total amount, the bioavailability of each metal in the environmental medium and the route of uptake [5]. Most of the evidence derived from laboratory studies indicates that uptake of heavy metals from water is the predominant route of exposure [6]. However, several recent studies have suggested that dietary accumulation may contribute significantly to total body burdens of heavy metals in organisms [7]. Cadmium (Cd) is a biotoxic element and one of the major metals that are ubiquitously distributed in aquatic systems [8]. It's also a widely used heavy metal in the industry [9,10] in recent years; serious Cd pollution in aquatic systems has become a serious problem [7]. Anthropogenic input is considered to be the main source of Cd contamination in aquatic environments [11] and is known to be toxic to zooplankton [12,13]. This metal can be accumulated by the organisms, so affecting their survival, growth, reproduction and antipredator behavioural [2,14]. Pelagic copepods are the dominant secondary producers of pelagic ecosystems, and play a considerable role in the transfer of energy and organic matter from primary producers to higher trophic levels [15]. The reproduction of pelagic copepods constitutes an important role in the control of fish and shellfish recruitment since eggs and nauplii constitute the main prey of many species of fish and fish larvae fish [16]. Copepods are emergent organisms in ecotoxicology; to date their employment is increasing [17]. For their ubiquitus distribution, small size and for their key-position in the food web, scientists have been suggesting them as test-species in ecotoxicology and the obtained results seem to be promising. Few studies exist on the copepod *C. ponticus*, particularly in the areas of its ecology, geophylogeny, and its behavioural [18-21]. It has a wide geographical distribution along the coast of the Mediterranean [22]. *C. ponticus* has several advantages as a model test species. Hence, its distinct developmental stages, dimorphic sexes, high fecundity, short life-cycle, and easy to culture permit detailed ecotoxicological studies on the effects of toxicants.

This study was conducted in response to the need for costal and marine species to be used in routine ecotoxicology and toxicological risk assessment. In order to develop the copepod *Centropages ponticus* as a model species, information on its sensitivity to pollutants is, therefore, highly desirable. Moreover, laboratory studies addressing acute and sublethal effects of heavy metals on physiological response of copepods are considered important to understand how copepod populations cope with metal pollution and to conclude a holistic risk assessment of metal pollution. Thus, to assess the toxicity of cadmium, we exposed copepods to a range of concentrations in 24-hour static mortality tests. From these results we focused to assess the effect of sublethal exposure on egg production and egg hatching success.

Materials and Methods

Chemicals

All reagents were of analytical grade, and all laboratory glassware was soaked in 10% HNO_3 for at least 48 h and rinsed 3 times with distilled water prior to use. Stock solutions were prepared by dissolving

***Corresponding author:** Ensibi Cherif, University of Carthage, Faculty of Sciences of Bizerte, Laboratory "Biodiversity and functioning of aquatic systems", 7021 Zarzouna, Bizerte, Tunisia, E-mail: cherif.bfsa@yahoo.fr

the toxicant in deionised water. Working solutions were prepared by dilution of the stock solutions.

Biological material

The copepod *C. ponticus* was collected during low tides from the Bizerte lagoon (37°14' N, and 9° 56'E) north east of Tunisia (Figure 1) by using plankton net with 200µm mesh (0.57 m mouth diameter). The Bizerte lagoon ecosystem was studied previously to determine the environmental variables controlling the planktonic copepods abundance, diversity and species composition [15,22,23]. The collected samples were immediately transported to the laboratory of Biodiversity and Functioning of Aquatics Systems (Faculty of Sciences of Bizerte, Tunisia). Copepods identification was based on morphological criteria following consultation of taxonomic keys and species descriptions [24] under a Leica binocular microscope. Animals were then acclimatised to laboratory and test conditions in the temperature controlled room at 18 ± 1°C for 24-48 h in 500 mL beakers containing seawater with continuous aeration and no additional food. Seawater used for acclimating copepods was collected 20 m offshore from Bizerte Baie, well away from any obvious contaminant inputs and was filtered throw 0.45 µm GF/F glass fiber filter (Whatman). After the acclimation period, only healthy and active ovigerous females that showed normal swimming ability were retained and used as test animals in the bioassay.

Experimental Procedures

Acute toxicity tests

In the bioassays, standard water quality parameters were as follows: salinity 32, dissolved oxygen 3.8 ± 0.2 mg L^{-1}, pH 8.2 ± 0.3 and temperature 18 ± 1°C. Toxicity tests were carried out according to the protocol Mohammed EH et al. [25], with modifications. Bioassays were performed with adult, ovigerous females of *C. ponticus*. Exactly 40 mL of filtered seawater (0.45 µm) was added to each glass beakers (5 cm height × 3.5 cm diameter).

Thirty ovigerous females were carefully placed in each of the glass beakers. Cadmium working solutions were then added to yield concentrations ranging from 0.1 to 200 µg $Cd.L^{-1}$. Experimental concentrations were chosen on the basis of preliminary trials and on data from the literature [13,25-27]. Both control and treatments had three replicates each. The bioassays lasted for 24 hours, because of the shorter time; copepods were not fed during the test period. Dead test organisms were counted and removed at each laps of times. Mortality was typically defined as a position on which the animal's urosome is perpendicular to its prosome [28]. In addition, no moving organisms were observed at the bottom of the recipients under a stereoscopic microscope after careful inspection. Copepods with no movement of appendices after an observation period of 30 seconds were also assumed to be dead.

Reproductive toxicity test

Experimental conditions were the same as acute toxicity tests. Cd concentrations used in the sublethal toxicity test were 0, 0.2, 0.4, 1 $\mu g.L^{-1}$ Cd (as nominal concentrations). Reproductive tests were conducted using a modified protocol described by [22], For each test concentration, 10 females of *C. ponticus*, bearing an egg sac, were individually transferred to a new three glass beakers per treatment equipped with 200 mm mesh false bottoms. Beakers were filled with 40 mL of 0.45 mm filtered seawater. These females were incubated for 3 days under a 13:11 h light-dark cycle (spring daily cycle). The spawned eggs (SE) were counted after 24, 48 and 72 h under Leica binocular microscope and all of them were gently transferred by micropipette into news glass beakers. The eggs were subsequently incubated for 48 h, after which the numbers of nauplii and unhatched eggs were recorded. The test conditions for egg hatching were the same as fecundity tests, three replicate experiments at each test concentration, including controls. Hatchling success (HS%) for the eggs produced was calculated as the number of nauplii present after 48 h / the number of eggs added at the start of the incubation x 100.

Statistical analysis

Statistical analyses were conducted using the SPSS version 12.0 statistical software. Differences in egg production rates, hatchling

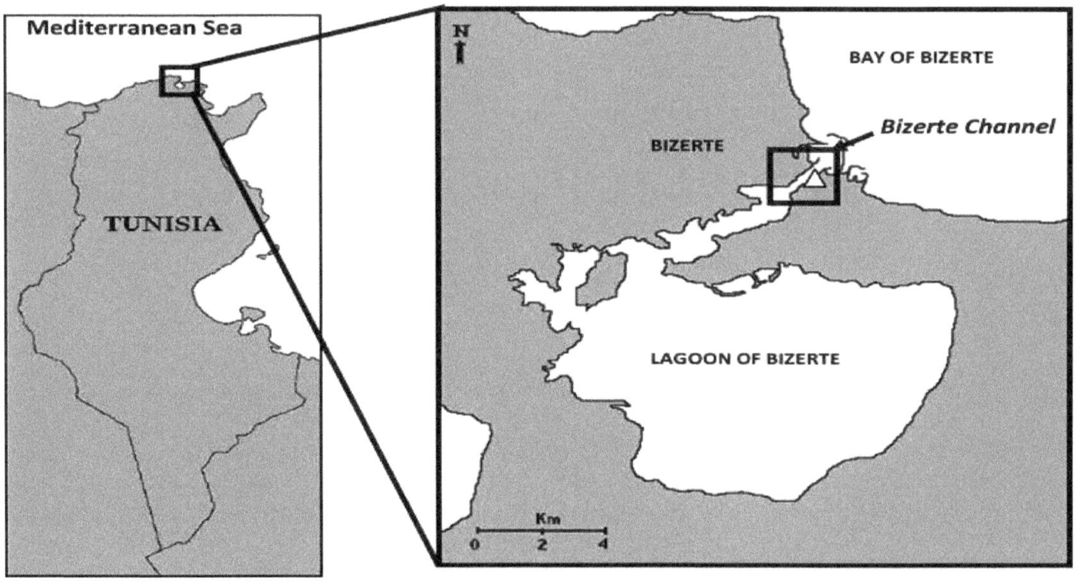

Figure 1: Map of the Bizerte Lagoon, North East of Tunisia (37°14' N, and 9°56' E), showing the sampling site from where copepods (*Centropages ponticus*) were collected.

success, and survival among treatments were tested for significance by means of one factor analysis of variance (ANOVA). When differences among groups were significant the Dunnet's test was employed to compare the control group and each of the experimental groups.

Results

Acute toxicity test

The trend of *C. ponticus* survival across cadmium concentrations shows that survival decreases as cadmium concentration increases, dropping from 97.77% and 83.34% in the control to 1.11% and 0% at the highest cadmium treatment after 12 and 24 hours respectively (Table 1). Survival analysis indicated that the effect on this trait was significant at cadmium concentrations equal or above 0.2 $\mu g.L^{-1}$ (Figure 2).

Reproductive toxicity test

Exposure of *C. ponticus* to Cd resulted in a significant reduction of its reproductive output. The spawned eggs decreased with increasing Cd concentrations and was significantly ($p<0.05$) reduced by 42.25, 68.76, and 77.48% at 0.2, 0.4 and 1 μgL^{-1} Cd, respectively compared to the controls (Figure 3). Copepod egg hatching success (EHS) ranged from 100 to 13.08% depending on the treatments. The exposure to Cd caused a significant reduction in egg hatching by 40.36, 60.75 and 86.91%, respectively for 0.2, 0.4 and 1 $\mu g.L^{-1}$, compared to the control (Figure 4).

Discussion

The present study demonstrates that cadmium exposure can have significant and detrimental effects on survival and fecundity of the planktonic copepod Centropages ponticus. The mortality increased significantly as cadmium concentration increased ($p<0.05$), with a mortality >90% at concentrations equal or above 0.2 $\mu g.L^{-1}$ after 24 hours. This is in agreement with others results obtained with the intertidal copepod Tigriopus japonicas. Our results showed that acute exposure to low and sublethal concentrations of cadmium produce a significant reduction in the reproduction rate and egg hatching success (Figures 3 and 4), which support previous studies on the effect of cadmium exposure on fecundity of some planktonic copepods [12]. The measurement of egg production rates and hutching success of adult female copepods has been proposed as an alternative method for estimating pelagic secondary production, since it constitutes a measure of potential population recruitment rate [22,29]. Moreover, egg production is an important aspect of the adult biology of copepods, which integrates a number of metabolic processes and therefore has high ecological relevance [15]. Thus, egg production and hatchling success of copepods has been previously shown to be a very sensitive biological response to several toxicants like heavy metals, crude oil [30-33], pesticides, PAHs and testosterone. In this study, the acute toxicity

Exposure time (hr)					12				
Treatment ($\mu g.L^{-1}$)	0	0.1	0.2	0.63	2	6.3	20	63	200
Mean value in %	2.22	12.22	32.22	46.66	61.11	72.22	78.88	92.22	98.88
SE	0.211	0.279	0.279	0.183	0.211	0.279	0.279	0.105	0.105
P Value					<0.0001				
Exposure time (hr)					24				
Treatment ($\mu g.L^{-1}$)	0	0.1	0.2	0.63	2	6.3	20	63	200
Mean value in %	16.66	24.44	97.77	100	100	100	100	100	100
SE	0.183	0.105	0.105	0	0	0	0	0	0
P value					0.0368				

Table 1: Mortality (%) of copepods during exposure to a range concentration of cadmium after 12 and 24 hours of exposure: mean value, standard error and *p* value.

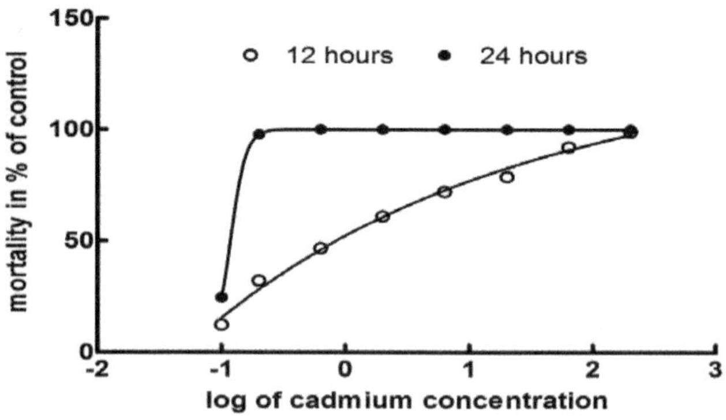

Figure 2: Effect of Cadmium on survival of *C. ponticus* after 12 and 24 hours of treatment. All points are means percentage of three replicates, 30 organisms per replicate.

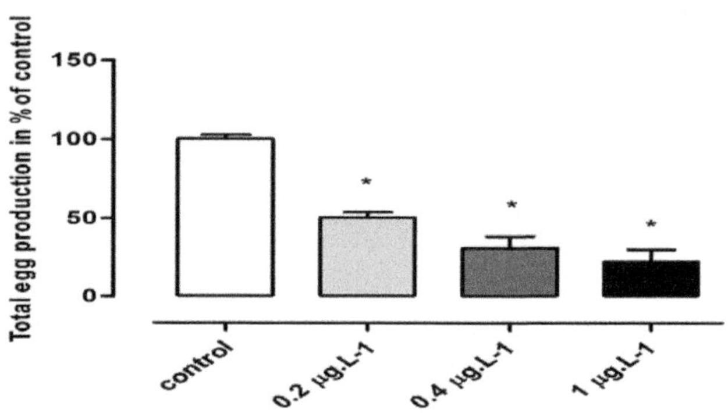

Figure 3: Eggs production of *C. ponticus* under different concentrations of dissolved cadmium in seawater after 48 hours of exposure. Data are described as mean ± error. Asterisks indicate significant differences from controls (ANOVA $p<0.05$).

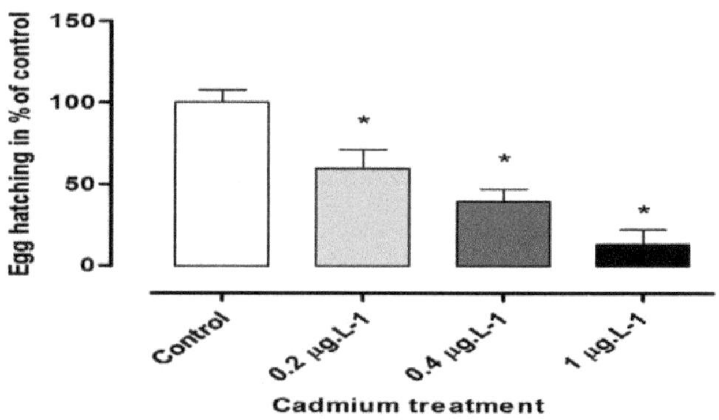

Figure 4: Eggs hatching of *C. ponticus* under different concentrations of dissolved cadmium in seawater after 48 hours of exposure. Data are described as mean ± error. Asterisks indicate significant differences from controls (ANOVA $p<0.05$).

of cadmium to adult female of *C. ponticus* was ascertained and the effect was time and dose dependent as ordinary xenobiotiques. Cadmium adversely affects on adult female survivability and fecundity at above 0.2 $\mu g.L^{-1}$. This study is the first to examine the effects of cadmium on the planktonic copepods *C ponticus* to evaluate its potential to serve as an appropriate test species. Thus, survival and reproduction could be effective physiological markers to monitor marine metal pollution.

Acknowledgements

This study was supported by the bilateral cooperation program between Tunisia and France (Jeunes Equipes AIRD; JEAI - ECOBIZ). Anonymous reviewers are acknowledged for their constructive remarks that resulted in significant improvement of this paper.

References

1. Ruilian Y, Xing Y, Yuanhui Z, Gongren H, Xianglin T (2008) Heavy metal pollution in intertidal sediments from Quanzhou Bay, China. J Environ Sci 20: 664-669.
2. DeForest D, Brix K, Adams W (2007) Assessing metal bioaccumulation in aquatic environments: The inverse relationship between bioaccumulation factors, trophic transfer factors and exposure concentration. Aquat Toxicol 84: 236-246.
3. Wang Q, Liu B, Yang H, Wang X, Lin Z (2009) Toxicity of lead, cadmium and mercury on 233 embryogenesis, survival, growth and metamorphosis of Meretrix meretrix larvae. Ecotoxicology 18: 829-837.
4. Naser HA (2013) Assessment and management of heavy metal pollution in the marine environment of the Arabian Gulf: A review Mar Poll Bull 72: 6-13
5. Rainbow PS (2002) Trace Metal Concentrations in Aquatic Invertebrates: Why and So What? Environmental Pollution 120: 497-507.
6. Williams DR, Giesy Jr JP (1978) Relative importance of food and water sources to cadmium uptake by Gambusia affinis. Environ Res 16: 326-332.
7. Wang Z, Yan C, Hyne RV (2010) Effects of dietary cadmium exposure on reproduction of saltwater cladoceran Moina monogolica daday: Implications in water quality criteria. Environ Toxicol Chem 29: 365-372.
8. El-Naga EHA, El-Moselhy KM, Hamed MA (2005) Toxicity of cadmium and copper and their effect on some biochemical parameters of marine fish Mugil seheli. Egyptian Journal of Aquatic Research 31: 60-71.
9. WHO (1992) Cadmium-environmental aspects: Environmental Health Criteria 135. World Health Organization, Geneva.
10. Jarup L, Akesson A (2009) Current status of cadmium as an environmental health problem. Toxicol Appl Pharmacol 238: 201-208.
11. Wright DA, Welbourn PM (1994) Cadmium in the aquatic environment: A review of ecological, physiological, and toxicological effects on biota. Environ Rev 2: 187-214.
12. Gnanamoorthy P, Manimaran K, Ashok Prabu V, Selvam T (2012) Cadmium toxicity study of copepod (Oithona similis) Int J Cur Tr Res 1: 8-12.
13. Faupel M, Ristau K, Traunspurger W (2012) The functional response of a freshwater benthic community to cadmium pollution. Environ Pollut 162: 104-109.
14. Sornom P, Gismondi E, Vellinger C, Devin S, Férard JF, et al. (2012) Effects of sublethal cadmium exposure on antipredator behavioural and antitoxic responses in the invasive amphipod Dikerogammarus villosus. PLoSONE 7.
15. Souissi A, Souissi S, Daly Yahia MN (2008) Temporal variability of abundance and reproductive traits of Centropages kroyeri (Calanoida; Copepoda) in Bizerte Channel (SW Mediterranean Sea, Tunisia). J Exp Mar Biol Ecol 355: 125-136.
16. Last JM (1980) The food of twenty species of fish larvae in the westcentral North Sea. Fisheries Research Technical Report, number 60. MAFF Directorate of Fisheries Research, Lowestoft.
17. Pane L, Agrone C, Giacco E, Somà A, Mariottini GL (2012) Utilization of Marine Crustaceans as Study Models: A New Approach in Marine Ecotoxicology for European (REACH) Regulation, Ecotoxicology, Dr. Ghousia Begum (Ed).
18. Soler ET, Del Rio JG, Vives F (1988) Morphological and taxonomical revision of Centropages ponticus Karavaev, 1895 (Copepoda, Calanoida). Crustaceana 55: 129-146.
19. Christou ED (1998) Interannual variability of copepods in a Mediterranean coastal area Saronikos Gulf, Aegean Sea. J Mar Syst 15: 523-532.
20. Riccardi N, Mariotto L (2000) Seasonal variations in copepod body length: a comparison between different species in the Lagoon of Venice. Aquat Ecol 34: 243-252.
21. Siokou-Frangou I, Zervoudaki S, Kambouroglou V, Belmonte G (2005) Distribution of mesozooplankton resting eggs in seabottom sediments of Thermaikos gulf (NW Aegean Sea, Greece) and possible effects of sediment resuspension. Cont Shelf Res 25: 2597-2608.
22. Neffati N, Ons DYK, Delphine B, Francois C, Mohamed Nejib DY (2012) Reproductive traits of two calanoid copepods: Centropages ponticus and Temora stylifera, in autumn in Bizerte Channel. J Plankton Res 1-17.
23. Afli A, Ayari R, Zaabi S (2008) Ecological quality of some Tunisian coast and lagoon locations, by using benthic community parameters and biotic indices. Estuarine, Coastal and helf Science 80: 269-280.
24. Razouls C, Durand J (1991) Inventaire des copépodes planctoniques méditerranéens. Vie Milieu 41: 73-77.
25. Mohammed EH, Wang G, Xu Z, Liu Z, Wu L (2011) Physiological response of the intertidal copepod Tigriopus japonicus experimentally exposed to cadmium. AACL Bioflux 4: 99-107.
26. Ferreira ALG, Loureiro S, Soares AMVM (2008) Toxicity prediction of binary combinations of cadmium, carbendazim and low dissolved oxygen on Daphnia magna. Aquat Toxicol 89: 28-39.
27. Fikirdeşici Ş, Ahmet A, Ebru O (2012) Investigation of acute toxicity of cadmium- arsenic mixtures to Daphnia magna with toxic units approach. Turk J Zool 36: 543-550.
28. Kwok KWH, Leung KMY (2005) Toxicity of antifouling biocides to the intertidal harpacticoid copepod Tigriopus japonicus (Crustacea, Copepoda): effects of temperature and salinity. Mar Pollut Bull 51: 830- 837.
29. Turner JT (2004) The Importance of small planktonic copepods and their roles in pelagic marine food webs. Zool Stud 43: 255-266.
30. Almeda Rodrigo, Baca S, Hyatt C, Buskey EJ (2014) Ingestion and sublethal effects of physically and chemically dispersed crude oil on marine planktonic copepods. Ecotoxicology 23: 988-1003.
31. Willis KJ, Ling N (2004) Toxicity of the aquaculture pesticide cypermethrin to planktonic marine copepods. Aquac Res 5: 263-270.
32. Bellas J, Thor P (2007) Effects of selected PAHs on reproduction and survival of the calanoid copepod Acartia tonsa. Ecotoxicology 16: 465-674.
33. Hossain K, Omura T, Takeda S, Takishi I (2005) Effect of testosterone on the calanoid copepod, Acartia omorii Bradford. La mer 43: 69-73.

Determination of some Heavy Metals in Sea-grass *"Poisdonia oceanica"(L.)* Delile of Eastern Libyan Coast "Mediterranean"

Abdulsalam A Benkhayal* and Khadija S El-turki

Chemistry Department, Faculty of Science, University of Benghaz, Libya

Abstract

The concentrations of heavy metals (Cr, Co, Zn, Cd, and Pb) were found in the samples of tissues of Sea-grass, *Posidonia oceanica (L.)* Delile, that were collected from eighteen different sites along 820 Km of the coast of eastern Libya on the Mediterranean sea (the first site is about 500 km to the east of Alexandria of Egypt). These samples were collected from the following 18 sites (east to west): Tobrouk, Ras-Etteen, Derna, Ras-Lehlal, Sousa, QasrLibya, Tulmaitha, Deryana, North Benghazi, Alssabri, Family Village (Qaria), Nairowz, Nakheil, Geminis, Sultan, Zuitina, Al-Briga and Ras-Lanouf. The technique used by this study was Flame Atomic Absorption Spectrometry (model novAA300). This study shows that elevated concentrations of Zn, Co, and Pb are found in most samples; particularly, those were collected near the discharge points of city sewages effluent and the power station sites. The concentrations of Cr and Co were very low in all the samples, except for those were collected near the city of Benghazi. The statistical analysis (one-way ANOVA) indicates that the Sea-grass" *Posidonia oceanica"(L.)* Del. tissues have different levels of metal bioaccumulation, which might be used as a bio monitors for trace metals. The higher values of MPI are recorded for BR and TP that is 121.18 and 124.29, respectively. Perhaps, the high values of MPI for these two sites are due to the high elevated values (387.56) µg /g and (186.40) µg/g of Chromium for both sites (BR & TP). The least values of MPI are obtained for the locations SP, ZU and RH (20.16, 20.96 and 22.20 µg\g), respectively, and this may be explained by the low levels of the five heavy metals (Cr, Co,Zn,Cd and Pb) in these sampled sites.

Keywords: Sea-grass; Metal bioaccumulation

Introduction

The major source of surface water pollution includes effluent discharges by industries, atmospheric depositions of pollutants and occasional accidental spills of toxic chemicals. Trace metals are regarded as serious pollutants of aquatic environment because of their toxicity. The persistence of their difficult biodegradability and their tendency to concentrate on aquatic organisms. They enter the marine environment through atmospheric and land-based effluent source [1,2]. There is currently a great interest in the use of living organisms as in aquatic ecosystem [3]. Given that the method used previously does not provide sufficient information on the bioavailability of the metals present in the environment. In the Mediterranean Sea, the endemic seagrass, *Posidonia oceanic (L.)* Delile, has been used as a metal bio indicator for several over decades [4,5]. Leaves *of P. oceanica* can give an indication of the metal concentration in the environment, a short time period (months) with good accuracy [6]. On the contrary, sheaths, which gave an indication of changes over long time periods (decades), seem to be less sensitive to variations in the metal concentration in the environmental. The use of *P.oceanica* as a bio indicator of water quality in relation to implementation of the water frame work directive [7-9]. The state of health of *P. oceanica* meadows is good indicator of the ecological status of a particular coastal water body. Therefore, *P. oceanica* has been proposed as a biological quality elements (BQES) for Mediterranean coastal water. Considerable work is being currently under taken to develop classification tools and metrics that would allow use of *P. oceanica* as a BQES for WFD (Water Frame work Directive). The *P. oceanica* eco-system is particularly threatened by anthropogenic coastal pollution and bio indication of heavy metal contaminations have already been used as such in the Mediterranean and elsewhere [10-12]. Thus, the aim of this study is to evaluate the state of metal contamination of the Benghazi coast in Libya metal contamination of the Benghazi coast in Libya using *Posidonia.*

Materials and Methods

Sampling and sample preparation

Seagrass Posidonia oceanica samples were collected in the summer of 2009, from 18 stations located along Eastern Libyan "Mediterranean" (Eastern of Libyan Coast; Figure 1). Dirt and foreign herbs from sea grass samples was removed and washed with deionized water to remove marine sediment, soil and others impurities. All sea grass samples were placed in a drying oven at about 65°C until most of the liquid has evaporated, then the temperature was increased to 80°C. Sea grass samples were placed in desiccators for about 48 hrs and then they were ground by a mill–Machine.

Trace metals analysis

Hundred milligrams of each samples was weighed, transferred into 25 ml beaker, dissolved in (8 ml 65% nitric acid) by heating to 70°C on a hotplate for about 8 hrs and the solution was then cooled, 2 ml of 30% hydrogen peroxide was added and heated again for another 4 hrs at 70°C, then filtered and diluted to the mark [13]. All samples solution were analyzed using Flame Atomic Absorption Spectrometer (FAAS), with detection limit of 0.0007 mg L^{-1} for Cr, 0.0013 mg L^{-1} for Co, 0.0009 mg L^{-1} for Zn, 0.0015 mg L^{-1} for Cd and 0.0023 mg L^{-1} for Pb. Good

***Corresponding author:** Abdulsalam A Benkhayal, Chemistry Department, Faculty of Science, University of Benghaz, Libya, E-mail: Absalamben @yahoo.com

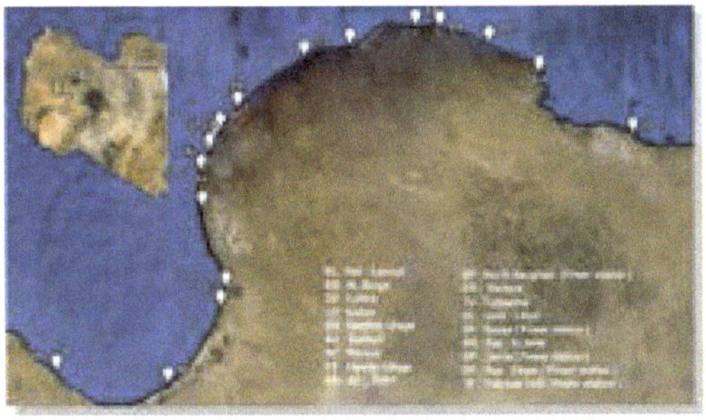

Figure 1: Map of eastern Libyan coast.

overview on sample preparation for atomic spectroscopic analysis was recently published [14]. Calibration standards solutions series were prepared from standard solutions of 1000 mg L^{-1} (Stock solution), to compare the total metal content at the different locations. The metal Pollution index (MPI) to compare the total metal content at the different stations, was used, it is obtained with the following equation: $MPI=(Cf_1 \times Cf_2 \ldots\ldots\ldots\ldots Cf_n)^{1/n}$; where Cf is the concentration of the metal n in the sample. For our study area, which is the multiplication of heavy metals concentration with their number

Statistical analysis

An explorative statistical analysis was carried out for all concentration values measured in all samples, and data were compared with other measures presented in the literature. The explorative parameter are mean value, range, standard deviation, relative standard error and 95% confidence interval of mean (Lower bound and Upper bound), and the results were analyzed using one - way analysis of variance (ANOVA) to examine statistical significantly of difference in the mean concentration of chromium, cobalt, zinc and lead metals ($P<0.05$). The mean concentration of cadmium metal shows no significantly ($P>0.05$), when determined in sea grass "Posidonia oceanica"(L.) Defile sample. Major species correlation matrices (ELOro) was used and showed a clear relation between locations and heavy metals (Cr, Co, Zn, Cd and Pb) (Table 2 data included as supplementary). There is strong relation between cobalt and chromium and zinc locations and lead metal (significance), and medium relation between cadmium and chromium, lead and zinc (significance), weak relation between locations and chromium, cobalt, zinc and cadmium.

Results

Determinationof metals concentration in seagrass "*Posidonia oceanica*" (L.) Defile

The mean concentration of heavy metals (Cr, Co, Zn, Cd and Pb) in *P. oceanica* samples from (18) sites in different location of Eastern of Libyan Oil Crescent are listed in Tables 1.1-1.6 (Data included as supplementary), respectively, the mean concentration of heavy metals were characterized by the following parameters: Mean, standard deviation and relative standard deviation of the trace metal concentration in *P. oceanica* are presented.

Discussion

This study was focused on the pollution by heavy metals, the study area extend about 820 Km along the eastern Libyan coast (Mediterraneansea); form Tobrouk city at the east to Ras-lanouf in the west. Seagrass "*Posidonia oceanica*" (L.), the sea plant known to be as a bio-indicator was chosen because of it is an ability to concentrate heavy metals.

This study have been proved that most expected contaminated sites were polluted with all the heavy metals (Cr, Co, Zn, Pb), with the exception of cadmium, all metals gave a significant indicator values and the average value was higher than the background level for each element. The values of MPI which represents the metal pollution index level, and that equal to an average of all heavy metals concentrations of this study multiplied by their number (five metals) were elevated in some sampling sites (TP, BR, SA, NA, DP, RP, RL, BP) (124.9, 121.18, 72..04, 59.86, 58.94, 51.88, 46.24, 43.68), respectively. The samples (TP, DP, RP, BP) were collected near power stations that uses heavy fuel oil, liquid fuel oil and fuel gas.

Wang et al. [15] states that the first two types of fuel contain zinc, (HFO: 1445 $\mu g/m^3$), (NG: 1479 $\mu g/Nm^3$), and lead: (HFO: 35.5 ng/m^3). These prove the leakage of such fuel into sea, in addition, the machines coolant oil that need to be changed may discarded without treatment into sea. The metal pollution index MPI values were found to be lower than that of previous samples, and it be calculated for the samples (SP, ZU, RH, TU) to be 20.16, 20.96,22.20,27.95, respectively. In order to support our results, a data function system was applied to link between the five heavy metals and their collection sites, this relation shows very strong link between (the sites and Pb metal, Cr and Co metals, Cr and Zn metals, Zn and Co metals), where their correlation values were (0.455**, 0.617**, 0.580**,0.529**), respectively. However, the correlation value was medium for Cr and Cd metals, Zn and Pb metals, that was found to be (0.356*, 0.297*), respectively, the weakest correlation value were recorded between the sites and the Cr, Co, Zn and Cd metals.

References

1. Lafabric C, Pergent-Martini C, Pergent G (2008) Metal contamination of Posidonia oceanica meadows along the corsican coastline (Mediterranean). Environ Pollut 151: 262-268.
2. Michel W, Renzo B, Ali T, Jean-Marie B, Michel J, et al. (1998) Distribution of heavy metals in the echinoid paracentrotus livldus from the Mediterranean Posidonia oceanica ecosystem: Seasonal and geographical variations. J Sea Res 39: 267-280.
3. Pergent-Martini C, Pergent G (2004) Marine phanerogams as a tool in the evaluation of marine trace-metal contamination: An example from the Mediterranean. Int J Environ Pollut 13: 126-147.
4. Markert BA, Breure AM, Zechmeister HG (2003) Bioindicators and Biomonitors principles, concepts and applications: Bioindication/biomonitoring of the environment trace metals and other contaminants in the environment 6: 3-40.
5. Islam MdS, Tanaka M (2004) Impacts of pollution on coastal and marine ecosystems including coastal and marine fisheries and approach for management: A review and synthesis. Mar Pollut Bull 48: 624-649.
6. Alfonso JA, Azocar J, LaBrecque JJ, Garcia B, Palacios D, et al. (2008) Trace metals in seageass collected from Venezuelan coastal sites. J Trop Biol 56: 215 - 222.
7. Pergent- Martini C, Leoni V, Pasqualini V, Ardizzone GD, Balestri E, et.al. (2005) Descriptors of Posidonia oceanica meadows: Use and application. Ecol Indicators 5: 213-230.
8. Ferrat L, Romeo M, Gnassia-Barelli M, Pergent-Martini C (2002) Effects of mercury on antioxidant mechanisms in the marine phanerogam Posidonia oceanica. Dis Aquat Organ 50: 157-160.
9. Lippmann M (2000) Environmental toxicants. Wiley-interscience, New York, USA.
10. Weiss MK (2005) Lead. In: Snell FD, Etle LS (Eds), Encyclopedia of Industrial Chemical Analysis 161-201.

11. Pergent-Martini C (1998) Posidonia oceanica: A biological indicator of past and present mercury contamination in the mediterranean sea. Mar Environ Res 45: 101-111.

12. Ledent G, Mateo MA, Wernau M, Temara A, Romero J, et al. (1995) Element losses following distilled water rinsing of leaves of the seagrass Posidonia oceanica (L.) Delile. Aquat Bot 52: 229 -235.

13. Manal.El-Gadi (2007) Determination of trace amount of lead in Benghazi–City coast using spectrophotometer and flame atomic absorption spectrometer. Garyounis University, Benghazi, Libya.

14. Nabil R. Bader, Zimmermann B (2012) Sample preparation for atomic spectroscopic analysis: An overview. Adv App Sci Res 3: 1733-1737.

15. Wang YF, Chao HR, Wang LC, Chang-Chien GP, Rsou TC (2010) Characteristics of heavy metals emitted from a heavy oil-fueled power plant in northern Taiwan. Aerosol Air Qual Res 10: 111-118.

9

Effects of Processing Methods on the Value of Bêche-de-mer from the Fiji Islands

Ravinesh Ram[1*], Roveena Vandana Chand[2] and Paul C. Southgate[1]

[1]*Centre for Sustainable Tropical Fisheries and Aquaculture, College of Marine & Environmental Sciences, James Cook University, Townsville, Queensland 4811, Australia*
[2]*School of Biological and Chemical Sciences, Faculty of Science Technology and Environment, University of the South Pacific, Suva, Fiji Islands*

Abstract

Sea cucumber harvesting, processing into bêche-de-mer (BDM) and exporting had been an important marine commodity trade for over two centuries. Sea cucumber trading has also become an important export earner for the Fiji Islands and a source of income generation for coastal communities involved in this trade. This study showed that the processing techniques being utilized for BDM in Fiji have changed little since they were developed in the 1800's. Fijian BDM (18 commercial species) is generally categorized as 'grade B' in Asian markets because of relatively poor product quality. The quality and value of bêche-de-mer are significantly impacted by negligence during processing by sea cucumber fishers and marine products agents. The loss of value due to poor quality of the final dried product in Fiji ranges from 10% to 50%. Sea cucumber fishers in Fiji need to adopt improved processing techniques that support production of better quality product, improved yield and greater income.

Keywords: Sea cucumber; bêche-de-mer; Value and processing

Introduction

The sea cucumber fishery is an important source of income for coastal communities in the Pacific [1]. Holothurians or sea cucumber are consumed as a delicacy and for their perceived medicinal properties, and are particularly sought by south-east Asian markets [2,3]. The major consuming countries are China, Hong Kong SAR, Taiwan PC, Singapore and Malaysia [4]. The global trade in sea cucumbers is based on the dried form known as bêche-de-mer (*iriko* in Japanese, *hai - som* in Chinese or *trepang* in Indonesian) [5-7]. From a total of approximately 1,200 species of sea cucumbers known today, around 58 are commercially exploited in Asian markets [1,5,8]. The majority of commercially exploited species belong to the Genera *Actinopyga* Bronn 1860, *Bohadschia* Jaeger 1833, *Stichopus* Brandt 1835, *Thelenota* Brandt 1835 and *Holothuria* Linnaeus 1758, with Asian buyers particularly targeting species from the genus *Holothuria* [8]. *Holothuria scabra* (Jaeger 1833) (Sandfish), *H. (Microthele) fuscogilva* (Cherbonnier 1980) (White Teatfish) and *H. (Microthele) whitmaei* (Bell 1887) (Black Teatfish) are among the highest value species [9] in Asian markets where well-dried 'A' grade product commands a price of $US 70-190 per kg depending on size and quality [5]. Papua New Guinea, Solomon Islands, Australia and the Fiji Islands are the leading suppliers of bêche-de-mer to Asian markets from the Pacific [4]. Approximately 98% of sea cucumber products imported by consuming countries are in the dried form with the remainder being brined, frozen or fresh [4].

Bêche-de-mer processing entails an uncomplicated sequence of actions resulting in a product that is non - perishable if stored in dry, dark conditions. The processing techniques currently used for bêche-de-mer in the Fiji Islands were developed in the 1800's and have changed little since. Post-harvest steps include first boiling, slitting and gutting, second boiling, smoking and finally sun drying [1,5,8-13]. Although these steps are uncomplicated, it requires continuous attention to obtain a standard dry product. Failure to do so can result in reduced quality and value of the final product [11,13].

Factors affecting the quality of bêche-de-mer have been poorly studied and there is limited literature in this field. There is a particular lack of data on the influence of processing methods on the value of bêche-de-mer in Asian markets [1,8,14-16]. The aim of this study was to document the potential losses in value (income) at different stages of bêche-de-mer processing in the Fiji Islands. Our results will help pinpoint critical points in the processing of bêche-de-mer, and provide a basis for improving current methods to maximise livelihood benefits from this important fishery.

Materials and Methods

A standard questionnaire was prepared to gather information on the perceptions of sea cucumber fishers (n=86 completed questionnaires), as well as marine product agents (n=5 completed questionnaire) and the middlemen (n=8 completed questionnaire) involved in the bêche-de-mer industry in the Fiji Islands (Figure 1). The marine product agents purchase sea cucumbers in a perishable state from fishers and process them for higher return from exporters. Comprehensive information on the issues and problems faced by marine product agents and fishers in Fiji was collected; it included information on problems related to sea cucumber harvesting and processing, socio–economic issues, species-specific issues, prices attained for sea cucumbers and the guidance provided to the fishers by the marine product agents relating to processing methods. Data were collected through formal and informal interviews held with the marine product agents and fishers. Data were also gathered on the socio–economic status of sea cucumber fishers, and for critical evaluation of the supply and value distribution chain of BDM in Fiji Islands. The information on supply and value chain was essential to understand the various channels within the bêche-de-mer trade in Fiji and the value received at each stage.

***Corresponding author:** Ravinesh Ram, Centre for Sustainable Tropical Fisheries and Aquaculture, College of Marine & Environmental Sciences, James Cook University, Townsville, Queensland 4811, Australia
E-mail: ravinesh.ram@my.jcu.edu.au; ravineshram@gmail.com

Figure 1: Sites visited in Fiji for collection of data (source: Ram, 2008)

Results and Discussion

Fiji Islands has exported a large quantity of bêche-de-mer to Asian markets for over two centuries. The major markets for Fijian bêche-de-mer products are now Taiwan, Hong Kong, China and Korea, as well as Australia, New Zealand, Canada and the USA [15]. Figure 2 shows official BDM export quantities from Fiji during the period from 1984 to 2012. Particularly large volumes of bêche-de-mer were exported in 1987 (>600 mt), 1988 (>700 mt) and 1996 (>600 mt). Declines in export volumes are notable following these peaks and these could be due to over fishing of sea cucumbers [17,18]. Subsequent peaks in export volumes of around 400 mt in 2005 and 2011 are considerably lower than those in the 1980s and 1990s. After 2005, the export trend levelled-off to average around 243 tonnes per year [19]. Sea cucumber harvests over recent years have generally declined for the high value species and fishing yield has now shifted from high value species to either medium or low value species [20]. Bêche-de-mer exporters now rely on the medium value species to make up a consignment for export. For example, data from Fiji's recent Annual Fishery Report [19] shows that the proportion of the sea cucumber fishery contributed by Hairy Blackfish (*Actinopyga miliaris,* Quoy and Gaimard 1833) has now declined to 6% of the harvest from around 90% of the harvest in 1988 [18,19]. This report further states that a decline in annual production of Black Teatfish (*H. whitmaei),* Hairy Blackfish, Curryfish (*Stichopus hermanni,* Semper 1868), Deepwater Redfish (*Actinopyga echinities,* Jaeger 1833), Greenfish (*Stichopus chloronotus,* Brandt 1835), Chalkfish (*Bohadschia marmorata,* Jaeger, 1833), Prickly Redfish (*Thelenota ananas,* Jaeger 1833), Lollyfish (*Holothuria atra,* Jaeger, 1833) and White Teatfish (*H. fuscogilva*) was observed between 2003-2012 whereas, a very low production of Dragonfish (*Stichopus horrens,* Selenka 1868), Flowerfish (*Pearsonothuria graeffei,* Semper 1868), Stonefish (*Actinopyga lecanora,* Jaeger 1835) and Pinkfish (*Holothuria edulis,* Lesson 1830) was noted, indicating that their populations may have collapsed [19].

Chinese agents still dominate the bêche-de-mer market in Fiji. China accepts the full range of product quality that is produced in Fiji and is primarily concerned about the size of products (the larger the size the better the price) [21]. The Hong Kong market, on the other hand, values product quality since bêche-de-mer bought from the Pacific islands is reprocessed and value-added for shipment to other markets such as Taiwan and Singapore [4,22]. White Teatfish is a high value product for Fiji exporters. It can earn as much as US$90 per kg in summer and up to US$140 per kg in winter. However, the current average rate paid to Fijian marine agents is only US$5–55 per kg because of the suboptimal quality of the product. A large quantity of lower grade BDM (grades "C" and "D") is generally exported from Fiji with a relatively small proportion of higher grade (grades "A" and "B") product (Table 1).

Table 1 summarizes the prices offered to marine product agents in Fiji by Asian markets. The results show that for high value species such as White Teatfish, the number of individuals required to make up one kg in the lower grades (e.g. grades C and D) is considerably higher than the number of individuals required to make up the same quantity in the superior grades (e.g. grades A and B) because of their smaller size. Lower grade BDM are also of lower value to Fijian bêche-de-mer agents (Table 1). Thus in order to receive greater income for bêche-de-mer from Asian markets, improving product quality will be a key issue for the Fijian BDM industry. The relatively low quality of Fijian BDM and fluctuations in product quality results in variations of 5–20% in the prices paid by international agents to Fijian agents. We identified the following issues that drove variations in market price for Fijian BDM in Asia:

Undersized bêche-de-mer (all species);

Products not cut and gutted properly resulting in the presence of gut contents that were still visible in the dried product;

Products did not have ideal cylindrical shape and appearance is a major grade determinant [11,13];

Burnt products; and

Products contaminated with sand and dust (poor hygiene).

From the survey, about 90% of responding fishers indicted that they do some processing before selling their products to the main agents. As a result some processing steps are usually missed or done too rapidly (e.g. smoking the BDM products over a fire for rapid drying compared to smoking only results in a burnt, poor quality product). Improving product quality begins with awareness of the factors that influence product quality from the time of harvest to storage and processing [11]. For example, inappropriate handling of sea cucumbers during and after harvest may causes skin lesions and damage to the 'teats' of high value species such as White Teatfish, which directly affects the value of the final product [11,13]. Furthermore, because local fishers generally lack the basic processing tools required to produce a good quality product, marine products agents in Fiji prefer to buy higher value species of sea cucumbers from fishers in a raw state so that they themselves can process to market quality grades that achieve a higher price.

The data collected during this study were used to generate estimates of the value lost at each stage of sea cucumber processing in Fiji by both

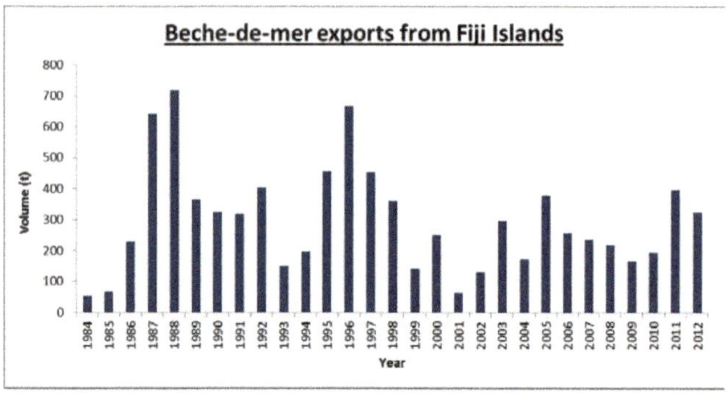

Figure 2: Bêche-de-mer exports from Fiji Islands from 1984-2012 (Source: Ministry of Fisheries and Forests: Fiji)

Common name	Grade	Grades of product	Number of pieces required for one kilogram	Size of sea cucumber in each grade (cm)	Price dry per kg ($USD) in Hong Kong markets*	Actual Price Dry per kg ($USD) offered by Asian marine product agents	Usual Price offered to marine product agents in Fiji depending on quality (USD/ Kg-dry)
White teatfish	High	A+	3-4	17-25	70-75	50-60	35-55
		B	5-6	13-20	50-60	40-45	25-35
		C	9-10	14-21	-	30-35	15-25
		D	14-15	6-15	-	20-25	5-10
Black teatfish	High	A	4-5	12-22	50-55	36-40	28-35
		B	9-10	10-17	40-45	26-30	20-25
		C	14-15	5-14	-	15-20	10-15
Sandfish	High	Banned	Banned	Banned	80-90	Banned	Still sold but price was unavailable from marine product agents
Prickly redfish	Medium	1 grade	3-4	9-23	35-40	26-30	20-25
Stonefish	Medium	1 grade	100-105	2-9	8-12	5-5.5	3-5
Surf Redfish	Medium	A	9-10	7-14	25-30	20-25	15-24
		B	13-15	5-10	20-25	15-18	15-16
Blackfish	Medium	A+	10-12	7-13	35-40	25-30	18-24
		B	18-20	6-11	-	20-24	15-20
		C	24-25	4-8	-	10-14	8-12
Elephant Trunkfish	Low	1 grade	3-4	10-25	5-6	5-5.5	4-5
Greenfish	Medium-High	1 grade	115-120	4-10	50-60	35-40	30-38
Curryfish	Low-Medium	A	9-10	8-17	33-38	25-28	20-25
		B	18-20	5-14	24-28	20-22	18-21
		C	25-30	4-13	-	14-16	10-15
		D (*Geci*)	-	2-5	-	-	-
Amberfish	Low	1 grade	4-5	8-30	6-8	5.5-6	4-6.5
Brown sandfish	Low	1 grade	30-35	5-13	10-12	7.5-9	7-8
Deep water redfish	Low-Medium	1 grade	7-8	6-40	25-35	10-27	15-25
Lollyfish	Low	A	38-40	4-10	6-8	4-5.5	3-5
		B	45-50	4-8	-	3-3.5	2.5-3
Tigerfish	Low-Medium	A	25-35	4-25	30-35	18-24	15-22
		B	50-70	6-15	15-20	8-15	5-11
Snake fish	Low	1 grade	28-30	5-17	6-8	6-7	5-6.5
Pinkfish	Low	1 grade	45-50	5-12	6-8	6-7	5-6.5
Sea cucumber	Low	1 grade	40-60	3-12	-	-	3-10

*Prices obtained from some marine product agents and the Ministry of Fisheries and Forests in Fiji Islands

Table 1: Prices offered to Fijian marine product agents for bêche-de-mer by Asian markets as of December 2008.

fishers and marine agents (Figures 3 and 4). Figure 3 shows that the key contributors to loss of value to fishers are mishandling after harvest (estimated value loss of 20%), overcooking (estimated value loss of 10%), burning of the product during smoke drying (estimated value loss of 40%) and improper drying prior to storage (estimated value loss of 20%).

Although the main marine agents may reprocess semi-dried product purchased from the fishers, considerable value is already lost because of initial damage to the semi-dried product (Figure 4). Figure 4 represents the processing steps conducted by the main BDM agents and shows that the losses generally occur when products have to be reprocessed to improve their overall appearance and shape (i.e. in an attempt to improve the quality of BDM purchased from fishers). In addition to these losses, BDM processing at the main agents is done with bulk quantities compared to the fishers who process very few sea cucumbers at a time. Boiling large numbers of sea cucumbers in a pot, however, leads to some sea cucumbers being overcooked resulting in bursting of the body-wall and/or semi-disintegration of the sea cucumbers. Instead of discarding such individuals, the agents remove them, dry them and class them as low grade product before exporting. Errors by the agents result in losses in income (Figure 4) which collectively account for a 5-20% overall loss based on the final quality of the product (Figure 5). Lower prices received by exporting agents vastly affect the distribution of revenue from the agents to the primary producers (fishers) (Table 2). Figure 6 shows the supply and value distribution chain for bêche-de-mer in the Fiji Islands and illustrates the relationships between fishers, middlemen and marine products agents (exporters). In some cases the involvement of middlemen in the supply chain further affects revenue distribution because the middlemen purchase items for less (Table 3) from the primary producers and demand higher prices [23] from the exporters or other middlemen involved in the supply chain (Figure 6).

Usually most Fijian sea cucumber fishers lack transportation to deliver their produce to the marine product agents and this situation is exploited by middlemen who buy sea cucumbers from fishers in a perishable state and process them to receive better value from the marine product agents. It seems that there are more middlemen operating in remote areas such as the outer islands where there are no marine product agents. Each fisher is only able to earn <$FJD 50 per

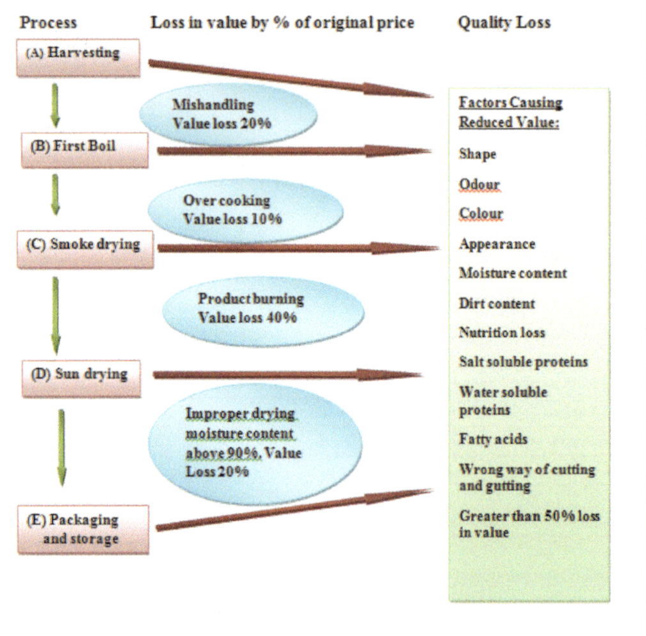

Figure 3: Schematic representation of bêche-de-mer processing done by the indigenous people of the Fiji Islands showing the various stages of processing with an estimation of the value lost at each stage, and the resulting impacts of product quality.

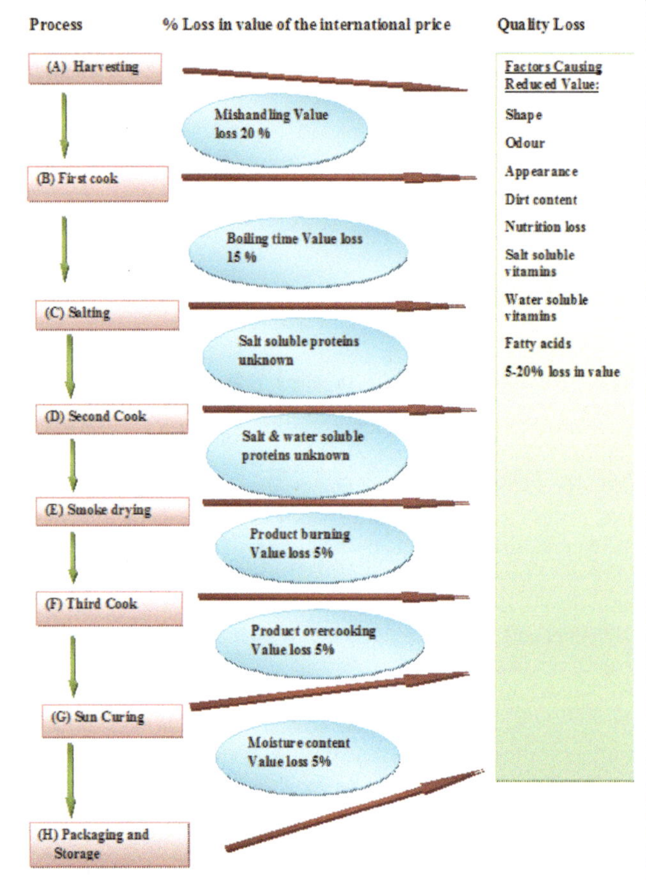

Figure 4: Schematic representation of bêche-de-mer processing done by marine agents of bêche de-mer in Fiji Islands showing the various stages of processing and estimated losses in income and the resulting impacts of product quality at each stage.

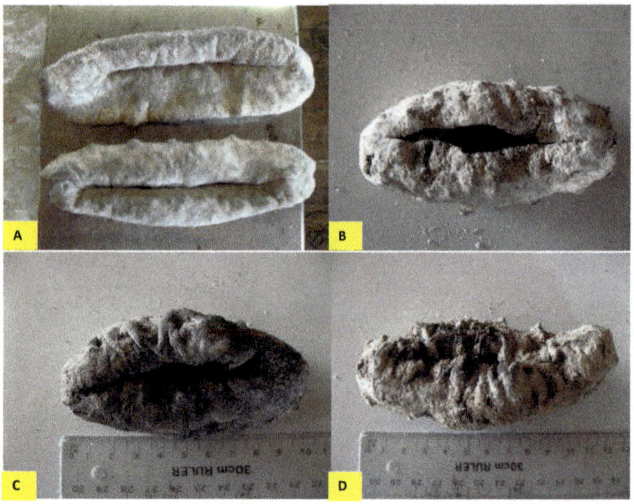

Figure 5: Different quality grades of a high value bêche-de-mer species, White Teatfish (*Holothuria fuscogilva*). Grade A (A), Grade B (B), Grade C (C) and Grade D (D)

week through middlemen, even though the product would contain an adequate number of high value species that could earn them greater income if sold directly to the main exporters who pay much more than the middlemen. Maximising the income of sea cucumber fishers in Fiji requires development of stronger links between them and the main exporters. The BDM exporters receive their income in US dollars and this income then filters down the chain to the fishers who receive relatively little (Figure 6). For example, the BDM exporters buy White Teatfish from the suppliers at (FJD$ 50-70 per piece), however the middlemen advise the fishers that the value of the product is only FJD$ 4-10 per piece. Middlemen encourage the fishers to deal with them directly, rather than any other buyer, telling them that they offer the best price. This is a successful strategy because fishers are often unaware of actual market prices. Dealings between the middlemen often result in a discrepancy of around FJD$ 10-15 per piece. 'Middlemen 2' (Figure 6) are sometimes Chinese nationals operating in smaller designated areas in Fiji who purchase the products from other middlemen and sell to the main exporters for higher revenue.

Price fluctuations in Fiji reflect those in the global export market. Such variations have a drastic effect on the fishers who receive varying income from the marine product agents as a result. During winter months when harvests are low, the price of BDM increases and in summer as supply improves, price falls. In Fiji the trading of marine invertebrate resource (e.g. trochus and bêche-de-mer) with international markets is an important source of foreign revenue. One problem is that there is improper stock monitoring of these fisheries and poor record keeping relating to production and export. Indigenous fishers in Fiji own much of the "*qoliqoli*" or traditional fishing ground, but they lack basic skills required for proper management of the supply of marine resource to the market [20,24]. This situation is exploited by middlemen and marine product agents in Fiji.

In addition, the issues of community-based resource management are generally complex in Fiji and there is a need to improve environmental awareness and sustainable management of marine resources. People need to actively participate in the management programs run at village level [25] and for better use of marine resources, resource-based education and awareness needs to be addressed allowing local people to acquire greater control over their resources

English name	Fijian name	Price received by fishers Dry per kg ($FJD)	Actual price that should be paid by marine product agents (dry/kg) ($FJD)	Price received by fishers Raw per piece ($FJD)	Actual price that should be paid by marine product agents. Raw per piece ($FJD)
White teatfish	*Sucuwalu*	25.00-$35.00	15.00-70.00	4.00-$25.00	5.00-55.00
Black teatfish	*Loaloa, lolo*	10.00-$20.00	20.00-45.00	5.00-$10.00	2.00-6.00
Sandfish*	*Dairo, tero*	8.00-$15.00	-	5.00-$10.00	-
Prickly redfish	*Sucudrau*	10.00-$30.00	25.00-35.00	3.00-$6.00	4.00-6.00
Stonefish	*Dri-vatu, Ki*	2.50-$10.00	10.00 -20.00	0.20-$0.30	0.10-0.30
Surf Redfish	*Tarase*	2.50-$12.00	18.00-25.00	1.00-$3.00	-
Blackfish	*Driloli*	2.00-$8.00	20.00-40.00	1.50-$3.00	-
Elephant Trunkfish	*Dairo-ni-cakau*	7.00-$15.00	7.00-10.00	1.00-$2.00	1.00-$2.00
Greenfish	*Tarasea*	20.00 -$30.00	40.00-52.00	-	-
Curryfish	*Laulevu*	10.00-$30.00	18.00-36.00	0.50-$1.00	0.50-2.50
Amberfish	*Dri-volavola*	7.00-$10.00	6.00-10.00	1.00-$2.00	-
Brown sandfish	*Vula*	2.50-$10.00	5.00-11.00	0.60-$0.80	0.50-2.00
Deep water redfish	*Dri-tabua*	15.00-$25.00	25.00-36.00	1.50-$3.00	4.00-10.00
Lollyfish	*Loliloli*	0.50-$5.00	5.00-9.00	-	-
Sea cucumber	*Mundra, midro*	3.00-$10.00	5.00-9.00	-	-
Snake fish	*Loliloli*	0.50-$4.00	5.00-9.00	-	-
Tiger fish	*Vula-ni-cakau*	6.00-$12.00	10.00-15.00	0.30-$0.80	2.00-5.00

*Species with fishing ban in Fiji Islands but priced for local consumption

Table 2: Prices paid by marine product agents to sea cucumber fishers in Fiji Islands in 2008.

English name	Fijian name	Raw per piece ($FJD) offered as buying price by middleperson	Prices received by the middleperson from the marine product agents ($FJD) when product sold dry
White teatfish	*Sucuwalu*	$4.00-$10.00	50.00-60.00
Black teatfish	*Loaloa, lolo*	$1.50-$2.50	22.00-35.00
Sandfish	*Dairo, tero*	Banned	-
Prickly redfish	*Sucudrau*	$.50-$2.00	30.00-40.00
Stonefish	*Dri-vatu, Ki*	$1.00-$2.00	30.00-38.00
Surf Redfish	*Tarase*	$1.00-$2.00	13.00-28.00
Blackfish	*Driloli*	$0.80-$2.00	20.00-30.00
Elephant Trunkfish	*Dairo-ni-cakau*	$0.20-$0.80	5.00-7.00
Greenfish	*Tarasea*	$0.20-$2.00	40.00-50.00
Curryfish	*Laulevu*	$0.50-$2.50	22.00-30.00
Amberfish	*Dri-volavola*	$0.10-$0.80	5.00-7.00
Brown sandfish	*Vula*	$0.60-$0.80	7.00-11.00
Deep water redfish	*Dri-tabua*	$1.50-$2.00	25.00-30.00
Lollyfish	*Loliloli*	$2.00-4.00/kg dry	5.00-6.00
Sea cucumber	*Mundra, midro*	$1.00-$2.00	4.50-5.50
Snake fish	*Loliloli*	$2.00-4.00/kg dry	5.00-5.50
Tiger fish	*Vula-ika*	$0.30-$0.80	14.00-16.00

Table 3: Prices paid by middlemen to Fijian fishers for raw sea cucumber product in Fiji Islands in 2008.

to ensure longer term economic benefits [8]. There is also a need for fair agreements relating to fisheries resources and proper management programs that could provide a greater and more sustainable source of revenue for the local communities [25].

This study has highlighted some important issues for sea cucumber management in the Fiji Islands and a number of recommendations are suggested on this basis:

(a) Sea cucumber fishers should be educated about the ecological importance of sea cucumbers through extension activities. Hands on training programmes on processing and handling techniques for sea cucumbers should be taught to the fishers to improve the quality and value of bêche-de-mer product [26-28].

(b) Seasonal closures or changing of harvesting sites would reduce pressures on wild sea cucumber stocks. Allowing sea cucumbers to spawn at least once prior to harvest and changes in the fishing grounds would support more sustainable harvesting practices [28,29].

(c) Bêche-de-mer fishery laws should be enforced properly since current regulations are not being followed by sea cucumber fishers and marine product agents in Fiji. Illegal sale of Sandfish, for example, still occurs without monitoring. Furthermore undersized sea cucumbers are harvested and sold.

(d) Total Export Limits (TEL) by species should also be instituted for each buyer in Fiji or for total exports from Fiji that could be in a range of 200-300 tonnes per annum. Once this limit is reached the export should cease until the following harvesting season.

(e) There should be restrictions on issuing SCUBA licenses and law enforcement on the non-usage of "hookah system" by sea cucumber fishers to reduce over-exploitation of stocks. SCUBA operator's exemption certificates have increased from 9 in 2010 to 25 in 2013 in Fiji and the use of SCUBA should be completely banned to avoid the

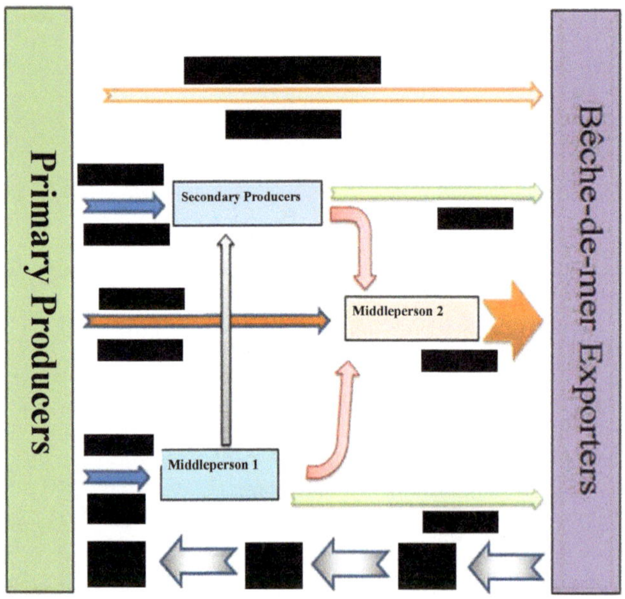

Figure 6: Supply and value distribution chain of bêche-de-mer in Fiji Islands showing the relationships between fishers, middlemen and marine products agents (exporters) and the values of transactions between them.

risk of over-exploitation [19].

(f) Further research is required on harvesting yields of sea cucumbers in the Fiji Islands and, based on the results obtained, group quotas, fisher's quota and industrial quotas should be allocated to protect the current stock from depletion [28].

Conclusion

The quality and value of bêche-de-mer are significantly impacted by negligence during processing by sea cucumber fishers and marine products agents in the Fiji Islands. The loss of value due to poor quality of the final dried product in Fiji ranges from 10% to 50% reflecting the broad number of people involved in the supply and distribution chain of sea cucumbers in Fiji. Sea cucumber fishers in Fiji need to adopt improved processing techniques that support production of a better quality product, improved yield and greater income. This could be achieved through a well-targeted extension program supported by extension materials such as processing manuals. In addition, communities involved in the sea cucumber fishery in Fiji need to be educated on the sustainable use of the marine resource and the ecological importance sea cucumbers in the marine ecosystem. Although aquaculture of tropical sea cucumbers [30-33] may offer an opportunity to improve supply of sea cucumbers for the BDM trade in Fiji while potentially reducing fishing pressures on wild stocks, in the short-term, a number of steps could be taken to improve the sustainability and Fiji sea cucumber fishery - these include enforcement of current fisheries legislation, seasonal closures or rotation of fishing sites, introduction of fishing quotas and export limits, and prohibition and enforcement relating the use of SCUBA and hookah equipment by sea cucumber fishers.

Acknowledgement

The authors would like to acknowledge village communities, sea cumber fishers and bêche-de-mer exporters of Fiji Islands, and the Ministry of Fisheries and Forests in Fiji for assisting with data collection.

References

1. Conand C (1990) The fishing resources of the Pacific Island Countries. Part 2: Holothuria, in Overview of Bêche-de-mer industry in Milne Bay Province, Papua New Guinea, J. Kinch, Editor SPC Bêche-de-mer information bulletin. p. 2-15
2. Bordbar S, Anwar F, Saari N (2011) High-value components and bioactives from sea cucumbers for functional foods--a review. Mar Drugs 9: 1761-1805.
3. Esmat AY, Said MM, Soliman AA, El-Masry KSH, Badiea EA, et al., (2013) Bioactive compounds, antioxidant potential, and hepatoprotective activity of sea cucumber (Holothuria atra) against thioacetamide intoxication in rats, in Nutrition (Burbank, Los Angeles County, Calif.). Elsevier B.V: United States. p. 258.
4. Ferdouse F (2004) World markets and trade flows of sea cucumber/bêche-de-mer, in Advances in sea cucumber aquaculture and management A. Lovatelli, et al. FAO Fisheries Technical Paper. No. 463: Rome; FAO 101-118.
5. McElroy S (1990) Beche-de-mer species of commercial species an update. SPC Bêche-de-mer information bulletin 2:2-7.
6. Bumrasarinpai R (2006) Overview of issues on concerned aquatic species under the convention on international trade in endangered species of wild fauna and flora (CITES), in Regional technical consultation on international fisheries related issues.WP03.
7. Ferdouse F (1999) Bêche-de-mer markets and utilisation. SPC Bêche-de-mer information bulletin.11:3-9.
8. Li X (2001) Fishery and resource management of tropical sea cucumbers in the islands of the South China Sea, in Advances in sea cucumber aquaculture and management. A. Lovatelli, et al. FAO Fisheries Technical Paper. No. 463: Rome; FAO 261-265.
9. Holland A (1994) The beche-de-mer industry in the Solomon Islands: recent trends and suggestions for management. SPC Bêche-de-mer information bulletin 6: p. 2-9.
10. Kinch J (2002) Overview of Bêche-de-mer industry in Milne Bay Province, Papua New Guinea. SPC Bêche-de-mer information bulletin 17: 2-15.
11. Sachithananthan K, Osman SS, Mlay M, Schoemaker R (1985) Report on the national workshop on fish handling in Zanzibar, in United Republic of Tanzania and report on the Tanzania/ SWIOP national workshop on Bêche-de-mer processing., National workshop on fish handling. 45-83.
12. Seeto J (1999) Bêche-de-mer processing – a little more effort to get much more money while saving precious resources. SPC Bêche-de-mer information bulletin 11: 2-3.
13. SPC, Sea cucumbers and beche-de-mer of the tropical pacific. Handbook No. 18. 1994, Noumea: SPC. 51.
14. Battaglene SC, Bell J.D (2004) The restocking of sea cucumbers in the Pacific Islands, in Advances in sea cucumber aquaculture and management, A. Lovatelli, et al. FAO Fisheries Technical Paper. No. 463: Rome; FAO. p. 176-178.
15. Conand C (2004) Present status of world sea cucumber resources and utilization. An international overview, in Advances in sea cucumber aquaculture and management, A. Lovatelli, et al., Editors. 2004, FAO Fisheries Technical Paper. No. 463: Rome; FAO 425.
16. Purcell SW (2014) Value, market preferences and trade of Beche-de-mer from Pacific Island sea cucumbers. PLoS One 9: e95075.
17. Ram R (2008) Impacts of harvest and post harvest processing methods on quality and value of bêche-de-mer in Fiji Islands. Master of Science Thesis. University of the South Pacific. 191.
18. Preston G (1990) Mass Bêche-de-mer production in Fiji SPC Bêche-de-mer information bulletin1: p. 4-5.
19. Pakoa K, Saladrau W, Lalavanua W, Valotu D, Tuinasavusavu I et al.(2013) The status of sea cucumber resources and fisheries management in Fiji. SPC. 51.
20. Friedman K, Eriksson H, Tardy E, Pakoa K (2011) Management of sea cucumber stocks: patterns of vulnerability and recovery of sea cucumber stocks impacted by fishing. Fish and Fisherie 12:75-93.
21. Subasinghe S (1992) Shark fin sea cucumber and jelly fish: a processor's guide. Malaysia INFOFISH Vol. 6:31.

22. Poh CS (2004) Fisheries trade and utilization of sea cucumbers in Malaysia, in Advances in sea cucumber aquaculture and management. A. Lovatelli et al. FAO Fisheries Technical Paper. No. 463: Rome; FAO 57-68.

23. Strehlow HV (2004) Economics and management strategies for restocking sandfish in Vietnam. NAGA, Worldfish center quarterly 27(3-4): 36-40.

24. Purcell S (2010) Managing sea cucumber fisheries with an ecosystem approach. A Lovatelli, M. Vasconcellos, Y. Yimin, FAO Fisheries and Aquaculture Technical Paper No. 520, Rome; FAO. p. 157.

25. Pomeroy RS, Rivera-Guieb R (2006) Fishery Co – management: A Practical Handbook, International Development Research Center, Cambridge, USA: CABI Publishing. 126-127.

26. Ram R, Friedman K, Sobey M.N (2010) Impacts of harvesting and post-harvesting processing methods on the quality and value of beche-de-mer in Fiji Islands. SPC Bêche-de-mer information bulletin 18: p. S19.

27. Purcell S.W, Ngaluafe P, Tamuera K (2012) Improving income of Pacific island fishers through better post-harvest processing of sea cucumber: scoping study. ACIAR/PARDI/2010/004 Project, Canberra, Australia.

28. Carleton C, Hambrey J, Govan H, Conand C (1990) The fishing resources of the Pacific Island Countries. Part 2: Holothuria, in Overview of Bêche-de-mer industry in Milne Bay Province, Papua New Guinea, J. Kinch, Editor., SPC Bêche-de-mer information bulletin. p. 2-15

29. Pakoa KM, Bertram I, Friedman KJ, Tardy E (2012) Sandfish (Holothuria scabra) fisheries in the Pacific region: present status, management overview and outlook for rehabilitation. Asia–Pacific tropical sea cucumber aquaculture p.168.

30. Hair C, Pickering T, Meo S, Vereivalu T, Hunter J et al. (2011) Sandfish culture in Fiji Islands. SPC Beche-de-mer information bulletin 31:(3-11).

31. Hair CA, Pickering TD, Mills DJ (2011) Asia-Pacific tropical sea cucumber aquaculture. In Proceedings of an international symposium held in Noumea, New Caledonia, 15-17 February. ACIAR Proceedings No. 136, ACIAR, Canberra.

32. Jimmy RA, Pickering TD, Hair C, Conand C (1990) The fishing resources of the Pacific Island Countries. Part 2: Holothuria, in Overview of Bêche-de-mer industry in Milne Bay Province, Papua New Guinea, J. Kinch, Editor. SPC Bêche-de-mer information bulletin. 2-15.

33. Battaglene SC (1999) Culture of Tropical Sea Cucumbers for Stock Restoration and Enhancement Naga. ICLARM. 22(4):4-11.

Commercially Important Crab Fishery Resources from Arukkattuthurai to Pasipattinam, South East Coast of India

D. Varadharajan* and P. Soundarapandian

Faculty of Marine Sciences, Centre of Advanced Study in Marine Biology, Annamalai University, Parangipettai-608 502, Tamil Nadu, India

Abstract

The documentation of crab landings especially commercially important crabs are need of the hour to get clear picture about total landings of crabs and also their seasonal availability for better management. So it is attempted to record the individual as well as total crab landings from Arukkattuthurai to Pasipattinam. The individual male, female and berried crabs were reported uniformly maximum in Mallipattinam and minimum in Pasipattinam (Mallipattinam > Manamelkudi > Jegathapattinam > Sethubavachatram > Pointcalimere > Arukkattuthurai > Muthupettai > Adirampattinam > Kattumavadi > Pasipattinam). However total landings of male (48976 kg), female (38745 kg) and berried (16533 kg) crabs were recorded maximum in Mallipattinam. However male (23932 kg), female (16087) and berried (5418) crabs were minimum in Pasipattinam. Among 12 crab species reported *P. pelagicus* (165297.8 kg) was maximum and *C.truncata* (18371.28 kg) was minimum for all sexes. Among various stations crabs were landed maximum in Mallipattinam (104254 kg) and minimum in Pasipattinam (45437 kg).

Keywords: Crab; *P. pelagicus*; *P. sanguinolentus*; Resource; Arukkattuthurai; Pasipattinam

Introduction

Fishery resources are renewable living resources that demand judicious exploitation for obtaining sustainable yields. The crustacean fisheries of India have been assumed significant in recent years. The marine crustaceans found along the Indian coasts, crabs rank third after shrimps and lobsters by virtue of importance as an esteemed seafood delicacy and also by the value of fishery they support. The haemolymph of the crab provides a valuable medical product critical to maintaining the safety of many drugs and devices used in medical care. Fresh and nutritious seafood of crab is always the spirit for various delicious dishes. Crabs are serving either as live, fresh, boiled, steamed, chilled, fried or baked and considered as popular dishes in many restaurants. Moreover, many crabs can serve as feed for fish cultures for turtles and domestic birds and other animals [1]. In duck farms, food mixed with crab powder not only stimulates growth of the ducks, but also raise their spawning rate. The data on the crab landings is essential to make out the contribution of crabs to the Indian economy. The area from different coast is supplying a significant quantity of seafood for the state, yet there is no comprehensive information on crab catches from this region, as they are very much useful for the development of marketing, management, research, export and policy making. Various authors have dealt with fishing methods and the landing of crabs from different parts in Indian coasts. [2,3] reported commercial fishing of crabs from Bombay waters. [4,5] have reported on the crab fishery from Goa. Reports on crab fishing and fishery were also available from Karnataka [6], Kerala [7], Tamil Nadu [8-10], Andhrapradesh [11] and Orissa [12]. From the previous crab landing reports the crab population is declining every year. So crab fishery resource study is essential to improve the population of the respective areas. Hence the present study was aimed to assess the crab fishery resources from Arukkattuthurai to Pasipattinam in south east coast of India for a period of one year.

Materials and Methods

Crabs were surveyed in different landing centres to estimate the landing of this important edible species in the Palk Bay coast of south east coast of India (Plates 3a-3e). The following stations were selected for the present study; Arukkattuthurai (St1), Pointcalimere (St2), Muthupettai (St3), Adirampattinam (St4), Mallipattinam (St5), Sethubavachatram (St6), Kattumavadi (St7), Manamelkudi (St8), Jegathapattinam (St9) and Pasipattinam (St10). The study was made from January 2010 to December 2010. In this study, weight in 5 grams accuracy was taken individually for males, females and berried females. For the purpose of crab catch estimation the observation of crab landings and counting of baskets was taken into consideration. The availability of the berried crabs was also recorded from the total crabs landed during the study period. The methodology for the above study is carried out by following the method of [10].

Results

The individual male, female and berried crabs were reported uniformly maximum in Mallipattinam and minimum in Pasipattinam (Mallipattinam > Manamelkudi > Jegathapattinam> Sethubavachatram> Pointcalimere> Arukkattuthurai> Muthupettai> Adirampattinam> Kattumavadi> Pasipattinam) (Tables 1-3). However total landing of male (48976 kg), female (38745 kg) and berried (16533 kg) crabs were recorded maximum in Mallipattinam. Similarly male (23932 kg), female (16087) and berried (5418) crabs were minimum in Pasipattinam. Total crabs were also followed similar trend as individual crabs (Table 4). Among 12 crab species reported *P. pelagicus* (16529.8 kg) was maximum and *C.truncata* (18371.28 kg) was minimum for all sexes. Among various stations crabs were landed maximum in Mallipattinam (104254 kg) and minimum in Pasipattinam (45437 kg) (Tables 1-5). The crabs landed in various stations showed significant difference (Table 6).

***Corresponding author:** D. Varadharajan, Faculty of Marine Sciences, Centre of Advanced Study in Marine Biology, Annamalai University, Parangipettai-608 502, Tamil Nadu, India, E-mail: heartvaradhan@gmail.com

S:No	Crabs	St 1	St 2	St 3	St 4	St 5	St 6	St 7	St 8	St 9	St 10	Total
1	*Scylla serrata*	4585	4055	2481	3477	7946	3726	2651	2073	2777	1904	35675
2	*S.tranquebarica*	2841	2938	2777	3144	8022	4961	2348	5092	3754	1695	37571.6
3	*Portunus pelagicus*	5116	6778	5773	6005	9949	8089	7325	14043	7658	6610	77346
4	*P.sanguinolentus*	3506	5021	5162	4494	7864	7888	5055	11452	8063	3409	61913.7
5	*P.gladiator*	1361	1857	1140	1092	2195	1608	1330	821.2	1786	1266	14456.1
6	*Podophthalmus vigil*	1175	2342	1972	1170	1874	2297	541.3	3050	1665	1154	17240.1
7	*Charybdis feriata*	1161	2467	1795	1119	2663	1872	1122	3161	1902	1375	18637.4
8	*C.natator*	1801	2591	1016	1892	2685	953	1575	2153	1865	1074	17604.8
9	*C.lucifera*	1619	1793	676.6	1586	1779	769.6	1539	1908	1301	1297	14267.7
10	*C.variegata*	1629	2654	643.4	1193	1432	606.2	963.4	1801	1387	1332	13640.8
11	*C.granulata*	1570	1582	1006	1310	1319	381.2	1029	1526	1837	1498	13058.6
12	*C.truncata*	664.6	1261	714	620.8	1249	1319	700.2	1433	1020	217	9218.8
	Total	27029	35340	25157	27103	48976	33389	26179	48512	35014	23932	330631

(St 1-Arukkattuthurai, St 2- Pointcalimere, St 3-Muthupettai, St 4- Adirampattinam, St 5- Mallipattinam, St 6-Sethubavachatram, St 7-Kattumavadi,St 8- Manamelkudi, St 9- Jegathapattinam, St 10-Pasipattinam)

Table 1: Total male crabs (Kg) landed from different stations during January 2010 to December 2010.

S:No	Crabs	St 1	St 2	St 3	St 4	St 5	St 6	St 7	St 8	St 9	St 10	Total
1	*Scylla serrata*	4707	2701	1648	2554	3313	3605	1626	1400	2649	1674	25876
2	*S.tranquebarica*	2851	2782	2773	2551	3696	4465	2550	4847	2698	1566	30779
3	*Portunus pelagicus*	5410	5814	5592	5432	10758	6936	6533	7641	7711	3019	64846
4	*P.sanguinolentus*	3804	3644	3799	3692	7858	6502	3908	10672	4043	3291	51212
5	*P.gladiator*	1123	1068	1363	771	1711	1380	1008	1143	1322	483.2	11372
6	*Podophthalmus vigil*	867.6	1877	1740	804.7	1983	1437	332.5	3717	1624	1025	15408
7	*Charybdis feriata*	428.9	1977	1591	1010	4265	1464	1023	1392	2016	768.5	15935
8	*C.natator*	1074	2065	585	765.2	1472	655	928	1519	1439	838	11340
9	*C.lucifera*	1143	1851	463	947.7	1489	494.7	1204	861.3	1319	572.9	10346
10	*C.variegata*	633.4	1958	570.6	1047	1197	512.5	743.7	2350	1939	945.5	11897
11	*C.granulata*	1058	1472	576	945.1	585	322.8	846.6	1812	1344	1498	10459
12	*C.truncata*	629	1210	313	587	418	407	510.1	664.8	1022	223.2	5984.1
	Total	23729	28419	21013	21107	38745	27997	19238	38018	29126	16087	263479

(St1-Arukkattuthurai, St2- Pointcalimere, St3-Muthupettai, St4- Adirampattinam, St5- Mallipattinam, St6-Sethubavachatram, St7-Kattumavadi, St8- Manamelkudi, St9- Jegathapattinam, St10-Pasipattinam)

Table 2: Total female crabs (Kg) landed from different stations during January 2010 to December 2010.

S:No	Crabs	St 1	St 2	St 3	St 4	St 5	St 6	St 7	St 8	St 9	St 10	Total
1	*Scylla serrata*	2529	1237	872	915	1202	1516	691	259.2	1074	358.3	10654
2	*S. tranquebarica*	1496	935.1	705	975	1482	2551	1043	994.7	1471	714	12367
3	*Portunus pelagicus*	1769	1279	1655	1658	6753	2515	1784	2356	1982	1355	23106
4	*P.sanguinolentus*	2076	1053	1500	1630	2534	2402	970.5	2005	2625	909.6	17704
5	*P.gladiator*	1290	471.6	444	443.1	766.3	994.2	440	354.7	674.7	190.6	6069.2
6	*Podophthalmus vigil*	458.5	779	665.2	359.5	765.2	1112	391	784.7	814.2	338.5	6467.7
7	*Charybdis feriata*	297.1	931.4	1565	393.9	1121	590	452	551.9	1008	319.5	7229.7
8	*C.natator*	460.6	832.6	927	383	590.6	340.4	483	363.4	1015	246	5641.6
9	*C.lucifera*	1116	750.6	635	459.3	427.9	289.5	485	323.5	631.9	240.6	5359.5
10	*C.variegata*	349	761.1	500.1	454.4	462.8	246	335.6	535.9	663.6	327.1	4635.6
11	*C.granulata*	461.2	624	676.5	368.9	266.3	207.9	529	301.3	709.4	164	4308.5
12	*C.truncata*	269.4	446.9	397.5	264	161.7	256	302	534.2	415.4	121.8	3168.4
	Total	12571	10101	10542	8304	16533	12885	7144	9364	13085	5418	105947

(St 1-Arukkattuthurai, St 2- Pointcalimere, St 3-Muthupettai, St 4- Adirampattinam, St 5- Mallipattinam, St 6-Sethubavachatram, St 7-Kattumavadi, St 8- Manamelkudi, St 9- Jegathapattinam, St 10-Pasipattinam)

Table 3: Total Berried crabs (Kg) landed from different stations during January 2010 to December 2010.

Discussion and Conclusion

Fisheries in India, particularly marine fisheries have shown a tremendous progress in recent times. Fluctuations have been noticed in the marine fish landings from year to year and these are due to fishery independent factors such as salinity, temperature and current and also fishery dependent factors such as rate of exploitation, season of exploitation and size exploited. The crustaceans are not only important scavengers of the sea but also valuable food for fishes. The data collected

[redacted]

Table 4: Total (Kg) male, female and berried female crabs landed from different stations during January 2010 to December 2010.

Crabs	[redacted]
Ss	
St	
Pp	
Ps	
Pg	
Pv	
Cf	
Cn	
Cl	
Cv	
Cg	
Ct	
Total	

(*Ss-Scylla serrata, St-S.tranquebarica, Pp-Portunus pelagicus, Ps-P.sanguinolentus, Pg-P.gladiator, Pv-Podophthalmus vigil, Cf- Charybdis feriata, Cn-C.natator,Cl-C.lucifera, Cv- C.variegata,Cg- C.granulata,Ct- C.truncata*)
[St1-Arukkattuthurai, St2-Pointcalimere, St3-Muthupettai, St4-Adirampattinam, St5-Mallipattinam, St6-Sethubavachatram, St7-Kattumavadi, St8-Manamelkudi, St 9- Jegathapattinam, St 10-Pasipattinam]

Table 5: Total crabs (Kg) landed from different stations during January 2010 to December 2010.

Source of Variation	**SS**	**df**	**MS**	**F**	**P-value**	**F crit**
Crab landed	269626389	9	29958488	1.153176	0.332315	1.966054
Within Groups	2857702142	110	25979110			
Total	3127328531	119				

Table 6: One way analysis of variance for the crabs landed in different landing centers from January-2010 to December-2010.

on landings of crabs contributing directly to support the mankind. Fisherman used different kinds of fishing gears to collect crabs such as crab nets, singhi valai, sembara valai, meen valai, koi valai, veral valai and oda valai. The trawl nets operated by mechanised boats and other powered vessels also bring appreciable amounts of crabs with other fishes all along the coastal waters [7]. In Indian coast commercial crab population is being overexploited has led to research on the life history, habitat requirements and population dynamics of this species [13]. To date, these factors and the overall health of the population have been poorly understood. Under these assumptions it is clear that though absolute estimates may vary over the period, the respective percentage contributions are expected to remain more or less the same [14].

Totally 12 (*S. serrata, S. tranquebarica, P. pelagicus, P. sanguinolentus, P. gladiator, P. vigil, C. feriata, C. natator, C. lucifera, C. variegata, C. granulata* and *C. truncata*) commercially important crabs were recorded in all stations of the present study. The crab species recorded was belonging to the family Portunidae includes the genera *Scylla, Portunus, Charybdis* and *Podophthalmus*. As in the present study, commercially important crab landings were already reported by different authors [10,15] in south east coast India. Tamil Nadu tops the list in crab landings all over India and the coastal belt from Tuticorin to Mallipattinam has been proven as the strongest potential of edible sea crabs [10,16]. In the present study not only total contribution of the crabs but also individual contributions of crabs were maximum in Mallipattinam than other stations. The total annual crab landings of the Parangipettai was 79,445.6 kg and Palk straight was 3, 88,196.032 kg as reported by [10,17] respectively. The total crabs landed in the present study are 7, 00057 kg, which is relative by more than last year as reported by [18]. The crab landings varied in the past two decades are due to many reasons [18]. Landing is abundant in some areas in India is due to large number of trawl operations. The important mechanised fishing centre is Mallipattinam, Sethubavachatram and Jegathapattinam but in other stations the fisherman using only the fiber boat, catamaran and small boats [19]. The number of mechanised trawling boats owned by local fishermen at Mallipattinam, Sethubavachatram and Jegathapattinam are 137, 82 and 118 respectively. This may be the reason for landing of crabs are maximum in Mallipattinam followed by Manamelkudi and Jegathapattinam.

The crabs like *P. pelagicus, P. sanguinolentus, S. serrata and S.tranquebarica* forming a major crab fishery along south east coast [10, 16, 17]. The above mentioned four species also contributed maximum throughout the study period of the present study in almost all the stations. This might be due to their continuous breeding activity and the fecundity up to 0.52 million eggs [15]. In the present study *P. vigil, P. gladiator, C. feriata, C. natator, C. lucifera, C. granulata, C. truncata* and *C. variegata* fishery is purely seasonal and contributes very less as far as total crab fishery is concerned. The previous report by [15] also supported the present findings among ten landing centers; *C. truncata* was less in Pasipattinam (582 kg), Muthupettai (1424.5 kg) and Adirampattinam (1471.8 kg). However, *P. gladiator* was available less in Pasipattinam (1939.8 kg) and Muthupettai (2947 kg). Similarly *C. lucifera* (1774.6 kg) and *C. granulata* (911.9) were available less in Sethubavachatram landing centre. Berried crabs of *P. pelagicus, P.sanguinolentus, S.serrata* and *S. tranquebarica* are available all the months of the year. In general the total numbers of berried crabs are maximum in Mallipattinam (16,533 kg) and minimum in Pasipattinam (5418 kg) which directly reflects the total number of crabs landed. Although, the commercial crabs are caught throughout the year at various regions of the coast, the main season varies from place to place [19]. On the west coast, the fishery is generally active in the latter half of the year. In the east coast, the peak season is observed in January to April and June to October and December. Landing of Portunid species have been estimated from other parts of the country as well [6, 7, 9, 11, 20, 21, 22].

There has been significant variation in the results obtained by different workers who studied fecundity of portunid crabs from various regions [5,23]. The number of eggs produced

by crustaceans varies widely [24]. According to [25] the fecundity can vary seasonally, annually and between areas. Among Brachyuran crabs, there is substantial variation in fecundity and is breeds round the year. Among various crab species reported in the present study, *P.pelagicus* and *P.sanguinolentus* were landed maximum in almost all the stations. This is also supported by various workers in different parts. *P. pelagicus* as a continuous breeder with maximum intensity during September - March near Mandapam on southeast coast of India [26,27] reported the occurrence of berried females of *P. pelagicus* throughout the year with pronounced abundance from January to March and September to December in Palk Bay and Gulf of Mannar. The high with definite peaks during December to January along the southwest coast of India suggesting peak breeding during that period [28]. The breeding of this species is from August to April with peak in January [5,29]. mentioned that the breeding of *P. pelagicus* takes place during February to March in the Zuari estuary in Goa. Two peaks in the incidence of gravid females of *P. pelagicus*, one in September and another in February were reported from Australian waters by [30]. The report on the biology of *P. sanguinolentus* observed that the species is breeding during February to April along the Malabar Coast [20,31] reported year round spawning with peak activity during March to April along the Mangalore coast, while [32] noticed that the crab is breeding throughout the year in the Bombay area [6] mentioned that this crab is breeding all through the season with maximum intensity in February and April to May along the South Kanara coast [33] indicated that this species spawn year round with a peak season from October to February [29] found that breeding is almost continuous throughout the year except July with peak activity in February along the southwest coast of India. The peak breeding seasons of *P. sanguinolentus* was from November to March and December to May and July to August and *C. feriata* from April to July in southwest coast [34,35] estimated the preponderance of berried *S. serrata* and *P. pelagicus* during post monsoon and in the months from August to October [36] along Tuticorin Bay observed that the maximum breeding activity in *P. pelagicus* during June followed by a gradual decrease in consecutive months and maximum in December. The portunid crab larval and juvenile abundance was at maximum level during late post monsoon and summer and minimum during December in the Vellar estuary [37]. The peak seasons of *S. serrata* were reported by [38]. For example *S. tranquebarica, S. serrata* have reported with comparatively higher fecundity [39-41]. The changes in the seasons of berried crab availability in two coasts might be due to different monsoonal periods, current patterns and environmental parameters.

According to the Annual Report 2009-2010 released by the Central Marine Fisheries Research Institute, the west coast remains more productive, contributing 56% of the total catch in the country, while the remaining 44 % is contributed by the east coast. However, the fish landing in Tamil Nadu, Gujarat and Andhra Pradesh had shown an increasing trend during the same period [19]. The fundamental economic laws of supply and demand can be applied to understand the market for commercial fisheries, but there are added complications due to the common property nature of crab resources [42]. This common property problem is the subject of most of the commercial fisheries economics literature analyzing the reasons for biological and economic overfishing [26,43]. The estimation of biodiversity values in monetary terms helps to promote and justify conservation actions at the public policy and decision making level and is essential to define the optimal level of conservation [22].

In the present study the Arukkattuthurai to Pasipattinam supports a major crab fishery along southeast coast of India. The occurrence of brachyuran zoea is fairly large numbers during January to June and throughout year in south east coast [21]. It is suggest that these crabs are not moving far away from the fishing ground for hatching their eggs. In contrast, the occurrence of juveniles of Portunid crabs (45-80 mm cw) in small numbers and the rarity of younger juveniles (25-50 mm cw) in trawl, mini trawl and shore seine catches throughout the year indicates that the berried crabs particularly the larger ones, probably move to more saline deeper waters for hatching. The low incidence of ovigerous crabs in comparison to that of portunid crabs can well be explained by the offshore migratory behaviour of this crab for hatching. The possible conservative measures to increase the crab landings and to make it available throughout the year are given as follows; The large scale destruction of young crabs would obviously have adverse effect on the crab resources; therefore a specific size (For example, *Scylla* spp. is 140 mm of carapace width) for harvest is to be fixed. The fishing areas should be closed or the fishing effort should be minimized for certain period of time or seasonal banning of the shipment of crabs should be done. The conservation measures for catching young, undersized and berried crabs should be done by educating the fishermen through audiovisual aids. Sea ranching should be promoted for over exploited species. And certain regulatory measures and laws should be enforced like, fixing legal minimum size (across the broadest part of carapace) and protection of ovigerous crabs. The law breakers should be fined or punished.

The existing crab fishery of India is not commercially developed as in other countries. But from a comparison of the present exploitation and the evaluation of the potential crab resources of different regions, it would appear that there is a scope for developing the fishery in to a major industry. Prerequisites for such an expansion from the commercial point of view are the extensive use of the efficient tackles to catch crabs, the development of quick transportation facilities of this extremely perishable commodity to the consumer centers and the development of a taste for crab meat among the population. From the scientific point of view, it is essential that intensive investigations on improved fishing technique, on farming of suitable varieties of crabs and on various biological aspects are to be carried out not only to harvest crabs from unexploited inshore and offshore regions, but also to safeguard the stock from injudicious exploitation, as it is not uncommon to see females in berry being extensively fished and sold in the markets. Various commercial crab species are reported as food by the local communities and future research should focus on establishing their nutritive food value, biology and ecological relationships in their marine habitats. The primary fishing is the lifeblood of many small coastal communities, often where other sources of jobs and incomes. So it is recommended the protection of the breeding stock and undersized commercial valuable crabs. As well as recommended returning back to the sea of juveniles, ovigerous females and recently moulted crabs as a way of preventing overfishing. The evident conclusion from this study is that coastal fishery landing reports are important contributors in country's economy, in terms of household income, employment and exports.

References

1. Tookwinas S, Srichantulk N, Kanchanavasite C (1992) Mud crab production in Thailand. In C.A. Angell (ed.). The mud crab. Report of the seminar on the mud crab culture and trade. Surat Thani, Thailand, November 5-8, 1991. BOBP/REP/51.Madras, India. 56-63.
2. Rai HS (1933) Shell fisheries of Bombay presidency(Part II). J Bombay Nat Hist Soc **36**: 884-897.
3. ChhapgarBF(1957)Onthemarinecrabs(Decapoda:Brachyura)ofBombayState. Rep marine zool **54**: 503-522.
4. Anzari ZA, Harkantra SN (1976) Crab resources of Goa. Sea Exp J **8**: 21-24.
5. Dhawan RM, Dwivedi SN, Rajamanickam GV (1976) Ecology of the blue crab *Portunus pelagicus* (Linnaeus) and its potential fishery in Zuari estuary. Indian J Fish **23**: 57-64.
6. Sukumaran KK, Telang KY, Thippeswamy D (1986) On the fishery and biology of the crab *Portunus pelagicus* (Herbst) along the south Kanara coast. Indian J Fish **33**: 188-200.
7. Bal DV, Rao KV (1984) The crabs, in marine fisheries. Tata Mcgraw- Publishing Company Ltd., New Delhi. pp. 375-385.
8. Menon MK (1952) A note on the bionomics and fishery of swimming crab, *Neptunus sanguinolentus* (Herbst) on the Malabar Coast. J Zool Soc India 4: 177-184.
9. Shanmugam S, Bensam P (1980) The fishery for the crab *Scylla serrata* (Forskal) at Tuticorin during 1974-75. Indian J Fish **27**: 102-110.
10. Varadharajan D, Soundarapandian P, Dinakaran GK, Vijakumar G (2009) Crab Fishery Resources from Arukkattuthurai to Aiyammpattinam, South East Coast of India. Curr Res J Biol Sci **1**: 118-122.
11. Lalitha Devi S (1985) The fishery and biology of crabs of Kakinada region. Indian J Fish **32**: 18-32.
12. Jones S, Sujasingani KH (1952) Notes on the crab fishery of Chilka Lake. J Bombay Nat Hist Soc **51**: 128-134.
13. Abdel-Razek FA (1987) Crab fishery of the Egyptian waters with notes on the biomomics of *Portunus pelagicus* (L.) adriat **28**: 143-154.
14. Michael JVDA, Dean LF (1984) Species Profiles: Life histories and environmental requirements of coastal fishes and invertebrates (South Atlantic) - blue crab. U .S. Fish Wildlife Serv **82**: 16.
15. Radhakrishnan CK (1979) Studies on portunid crabs of Porto Novo (Crustacea: Decapoda: Brachyura) Ph.D Thesis, Annamalai University India pp 1-129.
16. Sanil Kumar S (2000) New horizons in sea crab meat processing. Sea exp J **31**: 41-43.
17. John Samuel N, Thirunavukkarasu N, Soundarapandian P, Shanmugam A, Kannupandi T (2004) Fishery potential of commercially important portunid crabs along Parangipettai coast. In: Proceedings of Ocean Life Food & Medicine Expo pp 165-173.
18. CMFRI (2009) Annual Report 2009 -2010. Technical Report. CMFRI, Kochi.
19. CMFRI(2010)AnnualReport2010-2011.TechnicalReport.CMFRI,Kochi.Coast: A case study. J Mar Biol Ass India **45**: 99-107.
20. George PC, Nayak KR (1961) Observations on the crab fishing of Mangalore coast. Indian J Fish **8**: 44-53.
21. Rao P, Vedavyasa MM, Thomas, Sudhakara Rao G (1973) The crab fishery resources of India. In: Proceedinds of the symposium on living resources of the seas around India. CMFRI Sp Publ: 581-591.
22. Joel DR, Raj PJS (1987) Marine crab fisheries around Pulicat. Sea exp J **19**:18-21.
23. Potter IC, Chrystal PJ, Loneragan NR (1983) The biology of the blue manna crab *Portunus pelagicus* in an Australian estuary. Mar Biol **78**: 75-85.
24. Sastry A (1983) Ecological aspects of reproduction. In: FJ Vernberg and W Vernberg Biology of crustacea. Academic press: New York 8: 179-270.
25. Parsons DG, Tucker GE (1986) Fecundity of northern shrimp *Pandalus borealis* (Crustacea, Decapoda) in areas of North West Atlandic. Fish Bull US **84**: 549-558.
26. Prasad RR, Thampi PRS (1952) An account of the fishery and fishing methods for *Neptunus pelagicus* near Mandapam. J Zool Soc India **4**: 335-339.
27. Ameerhamsa KMS (1978) Fishery of the swimming crab *Portunus pelagicus* Linnaeus from Palk bay and Gulf of Mannar. Indian J Fish **25**: 229-232.
28. Pillai KK, Nair NB (1973) Observations on the breeding biology of some crabs from the south west coast of India. J Mar Biol Ass India **15**: 745-770.
29. Pillai K, Nair NB (1970) Observation on the reproductive cycles of some crabs from the southwest coast of India. J Mar Biol Ass India **10**: 384-386.
30. Campbell GR, Fielder DR (1986) Size at sexual maturity and occurrence of ovigerous females in three species of commercially exploited portunid crabs in S.E. Queensland. Proc R Soc Queens **97**: 79-87.
31. Menon MK (1953) A note on the bionomics and fishery of swimming crab, *Neptunus sanguinolentus* (Herbst) on the Malabar Coast. J Zool Soc India **4**: 177-184.
32. Chhapgar BF (1956) On the breeding habits and larval stages of some crabs. Rec Indian Mus **54**: 33-52.
33. Ryan EP (1967) Structure and function of the reproductive system of the crab *Portunus sanguinolentus* (Herbst) (Brachyura: Portunidae). I Male system Proc Symp Crustacea Mar Biol Ass India Part II: 506-521.
34. Saradha PT (1998) Crab fishery of the Calicut coast with some aspects of the population characteristics of *Portunus sanguinolentus*, *Portunus pelagicus* and *Charybdis cruciata*. Indian J Fish 45: 375-386.
35. Joel DR, Raj PJS (1982) Reproduction in portunid crabs *Scylla serrata* and *Portunus pelagicus* from Gulf of Mannar. In: Progress in Invertebrates Reproduction and Aquaculture. (Eds.) T Subramoniam and Sudha Varadarajan New Century Printers Madras 162-175.
36. Rajamani M, Manickaraja M (1998) A note on the fishery of the swimming crab *Portunus pelagicus* (Linnaeus) from Tuticorin bay. Indian J Fish **39**: 185-188.
37. Sethuramalingam S (1984) Studies on Brachyuran crabs from Vellar estuary, Killai backwater complex of Porto Novo coast. Ph.D Thesis, Annamalai University, India, p.243.
38. Hill BJ (1975) Abundance, breeding and growth of the crab *Scylla serrata* in two South African estuaries. Marine Biology 32: 119-126.
39. Prasad PN, Neelakantan B (1989) Fecundity of the mud crab, *Scylla serrata* (Forskal). Mahasagar **22**: 23-28.
40. Roberston WD, Kruger A (1994) Size at maturity, mating and spawning in the portunid crab *Scylla serrata* (Forskal) in Natal, South Africa. Estuar Coast Shelf Sci **39**: 185-200.
41. Srinivasagam S, Kathirvel M, Kulasekarapandian S (2000) Captive brood stock development induced breeding and larval stages of mud crabs (*Scylla* spp.) CIBA Bulletin **12**: 97.

42. Bueno PB (2005) Impacts of Tsunami on Fisheries, Coastal Resources and Human Environment in Thailand. In: 4th Regional Network of Local Governments Forum Bali. p 28.

43. Prasad RR, Thampi PRS (1953) A contribution to the biology of the blue swimming crab *Neptunus pelagicus* (Linnaeus), with a note on the zoea of *Thalamita crenata* Latrelle. J Bombay Nat Hist Soc **51**: 674-689.

11

Assessment of Heavy Metals in Some Marine Fish Species Relevant to their Concentration in Water and Sediment from Coastal Waters of Ondo State, Nigeria

Olusola JO* and Festus AA

Department of Chemistry, Federal University of Technology, Akure, Nigeria

Abstract

Due to oil exploration activities and urbanization as well as continuous industrial and agricultural growth in Ondo State, Nigeria. The coastal water has been heavily impacted by a number of pollutants originating from different sources including heavy metals. This investigation assessed and monitored accumulation levels of (Cr, Cd, Pb, Cu, Zn, and Ni) in different organs (Gill, Head, Bone, Muscle and Eye) of five fish species (*Pentanemus quinquarius, Pseudoltolithus senegalensis, Trichirus lepturus, Plectorhynchus meditarraneus* and *Pseudotolithus typus* together with water and sediment collected from the ocean shoreline in the coastal waters of Ondo State, Nigeria. Heavy metals concentrations in fish tissues, water and sediment samples were analyzed after treatment using atomic absorption spectrophotometer. Dissolve total metal in water (in mg/l) Zn 0.12-0.22 , Cu 0.26-0.27 and Ni BDL-0.01 recorded in the water samples in this study were low and within the maximum permissible level (MPL) recommended by WHO and USEPA, while higher concentrations above the MPL were recorded for Cr 0.31-0.34 Cd 0.08 -0.21 and Pb 0.57-0.79. The concentrations of these metals in sediment in (mg/kg) were in the range Cr (0.16-4.19, Cd 0.35-0.38, Pb 0.9-1.00, Cu 0.24-1.44, Zn 2.12-3.11 and Ni BDL-0.01. They are all lower than their probable effect concentrations (PEC) in sediment. Mean Concentrations of the heavy metals in fish species were of the order: Zn>Cu>Cr>Cd>Pb>Ni. Levels of heavy metals varied depending on different tissues in the fish species. Correspondingly, high concentrations of the metals were found in gills and eye compared to other tissues. While the concentration of Zn (0.34 mg/kg-1.29 mg/kg), Cu (BDL-1.88 mg/kg), Ni (BDL-0.01 mg/kg) , and Cr (BDL-1.28) in the fish tissues were within the maximum allowable level (MAL) for a food source, the findings of this study shows higher value above the allowable limit for Cd(BDL-1.14 mg/kg) and Pb (BDL-0.71 mg/kg), thus constituting potential health hazard to consumers of these fish species.

Keywords: Heavy metals; Fish; Sediment; Concentration factor; Coastal water

Introduction

The aquatic environment with its water quality is considered the main factor controlling the state of health and diseases in both cultured and wild fishes. Pollution of heavy metals in aquatic ecosystem is growing at an alarming rate and has become an important worldwide problem [1]. Increase in population, urbanization, industrialization and agricultural practices have further aggravated the situation [2]. As heavy metal cannot be degraded, they are deposited, assimilated or incorporated in water, sediments and aquatic animals and thus causing heavy metal pollution in water bodies [1]. Therefore, heavy metals can be bio accumulated and biomagnified via the food chain and finally assimilated by human consumers resulting in health risks [3]. As a consequence, fish are often used as indicators of heavy metals contamination in the aquatic ecosystem because they occupy high trophic levels and are important food sources [4]. Fishes are major part of the human diet and it is therefore not surprising that numerous studies have been carried out on metal pollution in different species of edible fish [5,6]. Predominantly, fish toxicological and environmental studies have prompted interest in the determination of toxic elements in seafood [7]. Analyzing pollutant in living organisms is more attractive and promising than analyzing pollutants of the abiotic environment, as living organisms provide precise information about the bioavailability of pollutants [8]. This may assist in predicting pollutants transfer exposure and its possible health consequence to humans. In addition, such information is crucial in making accurate risk assessment for seafood safety purposes. Report has shown that heavy metals have the tendency to accumulate in various organs of marine organisms, especially fish, which in turn may enter into the human metabolism through consumption causing serious health hazards [9]. In many cases, industry has main role in environmental crises especially in civil and energy projects. For example, more than 60 percent of annual greenhouse gas emissions are related to the industrial activities in the world (transportation fuels and distribution 25.3%, power stations 21.3%, industrial process 16.8%). Ondo State coastal water is a brackish water body located in southwestern Nigeria. The water is used by the local community for various purposes including fishery. Water and sediment quality within this coastal waters are increasingly polluted with contaminants especially heavy metals coming from oil spill from shell and chevron companies, sewage and run off-off from agricultural land [10]. Hence the main issue arises on whether the concentration of pollutants in the waters body represents a risk to human and aquatic biota. Generally, in environment field and related science, most previous studies have been carried out using classical methods and only the ways to deal with industrial pollution have been investigated. Researches about deal with challenges and using of new technologies have not been fully dealt with. However,

***Corresponding author:** Olusola JO, Department of Chemistry, Federal University of Technology, Akure, Nigeria, E-mail: layinkx@gmail.com,demolaktp@yahoo.co.uk

valipour et al used environmental flow diagram to identify pollutants of industrial units, the source of production, transmission and conversion processes, and the recipient thus, making decision about environmental pollution simple. The present study is undertaken to investigate the bioaccumulation potential of six selected heavy metals viz. Cadmium, Copper, Chromium, Nickel, Lead and Zinc in different tissues of the selected fish species, *Pentanemus quinquarius, Pseudoltolithus senegalensis, Trichirus lepturus, Plectorhynchus meditarraneus* and *Pseudotolithus typus* from the coastal waters with associated water and sediment samples.

Materials and Methods

Study area

The coastal waters in Ilaje local government area is one of the largest water bodies in ondo state, Nigeria. The research was carried out in the Atlantic Ocean shoreline from the coastal waters in Ilaje local government area. The Local Government Area lies between longitudes 6°12°E and 6°30°E of the Greenwich Meridian and between latitudes 4°10°N and 4°6°N of the Equator. It is bounded in the West by Ogun State, in the East by Irele Local Government Area and Edo State, in the North by Okitipupa Local Government Area and in the South by Bight of Benin and Atlantic Ocean (Figure 1). Due to oil exploration activities and urbanization as well as continuous industrial and agricultural growth in Ondo State, Nigeria. The coastal water has been heavily impacted by a number of pollutants originating from different sources including heavy metals. Such contamination must be an important issue regarding the health of the aquatic ecosystem and its animals and in turn, to human's health.

Sampling and sample analysis

Water and sediment samples were collected from randomly selected three sub location in each main location of the ocean shore line. Water samples were taken into pre-acid washed polyethene bottles. The water sample were acidified to a pH of less than 2 by adding 2 ml of analar grade conc. HNO_3 and kept in the refrigerator prior to analysis. Representative water sample (100 ml) from the three sub locations in each main location were digested by standard procedure. The concentrations of metals in each sample were determined against those of the blank and standard solutions using Buck scientific AAS. The sediment sample taken concurrently from the bottom surface using a sediment samples from the randomly selected three sub locations of each of the selected location were transported to the laboratory in a polythene bag that has been previously washed and leached with acid, and subsampled to obtain a representative sample for each location. The sediment was drained and kept in safe place for air drying prior to analysis. The air dried samples were ground and sieved through 2 mm aperture. Digestion of sediment samples was carried out by procedure described by Hulya [5]. Heavy metals in sediment sample was analyze using Atomic Absorption Spectrophotometer (AAS) Buck scientific model, 200A. Five to seven pieces each of (*Pentanemus quinquarius, Pseudoltolithus senegalensis, Trichirus lepturus, Plectorhynchus meditarraneus* and *Pseudotolithus typus.)* were caught from the water bodies with the assistance of local fisher men. The samples were transported to the laboratory, where the scales were removed where applicable, rinsed with distilled water to remove any adhering substance and each fish separated into eye, head, gills, bones and muscles with dissecting knife. The different fish organs were then separately oven-dried at 105 ± 2°C in a Gallenkamp moisture extraction oven until constant weight was obtained and then ground into powder. Extraction of metals from the fish samples was by mixed acid digestion. Heavy metal concentrations in water samples were determined having digested the samples in accordance with the standard procedure established by Parker [11]. The water samples (100 ml) were digested on a hot plate with the addition of 10 ml concentrated nitric acid. The solution was heated to almost dryness and the residue dissolved with 10% HNO_3. The resulting solution was filtered through Whatman number 1 filter paper into 50 ml standard flask and solution made to the mark. The concentrations of metals in each sample were determined against those of the blank and standard solutions using Buck scientific AAS. The sediment samples were air dried ground and sieved through 2 mm aperture. About two grams of the sediment sample in each case were put into 250 ml beaker and digested with 20 ml of mixed hydrochloric acid and nitric acid in the ratio 3:1 and 2 ml 35% H_2O_2 on hot plate in a fume cupboard for 2 hours [5]. On cooling, the digest was filtered into 50 ml standard flask and made to the mark with distilled water. The resultant solution was analyzed for heavy metals using Atomic Absorption Spectrophotometer (AAS) Buck scientific model, 200 A. Samples of various organs ranging from 0.5 g to 1.09 g were accurately weighed and digested using 10 ml of 65% HNO_3, 5 ml of 37% HCl and 2 ml of 35% H_2O_2 at 80 ± 50°C in a fume chamber until colorless liquid was obtained [12]. The solutions were allowed to cool at room temperature, filtered into a 50 ml standard flask. The filtrate obtained was made up to 50 ml mark with distilled water and used for heavy metals determination against those of the blank and calibration standards using Atomic absorption spectrophotometer, Buck Scientific model.

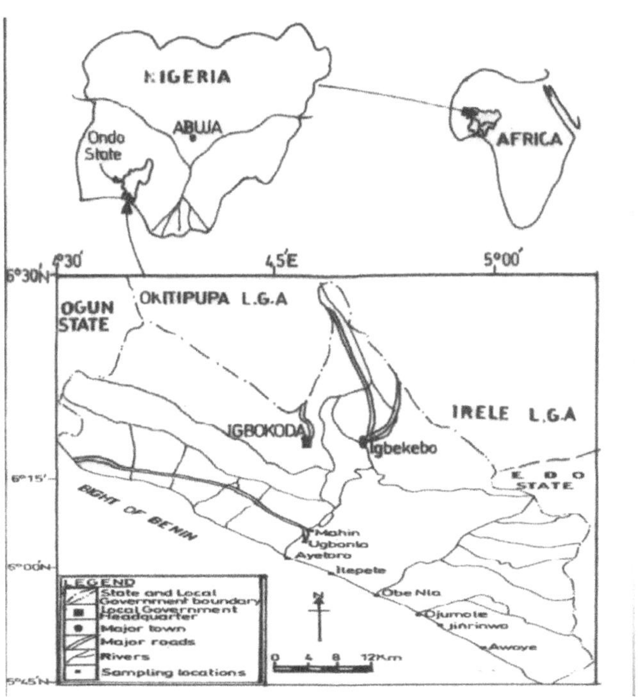

Figure 1: Map showing sampling location.

Result and Discussion

Heavy metal levels in water

Mean concentration and standard deviation of heavy metals level (mg/L) in the two sampled locations are presented in Table 1. The result were compare with federal ministry of Environment

Metals	Ocean shore L1	Ocean shore L2	USEPA Value	FME Limit	WHO
Cr	0.34 ± 0.001[a]	0.31 ± 0.001[a]	0.1	0.02-2.0	0.05
Cd	0.08 ± 0.016[b]	0.21 ± 0.014[a]	0.005	0.0002-0.0018	0.003
Pb	0.79 ± 0.001[a]	0.57 ± 0.142[b]	0.015(Action level)	0.001	0.01
Cu	0.27 ± 0.018[a]	0.26 ± 0.001[a]	1	2.0-4.0	2
Zn	0.22 ± 0.001[a]	0.12 ± 0.001[b]	2	50	3
Ni	0.01 ± 0.001[a]	0.01 ± 0.001[a]	-	0.0017	0.02

Data are mean of duplicate (2) determination. BDL: Below instrument detection limit. Letters a and b show differences among the different locations. Data shown with different letters are statistically different at P<0.05 level.

Table 1: Mean Concentration of heavy metals in the present studies and some international standard.

water standards for aquatic life fresh (FWA), USEPA and World Health Organization standard for permissible limit of heavy metals in water [13,14]. Heavy metal pollution in the marine environment is determined by measuring its concentration in water, sediment and living organisms. Levels of dissolved metals in water from the two main sampled locations at the Ocean shore line are presented in Table 1. No significant location-related differences were seen with respect to the levels of Cr, Cu and Ni in water. Metals concentrations in water were found in the following order: Pb > Cr > Cu > Zn >Cd >Ni in Location1, whereas they follow the order of Pb > Cu > Cr > Cd >Zn > Ni in Location 2. Of the six metals monitored, Pb and Cr were the most abundant. High concentration of Pb (0.57-0.79 mg/l) observed in the water may be associated to deposit of air born particulate matter as well as spill from fishing boat used as means of transportation. The mean concentrations of the metals in the coastal water were compared with guideline values recommended by USEPA and FME Table 1, Showed compliance with Cu, Zn and Ni, while Cr (0.31-0.34 mg/l) Cd (0.08-0.21 mg/l) and Pb (0.57-0.79 mg/l) were higher than the permissible recommended limits. Similar observation was made with the federal ministry of Environment water standards for aquatic life to which the metal Cr, Cd and Pb were also higher [15]. It is expected that the equilibrium of analyte between particular environmental components depend on the critical parameters responsible for kinetic processes that lead to this state. For the studied system the critic parameter should be as ordered as follows: the composition and physicochemical properties of neighboring rivers flowing into the coastal waters, type of the floated soil/sediment forming bottom sediment, current of flowing streams/rivers, seasonal parameters, and biological ability of particular environment al components for concentration of heavy metals.

Heavy metal levels in sediment

The mean results and standard deviation of heavy metals (mg/kg) in sediment samples from the two sampled locations are presented in Table 2. The results were compared with probable effect concentration of heavy metals in sediment as propose by McDonald et al., [16]. Sediments are essential component of aquatic ecosystem acting as carriers sink and potential sources of contaminants [17]. Heavy metals contamination in sediment can affect the water quality and bioaccumulation of metals in aquatic organisms, resulting in potential long term implication on human health and ecosystem [18]. The mean concentration of heavy metals in sediment has been found to reflect the degree of pollution in an area. The mean results of heavy metals recorded in sediment samples taken from the two sampled locations are presented in Table 2. Total metal concentrations of the bottom sediment varied slightly between the sampled locations especially in the values of Cr, Zn and Cu, but no significant differences (P<0.05) were observed in Pb, Cd and Ni concentration among the two locations studied. The value of Cr (0.02-4.19 mg/kg) and Zn (2.12- 3.11 mg/kg) was higher than those of other metals, while Ni remained within a relatively narrow range from BDL to 0.01 mg/kg. Cd and Pb were found to share a trend in their level at the two locations. The level of Cd and Pb in sediment of the coastal water could be attributed to industrial and agricultural discharges as well as from possible spill of petrol from fishing boat over the year. High level of metal at the Ocean shoreline could also be attributed to changes in velocity within the ocean which may lead to abundance of heavy metal at the ocean shoreline. Although the pH value of the sediment was not carried out, it has been show from literature that heavy metal are more accumulated in sediments with high pH value, which is one of the characteristics of the coastal water [10]. The order of abundance of these metals in sediments from the two locations were Cr> Zn > Cu > Pb > Cd > Ni (Table 3). Concentration of some metals investigated in the sediment were similar to those reported for River Orogodo in Delta State [17], but were lower compared to previous study on Ondo State Coastal Water (Tables 4a-4e) [18] and those reported for some communities in Ilaje local government area of Ondo State (Table 4.4) [10]. The difference in heavy metals concentrations in sediment may be attributed to seasonal variation due to sampling periods. Also, the constituents of sediment dictate their ability to concentrate heavy metals. Report has shown that

Metals	Ocean shore L1	Ocean shore L 2	PEC*
Cr	0.02 ± 0.001[b]	4.19 ± 0.038[a]	111
Cd	0.35 ± 0.021[a]	0.38 ± 0.025[a]	5
Pb	0.95 ± 0.023[a]	1.00 ± 0.025[a]	128
Cu	0.24 ± 0.007[b]	1.44 ± 0.028[a]	149
Zn	3.11 ± 0.110[a]	2.12 ± 0.035[b]	159
Ni	0.01 ± 0.003[a]	0.01 ± 0.007[a]	-----

Data are mean of duplicate (2) determinations. BDL: Below instrument detection limit.
Letters a and b show differences among the different locations. Data shown with different letters are statistically different at P<0.05 level.
*PEC- Probable effect concentration 'L1 - Location One 'L2 - Location Two

Table 2: Mean results and standard deviation of heavy metals (mg/kg) in sediment samples from the two sampled locations.

Metal	Present Study	Ondo Coastal water	Ilaje LGA	River Orogodo
Cr	0.02–4.19	9.20–21.20	2.01– 44.09	0.30–0.32
Cd	0.35 -0.38	4.10–13.10	0.01–2.99	0.10 - 0.27
Pb	0.95–1.00	1.20–10.10	10.00 -32.06	0.30–0.84
Cu	0.24–1.44	3.20–21.40	1004–2879	0.20–0.84
Zn	2.12 -3.11	4.00–26.70	0.02- 0.29	1.20–1.91
Ni	0.01	7.50–38.60	3.40–5.90	0.60–0.65

Table 3: Comparison of heavy metal levels (mg/kg) in Sediment of current and previous studies.

Metal	Muscle	Head	Eye	Gill	Bone	Overall mean
Cr	BDL	BDL	0.48[c]	1.19[a]	0.35[b]	0.4
Cd	0.33[b]	BDL	0.39[b]	1.07[b]	0.33[a]	0.42
Pb	0.09[c]	BDL	0.03[d]	0.18[a]	0.11[b]	0.08
Cu	BDL	BDL	BDL	1.86[a]	0.56[a]	0.48
Zn	0.25[c]	0.37[b]	0.43[b]	0.01[d]	0.35[a]	0.28
Ni	BDL	BDL	0.01[b]	0.01[b]	BDL	<0.01

Data are mean of replicate (2) determinations. BDL: Below instrument detection limit. Letters a, b, c and d show differences among the different tissues. Data shown with different letters are statistically different at P<0.05 level.

Table 4a: Level of heavy metal in *Trichiurus lepturus* organs (mg/kg).

Metal	Muscle	Head	Eye	Gill	Bone	Overall mean
Cr	BDL	0.32^{c}	0.35^{b}	0.49^{a}	0.35^{b}	0.3
Cd	BDL	0.21^{c}	0.33^{b}	0.48^{a}	0.32^{b}	0.27
Pb	0.09^{c}	0.20^{a}	0.20^{a}	0.15^{b}	0.12^{c}	0.15
Cu	BDL	BDL	BDL	0.87^{a}	0.59^{b}	0.29
Zn	0.34^{d}	0.38^{c}	0.38^{bc}	0.51^{a}	0.10^{b}	0.4
Ni	BDL	0.02^{d}	0.04^{d}	0.03^{d}	BDL	0.02

Data are mean of replicate (2) determinations. BDL: Below instrument detection limit. Letters a, b, c, and d show differences among the different tissues. Data shown with different letters are statistically different at P<0.05 level.

Table 4b: Level of heavy metal in *Pentanmius guigarius* organs (mg/kg).

Metals	Muscle	Head	Eye	Gill	Bone	Overall mean
Cr	BDL	0.37^{c}	0.62^{b}	1.28^{a}	0.38^{c}	0.53
Cd	BDL	0.26^{c}	0.49^{b}	1.14^{a}	0.32^{c}	0.44
Pb	0.10^{e}	0.34^{b}	0.14^{c}	0.40^{a}	0.11^{d}	0.22
Cu	0.58^{b}	BDL	BDL	1.88^{a}	0.59^{b}	0.61
Zn	0.35^{c}	0.36^{c}	0.65^{b}	1.29^{a}	0.39^{c}	0.61
Ni	BDL	0.01^{d}	0.01^{d}	0.05^{d}	BDL	0.01

Data are mean of replicate (2) determinations BDL: Below instrument detection limit. Letters a, b, c, and d show differences among the different tissues. Data shown with different letters are statistically different at P<0.05 level.

Table 4c: Level of heavy metal in Pseudotolithus senegalensis organs (mg/kg).

Metal	Muscle	Head	Eye	Gill	Bone	Overall mean
Cr	BDL	0.26^{c}	0.32^{b}	0.86^{c}	0.38^{b}	0.38
Cd	BDL	0.27^{b}	0.29^{b}	0.81^{a}	0.31^{b}	0.34
Pb	0.10^{a}	0.71^{a}	0.04^{a}	0.27^{a}	0.12^{a}	0.28
Cu	0.46^{b}	0.38^{b}	0.39^{b}	0.94^{a}	0.39^{b}	0.51
Zn	0.36^{b}	0.38^{b}	0.39^{b}	0.94^{a}	0.39^{b}	0.49
Ni	0.02^{c}	0.12^{b}	0.08^{c}	0.07^{c}	0.02^{c}	0.06

Data are mean of replicate (2) determinations. BDL: Below instrument detection limit. Letters a, b, c, and d show differences among the different tissues. Data shown with different letters are statistically different at P<0.05 level.

Table 4d: Level of heavy metal in *Pseudotolithus typus* organs (mg/kg).

Metal	Muscle	Head	Eye	Gill	Bone	Overall mean
Cr	0.36^{c}	0.23^{d}	0.37^{ab}	0.37^{bc}	0.37^{a}	0.34
Cd	0.31^{b}	0.28^{d}	0.30^{c}	0.32^{a}	0.32^{b}	0.32
Pb	0.07^{c}	0.16^{a}	0.01^{d}	0.12^{b}	0.11^{b}	0.09
Cu	BDL	BDL	BDL	0.62^{a}	0.59^{a}	0.24
Zn	0.36^{d}	0.38^{c}	0.39^{b}	0.42^{a}	0.38^{b}	0.39
Ni	BDL	0.01^{c}	0.04^{c}	BDL	0.01^{c}	0.01

Data are mean of replicate (2) determinations. BDL: Below instrument detection limit. Letters a, b, bc, and c show differences among the different tissues. Data shown with different letters are statistically different at P< 0.05 level.

Table 4e: Level of heavy metal in *Plectorhynchus mediterraneus* organs (mg/kg).

sediments containing more sand and lower values of organic matter exhibit low metals enrichment. Also, the concentration of heavy metals in sediment increases as the amount of organic material increases [19]. It has also been reported that metals such as Cd, Cu, Pb and Zn have high tendency to binding tenaciously to organic matter contained in the soil, sediment and suspended particles within the water column, while sediment with high clay content has ability to concentrate most heavy metals compared to others such as carbonate which are selective [20,21]. Therefore the variation in heavy metals in the sediment samples analyzed may be due in part to contribution from these factors. By comparing the concentrations of heavy metals analyzed in the water and sediments samples, it can be concluded that heavy metals are highly accumulated in sediments than water corroborating what has earlier been reported that sediments act as reservoir for all contaminants and dead organic matter [17,22]. Probable effect concentrations (PEC) of sediments metal levels Pb 128, Cd 5, Cr 111, Cu 149, Zn 159 (mg /kg) has been proposed by McDonald et al. [16]. The results of the present study show that the concentrations of Cd, Cu, Cr, Pb and Zn in the sediment samples were much below the probable effect concentration of sediment metals levels. Above this level, metals may have adverse effects on sediment dwelling organisms [16].

Heavy metal concentration in fish

Mean concentrations of chromium, cadmium, lead, copper, zinc and nickel in muscle, head, gill, eye and bone of *Pentanemus quinquarius, Pseudoltolithus senegalensis, Trichirus lepturus, Plectorhynchus meditarraneus and Pseudotolithus typus* from the coastal waters are shown in Tables 4 (a-e). Measure concentration of the six tested metals in the muscle, gill, eye, bone and head of the five fish species (*Pentanemus quinquarius, Pseudoltolithus senegalensis, Trichirus lepturus, Plectorhynchus meditarraneus and Pseudotolithus typus*) under investigation are presented in Tables 4a-e. Zinc was the most abundant in all the examined tissues. Tissues specific comparison of the accumulation of the investigated metals showed that the metals concentration were significantly high in the gill tissues compared to the muscle tissues. Jobling [23] attributed the high accumulation of heavy metals in gills tissues to the metallothionein proteins which are synthesized in liver and gills tissues when fishes are exposed to heavy metals and detoxify them. These proteins are thought to play an important role in protecting them from damage by heavy metal toxicants. Also, gills are the site directly exposed to the ambient conditions and also are known for their excretory function [24]. Chromium, copper and Zinc are essential element and are regulated by physiological mechanisms in most organism. However, occurrence of excessive levels of them is regarded as potential hazard which can endanger both animal and human health [25]. The level of these essential metals Cr, Cu and Zn in the muscle, gill, bone, eye and head are presented in Tables 4a-4e. Cr was not detected in the muscle of all examined fish species except *Plectorhynchus meditarraneus* (0.36 mg/kg) this may be associated to their large size and long exposure period. Cr concentration recorded comparable value in most tissues of the examined fish species. The level of Cr in muscle and other organs were lower than USEPA limit 8 mg/kg [14].The levels of Cu and Zn in edible muscle of the fish species ranged from BDL-0.58 mg/kg and 0.25-0.36 mg/kg respectively. The USEPA limit for Cu and Zn in fish food for human consumption is 120 and 50 mg/kg respectively [14]. The level of Cu and Zn in edible muscle in all fish species were low and below the international limits. Hence, Cu and Zn levels have posed no threat for consumption of the fish muscle from the coastal water. However, because a metal concentration in the aquatic environment is low and considered to be naturally occurring or background does not mean that the concentration could not cause adverse ecological effects [26]. The presence of one metal can significantly affect the impact that another may have on an organism. The effect may be synergistic, additive or antagonistic. The low level of Cu and Zn in muscle could be associated to lower level of binding protein metallothioneins. Metallothioneins are a family of low molecular weight cysteine rich proteins that have been reported in vertebrate and several invertebrate. Their synthesis can be induced by a wide variety of metal ions including Cu, Zn and Cd. Hence metallothioneins has been proposed as biomarkers to indicate the presence of high metals in the environment [8]. Although Cu and Zn play a crucial role in several enzymatic processes, they are tentatively classified as highly toxic metals by Hellawell [27] and are bioaccumulated in aquatic organism. The concentration of nickel in the tissues of all examined species is relatively low ranges from BDL-0.12 mg/kg. The result confirms that most fish species do not bio accumulate the metal contrary to what was previously reported by Hulya and Mendil [5,28]. The mean concentration of nickel in different fish species observed in this present study was below the permissible limit of 1.0 mg/kg prescribed by United State Environmental Protection Agency [14]. Lead and cadmium are non-essential elements which are accumulated in human tissue and harmful to human health. Comparison of Cd and Pb accumulation in the various organs shows that the highest concentration of Pb was found in head of *Pseudotolithus typus* (0.71 mg/kg) (Table 4d) and gill of *Pseudoltolithus senegalensis* (0.40 mg/kg) (Table 4c). While highest Cd concentration was found in gill of *Pseudoltolithus senegalensis* (1.14 mg/kg) (Table 4c) and gill of *Trichirus lepturus* (1.07 mg/kg) (Table 4a). The level of Pd in the edible muscle of the fish species ranges from 0.07-0.10 mg/kg. While Cd was observed only in the muscle of *Plectorhynchus meditarraneus* (0.07 mg/kg) (Table 4e) and *Trichirus lepturus* (0.09 mg/kg) (Table 4a). The accumulation of Pb and Cd recorded in fish bone follows similar pattern in all the species ranging from 0.11-0.12 mg/kg and 0.31-0.33 mg/kg respectively. The accumulation of these metals in gills, bones and head of fish do not directly affect human health because these are not edible parts. Nevertheless, the predatory animals such as birds who consume the whole fish from coastal waters are at risk of excess metal contamination. Moreover, fish head and bone are used in production of food additives for livestock feed. Therefore, its health quality is also very important. The maximum allowable levels of lead and cadmium in fish for human consumption specified by European Union are 0.2 and 0.05 mg/kg respectively. Of the different metals levels detected in edible muscle of all the examined fish species, the level of Cd in *Plectorhynchus meditarraneus* and *Trichirus lepturus* exceed the maximum allowable limit.

Concentration Factor

Concentration factors of the metals in fish relative to superficial water and sediment are presented in the tables below (Tables 5 and 6). Estimated concentration factor of heavy metals from water were greater than from sediment, suggesting bioaccumulation of the metals by the fish from water column. Fish has been reported to accumulate metals from water by diffusion via skin and gills as well as oral consumption/drinking of water [29]. Similar observation has been reported by many workers in Nigeria [18,30-32]. Figures 2 and 3 Show the comparisons of the ratio of investigated heavy metals levels in fish/water and fish sediment respectively.

	Cr	Cd	Pb	Cu	Zn	Ni
P.q	1.57	7.01	0.34	2.09	2.89	26
P.s	0.89	3.48	0.19	1.09	1.79	19
T.l	1.58	5.7	0.28	2.31	2.73	14
P.m	1.12	4.36	0.37	1.93	2.2	62
P.t	1.01	4.09	0.12	0.9	1.74	13

Table 5: Concentration factor from water to fish.

	Cr	Cd	Pb	Cu	Zn	Ni
P.q	0.13	1.42	0.28	0.39	0.3	2
P.s	0.07	0.71	0.16	0.2	0.19	1.46
T.l	0.13	1.16	0.22	0.42	0.29	1.08
P.m	0.09	0.89	0.29	0.36	0.23	4.77
P.t	0.08	0.83	0.09	0.17	0.18	1

Pq: Pentanemus quinquarius; P.s: Pseudoltolithus senegalensis; T.l: Trichirus lepturus; P.m: Plectorhynchus meditarraneus; P.t: Pseudotolithus typus

Table 6: Concentration factor from sediment to fish.

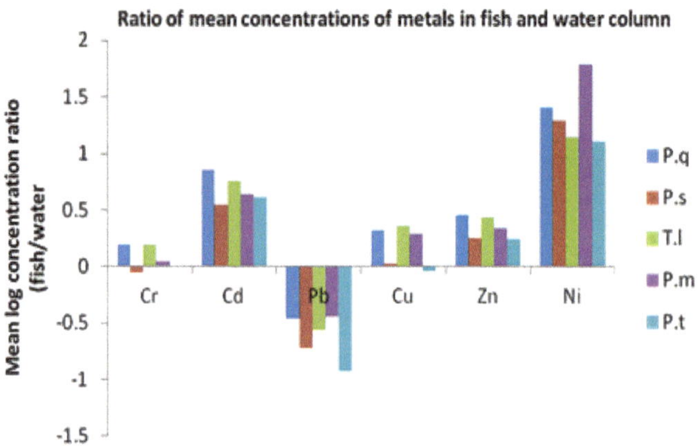

P.q: *Pentanemus quinquarius*; P.s: *Pseudoltolithus senegalensis*; T.l: *Trichirus lepturus*; P.m: *Plectorhynchus meditarraneus*; P.t: *Pseudotolithus typus*

Figure 2: A plot of the ratio of mean concentrations of metals in fish and water column.

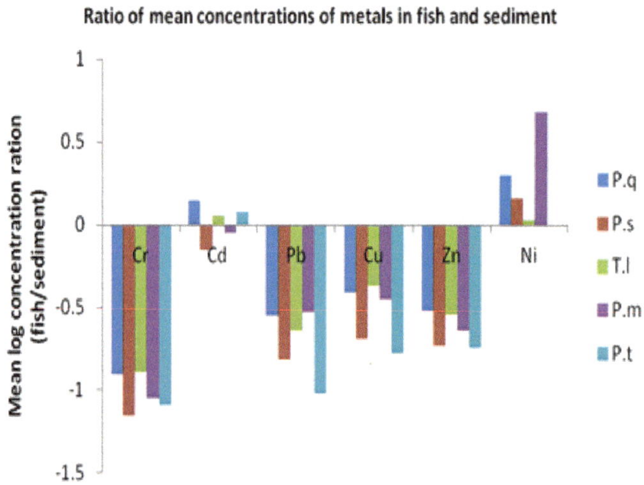

P.q= Pentanemus quinquarius; P.s: Pseudoltolithus senegalensis; T.l: Trichirus lepturus; P.m: Plectorhynchus meditarraneus; P.t: Pseudotolithus typus.

Figure 3: A plot of the ratio of mean concentrations of metals in fish and sediment.

Conclusion

The health safety of Ondo state coastal water for some fish species has been assessed, though there have not been reported cases of metal toxicity arising from direct use of water or consumption of fish from the area. The present results showed that metal concentration were lowest in muscle and highest in gill due to physiological roles in fish metabolism where the target tissues of heavy metals are the metabolically active ones. Therefore heavy metals in these active tissues are higher compared to muscle where metabolic activity is relatively lower. The result of these research has provided baseline information of heavy metals in water, sediment and tissues of some fish species from the coastal waters of Ondo state, Nigeria with which future assessment program can be compared.

References

1. Malik N, Biswas AK, Qureshi TA, Borana K, Virha R (2010) Bioaccumulation of heavy metals in fish tissues of a freshwater lake of Bhopal. Environmental Monitoring Assessment 160: 267-276.
2. Gupta A, Rai DK, Pandey RS, Sharma B (2009) Analysis of some heavy metals in the riverine water, sediments and fish from river Ganges at Allahabad. Environmental Monitoring Assessment 157: 449-458.
3. Agah H, Leermakers M, Elskens M, Fatemi SMR, Baeyens W (2009) Accumulation of trace metals in the muscles and liver tissues of five fish species from the Persian Gulf. Environmental Monitoring Assessment. 157**:** 499-514.
4. Blasco J, Arias AM, Sa Enz V (1999) Heavy metals in organisms of the River Guadalquivir estuary: possible incidents of the Aznalcollar disaster. The Science of the Total Environment 242: 249-259.
5. Hulya K, Erhan U (2000) Concentrations of heavy metals in water, sediment and fish species from Ataturk Dam Lake (Eupharates),Turkey. Chemosphere 41: 1371-1376.
6. Prudente M, Kim EY, Tanabe S, Tatsukawa R (1997) Metal levels in some commercial fish species from Manila Bay, Philippines. Marine Pollution Bulletin 34: 671- 674.
7. Waqar A (2006) Levels of selected heavy metals in tuna fish. The Arabian Journal for Science and Engineering 31: 89-92.
8. Peakall D and Burger J (2003) Methodologies for assessing exposure to metals: speciation, bioavailability of metals and ecological host factors. Ecotoxicology and Environmental Safety 56: 110-121.
9. Puel D, Zsuerger N, Breittmayer JP (1987) Statistical assessment of a sampling pattern for evaluation of changes in Hg and Zn concentration in Patella coerulea. Bull Environ Contam Toxicol 38: 700-706.
10. Ololade IA, Lajide I, Amoo IA (2007) Enrichment of Heavy Metals in Sediments as Pollution Indicators of the Aquatic Ecosystem. Pakistan Journal of Scientific Research 50: 2-35.
11. Parker RC (1972) In: E. I. Adeyeye (Editor) Water analysis by atomic absorption spectroscopy. Varian techtron, Switzerland. Determination of trace heavy metals in Illisha Africana fish and in associated water and sediment from some fish ponds. International Journal of Environmental Studies 45: 231-238.
12. Khansari FE, Khansari MG, Abdollahi M (2005) Heavy metals content of canned tuna fish. Food Chemistry 93: 293-296.
13. World Health Organization (1993) Guidelines for drinking water quality Geneva.
14. USEPA (2002) Toxic metals sources and specific effects retrieved on 30/10/2004 from http:www.extremehealth.com
15. Federal Ministry of Environment: National Guidelines and standard for water quality in Nigeria (2001) 114.
16. McDonald DD, Ingersoll CG, Berger TA (2000) Development and evaluation of consensus based sediment quality guidelines for freshwater ecosystems. Archives of Environmental Contamination and Toxicology 39: 20-31.
17. Isah BR, Arimoro FO, Ibrahim M, Birma GI, Fadairo EA (2011) Assessment of sediment contamination by heavy metal in river orogodo Delta state, Nigeria. Current World Environment 10: 29-38.
18. Asaolu SS, Ipinmoroti KO, Adeeyinwo CE, Olaofe O (1997) Interrelationship of heavy metal contamination in water, sediment and fish of Ondo State coastal waters. African Journal of Science 1: 55-60.
19. Tsai LJ, Ho ST, Yu KC (2003) Correlation of extractable heavy metals with organic matters in contaminated rivers sediments. Water Science and Technology 47: 101-107.
20. Aiyesanmi AF (2008) Baseline heavy metals concentration in river sediments within Okitipupa South East belt of Nigeria bituminous sand field. Journal of Chemical Society of Nigeria 33: 29-41.
21. Aiyesanmi AF, Idowu GA (2012) Levels of heavy metals in leafy vegetables grown n around waste dump sites in Akure, Southwestern Nigeria. FUTA Journal of Research in Sciences 8: 27-35.
22. Nguyen H, Leermakers M, Osan J, Tfrfk S, Baeyens W (2005) Heavy metals in Lake Balaton: water column, suspended matter, sediment and biota. Science of the Total Environment. 340: 213-230.
23. Jobling m (1995) Environmental Biology of Fisheries 1st Edition Printed in Great Britain. Chapman and Hall, London.
24. Mathiessen P, Brafield AE (1977) Uptake and loss of dissolved Zinc by Stickle back *Gasterosteus aculeatus* (L). Journal of Fisheries Biology 10: 399-410.
25. Wright DA, Welbourn P (2002) Environmental Toxicology, Cambridge press Cambridge.
26. United State Environmental Protection Agency (USEPA) (1976) Quality criteria for water, USEPA, Washington DC 440: 76-123.
27. Hellawell JM (1986) Biological Indicators of Freshwater Pollution and Environmental Management. Elsevier Applied Science Publishers Ltd, London and New York. 546.
28. Mendil D, Demirci Z, Tuzen M, Soylak M (2010) Seasonal investigation of trace element contents in commercially valuable fish species from the Black Sea, Turkey. Food and Chemical Toxicology 48: 865-870.
29. Nussey G, Van Vuren JHJ, Du Preez HH (2000) Bioaccumulation of chromium manganese, nickel and lead in the tissues of the moggel, Labeo umbratus (Cyprinidae), from Witbank Dam, Mpumalanga. South Africa. Water Safety 26: 269-284.
30. Adefemi OS, Olaofe O, Asaolu SS (2004) Concentration of Heavy metals in water sediment and fish parts (Illisha africana) from Ureje dam, Ado-Ekiti, Ekiti State. Nigerian Journal of Biology and Physical Sciences 3: 111-114.
31. Valipour M, Mousavi SM, Valipour R, Rezaei E (2013) Deal with Environmental Challenges in Civil and Energy Engineering Projects Using a New Technology. Journal of Civil and Environmental Engineering 3: 1.
32. Valipour M, Mousavi SM, Valipour R, Rezaei E (2012) Air, Water, and Soil Pollution Study in Industrial Units Using Environmental Flow Diagram. Journal of Basic and Applied Scientific Research 2: 12365-12372.

12

Embryonic Development of Commercially Important Portunid Crab, *Charybdis feriata* (Linnaeus)

Soundarapandian P[1]*, Ilavarasan N[2], Varadharajan D[1], Jaideep Kumar[1] and Suresh B[1]

[1]*Faculty of Marine Sciences, Centre of Advanced Study in Marine Biology, Annamalai University, India*
[2]*Department of Zoology, Government Arts College, Karur, India*

Abstract

The quality and survival of newly hatched zoea will depend primarily on the embryonic development. So an attempt has been made to study the embryonic development in the portunid crab, *Charybdis feriata*. The embryonic development of *C. feriata* is divided into six stages, *viz.* blastula, gastrula, eye placode, pigment; heart beat and freshly hatched first zoea. In blastula stage, the eggs were round, golden yellow in colour and the diameter of the egg was 0.36 mm-0.37 mm. In gastrula stage, the eggs were round and deep yellow or yellowish orange in colour, and the diameter of the egg was 0.38 mm-0.39 mm. In eye placode stage, the eggs were round orange in colour and the diameter of egg was 40 mm-41 mm. In pigment stage, the eggs were brown in colour and the diameter of the egg was 0.42 mm-0.43 mm. In hear beat stage, the eggs were dark brown or black in colour, and the diameter of the egg was 0.44 mm-0.45 mm. The freshly hatched I zoea moved freely in the water, and its carapace length ranges from 1.05 mm-1.25 mm.

Keywords: Embryonic development; Blastula; First zoea; Crucifix crab; Survival

Introduction

After the disease outbreak in shrimp farming, crabs have become a candidate species for aquaculture because of their export potential [1-7]. In order to develop a hatchery technology, the crab larval biology should be thoroughly studied to produce good quality eggs and healthy zoea from the mother crab [1,2,4,5]. So far, there is no reliable hatchery for crabs because of the mass mortality of the zoeal stages. Hence, embryological study forms a base line to get healthy first zoea from developing eggs. So, in the present study, an attempt has been made to study the embryonic development of *C. feriata.*

Materials and Methods

Gravid females of C. *feriata* with early broods (yellowish orange coloured eggs) were collected from the Parangipettai coastal waters and retained in separate tanks containing sea water (salinity-35 ± 1%; temperature-28-31°C; dissolved oxygen was up to the saturation level and photoperiod-12 L: 12 D). The crabs were fed with mussel and clam meat once in a day. Every day, the excess food, excreta and shed out eggs were siphoned out. Continuous aeration was given throughout the incubation period and the development of the egg was closely observed. Daily colour changes of the eggs during incubation period could be noted. Small clumps of eggs were snipped from random locations in each clutch using sharp scissors. All the developing embryos were examined with a MEIJI binocular dissecting microscope (100 X), to ensure that only viable embryos were sampled, and the colour change corresponding to the development and length of incubation period was noted [1,8]. The time course of embryonic development, as indicated by the appearance of specific morphological features, including the development of the compound eye, initiation of the heart beat, development of the limb pigmentation and initiation of limb twitches were monitored. The gradual change in the embryonic development and increase in the size of eggs were recorded to understand the different developmental stages [9]. All these developmental stages were photographed using the digital camera (Nikon, COOLFIX 990, and Japan), attached with the microscope.

Results

The eggs of C. *feriata* went through different colours with its gradual development. The newly spawned eggs were bright yellow, and the number of eggs attached to each seta of the pleopod was not definite in number. The eggs were spherical and surrounded by two membranes, an inner and outer membrane. Both membranes were transparent, and the yolk was visible as yellow granules with polygonal areas. Owing to the large size of the egg mass, the abdomen was almost straight, continues with the cephalothorax, and the telson was slightly tilted upwards. The newly oviposited eggs contain all the necessary material for synthetic processes associated with embryogenesis and morphogenesis, and all of the compounds required for oxidative metabolism and energy production. The egg contains nutritive reserves, in the form of proteinaceous yolk and lipid vesicles scattered throughout the cytoplasm. The newly spawned eggs were bright yellow, as the yolk contains carotenoid pigments. As the development progresses, the bright yellow colour changes to dull yellow and finally, to dark grey, just one day before hatching. At this stage, the developing larvae with its occasional twitching movements were observed under the microscope. During this period, there was considerable increase in the egg size also. The total days of incubation varied between 8-11 days. The eggs at the time of oviposition were quite distinct and large. They could be divided into six stages, *viz.* blastula, gastrula, eye placode, pigment, heart beat and freshly hatched first zoea [10].

***Corresponding author:** Soundarapandian P, Faculty of Marine Sciences, Centre of Advanced Study in Marine Biology, Annamalai University, Parangipettai-608 502, Tamil Nadu, India, E-mail: soundsuma@gmail.com

Stage-I-Blastula

Eggs were round, golden yellow in colour and were undeveloped, and mass of undifferentiated cells are found. Yolk granules were denser. Cleavage and gastrulation were not clear. The diameter of the freshly laid *C. feriata* egg was 0.36 mm-0.37 mm.

Stage-II-Gastrula

Eggs were round and deep yellow, or yellowish orange in colour. The space between the egg wall and the inner developing embryo was visible. The diameter of the egg was 0.38mm-0.39 mm.

Stage-III-Eye placode

Eggs were round orange in colour. Yolk granules were not denser. Segmentation and organogenesis were distinct. The eyespots were appeared as scarlet crescent. The diameter of egg was 40 mm-41 mm.

Stage-IV-Pigment

Eggs were brown in colour with slightly elliptical in shape. Appendages of embryonic larvae were pigmented. The diameter of the egg was 0.42 mm-0.43 mm.

Stage-V-Heart beat

The eggs were dark brown or black in colour. Eyes were round in shape. Heart starts to beat vigorously. The diameters of *C. feriata* eggs were 0.44 mm-0.45 mm.

Stage-VI-Newly hatched first zoea

The freshly hatched I zoea moved freely in the water, and its carapace length ranges from 1.05 mm-1.25 mm. The hatching success of freshly hatched I zoea is 95%.

Discussion

In decapod crustaceans, the embryos develop in broods that are carried by the females, and experience the parental environmental conditions. In crabs, during mating, the sperms are transferred to the seminal receptacle, which act as a storage organ. Viable sperms are utilized during subsequent spawnings of that particular intermoult period. As the eggs are laid, they adhere to one another and to the setae of the endopodites of the abdominal segment, and the maturing egg mass/sponge is held between the reflexed abdomen and venter of the cephalothorax. The abdominal chamber acts as an incubation chamber for the developing eggs. Hamasaki et al. [11] reported embryogenesis in *P. trituberculatus*, in which, females extrude their eggs from gonopores onto a bottom substratum such as sand, where they bury themselves; then they attach the extruded eggs to the ovigerous setae of the endopods of the pleopods, which are moving forward and backward.

Needham [12] classified the animal eggs into two categories based on the major substrate (fat or protein) utilized during embryogenesis. They are terrestrial or cledoic eggs, utilizing fat during embryogenesis, and aquatic or non-cledoic eggs utilizing proteins for oxidation. However, after working with several marine crustaceans discussed the merits and demerits of classification of eggs [12,13]. Pandian [13] classified the eggs according to the habitat of the species and the main substrate metabolized. They are: terrestrial eggs-protein metabolism is greatly suppressed and oxidation of fat is geared up; freshwater eggs-protein metabolism is predominant and limited fat metabolism; marine eggs-fat depletes heavily and considerable suppression of protein metabolism. Based on the egg size, classified the crustacean embryogenesis into two groups as the larger eggs utilize more lipid than protein, and the smaller ones utilize more protein than lipid [14]. In most marine invertebrates, the newly laid eggs contain all the energy and reserves for embryonic development [15,16]. Female nutritional and reproductive condition, temperature and salinity may affect oogenesis, embryogenesis and larval quality [17-19].

As the embryo develops, the size of the eggs, increases gradually. The eggs swell as they develop, so that by the time they are ready to hatch, they are roughly double their new-laid volume. The egg size of nine species of brachyuran crabs did not vary significantly within species and mean egg size was not correlated with female body size [20]. The same trend was observed in *S. serrata,* by where egg diameter was not related in any way to female size, and also the egg diameters increased at a relatively steady pace throughout ontogeny [21]. There was an accelerated increase in egg diameter, when the embryonic heartbeat was first observed. Similar observations were made by for green crab (*C. maenas*), the nut crab (*Ebalia turberosa*), and some lobsters (*Galathea dispersa* and *G. squamifera*) [22]. Under constant environmental conditions, the variability in egg size and biomass has been attributed to variation in female size or age and genetic factors [23,24].

Hamasaki et al. [25] studied batch fecundity in *P. trituberculatus*, in which they emphasized that the size of eggs decreased with increasing temperatures, the number of first zoeae showed no fluctuation in the same-size females throughout the breeding season. They suggested that three patterns of reproductive characters may cause this phenomenon: (1) Females invest the same amount of energy in the reproductive output throughout the breeding season, so that they increase egg number with decreasing egg size dependent on the trade-off between egg size and number, but lose more eggs between oviposition and hatching as the breeding season advances [26]. (2) Females produce similar egg numbers, and decrease egg size and energy investment for reproductive output as the breeding season advances; and (3) Females decrease both egg size and number, and egg loss rate decreases as the breeding season advances. Seasonal and regional variability in egg size and number is known for a few brachyuran species. In the Japanese mitten crab, *Eriocheir japonica*, egg size varies within the breeding season; large eggs are spawned and developed at low temperature and small eggs at high temperature [27]. Brante et al. [28] reported the egg number and size for the cancrid crab, *C. setosus,* distributed along the Chilean coast. Egg numbers produced by females showed no significant difference along a latitudinal (temperature) gradient, but egg size and reproductive output decreased in Northern Chile (high temperature). In this study, the number of eggs was examined using ovaries that developed during the overwintering period.

During the development, the colour of the egg changes through brown to grey, as the yolk is used up and the outline of the embryo become visible. The eyes and pigment spots appear first, followed by the outlines of the abdomen and cephalothorax [29]. The eggs when deposited are yolky with bright yellow or yellowish orange, but they become brown and then dark brown or black before hatching. The colour change was caused by absorption of the yellow yolk and development of dark pigment in the eyes [30,31]. Veera [30] reported that as the progression of development occurs, the embryo decreases in dry weight stage by stage, as it utilizes the yolk material. Subramoniam [32] studied the yolk utilization during embryogenesis in *Emerita asiatica* and reported that the water content steadily increased; conversely, protein content showed a steady decline; protein-bound carbohydrates declined, where as the concentration of free carbohydrates and glycogen exhibited an increasing trend; lipid level remained unaltered almost up to stage V, thereupon, the value fell precipitously, reaching the minimum in stage IX.

The embryonic development includes the stages of newly spawned egg, multicell stage, eye stage, pigment stage, heart beat stage and prehatching stage. The course of embryonic development includes cleavage, blastula, gastrula, segmentation, organogenesis, formation of appendages, formation and functioning of heart, and formation of chromatophores all around the body. Similar pattern of embryonic development was reported for many portunid crab species [5,30,31,33].

Many workers have divided the crustacean egg stages, based on the appearance of distinctive morphological features such as the eye, heart beat and appendage formation. However, such morphological characters only begin to appear mid-way during embryonic development. Cellular differentiation starts soon after gastrulation and requires enormous energy expenditure. Therefore, Subramoniam [32] emphasized the importance of a detailed classification of early development of decapod crustaceans, to understand the changes in the metabolic pathways involving interconversion of already stored substrates within the closed system of egg development. In the embryology of *E. asiatica*, divided the egg development into nine stages based on colour change and other concomitant morphological features of the embryo. His study reports on the biochemical alterations in the major organic substrates, as well as the activity of non-specific esterases during egg development, leading to the release of the zoea.

Subramoniam [32], while studying the embryonic development in *S. serrata* observed different stages, which includes the stages of cleavage, blastula, gastrula, egg-nauplius, embryo with five pairs of appendages, embryo with seven pairs of appendages, embryo with compound eye pigments formation, and embryo almost ready to hatch. The cleavage evolves those from the spiral type to the superficial type was studied by Sun and Yu [34]. The dividing furrow can be seen from 2 cells. The gastrula is formed through invagination. Xue et al. [35] studied the histology of embryonic development in *P. trituberculatus* and reported 5 stages, i.e. two egg-nauplius stages and three egg-zoea stages. The egg-nauplius stage could be divided into egg-nauplius I and egg-nauplius II. The egg-nauplius I includes the formation of optic lobe, antennule, antenna and mandible with the cleaving cells. These appendages are unsegmented; the thoracoabdominal process formed abdomen with proliferating cells; the labrum rudiment and labium rudiment formed in the tip of the stomodeum. The maxillule, maxilla, cavity abdomen, optic ganglia and antennule ganglia formed during the egg-nauplius II. The segmentation emerged between thorax and abdomen. The egg-zoea stages were divided into egg-zoea I, egg-zoea II and egg-zoea III. The carapace, stomach, hind gut, compound eye and thoracic ganglia formed in the egg-zoea I. The antenna ganglia keep in touch with each other; each of the optic ganglia was independent; the abdomen was divided into 6 segments. The appendages segmented during the egg-zoea II. The ends of the appendages were divergent with setae; the parts of the compound eye have formed; the shape of the yolk sac was like butterfly. The pigment cells, maxilliped I, maxilliped II, heart, gonad and brain have formed. The mesoderm formed muscle, and arranged themselves as beads in the next stage. The digestive system has formed, except hepatopancreas. Compared with the egg-zoea I and egg-zoea II, the brain and carapace further developed in the egg-zoea III. The muscle and the hepatopancreas formed at the same time.

In the present study, the duration of embryonic development in *C. feriata* lasts for 8-11 days. Almost similar duration was also reported in *P. sanguinolentus* [5]. Whereas in *P. pelagicus,* the period reported was 8-10 days [36]. In general, the duration of embryonic development in the portunid crabs varies between 8 to 14 days depending on the environmental conditions. Sun and Yu [34] have reported that the duration of embryonic development could be shortened by raising temperature in the range of 18-25°C. In most crustaceans, the incubation period is highly dependent on temperature [37]. Incubation periods of Moreton Bay (Australia) population of *S. serrata* are usually in early spring, at water temperatures of 18-20°C [37]. Water temperatures in estuaries and coastal waters along the north coast of South Africa range between 17-22°C in winter and 23-30°C in summer [38]. The egg incubation period of *S. serrata* exponentially decreased with increasing temperature [39].

Salinity also holds an important role in the embryonic development and egg hatching. During the course of current study, while maintaining the brooders of *C. feriata* at different salinities, it was observed that the hatching of eggs and release of larvae occurs in the waters of salinity ranging between 30-35%. Campbell and Fielder [40] opined that, in *P. sanguinolentus*, the occurrence of prezoeae increased when eggs were hatched at salinities below the oceanic salinities that this species normally encounters in nature. Similarly, found that prezoeae were more prevalent in *C. sapidus*, when experimental salinities were below 20% [41].

Among crustaceans; brachyuran crabs carry the embryos for extended periods of time [22]. During brooding, female crabs exhibit complex behaviors, which seem to be mostly directed towards providing embryos with oxygen [15,42-45]. Since oxygen consumption of the embryos increases with progressing development, brooding females increase ventilation frequency, providing oxygen to the embryos according to their demand [44]. This change in brooding behaviour is related to a 2-fold increase in oxygen consumption of brooding females (after discounting oxygen consumption of the embryo mass), when compared to nonbrooding females of similar size [44]. This suggests that oxygen provision to the brood may account for a substantial fraction of total reproductive costs.

Hatching usually occurs in the early hours of the day in *C. feriata,* as all the portunid crabs generally hatch. During the hatching process, the fully developed first zoea hatch out of the egg cases and swims freely in the water column. Davis [45] reported the process of hatching, a period of swelling of the eggs followed by osmotic swelling of the inner egg membrane at the start of hatching. The swelling inner egg membrane then ruptures the chorion by pressure, from within the larvae plays no part. The inner membrane is subsequently ruptured by mechanical action of the larval abdomen.

Water quality and feed are the important criterion in rearing the berried crabs. If they are not maintained properly, diseases will attack, which leads to the hatching of unhealthy larvae and their mortality in the early stages itself. Iin SD et al. [46] observed that the egg quality, egg hatching rate and the quality of the newly hatched zoea are very good when berried crabs fed with formulated feeds compared than normal diets. The mortality of the eggs has been attributed to fungus, predation and suffocation in fouled water and changes in temperatures. On the average, only one out of every million eggs survives to become a mature adult [47]. The quality of the eggs produced by individual females has also been highly variable. In order to select the best eggs for incubation and subsequent rearing, an estimate of egg quality had to be established, and this could be achieved immediately after extrusion. Giles [21] reported that, the newly hatched larvae could be subjected to a variety of stress tests, including ammonium, salinity, formalin and starvation stress tests, and the results of these tests would ultimately be the deciding factors in classifying eggs as good or poor quality. The study on embryonic development in crabs is more important, since the health of the newly hatched zoeae rely on the quality and health of the brooded eggs. The size and health of the ovigerous crabs along with good water quality parameters, like temperature and oxygen will play a major role in the production of healthy eggs and larvae which in turn is a prerequisite for successful production of crab juveniles in any hatchery. The water quality and feed are the important criteria in rearing the berried crabs, and if they are not maintained properly diseases will attack which leads to the hatching of unhealthy larvae and their mortality in the early stages itself.

References

1. John Samuel N, Soundarapandian P (2009) Embryonic Development of Commercially Important Portunid Crab *Portuns sanguinolentus* (Herbst) Inter J Ani Veter Advan 1: 32-38.
2. Dinakaran GK, Soundarapandian P (2009) Mating behaviour and broodstock development of commercially important blue swimming crab, *Portunus sanguinolentus* (Herbst). Indian J Sci Tech 2: 71-75.
3. Varadharajan D, Soundarapandian P, Dinakaran GK, Vijakumar G (2009) Crab Fishery Resources from Arukkattuthurai to Aiyammpattinam, South East Coast of India. Curr Res J Biol Sci 1: 118-122.
4. Anand T, Soundarapandian P (2011) sea ranching of commercially important blue swimming crab *Portunus pelagicus* (linnaeus, 1758) in Parangipettai coast. Inter J Sci Nat 2: 215-219.
5. John Samuel N, Soundarapandian P (2010) Embryology of commercially important portunid crab *Scylla serrata* (Forskal). Asian J. Exp. Biol. Sci 1: 178-182.
6. Varadharajan D, Soundarapandian P, Pushparajan N (2013) The global science of crab biodiversity from Puducherry coast, south east coast of India. Arthropods 2: 26-35.
7. Varadharajan D, Soundarapandian P (2012) Commercially Important Crab Fishery Resources from Arukkattuthurai to Pasipattinam, South East Coast of India. J Marine Sci Res Dev 2: 2-6.
8. Srinivasagam S, Kathirvel M, Kulasekarapandian S (2000) Captive broodstock development induced breeding and larval stages of mud crabs (*Scylla* spp). CIBA Bulletin.
9. Quinitio ET, Parado-Estepa FD (2003) Biology and hatchery of mud crabs Scylla spp. SEAFDEC Aquacult Exten Manu No 34 Iloilo Philippines.
10. Amsler MO, George RY (1984) Seasonal variation in the biochemical composition of the embryos of *Callinectes sapidus* Rathbun. J Crust Biol 4: 546-553.
11. Hamasaki K, Fukunaga K, Nogami K, Maruyama K (1991) Egg-laying and process of egg attachment to the pleopods in the swimming crab *Portunus trituberculatus*. Saibai Giken 19: 85-92.
12. Needham J (1942) Biochemistry and morphogenesis. University Press, Cambridge. 86: 585-590.
13. Pandian TJ (1970) Cleidioc properties of marine demarsal eggs. Indian J Exp Biol 8: 340-342.
14. Shakuntala K, Reddy SR (1982) Crustacean egg size as an indicator of egg fat/protein reserves. Int J Invertebr Reprod 4: 381-384.
15. Hoagland K (1979) The behavior of three sympatric species of *Crepidula* form the Atlantic, with implications for evolutionary ecology. Nautilus 93: 143-149.
16. Jaeckle W (1995) Variation in the size, energy content, and biochemical composition of invertebrate eggs: correlates to the mode of larval development. In: McEdward, L. (Ed.), Ecol Mar Invert Lar CRC Press, Boca Raton 49-78.
17. Harrison K (1990) The role of nutrition in maturation, reproduction and embryonic development of decapod crustaceans: a review. J Shellfish Res 9: 1-28.
18. McLaren I, Corkett C, Zillioux E (1969) Temperature adaptations of copepod eggs from the arctic to the tropics. Biol Bull 137: 486-493.
19. Crisp D, Costlow J (1963) The tolerance of developing cirripede embryos to salinity and temperature. Oikos 14: 22-34.
20. Hines AH (1991) Fecundity and reproductive output in nine species of cancer crabs (*Crustacea, Brachyura, Cancridae*). Can J Fish Aquat Sci 48: 267-275.
21. Giles C (2003) Selecting criteria to evaluate the quality of mud crab, *Scylla serrata* eggs in South Africa. In: Workshop on mud crab culture, ecology and fisheries, Can Tho University, Vietnam.
22. Wear RG (1974) Incubation in British Decapod crustacea and the effects of temperature on the rate and success of embryonic development. J Mar Biol Ass UK 54: 745-762.
23. Qian PY, Chia FS (1992) Effect of aging on reproduction in a marine polychaete, *Capitella sp.* J Exp Mar Biol Ecol 156: 23-38.
24. Eyster L (1979) Reproduction and developmental variability in the opistobranch *Tenellia pallida*. Mar Biol 51: 133-140.
25. Hamasaki K, Fukunaga K, Kitada S (2006) Batch fecundity of the swimming crab *Portunus trituberculatus* (Brachyura: Portunidae). Aquacul 253: 359-365.
26. Hines AH (1982) Allometric constraints and variables of reproductive effort in brachyuran crabs. Mar Biol 69: 309-320.
27. Kobayashi S, Matsuura S (1995) Egg development and variation of egg sizes in the Japanese mitten crab *Eriocheir japonica* (de Hann). Benthos Res 48: 29-39.
28. Brante A, Fernandez M, Eckerle L, Mark F, Portner HO, et al. (2003) Reproductive investment in the crab, *Cancer setosus*, along a latitudinal cline: egg production, egg losses and embryo ventilation. Mar Ecol Prog Ser 251: 221-232.
29. Warner GF (1977) The Biology of Crabs. Elek Science, London, UK 202.
30. Veera RA (1994) Biochemical changes during embryonic and larval development of the edible portunid crab *Charybdis lucifera* (Fabricius). PhD Thesis, Annamalai University, India.
31. Parimalam K (2001) Embryonic and larval development of the hermit crab *Clibanarius longitarsus* (De Haan) (Crustacea: Decapoda: Anomura). M. Phil Thesis, Annamalai University, India.
32. Subramoniam T (1991) Yolk utilization and esterase activity in the mole crab *Emerita asiatica* (Milne Edwards). In: Crustacean egg production (Eds. A. Wenner and A. Kuris). Crusta 7: 19-30.
33. Thirunavukkarasu N (2005) Biology, nutritional evaluation and utilization of mud crab *Scylla tranquebarica* (Fabricius, 1798). Ph.D. Thesis, Annamalai University, India:1- 126.
34. Sun Y, Yu Yan (1989) A preliminary study of artificial rearing of juvenile *(Charybdis japonica)*. Mar Sci Haiyang Hexue 3: 43-47.
35. Xue JZ, Du SN, Lai W (2001) Morphology of egg-larva of swimming crab *(Portunus trituberculatus)* during embryonic development. Acta Zool Sinic 47: 447-452.
36. Josileen J (2002) Studies on reproductive biology and larval culture of blue crab *Portunus pelagicus* (Linnaeus). Ph.D. Thesis, Cochin University of Science and Technology, India: 260.
37. Heasman MP, Fielder DE (1983) Laboratory spawning and mass rearing of the mangrove crab *Scylla serrata* (Forskal), from first zoea to first crab stage. Aquacult 34: 303-316.
38. Robertson WD, Kruger A (1994) Size at maturity, mating and spawning in the portunid crab, *Scylla serrata* (Forskal) in Natal, South Africa. Est Coast Shelf Sci 39: 185-200.
39. Hamasaki K, Fukuanga K, Maruyama K (2003) Egg development and incubation period of the swimming crab *Portunus trituberculatus* (Decapoda: Portunidae) reared in the laboratory. Crust Res 32: 45-54.
40. Campbell GR, Fielder DR (1987) Occurrence of a prezoea in two species of commercially exploited portunid crabs (Decapoda, Brachyura). Crust 52: 202-206.
41. Sandoz M, Rogers R (1944) The effect of environmental factors on hatching, moulting and survival of zoea larvae of the blue crab *Callinectes sapidus* Rathbun. Ecol 25: 216-228.
42. Hartnoll RG, Paul RKG (1982) The embryonic development of attached and isolated eggs of *Carcinus maenas*. Int J Invert Reprod 5: 247-252.

43. Fernandez-Vergaz V, Lopez Abellan LJ, Balguerias E (2000) Morphometric, functional and sexual maturity of the deep-sea red crab *Chaceon affinis* inhabiting Canary Island waters: chronology of maturation. Mar Ecol Prog Ser 204: 169-178.

44. Baeza JA, Fernandez M (2002) Active brood care in *Cancer setosus* (Crustacea: Decapoda): The relationship between female behaviour, embryo oxygen consumption and the cost of brooding. Func Ecol 16: 241-251.

45. Davis CC (1965) A study of the hatching process in aquatic invertebrates: XX. The blue crabs, *Callinectes sapidus*, Rathbun. Chesapeake Sci 6: 201-208.

46. Iin SD, Wille M, Kontara EK, Sorgeloos P (2003) Reproductive performance and offspring quality in mud crab *(Scylla paramamosain)* broodstock fed different diets. Aquacult Internat 11: 3-15.

47. Van Engel WA (1958) The blue crab and its fishery in Chesapeake Bay. Part I. Reproduction, early development, growth and migration. Commer Fish Rev 24: 1-10.

Comparative study of Condition factor, Stomach Analysis and Some Aspects of Reproductive Biology of Two Land Crabs: *Cardiosoma armatum* (Herklots, 1851) and *Cardiosoma guanhumi* (Latreille, 1825) from a Mangrove Swamp Ecosystem, Lagos – Nigeria

E Isa Olalekan[1]* and Aderonke Omolara Lawal-Are[2]

[1]*Environmental and Resource Management, Brandenburg University of Technology, Cottbus, Germany*
[2]*Department of Marine Sciences, University of Lagos, Lagos, Nigeria*

Abstract

A total samples of 858 of *Cardisoma armatum* and *Cardisoma guanhumi*) collected from the Lagos Lagoon mangrove area of the University of Lagos were studied for their frequency, distribution, growth and sex ratio and a comparative analysis was done on both crabs. Investigation into their length-weight relationships, growth pattern, condition factor, food and feeding habits were carried out. The calculated Chi-square (χ^2) test showed that male crabs of *Cardisoma armatum* and *Cardisoma guanhumi* were significantly ($p<0.05$) more abundant than the female crabs. The Condition factor K values for the *Cardiosoma armatum* and *Cardiosoma guanhumi* ranged from 3.3 to 30.7 and 3.1 to 28.6, respectively. This research study indicates almost similar biological features for both species.

Keywords: Reproductive biology; Land crabs; Mangrove swamp ecosystem; Cardisoma armatum

Introduction

Crabs belong to the brachyuran infra order family comprising more than 6,793 species peculiarly known for their ten legged creature (decapod) Hosseini et al. [1]. Crabs have flourished to be a predominant icon in the invertebrate fauna because of its ubiquitoes existence in almost all part of the world oceans including freshwater, marine even on land [2]. Crabs are least exploited amongst other crustacean and crabs have been commonly found in West Africa. The Gecarcinidae currently consists of 20 species and has been recognized to include 4 [3] or 6 genera [4,5].

The eggs of crabs have to hatch in the sea, where the larvae undergo typical planktonic development [3]. The mass migration of reproductive individuals to the surf for larval release has been reported for many crab species [6]. The aims of this research was to provide baseline data on condition factor, feeding pattern and sex ratio and to make comparison of populational differences based on morphological analyses of the two crabs species: *Cardiosoma armatum* [7] and *Cardiosoma guanhumi* [7].

Materials and Methods

Description of study site

The study was carried out in the coast/Mangrove area of University of Lagos Lagoon front which is located opposite the Lagos Lagoon on the geographical platform of 6°26'N and 6°39'N and longitude 3°29'E and 3°50'E (Figure 1). The lagoon is the largest of the four lagoon systems of Gulf of Guinea and is located at South Western Nigeria. The mangrove swamp connects to the Lagos Lagoon by tidal creek.

Collection of specimens

Specimens were collected at the mangrove part of the Lagos Lagoon of the University of Lagos. They were caught with hand at the same time and place between 7 pm-11 pm to allow for precise readings and analysis of the samples. The collection was done randomly and was collected over a period of six months on weekly bases between February and July, 2012. The crabs were collected in 2 different stations within the mangrove swamp. A total of 300 crabs were collected from the site and were preserved immediately in a deep freezer in the laboratory prior to examination.

Laboratory procedure

The crabs were removed from the freezer and allowed to thaw. Excess water was removed from the specimens using filter paper.

Condition factor

This is the condition of general well-being of a crab. It was studied in relationship to size. The equation for condition factor as follows:

$K=100W/L^3$

Where K=condition factor

W=weight of the crab (g)

L=length of the crab (cm)

It is defined as a condition representing how lean or fat the crab.

Food analysis

The crabs were dissected and the stomachs removed for food analysis. Each stomach was studied as a unit in order to provide information on individual variation. The stomach is greenish in color,

***Corresponding author:** Isa Olalekan, Environmental and Resource Management, Brandenburg University of Technology, Cottbus, Germany
E-mail: isaelegbede@gmail.com

Figure 1: Map showing the sampling location.

located underneath the carapace and it is divided into four parts. The stomach contents were examined and scored with regards to whether they were empty, one-quarter full (¼), half full (½), three-quarter full (¾) or full stomach (4/4). The methods of food analysis used were the numerical method and frequency of occurrence method.

Numerical method: The number of individuals of each food items in each stomach was counted. They were summed up to give totals for each kind of item in the whole sample. Then a grand total of all food items were obtained and each food item was expressed as a percentage of the total number of food found in all crabs examined.

Frequency of occurrence method: In this method, stomach content was examined and individual food organisms sorted and identified. The number of stomachs in which each item occur was recorded and expressed as a percentage of the total number of stomach with food. The method gives information only on the organisms fed on. Its main disadvantage is that it does give information on quantities or numbers;

Reproductive biology

Sex ratio: The crabs were sorted out and sexed using gonopod (a thin abdominal segment) and the gonophores (a broad abdominal segment). These structures were used to identify the male and female respectively. In the male, the abdominal segment is present only on the first and second abdominal somites and is modified to form copulatory organs. The female differs by having all the somites freely moveable and there is one pair of appendages on each of the 2nd, 3rd, 4th and 5th somites, these form the swimmerets to which the eggs are attached in ovigenous crabs.

Fecundity: This is defined as the number of ripe eggs in the female prior to the next spawning [7]. The egg mass was carefully removed from the pleopods using tweezers and washed in running water. The eggs were placed in a 50 ml beaker and filled with seawater. Egg diameter was determined using an ocular micrometer before the eggs were separated.

Results

418 and 440 specimens of *Cardiosoma armatum* (Plate 1) and *Cardiosoma guanhumi* (Plate 2) were studied respectively making a total of 858 species of crabs collected and studied. The specimens were studied for the length and width frequency distributions between the months of February to July, 2012.

Condition factor of *Cardiosoma Armatum* and *Cardiosoma Guahunmi*

The condition factor (CF) or coefficient of condition is referred to as the K factor which indicates the state or overall wellbeing of the *Cardiosoma armatum* and *Cardiosoma guanhumi* was calculated for the 418 *C. armatum* (combined sex) and 440 *C. guanhumi* (combined sex) in relation to size. The K values for the *Cardiosoma armatum* ranged from 3.3 in size group 8.5-9.4 and 30.7 in size group 2.5-3.4. For the *Cardiosoma guanhumi* the K value ranged from 3.1 in size group 8.5-9.4 and 28.6 in size group 2.5-3.4. This is illustrated in the Tables 1 and 2.

Plate 1: *Cardiosoma armatum*.

Plate 2: *Cardiosoma guanhumi*.

Food items	Numerical method	Frequency of occurrence	Food items	Numerical method
	Number	% frequency	Number	% frequency
frequency				
Plant material	9,945	64.8	320	90.00
Crustaceans	2,390	15.6	110	31.25
Fish fragment				
(bones and scales)	789	5.1	90	25.5
Sand grains	-	-	305	86.7
Unidentified masses	-	-	290	82.4
Total	15,356	100		

Table 1: Stomach Food Content of *Cardiosoma armatum* from the Lagos Lagoon mangrove Swamps.

Period	Number	Number of crabs With empty stomach	% empty stomach
February	60	9	15.00
March	75	7	9.33
April	61	5	8.20
May	78	11	14.10
June	80	8	10.00
July	86	13	15.12
Total	440	53	12.05

Table 2: Stomach analysis of *Cardiosoma guanhumi* from the Lagos Lagoon mangrove swamps (February- July, 2012).

Sex ratio of *Cardiosoma Armatum* and *Cardiosoma Guahunmi*

The Carapace length of the male *Cardiosoma armatum* showed significant value as against the Carapace length of the female *Cardiosoma armatum* from the null hypothesis (H_0) of the expected ratio of 1:1 in the Chi-square test shown in Table 3. The chi-square test carried out to test if there is any significant difference in the sex ratio of *Cardiosoma armatum* gave 1.38 for 1 *df* at 5% level of significance. This is less than the tabulated value of 3.84 for 1 *df* at 5% level of significance (Table 4).

$$x^2 = \frac{(221-209)^2}{209} + \frac{(197-209)^2}{209}$$

$$x^2 = \frac{(12)^2}{209} + \frac{(12)^2}{209}$$

$$\frac{144}{209} + \frac{144}{209} = \frac{288}{209}$$

$$0.6889 + 0.6889$$

$=1.378$

Calculate x^2 (1df.5%)=1.378 (Non-significant)

Tabulated x^2 (1df.5%)=3.84

Hence, Tab X^2 > Cal.X^2

No significance

Therefore, there is no significance between the male and female of *Cardiosoma amatum* from the Lagos Lagoon mangrove swamps (February- July, 2012).

$$x^2 = \frac{(235-220)^2}{220} + \frac{(205-220)^2}{220}$$

$$x^2 = \frac{(15)^2}{220} + \frac{(-15)^2}{220}$$

$$\frac{225}{220} + \frac{225}{220} = \frac{450}{220}$$

$=2.045$

Calculated x^2 (1df.5%)=2.045 (No significant)

Tabulated x^2 (1df.5%)=3.84

Hence Tab x^2>Cal.x^2

No significance

Therefore, there is no significance between the male and female of *Cardiosoma guanhumi* from Lagos Lagoon mangrove swamps between February and July, 2012 (Table 5).

Discussion and Conclusions

In studies of population dynamics, high condition factor values shows favorable environmental conditions such as habitat and prey availability Moradinasab et al. [8] this assertion shows relevance to this research work, the condition factor for the *Cardiosoma armatum* and *Cardiosoma guanhumi* has a higher k values of 28.60 and 30.75 for both crabs respectively, though *Cardiosoma guanhumi* had a higher condition factor k than *Cardiosoma armatum*, this is obviously related to the relative difference in habitat condition and adequate prey inclusion. This is also supported by the works of Lawal-Are and Nwankwo [9] with k-values of *Sersema huzadii* from a tropical estuarine lagoon.

The stomach content analysis carried out on *Cardiosoma armatum* and *Cardiosoma guanhumi* from the Lagos Lagoon, Unilag Water front, indicated that the percentage empty stomach of *Cardiosoma armatum* and *Cardiosoma guanhumi* were 66(5.79%) and 53(2.05%) respectively. The result was in conformity with Lawal-Are and Bilewu [10] for *Portunis validus* off Lagos's coast Nigeria, the percentage empty stomach content was lowest in March and April for both *C. armatum* and *C. guanhumi,* this is due to the low environmental condition at the

	Observed number	Expected number	Calculated x^2	Tabulated x^2
Male	221	209	1.38	3.84
Female	197	209		
Total	418	418		

Table 3: Chi square Test on Sex ratio of *Cardiosoma armatum* from the Lagos Lagoon mangrove swamps (February- July, 2012).

Food items	Numerical method		Frequency of occurrence	
	Number	% frequency	Number	% frequency
Plant materials	8,786	39.5	302	78.03
crustaceans	2,857	12.9	95	25.00
Fish fragments				
(bones and scales)	750	3.4	83	21.44
Sand grains	-	-	287	74.16
Unidentified masses	-	-	245	63.31
Total	22,225	100		

Table 4: Stomach Food Content of *Cardiosoma guanhumi* from the Lagos Lagoon mangrove swamps (February- July, 2012).

Sex	Observed number	Expected number	Calculated x2	Tabulated x2
Male	235	220	2.045	3.84
Female	205	220		
Total	440	440		

Table 5: Chi square Test on Sex ratio of *Cardiosoma guanhumi* from the Lagos Lagoon mangrove swamps (February- July, 2012).

	Cardiosoma armatum			*Cardiosoma guanhumi*			Chi-Square Significant Level
	Female	Male	Sex ratio	Female	Male	Sex ratio	
February	26	30	1:1.15	26	30	1:1.15	0.147
March	29	33	1:1.14	29	33	1:1.14	0.997
April	33	37	1:1.12	33	37	1:1.12	0.015
May	31	38	1:1.22	31	38	1:1.22	0.835
June	39	39	1:1	39	39	1:1	0.794
July	39	44	1:1.12	39	44	1:1.12	0.850

Table 6: Sex Ratio of *Cardiosoma armatum* and *Cardiosoma guanhumi* from the Lagos Lagoon mangrove swamps (February- July, 2012).

period of collection.

Both crabs showed leaf preference because of the flora associated to their habitat, they showed high level of omnivorous feeding habit, as shown in the stomach content analysis indicated that they both feed on plant materials, crustaceans, fish fragments (bone and scales), sand grains and unidentified items, this support the work of Micheli et al. [11] for *Cardiosoma carnifex* and *Sesarma mainerti*. Fish fragments and crustacean found in their stomach content was attributed to the inter migration to shallow part of coastal water. The wide opportunistic feeding pattern of *Cardiosoma armatum* and *Cardiosoma guanhumi* was due to their accidental predatorship [12]. The large amount of sand grains discovered was attributed to the burrowing nature of the crabs and inherent soil habitat.

The cumulative sex ratio of both crabs *Cardiosoma armatum* and *Cardiosoma guanhumi* showed that males are higher than female; the large number of males in both crab species conforms with [9] for *Sersema* huzadii which is a mangrove crab (Table 6). According to De-Rivera et al. [13] in a population of the California fiddler crab, *Uca crenulata*. Mensurative studies revealed there were almost twice as many adult males as females, mating occurred across half of the days within the breeding season, and females had much longer individual reproductive cycles than males. Hence more males than females were available for mating on each breeding day. Perhaps as a consequence, males spent a large proportion of their time fighting with neighbors and rapidly waving their large claws when females passed by. Statistically the chi-square for male female ratio of both crabs showed no significance for male of both crab species and the females, based on the research of Male crabs were more abundant than females [14].

Thus, the conclusion was that the male *Cardiosoma armatum* were significantly more numerous than the female in the total number of crab sampled from the Lagos lagoon Unilag water front. There was no ripe female in the sample that could be used for fecundity study.

References

1. Hosseini M, Vazirizade A, Parsa Y, Mansori A (2012) Sex Ratio, Size Distribution and Seasonal Abundance of Blue Swimming Crab, Portunus pelagicus (Linnaeus, 1758) in Persian Gulf Coasts, Iran. World Appl Sci J 17: 919-925.
2. Akin-Oriolam G, Anetekhai M, Olowonirejuaro K (2005) Morphometric and Meristic Studies in Two Crabs: Cardiosoma armatum and Callinectes pallidus. Turkish J Fisheries Aquatic Sci 5: 85-89.
3. Hartnoll RG (1988) Evolution, systematics, and geographical distribution. In WW Burggren, BR McMahon, eds. Biology of the land crabs. Cambridge, UK: Academic Press, 6-54.
4. Ng PKL, Guinot D (2001) on the land crabs of the genus Discoplax A. Milne Edwards, 1867 (Crustacea: Decapoda: Brachyura: Gecarcinidae), with description of a new cavernicolous species from the Philippines. Raffles Bull Zool 49: 311-338.
5. Liu HC, Jeng MS (2005) Reproduction of *Epigrapsus notatus* (Brachyura: Gecarcinidae) in Taiwan. J Crustacean Biol 25: 135-140.
6. Adamczewska AM, Morris S (2001) Ecology and behavior of *Gecarcoidea natalis*, the Christmas Island red crab, during the annual breeding migration. Biol Bull 200: 305-320.
7. Mantelatto FLM, Fransozo A (1996) Size at maturity in *Callinectes ornatus* (Brachyura, Portunidae) from the Ubatuba Region (SP), Brazil. Nauplius 4: 29-38.
8. Moradinasab GH, Raeisi H, Paighambari SY, Ghorbani R, Bibak Z (2012) Length-weight relationships, relative condition factor and relative weight of three fish species from beach seine fishing grounds in Iranian coastal waters of Caspian Sea. Academic Journals 7: 1809-1812.
9. Lawal-Are AO, Nwankwo H (2011) Biology of the Hairy Mangrove Crab, *Sersema huzardii* (Decapoda: Graspidae) from a Tropical Estuarine Lagoon. Nat and Sci 9: 58-164.
10. Lawal-Are AO, Bilewu B (2009) The Biology of the Smooth Swim Crab, *Portunus validus* (Herklots) off Lago`S Coast, Nigeria. Eur J Sci Res 30: 402-408.
11. Micheli F, Gherardi F, Vannini, M (1991) Feeding and burrowing ecology of two East African mangrove crabs. Mar Biol 111: 247-254.
12. Lawal-Are AO (2009) Food and feeding habits of the blue crabs, *Callinectes amnicola* (De Rocheburne) from three different interconnecting lagoons in South-West, Nigeria. Eur J Sci Res 32: 89-95.
13. De-Rivera CE (2003) Causes of male-biased operational sex ratio in the fiddler crab *Uca crenulata*. J Ethol 21: 137-144.
14. Harding JM, Mann R (2010) Distribution, size, and sex ratio of mature blue crabs *Callinectes sapidus* Rathbun 1896, from a Chesapeake Bay tributary in relation to habitat type and environmental factors. Bull. Mar Sci 86: 75-91.

A Study on the Zoogeography of Indian Penaeidae

Angsuman Chanda*

Post Graduate Department of Zoology, Raja NL Khan Women's College, Gope Palace, Midnapur, Paschim Medinipur-721 102, WB, India

Abstract

From the distributional point of view of Penaeid prawn, Indian region in the present study is divided into three main sub-regions viz., East Coast, West Coast including Laccadive-Minicoy Islands and Andaman Islands. Present study also records existence of 79 species in the coastal water of India. Out of 79 species 9 are endemic, 24 are common in east coast, west coast and Andaman Islands; 16 species are restricted to East coast, 13 are restricted to west coast, 2 are restricted to Laccadiv Islands, 3 species are restricted to Andaman Islands only. Global Analysis shows that most species are distributed within Indo-West Pacific region and only 6 species are distributed up to east Atlantic region. A statistical analysis based on the distribution of prawn shows more moderate similarity between east coast and west coast in India and strong similarity between sub-region 1 and 4 of Indo-West Pacific indicates that the Indian species are more common with Indonesian and Indo-Malaysian than East African species.

Keywords: Distribution; Penaeid prawn; Indian region; Global

Introduction

Shrimps and Prawns of various kinds have certainly been a source of protein for human consumptions from very early times. Within historical times reference is made to prawn in ancient Chinese and Japanese literature. Usage of the term 'Prawn' and 'Shrimp' are somewhat confusing. In some western literature the term 'Shrimp' is applied for Penaeoidea and Sergestoidea, but in the east these are called 'Prawn'. Holthuis [1] discussed the contradiction but did not arrive at any conclusion. In the Prawn Symposium of the Indo-Pacific Fisheries Council held at Tokyo in 1955 it was decided that the word 'Prawn' should be applied to the Penaeids, Pandalids and Palemonids while 'Shrimp' to the smaller species belonging to the other families [2]. As such in the present study the term 'Prawn' is used for all the species belonging to family Penaeidae. Among a variety of edible decapod crustaceans, prawns contribute largely to the fishery wealth of many nations. Exploitation of prawn resource from the seas around each country is playing increasingly significant role in furthering their national economy. In recent years, in spite of some ecological hazards, the demand for prawns and prawn products has increased so much that every country is making efforts to utilize hitherto unknown but usable stocks and expansion of prawn fisheries and industries near coast line is rightly being given the maximum encouragement in the development programme of each nation. Therefore, the distribution of different species gradually extending their range of distribution and newer species were recorded throughout coastal region of each nation.

According to distribution of penaeid prawns, Indian region has been divided into three main sub-regions viz., Eat coast, West coast including Laccadive-Minicoy Islands and Andaman Islands (Table 1). The sub-region wise distribution of penaeid prawn has been shown in Table 2. Sorensen's quotient of similarity has been included in Table 3. Shallow and warm water marine areas of the world were classified into four zoogeographic regions namely Indo-West Pacific, Eastern Pacific, Western Atlantic and Eastern Atlantic regions [3]. Each region is further divided into several sub-regions. Since penaeid prawns are mostly distributed in these regions and sub-regions Dall et al., [3] have designated those as penaeid zoogeographical regions and sub-regions.

Materials and Methods

The present study is mainly based on the specimens collected by the author from different commercial fish landing centres throughout Indian coast line (Figure 1). In addition to this penaeid prawns preserved in the National Collection of the Zoological Survey of India, Kolkata, India; Central Marine Fishery Research Institute, Cochin, Kerala and its regional stations at Mandapam, Tamil Nadu and National Institute of Oceanography have also been studied.

The collected materials were preserved in rectified spirit (90%) and body parts of taxonomic importance have been dissected and studied under a stereoscopic binocular microscope and identified. All the materials were registered and preserved in the national collection of Zoological Survey of India, Kolkata.

Results

Indian distribution

The present study records existence of 79 species of penaeid prawn in the coastal water of India. A look into the Table 2 reveals that East coast has 58 species, West coast 55 and Andaman Island has 34 species. Out of the 79 species recorded in this dissertation 24 species namely *Penaeus indicus* H. Milne Edwards, 1837, *P. merguiensis* De Man, 1888, *P. monodon* Fabricin 1798, *P. semisulcatus* De Haan, 1844 *P. japonicus*

Sl. No.	Indian Subregions	Total No. of species	Endemic species
1.	East Coast: West Bengal, Orissa, Andhra Pradesh, Tamil Nadu	58	3
2.	West coast including Laccadive Islands: Gujarat, Maharastra, Goa, Karnataka, Kerala	55	4
3.	Andaman Islands.	34	2

Table 1: Penaeid species including endemic species found in various Zoogeographical subregions of India.

***Corresponding author:** Angsuman Chanda, Post Graduate Department of Zoology, Raja N. L. Khan Women's College, Gope Palace, Midnapur, Paschim Medinipur-721 102, W.B, India, E-mail: angsumanchanda@yahoo.in

Sl. No.	Name of genera, species & subspecies	East coast (EC)	West coast including Laccadive (WC)	Andaman Islands (AI)	Endemic
	Alcockpenaeopsis Chanda, 2016				
1	*Alcockpenaeopsis uncta* (Alcock, 1905)	+	+	-	-
	Atypopenaeus Alock, 1905				
2	*A. compressipes* (Henderson, 1893)	+	+	-	-
3	*A. stenodactylus* (Stimpson, 1860)	-	+	-	-
	Batepenaeopsis Chanda, 2016				-
4	*B. acclivirostris* (Alcock, 1905)	+	+	-	-
5	*B. tenella* (Bate, 1888)	+	-	-	-
	Funchalia Johnson, 1867				
6	*F. danae* Burkenroad, 1940	-	+	-	-
7	*F. villosa* (Bouvier, 1905 b)	-	+	+	-
8	*F. woodwardi* Johnson, 1867	-	+	+	-
	Helleropenaeopsis Chanda, 2016				
9	*H. cultirostris* (Alcock, 1906)	+	-	-	-
10	*H. hardwickii* Miers,1878)	+	+	-	-
11	*H. indica* (Muthu, 1972)	+	-	-	-
12	*H. sculptilis* (Heller, 1862)	+	+	+	-
	Kishinouyepenaeopsis Chanda,2016				
13	*K. cornuta* (Kishinouye, 1900)	+	+	+	-
14	*K. maxillipedo* (Alcock, 1906)	+	+	-	-
	Megokris Pérez Farfante and Kensley, 1997				
15	*M. granulasus* (Haswell, 1879)	+	-	-	-
16	*M. pescadoreensis* (Schmitt, 1931a)	+	+	-	-
17	*M. sedili* (Hall, 1961)	+	+	-	-
	Metapenaeopsis Bouvier, 1905				
18	*M. andamanensis* (Wood-Mason, 1891)	-	+	+	-
19	*M. barbata* (De Haan, 1844)	+	-	-	-
20	*M. commensalis* Borradaile, 1898	-	+	-	-
21	*M. coniger* (Wood-Mason, 1891)	+	+	+	-
22	*M. ceylonica* Starobogatov, 1972	+	-	-	-
23	*M. gaillardi* Crosnier, 1991	+	+	-	-
24	*M. gallensis* (Pearson, 1905)	+	-	-	-
25	*M. hilarula* (De Man, 1911)	+	+	-	-
26	*M. mogiensis* (Rothbun, 1902)	+	+	+	-
27	*M. novaeguineae* (Haswell, 1879)	+	-	-	-
28	*M. palmensis* (Haswell, 1879)	+	-	-	-
29	*M. philippii* (Bate, 1881)	-	+	-	-
30	*M. stridulans* (Alcock, 1905)	+	+	+	-
31	*M. toloensis* Hall, 1962	+	-	+	-
	Metapenseus Wood-Mason, 1891				
32	*M. affinis* (H. Milne Edwards, 1837)	+	+	+	-
33	*M. alcocki* George and Rao, 1968	-	+	-	+
34	*M. brevicornis* (H. Milne Edwards, 1837)	+	+	+	-
35	*M. bengalyensis* Tirmizi, 1971	+	-	-	+
36	*M. dobsoni* (Miers, 1878)	+	+	+	-
37	*M. eboracensis* Dall, 1957	+	-	-	-
38	*M. elegans* De Man, 1907	+	-	+	-
39	*M. endeavouri* (Schmitt, 1926a)	+	-	-	-
40	*M. ensis* (De Haan, 1844)	+	+	+	-
41	*M. intermedius* (Kishinouye, 1900)	-	-	+	-
42	*M. krishnatrii* (Silas and Muthu, 1976)	-	-	+	+
43	*M. kutchensis* George, George and Rao, 1963	-	+	-	+
44	*M. lysianassa* (De Man, 1888)	+	+	+	-
45	*M. monoceros* (Fabricius, 1798)	+	+	+	-
46	*M. moyebi* (Kishinouye, 1896)	+	+	+	-
47	*M. stebbingi* Nobili, 1904	-	+	-	-
	Miyadiella Kubo, 1949				
48	*M. podophthalmus* (Stimpson, 1860)	-	+	-	-
	Parapenaeopsis Alcock 1901				
49	*P. longirostris* Chanda and Bhattacharya,2002.	+	-	-	+
50	*P. nana* Alcock, 1905	+	-	-	-
51	*P. stylifera cochinensis* George, 1975	-	+	-	-
52	*P.* stylifera *coromandelica* Alcock, 1906	+	+	-	-
53	*P. stylifera stylifera* (H. Milne Edwards, 1837)	+	+	+	-
	Perapenaeus Smith, 1885				
54	*P. fissuroides indicus* Crosnier, 1985	-	+	-	-
55.	*P. fissurus* (Bate, 1881)	+	-	+	-
56	*P. investigatoris* Alcock and Anderson, 1899	+	+	+	-
57	*P. longipes* Alcock, 1905	+	+	-	-
58	*P. sextuberculatus* Kubo, 1949	-	+	-	-
	Pelagopenaeus Perez Farfante & Kensley, 1997				
59	P. *balboae* (Faxon, 1893)	-	+	-	-
	Penaeopsis Bate, 1881				
60	*P. edwardoi* Perez Farfante, 1977	+	-	-	-
61	*P. jerryi* Perez Farfante, 1979	+	+	+	-
62	*P. rectacuta* (Bate, 1881)	+	+	+	-
	Penaeus Fabricius, 1798				
63.	*P. canaliculatus* (Oliver, 1811)	+	+	+	-
64.	*P. hathor* Burkenroad, 1959	-	+	-	-
65.	*P. indicus* (H. Milne Edwards, 1837)	+	+	+	-
66.	*P. japonicus* (Bate, 1888)	+	+	+	-
67.	*P. konkani* (Chanda and Bhattacharya,2003)	-	+	-	+
68.	*P. latisulcatus* (Kishinouye, 1896)	+	+	+	-
69.	*P. marginatus* (Randall, 1840)	+	+	-	-
70.	*P. merguiensis* (De Man, 1888)	+	+	+	-
71.	*P. monodon* Fabricius, 1798	+	+	+	-
72.	*P. penicillatus* (Alcock, 1905)	+	+	-	-
73.	*P. semisulcatus* De Haan, 1844	+	+	+	-
74.	*P. silasi* Muthu & Motho, 1979	+	-	-	+

75.	*P. similis*.(Chanda and Bhattacharya,2000.)	-	-	+	+
	Trachypenaeopsis Burkenroad, 1934				
76.	*T. minicoyensis* Thomas, 1972	-	+	-	+
	Trachysalambria Burkenroad, 1934				
77.	*T. aspera* (Alcock, 1906)	+	-	+	-
78.	*T. curvirostris* (Stimpson, 1860)	+	+	+	-
79.	*T. fulva* (Dall, 1957)	+	-	-	-
	TOTAL	**58**	**55**	**34**	**9**

Table 2: Distribution of penaeid prawns in Indian subregion.

Region	Q/S Value	Ramarks
EC vs WC	0.673	Moderately similar.
EC vs AI	0.608	Moderately similar.
WC vs AI	0.606	Moderately similar.

Table 3: Sorensen's quotient of similarity based on the distribution of penaeid fauna in India.

Bate, 1888, *P. canaliculatus* Olivier, 1811, *P. latisulcatus*; *Metapenaeopsis coniger*, *M. mogiensis*, *M. stridulaus* [4] *Metapenaeus affinis*, *M. brevicornis*, *M. dobsoni*, *M.* ensis, M. *lysianassa* , *M. monoceros*, *M. moyebi*, *Helleropenaeopsis sculptilis*, *Kishinouyepenaeopsis cornuta*, *Parapenaeopsis stylifera*, *Parapenaeus investigatoris* Alcock and Anderson [4], *Panaeopsis jerryi* Pérez Farfante, *P. rectacuta* and *Trachysalambria curvirostris* are found both in East coast, West Coast and Andaman Islands. 10 species namely *Atypopenaeus compressipes*, *Penaeus penicillatus* [4], *P. marginatus*, *Megokris sedili*, *Metapenaeopsis hilarula*, *Batepenaeopsis acclivirostris* [4], *Alcockpenaeopsis unclta* [4], *Helleropenaeopsis hardwickii*, *Kishinouyepenaeopsis maxillipedo* Alcock [4] and *Parapenaeus longipes* Alcock [4] are found in East and West coast but not in Andaman Islands. 4 species namely *Metapenaeopsis toloensis*, *Metapenaeus elegans* De Man, 1907, *Parapenaeus fissurus* and *Trachysalambria aspera* [4] are common in East coast and Andaman Islands. 2 species namely *Funchalia woodwardi* Johnson, 1867 and *Metapenaeopsis andamanensis* are common in West coast and Andaman Islands. 12 species namely *Megokris granulosus*, *Metapenaeopsis barbata*, *M. gallensis*, *M. novaeguineae*, *M. palmensis*, *Metapenaus bengalensis* Tirmizi, *M. eboracensis* Dall [3], *Batepenaeopsis tenella* (Bate, 1888), *Helleropenaeopsis cultirostris* Alcock [4], *Parapenaeopsis longirostris* Chanda and Bhattacharya, 2004; *P. nana* Alcock [4] and *Penaeopsis eduardoi* Pérez Farfante, 1977 are restricted to East coast only. 10 species namely *Atypopenaeus stenodactylus*, *Penaeus Konkani* [5], *Metapenaeopsis commensalis* Borradaile, 1898, *M. Philippii*, *Metapenaeus alcocki*, *M. kutchensis* George, George and Rao, *M. stebbingi* Nobili, 1904, *Miyadiella podophthalmus*, *Parapenaeus sextuberculatus*, and *Trachypenaeopsis minicoyensis* Thomas, 1972 are

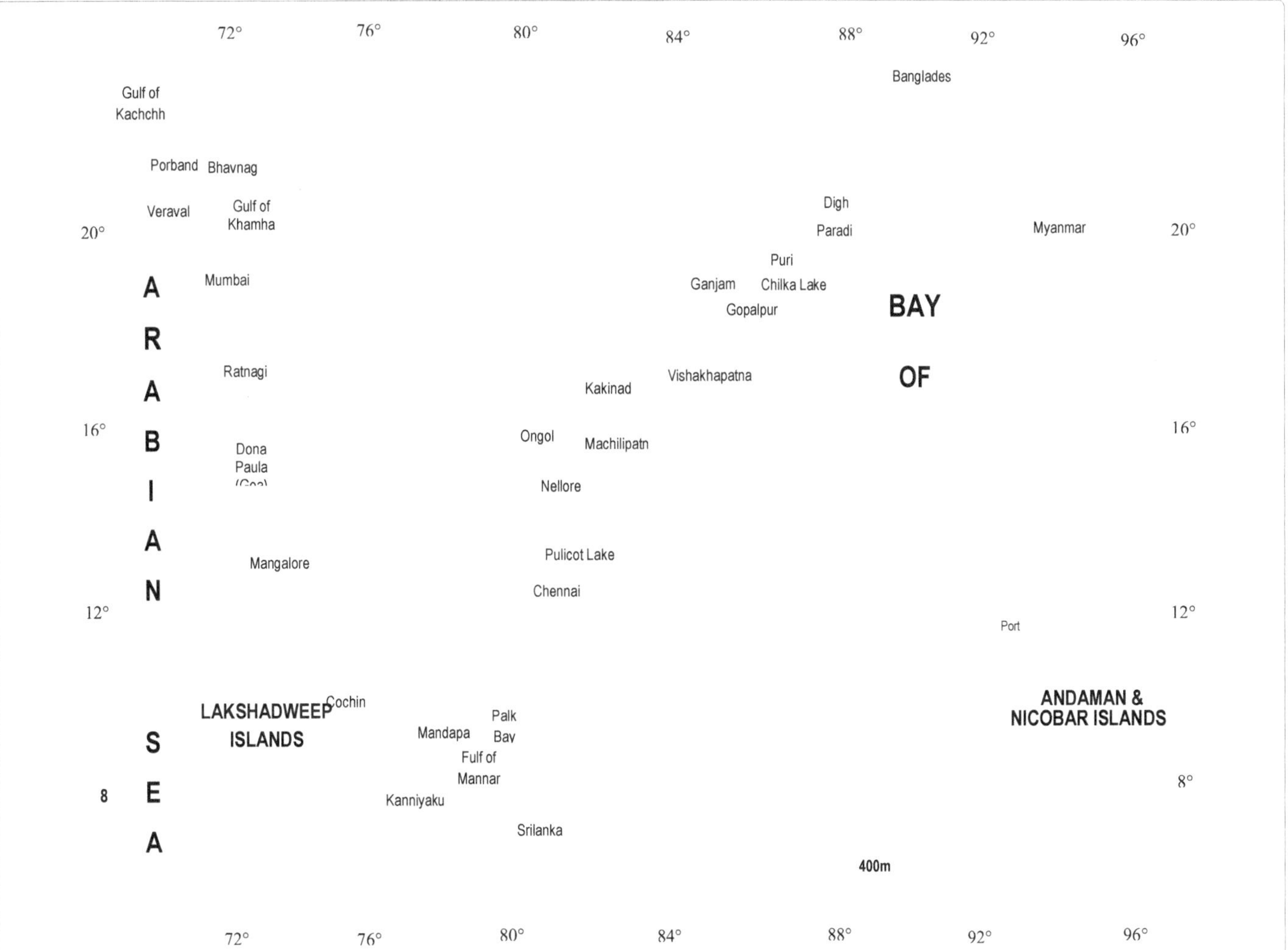

Figure 1: Study area and collection spots along Indian Coast.

restricted to wast coast of which *M. commensalis* Borradaile, 1898 and *T. minicoyensis* Thomas, 1972 are restricted to Laccadive Islands. 3 species namely *Penaeus similis* [5]. *Metapenaeus intermedius* and *M. krishnatrii* Sisal and Muthu, 1976 are restricted to Andaman Island only. Two species have been recorded for the first time by Chanda [6] from India namely *Metapenaeopsis palmensis* and *Metapenaeus eboracensis* Dall, 1957, both from Andhra Pradesh in the east coast of India.

The present study also extends the limits of distributions of several earlier recorded species. Some of the species like *Metapenaeopsis toloensis* Hall, 1962 and *Metapenaeus elegans* De Man, 1907 which were previously recorded only from Andaman Sea have now been recorded from the eastern coast of mainland, India. In contrast to this *Penaeus indicus* H. Milne Edwards, 1837 which was earlier reported from east and west coast of main land, India [7] has now been recorded from Andaman Sea.

Penaeus latisulcatus Kishninouye, 1896 which was previously recorded only from west coast has now been recorded from east coast also. Likewise *Metapenaeus lysanassa* and *Penaeopsis acclivirostris* [4] was previously recorded only from east coast have now been recorded from west coast also.

In case of some species the limits of distributions have been extended either in east or west coast. For example northern limit of distribution in west coast has been extended upto Gujarat for *Atypopenaeus stenoductylus, Metapenaeopsis stridulans* [4], *Metapenaeus ensis* and *Helleropenaeopsis hardwickii.* Similarly northern limit of the distribution in east coast has been extended for *Metapenaeus ensis* and *Parapenaeopsis stylifera* up to Ganjatic delta and West Bengal and up to Andhra Pradesh for *Kishinouyepenaeopsis cornuta.* Present study also suggest that *Penaeus penicillatus* [4] has a broad range of distribution in both east and west coast and not restricted to Maharastra and Orissa coast as thought earlier by George [8].

Of these reported 79 species ,9 species are endemic to India and 4 of these are restricted to West coast including 2 in Laccadive Island, 2 are restricted to each coast and remaining 3 are restricted to Andaman Islands. Analysis of the Sorensen's quotient of similarity based on the distribution in three sub-regions of India revealed that penaeid fauna is moderately similar (0.61-0.70) among different sub-regions. As expected similarity maximum between East and West coast (0.673) due to continuity of coast line and least between West coast and Andaman Island (0.606) due to distance.

Global distribution

Some penaeid prawns found in India (Indo-West Pacific region) are also distributed in Eastern Atlantic zoogeographic region but no species of this region is found in Eastern Pacific and Western Atlantic. The region and sub-regions wise distribution of Indian penaeids have been computed and shown in Tables 4 and 5. Of the 79 species reported here only 6 species are common with Eastern Atlantic region where they are found in the sub-region 2 (Mediterranean Sea) only. Most of the Indo-West Pacific Indian species are found in sub-region 1and4 (69 and 61 respectively). Some of the species are distributed as far as sub-regions 7, 8 and 9 [8-12].

Analysis of the Sorensen's quotient of similarity based on the distribution of species in different sub-regions of the Indo-West Pacific region (Table 6) revealed that as expected, penaeid species composition is strongly similar (>0.71) between sub-region 1 and 4. Slightly similar (0.51-0.60) species composition is seen between sub-region 1 and 2, 1 and 3 and 3 and 4. Species composition between sub-region 1 and 5, 2 and 4 and 4 and 5 are found to be moderately dissimilar. Species composition between sub-region 1 and 6, 1 and 7, 1 and 8, 1and 9 and between remaining other sub-regions it is found to be strongly dissimilar as compare to sub-region 4.

Such pattern of distribution is quite expected and can be correlated with the distance. Coastal continuity between East and West coast of India is the cause of the moderate to strongly (·673 ≈ ·7) similar species composition. Relative proximity between the East African coast and the West Coast of India has resulted into the slightly similar species composition between these two sub-regions but a moderately dissimilar species composition has been observed between East coast of India and East African coast. As such distributional tendency of Indian penaeids towards east is more than west. Only 6 species of India has been found to be distributed outside Indo-West Pacific region and that too only found in Mediterranean sea. Such distribution in all probability is a recent phenomenon and might have arisen due to the dispersal of species through Suez canals which is also suggested by Dall [3].

		Global penaeid zoogeographical area	Total No. of species	Endemic species
I.	**Indo – West Pacific**			
	Indian Ocean and Western part of Pacific Ocean are under the area		79	9
	Sub-regions:			
	1.	Indo-Malaysian, Sri Lanka, Bay of Bengal, Philippines, Gulf of Tonkin, South Taiwan, Indonesia including northern West Irian, Northern Papua, New Guinea, Solomon Islands	69	4
	2.	Tropical Australia – Torres Strait to Shark Bay. Gulf of Papua, wide Bay, Queensland.	28	
	3.	Sino-Japanese, Gulf of Tonkin to Yellow Sea and the Inland Sea of Japan	28	
	4.	Arabian Sea, West Coast of India through Arabian Sea, Gulf of Iran and Red Sea	61	5
	5.	East African Coast, Cape Guardafui to Durban	21	
	6.	South Africa, Durban to Swakopmunad	11	
	7.	Southeastern Australia. Wide Bay to Bass strait	4	
	8.	South Western Australia. Shark Bay of Vincents Gulf	5	
	9.	Oceania. South and Central Pacific islands, including the Hawaiian Islands	12	
II.	**Eastern Atlantic**			
	Eastern part of Atlantic Ocean starting from Mediterranean Sea to South Angola			
	Sub-regions:			
	1.	Eastern Atlantic, Lisbon, Portugal, 40°N, to Porto Alexandre, South Angola, 16°S.	0	
	2.	Mediterranean Sea	6	

Table 4: Distribution of Indian penaeids including endemic species in global sub-regions.

Sl. No	Name of Genera & Species	Indo-West Pacific									Eastern Atlantic	
		1	2	3	4	5	6	7	8	9	1	2
	Alcockpenaeopsis Chanda, 2016											
1.	*Alcockpenaeopsis uncta* (Alcock, 1905)	+	-	-	+	-	-	-	-	-	-	-
	Atypopenaeus Alcock, 1905											
2.	*A. compressipes* (Henderson, 1893)	+	+	-	+	-	-	-	-	-	-	-
3.	*A. stenodactylus* (Stimpson, 1860)	+	+	+	+	-	-	-	-	-	-	-
	Batepenaeopsis Chanda, 2016											
4.	*B. acclivirostris* (Alcock, 1905)	+	-	-	+	+	+	-	-	-	-	-
5.	*B. tenella* (Bate, 1888)	+	+	+	+	-	-	-	-	-	-	-
	Funchalia Johnson, 1867											
6.	*F. danae* Burkenroad, 1940	-	-	-	+	-	-	-	-	-	-	-
7.	*F. villosa* (Bouvier, 1905 b)	+	-	-	+	-	-	-	-	-	-	-
8.	*F. Woodwardi Johnson, 1867*	+	-	-	+	+	-	-	-	-	-	+
	Helleropenaeopsis Chanda, 2016											
9.	*H. cultirostris* (Alcock, 1906)	+	-	-	-	-	-	-	-	-	-	-
10.	*H. hardwickii* Miers,1878)	+	-	+	+	-	-	-	-	-	-	-
11.	*H. indica* (Muthu, 1972)	+	-	-	-	-	-	-	-	-	-	-
12.	*H. sculptilis* (Heller, 1862)	+	+	-	+	-	-	-	-	-	-	-
	Kishinouyepenaeopsis Chanda,2016											
13.	*K. cornuta* (Kishinouye, 1900)	+	+	+	+	-	-	-	-	-	-	-
14.	*K. maxillipedo* (Alcock, 1906)	+	+	-	+	-	-	-	-	-	-	-
	Megokris Pérez Farfante and Kensley, 1997											
15.	*M. granulasus* (Haswell, 1879)	+	+	-	+	-	-	-	-	-	-	-
16.	*M. pescadoreensis* (Schmitt, 1931a)	+	-	-	+	-	-	-	-	-	-	-
17.	*M. sedili* (Hall, 1961)	+	-	-	+	+	-	-	-	-	-	-
	Metapenaeopsis Bouvier, 1905											
18.	*M. andamanensis* (Wood-Mason, 1891)	+	-	+	+	-	-	-	-	-	-	-
19.	*M. barbata* (De Haan, 1844)	+	-	+	+	-	-	-	-	-	-	-
20.	*M. commensalis* Borradaile, 1898	+	+	-	+	-	-	-	-	+	-	-
21.	*M. coniger* (Wood-Mason, 1891)	+	-	+	+	-	-	-	-	-	-	-
22.	*M. ceylonica* Starobogatov, 1972	+	-	-	-	-	-	-	-	-	-	-
23.	*M. gaillardi* Crosnier, 1991	+	-	-	+	-	-	-	-	-	-	-
24.	*M. gallensis* (Pearson, 1905)	+	+	+	+	-	-	-	-	-	-	-
25.	*M. hilarula* (De Man, 1911)	+	+	-	+	+	+	-	-	-	-	-
26.	*M. mogiensis* (Rothbun, 1902)	+	-	+	+	+	-	-	-	-	-	-
27.	*M. novaeguineae* (Haswell, 1879)	+	+	-	-	-	-	-	-	-	-	-
28.	*M. palmensis* (Haswell, 1879)	+	+	+	-	-	-	-	+	-	-	-
29.	*M. philippii* (Bate, 1881)	+	-	-	+	-	-	-	-	-	-	-
30.	*M. stridulans* (Alcock, 1905)	+	-	-	+	-	-	-	-	-	-	-
31.	*M. toloensis* Hall, 1962	+	+	-	+	-	-	-	-	-	-	-
	Metapenseus Wood-Mason, 1891											
32.	*M. affinis* (H. Milne Edwards, 1837)	+	-	-	+	-	-	-	-	+	-	-
33.	*M. alcocki* George and Rao, 1968	-	-	-	+	-	-	-	-	-	-	-
34.	*M. brevicornis* (H. Milne Edwards, 1837*)*	+	-	-	+	-	-	-	-	-	-	-
35.	*M. bengalyensis* Tirmizi, 1971	+	-	-	-	-	-	-	-	-	-	-
36.	*M. dobsoni* (Miers, 1878)	+	-	-	+	-	-	-	-	-	-	-
37.	*M. eboracensis* Dall, 1957	+	+	-	-	-	-	-	-	-	-	-
38.	*M. elegans* De Man, 1907	+	-	-	-	-	-	-	-	+	-	-
39	*M. endeavouri* (Schmitt, 1926a)	+	+	-	-	-	-	-	-	-	-	-
40.	*M. ensis* (De Haan, 1844)	+	+	+	+	-	-	-	-	+	-	-
41.	*M. intermedius* (Kishinouye, 1900)	+	-	+	-	-	-	-	-	-	-	-
42.	*M. krishnatrii (*Silas and Muthu, 1976)	+	-	-	-	-	-	-	-	-	-	-
43.	*M. kutchensis* George, George and Rao, 1963	-	-	-	+	-	-	-	-	-	-	-
44.	*M. lysianassa* (De Man, 1888)	+	-	-	+	-	-	-	-	-	-	-

45.	*M. monoceros* (Fabricius, 1798)	+	-		+	+	+	-	-	-	-	+
46.	*M. moyebi* (Kishinouye, 1896)	+	-	+	+	-	-	-	-	-	-	-
47.	*M. stebbingi* Nobili, 1904	-	-	-	+	+	+	-	-	-	-	+
	Miyadiella Kubo, 1949											
48.	*M. podophthalmus* (Stimpson, 1860)	+	-	+	+	-	-	-	-	-	-	-
	Parapenaeopsis Alcock 1901.											
49.	*P. longirostris* Chanda and Bhattacharya,2002.	+	-	-	-	-	-	-	-	-	-	-
50.	*P. nana* Alcock, 1905	+	-	-	-	-	-	-	-	-	-	-
51.	*P. stylifera cochinensis* George, 1975	-	-	-	+	-	-	-	-	-	-	-
52.	*P. stylifera coromandelica* Alcock, 1906	+	-	-	+	-	-	-	-	-	-	-
53.	*P. stylifera stylifera* (H. Milne Edwards, 1837)	+	-	-	+	-	-	-	-	-	-	-
	Perapenaeus Smith, 1885											
54.	*P. fissurus* (Bate, 1881)	+	-	+	+	+	-	-	-	-	-	-
55.	*P.fissuroides indicus* Crosnier, 1985	-	-	-	+							
56.	*P. investigatoris* Alcock and Anderson, 1899	+	-	+	+	+	-	-	-	-	-	-
57.	*P. longipes* Alcock, 1905	+	-	+	+	+	-	-	-	-	-	-
58.	*P. sextuberculatus* Kubo, 1949	+	+	+	+	+	-	+	-	-	-	-
	Pelagopenaeus Perez Farfante & Kensley, 1997											
59.	*P. balboae* (Faxon, 1893)	-	-	-	+	-	-	-	-	-	-	-
	Penaeopsis Bate, 1881											
60.	*P. edwardoi* Perez Farfante, 1977	+	+	+	-	+	-	-	-	+	-	-
61.	*P. jerryi* Pérez Farfante, 1979	+	-	-	+	+	-	-	-	-	-	-
62.	*P. rectacuta* (Bate, 1881)	+	-	+	+	+	-	-	-	+	-	-
	Penaeus Fabricius, 1798											
63.	*P. canaliculatus* (Oliver, 1811)	+	+	+	+	-	-	-	-	+	-	-
64.	*P. hathor* Burkenroad, 1959	-	-	-	+	-	-	-	-	-	-	-
65.	*P. indicus* (H. Milne Edwards, 1837)	+	+	+	+	+	+	-	-	-	-	-
66.	*P. japonicus* (Bate, 1888)	+	+	+	+	+	+	-	-	-	-	+
67.	*P. konkani* (Chanda and Bhattacharya,2003)	-	-	-	+	-	-	-	-	-	-	-
68.	*P. latisulcatus* (Kishinouye, 1896)	+	+	+	+	+	+	-	+	+	-	-
69.	*P. marginatus* (Randall, 1840)	+	+	+	+	+	+	-	-	+	-	-
70.	*P. merguiensis* (De Man, 1888*)*	+	+	-	+	-	-	-	-	+	-	-
71.	*P. monodon* Fabricius, 1798	+	+	+	+	+	+	+	+	+	-	-
72.	*P. penicillatus* (Alcock, 1905)	+	-	-	+	-	-	-	-	-	-	-
73.	*P. semisulcatus* De Haan, 1844	+	+	+	+	+	+	+	+	+		+
74.	*P. silasi* Muthu & Motho, 1979	+	-	-	-	-	-	-	-	-	-	-
75.	*P. similis.(*Chanda and Bhattacharya,2002).	+	-	-	-	-	-	-	-	-	-	-
	Trachypenaeopsis Burkenroad, 1934											
76.	*T. minicoyensis* Thomas, 1972	-	-	-	+	-	-	-	-	-	-	-
	Trachysalambria Burkenroad, 1934											
77.	*T. aspera* (Alcock, 1906)	+	-	-	-	-	-	-	-	-	-	-
78.	*T. curvirostris* (Stimpson, 1860)	+	+	+	+	+	+	+	+	-	-	+
79.	*T. fulva* (Dall, 1957)	+	+	-	-	-	-	-	-	-	-	-
	TOTAL NUMBER OF SPECIES	**69**	**28**	**28**	**61**	**21**	**11**	**4**	**5**	**12**	**0**	**6**

Table 5: Zoogeographic distribution of penaeid species found in India.

***Subregion**	**1**	**4**
1		
4	.785	
3	.577	.584
2	.577	.517
5	.444	.488
9	.296	.329
6	.275	.306
8	.135	.121
7	.110	.123

*Reference table number 5.

Table 6: Sorensen's quotient of similarity (Q.S).

Acknowledgements

Author is extremely indebted to Professor Tanmoy Bhattacharya, Ex-Emeritus Professor, Department of Zoology, Vidyasagar University; Dr. Probodh Kumar Maity, Emeritus scientist-SF, ZSI, Kolkata and Dr. Tusherendu Roy, Scientist-C, ZSI, Kolkata for their endless guidance and inspiration. Author expressed his gratitude to Prof. Samir Kumar Benerjee, Prof. of Zoology, Calcutta University for his inspiration during preparation of manuscript in present format. Author is also grateful to the Director, Zoological Survey of India for granting him a fellowship during which the work has been completed.

References

1. Holthuis LB (1980) FAO Species Catalogue-Shrimps and prawns of the world. *FAO Fish Synop* (125), 1: 1-271.
2. Kurian CV, Sebastian OV (1993) Prawns and Prawn Fisheries of India. Hindustan Publishing Corporation Delhi 1-267.
3. Dall W, Hill BJ, Rothlisberg PC, Sharples DJ (1990) The biology of the Penaeidae. Adv Mar Biol 27: 1-489.
4. Alcock A (1906) Catalogue of the Indian Decapod Crustacea in the collection of the Indian Museum. Part III. Macrura. Fasciculus I. The prawns of the *Peneus* group. Indian Museum, Calcutta: 1-55.
5. Chanda A, Bhattacharya T (2002) Penaeiod shrimp of Digha and adjacent coast of Midnapore, West Bengal, India. Vidyasagar University J of Bio Sci 8: 1-22.
6. Chanda A (2014) First Record of Two Australian Species : One under Metapenaeopsis and Another under Metapenaeus from Indian Water, Their Diagnosis and Distribution. J Ento Zoo Studies 2: 18-20.
7. George MJ (1972) On the zoogeographic distribution of Indian Penaeidae. Indian J Mar Sci 1: 89-92.
8. George MJ (1979) Taxonomy of Indian prawns (Curstacea, Decapoda, Penaeidae). "In contribution to Marine Science", dedicated to Dr CV Kurian: 21-59.
9. Chanda A (2016) A Study on Newly Described Genera Alcockpenaeopsis, Batepenaeopsis, Helleropenaeopsis, Kishinouyepenaeopsis and Parapenaeopsis from Indian Water. *Poult Fish Wildl Sci*, 4: 147.
10. Chanda A, Bhattacharya T (2003) Fenneropenaeus konkani, a new species of prawn (Decapoda: Penaeidae) from Indian Coast. Science and Culture 69: 229-230.
11. Gorge MJ, Suseelan C (1982) Distribution of species of prawn in the backwaters and estuaries of India with reference to coastal aquaculture. *Proc Symp Coast Aqua* 1: 273-284.
12. Sorensen T (1948) A Method of Establishing Groups of Equal Amplitude in Plant Sociology Based on Similarity of Species Content and Its Application to Analysis of the Vegetation of Danish Commons. Biol Skr 5: 1-34.

15

Effect of Liver Biotoxins of Certain Marine Fishes on Mouse Cell Culture

Yogesh Kumar and Pani Prasad Kurcheti*

Division of Aquatic Environment and Health Management, Central Institute of Fisheries Education, (Deemed University), Mumbai-400 061, India

Abstract

Biotoxins from certain marine fishes are not only harmful to marine animals and their native environment, but also have pharmacological importance, and can be used as medicines. An investigation was carried out to study the effects of liver biotoxins of marine fish (*Chelonodon patoca*), catfish (*Osteogeneiosus militaries*), and mackerel (*Rastrelliger kanagurta*) on mouse, and their established kidney and liver cell lines. Among the three toxins, puffer liver toxin was the most toxic in mouse, with 5 mouse units/ml, followed by 2.5 mouse units/ml and 0.67 mouse units/ml of catfish and mackerel, respectively. The morphological rigidity, viability, swelling and syncitia formation, followed by cell death, are the common features recorded in both cell lines. The Tissue Culture Infective Dose ($TCID_{50}$) was the highest for puffer liver toxin in both cell lines, followed by catfish and mackerel liver toxin.

Keywords: Biotoxin; Puffer fish; Catfish; Mackerel; $TCID_{50}$

Introduction

The miraculous oceanic water world is inhabited by a large variety of aquatic animals and fishes. The fishes are the most diversified group of the animal kingdom. There are a number of marine organisms, including some group of fishes that they are dangerous and deadly poisonous. Among all poisonous and venomous fishes, puffer fishes are considered to be the most poisonous and are found in all the oceans of the world. Although these are deadly poisonous, they command the highest price as food fish in several East Asian countries, particularly in Japan. There are many marine and freshwater fishes that contain toxins in their body. The two most common marine toxins are tetrodotoxin and saxitoxins. Tetrodotoxin is present predominantly in puffer fish [1]. The toxin is concentrated mostly in ovaries and liver, with small amount in intestine, skin, muscle and blood [2].

Despite serious public health threat by poisonous fishes to the tropical and subtropical regions, very little attention has been paid to the study of toxicological aspects of different fishes. They are dangerous to several other species, including human, by virtue of their ability to bite, lacerate or sting, or toxin substances contained in their body or organs. The problem of poisonous fishes is rapidly becoming recognized as one of the significant public health hazards, and one which can't be circumvented of, if we were to harvest enormous food resources of our oceans. Some toxins extracted from these fishes have pharmacological importance. Hence, the scientists are continuously using these fishes as a valuable experimental animal/model.

Biotoxic fishes are those fishes which when digested, causes biotoxication in humans and animals due to the bioactive substance present in the fish. Fish biotoxins hold the promise of many medical compounds and substances useful for research in human and animal health problems. After intake of these fishes, humans may develop headache, vomiting sensation, and if concentration of toxin is high, death may occur. Therefore, the present investigation is to emphasize the toxicity of certain toxins extracted from different fishes, with the help of cell lines, to study the toxic effect of biotoxins extracted from liver tissues of three marine fish, like puffer fish (*Chelonodon patoca*), catfish (*Osteheneiosus militaries*), and mackerel (*Rastrelliger kanaguta*) on mouse cell lines, for development of a suitable alternate to mouse bioassay study. Moreover, this method is very sensitive and even small changes at the cellular level that may not result in the death of animal, are easily detected and demonstrated. This method prevents killing of hundreds of innocent animals for the sake of research, thereby making research more economical and ecofriendly.

Materials and Methods

Different fishes like puffer fish (*Chelonodon patoca*), catfish (*Osteheneiosus militaries*), and mackerel (*Rastrelliger kanaguta*) in a live condition were collected from Versova landing centre, Mumbai, India. The fish were maintained under laboratory conditions, with proper aeration and artificial diet.

Mice

The Swiss albino male mice (*Kasauli* strain) of 20+2 g procured from Haffkine Biopharma Corporation Ltd., Mumbai, India, were used for cell culture and mouse bioassay studies.

Extraction of toxin

The toxin was extracted from the liver tissues of the above fishes, following procedures with necessary modification [3-5]. In brief, different tissues were excised, minced and extracted, with an equal volume of 80% methanol and 1% acetic acid, in water bath maintained at 60°C for 30 minutes (min). Supernatant was collected after 30 min, and the residue was again processed twice in a similar manner. Finally, the supernatant was pooled. The supernatant was defatted with equal volume of dichloromethane. The aqueous layer was then separated and evaporated to dryness. The resulting residue was used as crude toxin.

Preparation of cell culture

Liver and kidney cell culture of mice was done, as per the procedure of Wenzel [6]. The Eagles' minimum essential medium (Sigma Aldrich

***Corresponding author:** Pani Prasad Kurcheti, Division of Aquatic Environment and Health Management, Central Institute of Fisheries Education, (Deemed University), Mumbai-400 061, India, E-mail: kpaniprasad@cife.edu.in

USA), calf bovine serum (Gibco, USA), and L-glutamine were used for the cell culture.

Tissue culture assay

Toxicity assay was done, as per the following procedure [7]. The 96 well microtitre plates were seeded with 200 µl of culture medium containing 1.0×10^5 cells/ml, and incubated for 2 days for monolayer formation. Then 10 µl of toxin adjusted to pH 7.4, pre-filter sterilized through 0.22 µm membrane, were serially diluted 10 fold with Dulbecco's PBS buffer. The study was conducted in triplicate. Cells with swollen or lost morphological rigidity were considered as dead. Atleast 200 cells in each well were randomly chosen, and the 50% tissue culture infected dose by Reed and Muench formula was calculated.

Bioassay study

The extracted crude toxin of each tissue was dissolved individually in 1 ml of distilled water, so that 1 ml of test solution was equivalent to 1 gm of tissue. The amount of toxin solution for injection and the dilution that gives death time of 1-2 min was determined, and referred to as one mouse unit of the toxin.

Results

Mouse bioassay

The number of mouse units per ml of toxin was found to be 5.0, 2.5 and 0.67 for puffer, catfish and mackerel, respectively (Table 1). The puffer liver toxin injected mouse showed paralysis of hind limbs, urination, tonic spasm, gasping, hyperirritability, sluggish moment, lying in prostate condition, and during the terminal stage, micturation and defecation were recorded; whereas the mackerel liver toxin injected mouse showed sluggish movement, shivering, gasping paralysis of hind limbs, and finally death. Similar type of symptoms as in the case of mackerel, followed by death, was recorded by catfish liver toxin.

Tissues culture assay

Table 2 indicates the $TCID_{50}$ values of puffer fish, catfish, and mackerel liver extracts on kidney and liver cell lines of mouse. Puffer liver toxin was found to be most toxic for both kidney and liver cell line, followed by catfish and mackerel toxins.

Effect of toxins on kidney cell line

After treatment with puffer toxin, kidney cells undergo morphological deformities, and began to lose rigidity, viability, and got detached from each other within 2 min (Figure 1). The cells started degeneration, syncitia formation with dispersed cytoplasm and nuclear materials after 40 min, while some cells were found to be swollen (Figure 2). The nuclear and cytoplasmic demarcation and cell death were found to be within 1 hr and 11 min (Figure 3). The $TCID_{50}$ was found to be 1.585×10^5/ml. Similarly, in case of catfish liver toxin, swelling, rounding, stretching and detaching of cells were observed after 44 min and 05 sec (Figure 4), and within 1 hr and 32 min, most of the cells degenerated with fusion of nuclear and cytoplasmic materials and syncitia formation (Figure 5). The $TCID_{50}$ was found to be 2.512×10^3/ml.

The effect of mackerel toxin was slower than the other two toxins. The fusion and swelling of cells started after 23 min (Figure 6), and after 45 min, most of the cells were clubbed and fused with degenerating cells (Figure 7). All the cells degenerated within 5 hrs, and the $TCID_{50}$ value was found to be 2.215×10^2/ml.

Dose in ml	Mean death time in hour : min : sec			Remark
	Puffer	Cat Fish	Mackerel	
0.1	00:05:15	02:40:15	05:30:05	Lethal
0.2	00:00:45	00:04:20	00:10:37	Lethal
0.3	00:00:29	00:03:46	00:05:20	Lethal
0.4	00:00:20	00:01:30	00:03:10	Lethal
0.5	00:00:32	00:02:25	00:01:04	Lethal

Table 1: Toxicity of liver toxins of three fish on mice.

Toxin	Mouse unit/ml	$TCID_{50}$ in kidney cell line	$TCID_{50}$ in liver cell line
Puffer	5	1.585×10^5	2.215×10^4
Catfish	2.5	2.512×10^3	1.585×10^3
Mackerel	0.67	2.512×10^2	1.585×10^2

Table 2: Mouse units/ml of toxin with their $TCID_{50}$ values in liver and kidney Cell lines of mouse.

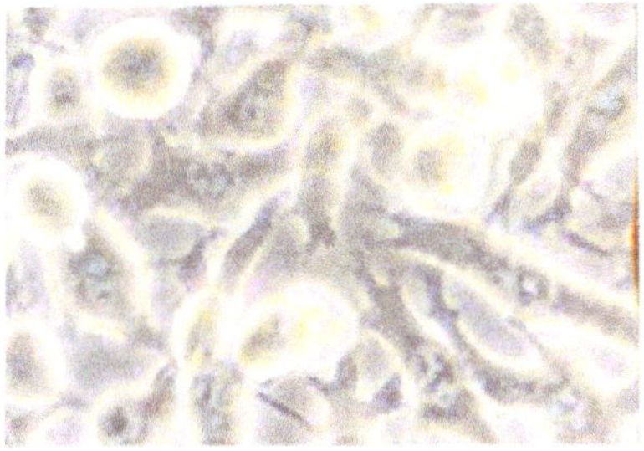

Figure 1: Mouse kidney cell line after 3 min of treatment with puffer liver toxin.

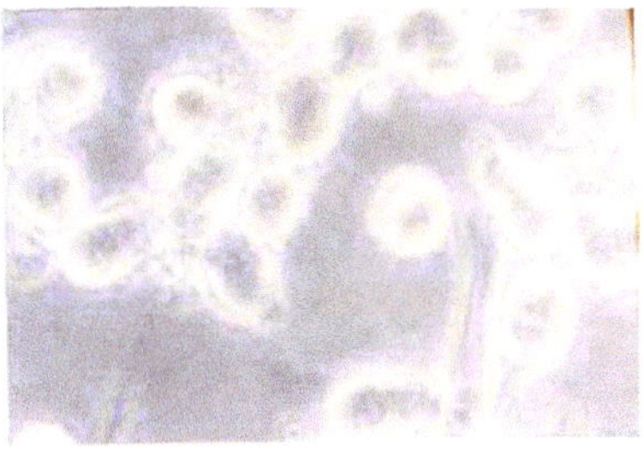

Figure 2: Mouse kidney cell line after 40 min of treatment with puffer liver toxin.

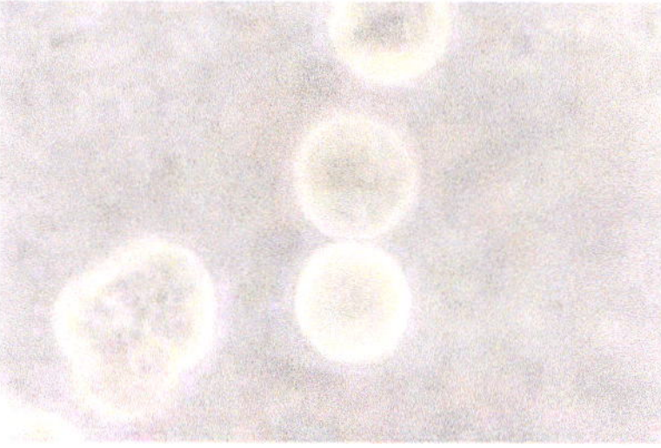

Figure 3: Mouse kidney cell line after 1 hr of treatment with puffer liver toxin.

Effects of different toxin on liver cell line

The swelling of cells and breaking of intracellular connection started after 3 min in puffer toxin, and the nuclear changes occurred within 16 min. The cells lost their morphology with degeneration of nucleus and loss of cell membrane, and bursting of cells (Figures 8 and

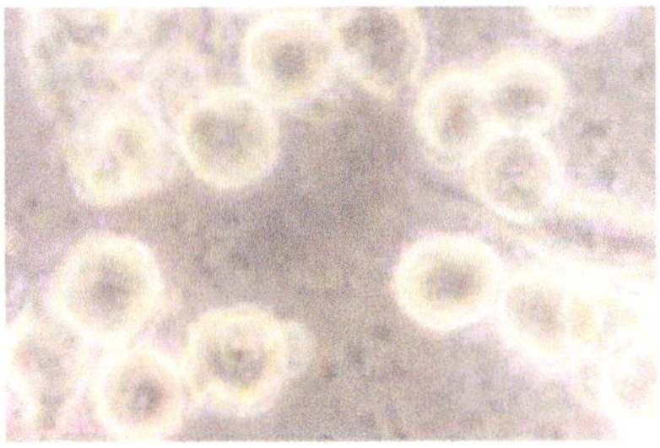

Figure 4: Mouse kidney cell line after 45 min of treatment with puffer liver toxin.

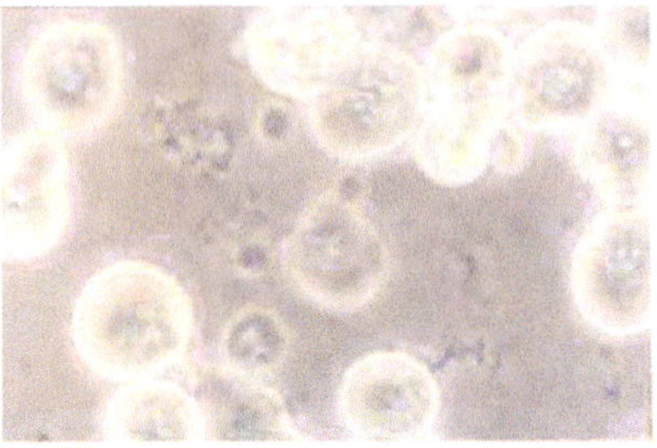

Figure 5: Mouse kidney cell line after 1 hr of treatment with catfish liver toxin.

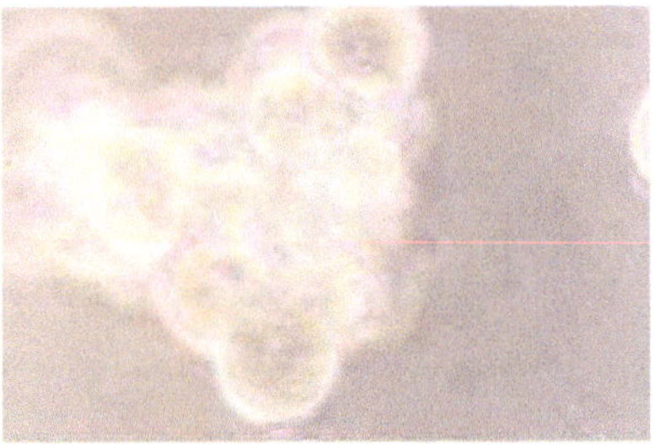

Figure 6: Mouse kidney cell line after 6 min of treatment with mackerel liver toxin.

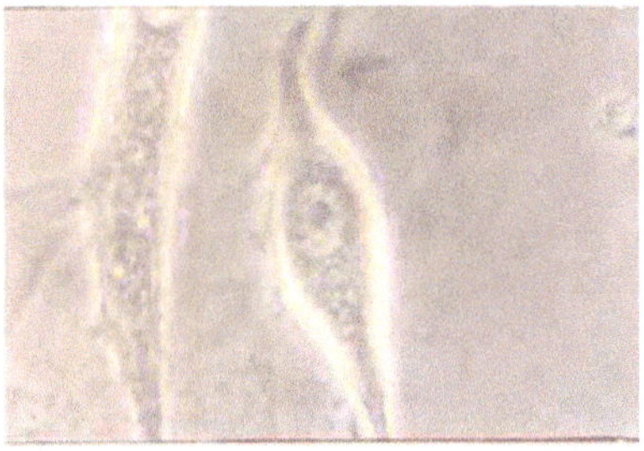

Figure 7: Mouse kidney cell line after 45 min of treatment with mackerel liver toxin.

9). After 30 min, the medium became colloidal with death of all the cells. The $TCID_{50}$ value was found to be 2.512×10^4/ml.

In the case of catfish liver toxin, morphological changes started after 6 min of toxin treatment with swollen, rounded and detached cells. Cell death was found to occur within 21 min, which was confirmed by trypan blue staining of the cells. The $TCID_{50}$ was found to be 1.585×10^3/ml, whereas the effects were slow in case of mackerel toxin, which started after 30 min of treatment (Figure 10). After 30 min of syncitia formation, degenerated cells appeared which turned the growth medium to colloidal solution. The $TCID_{50}$ value was found to be 1.585×10^2/ml.

Discussion

Marine toxins are those observed and extracted from certain marine fishes, are very complex in their nature. They have the ability to destroy life by complicated, and sometimes, undetermined means. Their investigation has invited speculation, and sometimes, pure fantasy [1].

Serious health problems are often reported due to marine and freshwater animals. As a result of which, continuous monitoring of the toxins are regularly done through mouse bioassay [8,9]. Detection of toxins by ELISA and HPLC are often restricted to sophisticated laboratory only [10].

The toxins are of public health importance, but the studies on human being are not possible. A simple and sensitive bioassay was used to detect the toxins also, to evaluate their toxicity. Thus, cell culture is

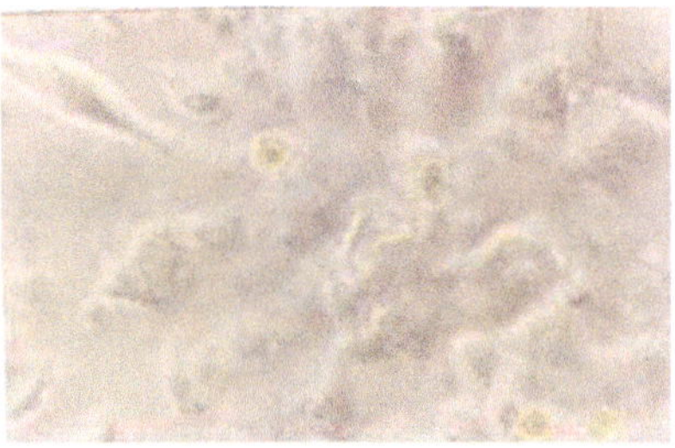

Figure 8: Mouse kidney cell line after 3 min of treatment with puffer liver toxin.

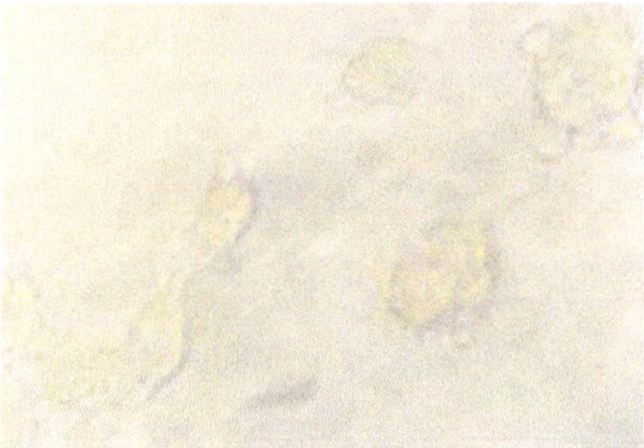

Figure 9: Mouse kidney cell line after 16 min of treatment with mackerel liver toxin.

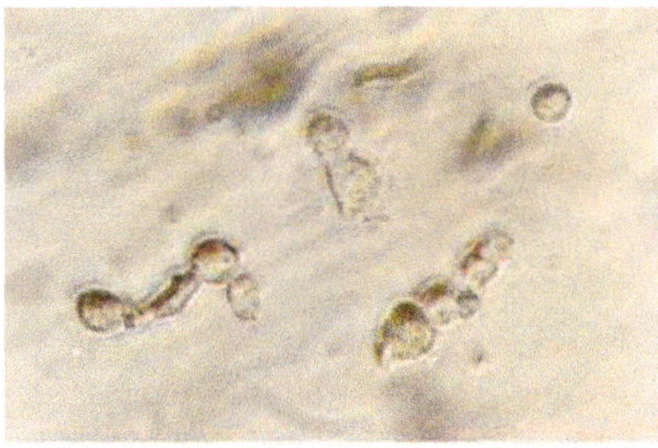

Figure 10: Mouse kidney cell line after 30 min of treatment with mackerel liver toxin.

used to evaluate the toxic effect of different fish toxins, and correlated with mouse bioassay.

The puffer liver toxin was found to be highly toxic, when compared to the other toxins.

The mean death time of one mouse unit was found to be 45 minutes and 1 ml of toxin was containing 5 mouse units. The present findings were in accordance with the result obtained by Hwang et al. [3] and Reddy [11], with subtle differences. Reddy [11] reported 6.25 mouse unit gram of puffer toxin. The puffer liver toxin was found to be most toxic in both liver and kidney cell lines of mouse. The morphological changes like swelling, syncitia, dispersed cytoplasm and nuclear material, were the common toxic effects.

Pharmacological studies on puffer poison were initiated by Murtha et al. [12]. According to them, a small dose of puffer poison can cause neuromuscular paralysis, respiratory failure, and marked hypertension to cat, dogs and rabbits. More studies have been done by various workers [13-17].

The result of mouse bioassay and Culture Infective Dose ($TCID_{50}$) can be correlated. 5 mouse unit/ml of puffer liver toxin was equal to 1.585×10^5 to 1.585×10^4 $TCID_{50}$ on mouse kidney and liver cell lines. Similarly, 2.5 mouse unit/ml of catfish liver extract was equal to 2.512×10^3 and 1.585×10^3 $TCID_{50}$ of kidney and liver cell lines respectively (Table 2).

It is evident from this study that use of cell lines for toxicological studies is more sensitive and less quantity of sample is required for the assay. The changes at the cellular level were also studied simultaneously, which was possible by mouse bioassay studies. In conclusion, the morphology and viability changes induced in the mouse liver and kidney cells provided a new sensitive bioassay for puffer liver toxin, catfish liver toxin, and mackerel liver toxin.

Many researchers have worked on the effect of the marine toxins on cell lines *viz.* Pichardo et al. [18] used fish cell lines RTG-2 and PLHC-1, to compare the toxic effects produced by microcystins LR and RR of different concentrations. The embryotoxic effects of the marine biotoxin, okadaic acid on murine embryonic stem cells, was determined by Ehlers et al. [19]. No research has been undertaken to compare the mouse bioassay and $TCID_{50}$ values.

Finally, it could be concluded that comparing and correlating the mouse bioassay and $TCID_{50}$ values, the effect or lethality of toxins would be calculated as cell lines are sensitive and expensive. A large number of samples could be detected and quantified within a limited time. Also, the use of laboratory animals like mice can be minimized in toxicological research, and this assay is recommended for detecting and quantifying these toxins. The present study confirmed the more toxic nature of puffer liver toxin, followed by catfish liver toxin, while mackerel was found to be least toxic.

Acknowledgements

The authors are thankful to the Director, Central Institute of Fisheries Education, for his kind support and cooperation during the entire course of investigation.

References

1. Russell FE (1965) Marine toxins and venomous and poisonous marine animals. Adv Mar Biol Academic Press, London, New York, USA.
2. Halstead BW, Courville DA (1965) Poisonous and venomous marine animal of the world. U.S. Government Printing Office, Washington DC, USA.
3. Hwang DF, Noguchi T, Arakawa O, Abe T, Hashimoto K (1988) Toxicological studies on several species of puffer in Taiwan. Bulletin of Japanese Society of Scientific Fisheries 54: 2001-2008.
4. Hwang DF, Chuch CH, Jeng SS (1990) Susceptibility of fish crustacean and molluscs to tetrodotoxin and paralytic shellfish poison. Bulletin of Japanese Society of Scientific Fisheries 56: 337-343.
5. Sato S, Komaru K, Ogata T, Kodama M (1990) Occurrence of tetrodoxin in cultured puffer. Bulletin of Japanese Society Fisheries 56: 1129-1131.
6. Wenzel P (1970) Establishment of cell culture model with heart obtained from 2-4days old mice. Methods in Toxicology, Academic Press Inc, USA.
7. Cunnighum CH (1966) A laboratory guide in virology. (4th edn), Minneapolis Minn, Burgees Publishing Company, USA.
8. Evans MH (1972) Tetradotoxin, saxitoxin and related substances: their applications in neurobiology. Int Rev Neurobiol 15: 83-166.
9. Kawabata T (1978) Puffer toxin. The manual for the methods of food sanitation test. Food Hygiene Association, Tokyo, Japan 2: 232.
10. Davio SR, Hewetson JF, Beheler JE (1985) Progress towards development of monoclonal antibodies to saxitoxin: antigen preparation and antibody detection. In: Toxin Dinoflagellats, Elsevier, New York, USA.
11. Reddy BJC (1997) Variation in the space in the toxicity of the puffer *Lagocephalus linaries* of Mumbai coast. M.F.Sc. Dissertation. Submitted to Central Institute of Fisheries Education, Mumbai, India.
12. Murtha EF, Stabile DE, Wills JH (1958) Some pharmacological effect of puffer poison. J Pharmacol Exp Ther 122: 247-254.
13. Kao CY (1966) Tetrodotoxin, Saxitoxin and their significance in the study of excitation. Pharmacol Rev 18: 997-1049.
14. Russell FE (1967) Comparative pharmacology of some animals toxins. Fed Proc 26: 1206-1224.
15. Mardersin AD (1968) Drigs from the sea. Marine Technology Society, Washington 317.
16. Halstead BW (1965) Poisonous and venomous marine animals of the world. Vol. I Invertebrates. U.S. Government Printing Office, Washington, USA.
17. Venkateshvaran K, Pani Prasad K (1997) Comparative pharmacology of marine biotoxins. Advances technology in marine biotoxicology. CAS in Fishery Science, Central Institute of Fisheries Education, Mumbai, India.
18. Pichardo S, Jos A, Zutira JL, Salguero M, Camean AM, et al. (2005) The use of the fish cell lines RTG-2 and PLHC-1 to compare the toxic effects produced by microcystins LR and RR. Toxicol *In Vitro* 19: 865-873.
19. Ehlers A, Stempin S, Al-Hamwi R, Lampen A (2010) Embryotoxic effects of the marine biotoxin okadaic acid on murine embryonic stem cells. Toxicon 55: 855-863.

A Submersible Holographic Microscope for 4-D In-Situ Studies of Micro-Organisms in the Ocean with Intensity and Quantitative Phase Imaging

Lina M. Rotermund[1], John Samson[2] and Hans Jürgen Kreuzer[3]*

[1]Department of Physics & Atmospheric Science, Dalhousie University, Halifax, NS B3H 4R2, Canada
[2]4Deep Inwater Imaging, 6589 Chebucto Rd, Halifax, NS B3L 1L9, Canada
[3]Department of Physics & Atmospheric Science, Dalhousie University, Halifax, NS B3H 4R2, Canada

Abstract

Digital holographic reconstructions yield the complex wave amplitude throughout a 3-D volume from which intensity and quantitative phase images can be obtained. A submersible digital holographic microscope with a point source has been used to demonstrate the feasibility of continuous monitoring of micro-organisms such as algae at any depth in oceans, lakes and rivers. The local refractive index can be obtained throughout the organisms. A tracking algorithm is used to study their mobility.

Keywords: 4-D imaging; In water; Intensity images; Quantitative phase imaging

Introduction

With the steady increase of aquaculture sites in coastal areas monitoring water quality becomes mandatory to avoid pollution and resulting economic setbacks. Standard procedure to monitor water quality is to take samples which are then analyzed in the lab for chemical pollutants and for micro-organisms. This procedure is not only time consuming and costly but also leads to delays for any response to correct difficulties arising. This could be avoided by continuous in-situ remote monitoring. An instrument to do this for micro-organisms has been developed and is successfully used at the research level, namely the submersible holographic microscope [1] that not only is capable of registering and identifying micro-organisms and microscopic solids but also record variations as a function of time. It can be used at shallow depths but also at great depth such as in an exploration down to 6000 m in the South Atlantic [2], and even under extreme conditions such as the high Arctic [3]. In this feasibility study we have used such a microscope over the summer of 2015 to record micro-organisms in the North West Arm of Halifax Harbour in Nova Scotia, Canada. The aim is to show that this technique can be used for continuous in-situ monitoring of oceans, rivers and lakes.

Materials and Methods

A brief introduction to digital in-line holographic microscopy (DIHM) with a point source is in order. Illumination is done with a point source which is either a pinhole illuminated by a laser or the end of a fibre optic cable. Both must be of the order of the wavelength of the laser so that the emanating waves are spherical as sketched in Figure 1. These waves travel through the volume to be imaged and scatter off any objects therein such as algae or oil drops, etc. The interference of the scattered waves with the primary wave from the point source creates an interference pattern, called a hologram that is recorded by a CCD chip. This hologram has no resemblance to the objects imaged. Therefore, in a second step the hologram is used in numerical reconstruction to produce a 3D image of all the objects in the volume illuminated. The reconstruction is based on the Helmholtz-Kirchhoff transform

$$K(\mathbf{r}) = \int_{screen} d^2\xi \tilde{I}(\xi) \exp\left(ik\mathbf{r} \cdot \frac{\xi}{|\xi|} \right) \qquad (1)$$

The screen is located at coordinates $\xi = (X,Y,L)$, where L is the distance from the source to the screen centre, $k = 2\pi/\lambda$ is the wave number of the radiation and $\tilde{I}(\xi)$ is the (contrast) hologram. The algorithm is explained in detail in recent reviews, covering both the technical details and wide range of applications [4-6] and, in a second review [7], applications in the marine environment are discussed. What is important is the fact that K is a wave amplitude so that we obtain both intensity images, $I(r) = KK^*$, and phase images, $\varphi(r) = \tan^{-1}(im(K)/$

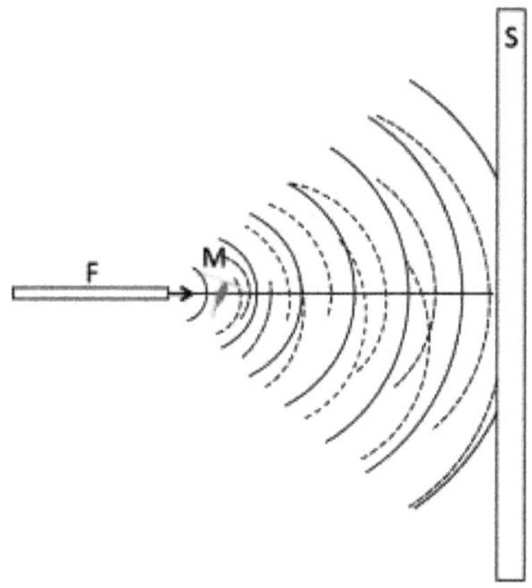

Figure 1: Schematic of (digital) holographic microscopy with a point source. The screen is a CCD chip connected to a computer for data transfer, reconstruction and further data analysis.

***Corresponding author:** Hans Jürgen Kreuzer, Department of Physics & Atmospheric Science, Dalhousie University, Halifax, NS B3H 4R2, Canada
E-mail: h.j.kreuzer@dal.ca

real(*K*)) where real(*K*) and *im*(*K*) are real and imaginary parts of *K*. As shown elsewhere [8] DIHM can determine the local phase throughout the object quantitatively. We will rely heavily on this duality of intensity and phase images. The schematics of the submersible microscope[1] are shown in Figure 2 together with an instrument.

The left pressure chamber contains the laser and fibre optic cable from which light shines through a window into the gap between the two cylinders which together with the light scattered by the objects enters through a window into the right chamber and is recorded by a CCD chip. The data are either recorded on a hard drive or sent via an Ethernet cable ashore for further processing. Software packages are available not only for the reconstruction of the images but also for data analysis such as obtaining the size of the objects and their number both of which are recorded automatically and continuously on histograms as a function of time. The overriding advantage of this instrument is that all objects in the volume illuminated by the light source are recorded whereas a conventional microscope can only image objects in a narrow focal plane of typically less than 10 μm for a high resolution instrument. The DIHM instrument has a spatial resolution of better than 1 μm if a blue laser is used, and a depth of field of several millimeters. This is achieved by two tricks: (1) because the holograms are formed in water the wavelength of the laser is reduced [9] from $\lambda = 0.408\mu m$ to $\lambda/n_w = 0.3\ \mu m$. (2) As shown elsewhere [10] two deconvolutions are used, one for the finite size of the source and a second for the finite size of the CCD chip. The temporal resolution is controlled by the frame rate of the CCD chip. The software also generates movies to study the motion of the micro-organisms in 3-D. This method of quantitative phase measurements has been used successfully to study cell death [11]. The instrument used here has several advantages: (1) it is a submersible microscope that allows in-situ, real time experiments. (2) the simplicity of the optical components (fibre optic cable, laser, CCD chip) with no lenses or mirrors as in off-line holography guarantee robustness of the instrument, and (3) the software allows real time reconstructions in a sub-second time frame.

Results and Discussion

This paper presents a brief survey of the microorganisms found in the North West Arm of Halifax Harbour using digital in-line holographic imaging. Most of the images were obtained with the submersible DIHM but some experiments were also done with the desktop version using water samples from the same location.

To demonstrate the quality of intensity images we show in Figure 3 a short chain of the diatom *Chaetoceros socialis* in panel A and a fresh water organism, *Dynobyrn duvergens*, in panel B. The latter were very abundant after heavy rains and were most likely washed into the ocean by run-off.

To examine the diatom further we show in Figure 4 its phase image. Two things are worth noticing: (1) the part that was black, i.e. not visible in the intensity plot can be seen in the phase image but only their outline indicating that they are empty of any material except for water. A suggestion would be that they are dead; yet they hold the whole structure together. In the enlargement of panel B we can also see the network of *setae* or *cileae*. To obtain further information we make a cross cut through one *cilea*, the short black line, depicted in panel C. The bell-like shape suggests that the cross section of the *cilea* is circular which results in a maximum phase shift relative through the center to the surrounding seawater

$$\Delta\phi = 2\pi\Delta n\frac{d}{\lambda} \qquad (2)$$

Here Δn is the difference in the refractive index of the *cilea* relative to water and *a* is its diameter. This shows clearly that our measurement of the index of refraction is a microscopic implementation of the Jamin-Lebedeff interferometer where the reference "beam" is simply that portion of the illuminating light that does not cross the sample [12]. The result is a local measurement of the index of refraction throughout the sample at the micron scale. A fit to the cross section gives *d* approximately 7 μm and Δn approximately 0.004 i.e. a refractive index of the *cilea* not much larger than water.

The cut through the cell in panel D is quite revealing. The cell is

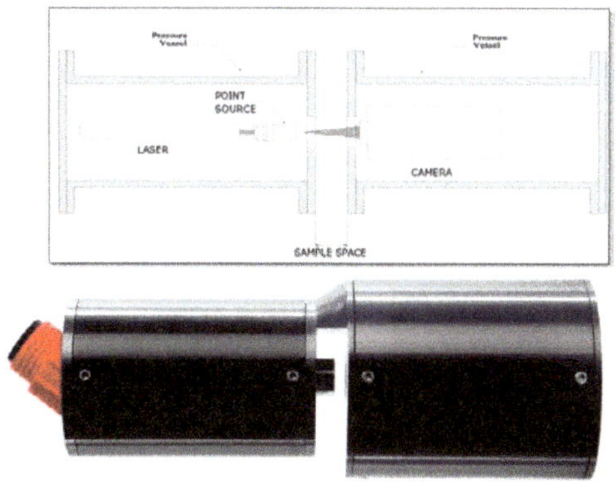

Figure 2: The submersible holographic microscope, courtesy of 4Deep Inwater Imaging.

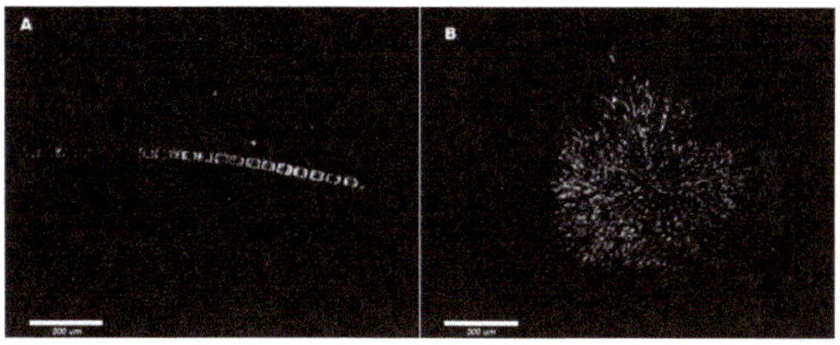

Figure 3: Intensity reconstruction of A) a chain of *Chaetoceros* sp., B) *Dinobyron divergens*.

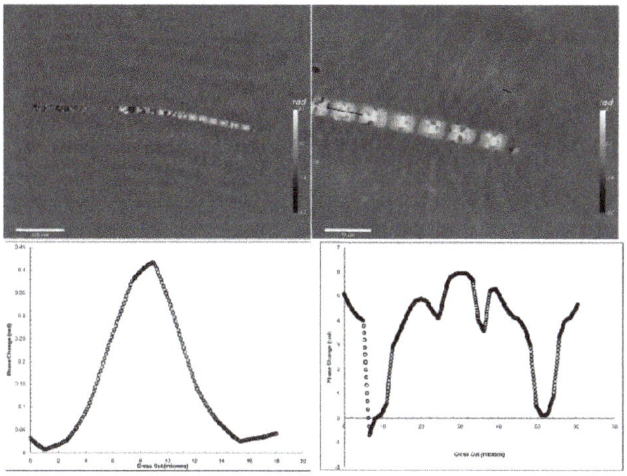

Figure 4: Phase image of a diatom *Chaetoceros socialis* and cuts through a *cilea* and the diatom.

roughly 40 μm long. The dips to zero phase shift (relative to water) is the gap between two cells which is about 4 μm as one would expect for a van der Waals attraction through a water layer of that thickness. The big peak in the middle is most likely the nucleus of the cell and the two smaller peaks are perhaps oil drops. The cell wall or frustule made of SiO_2, shows up as the break in the flanks of the profile.

Going to the macroscopic scale we show in Figure 5 large aggregates of cells abundant in early spring. The large setae or *cilea* of the diatoms are capable of collecting debris (the bright spots surrounding the organisms). Their purpose seems to be defensive i.e. to irritate the gills of larger marine organisms.

To demonstrate the use of both intensity and phase images we show examples for three organisms in Figure 6.

To dramatize phase imaging further we show in Figure 7 the phase image of Copepod Nauplius in a 3-D rendering which is a true representation of the optical thickness of the species. Of interest is the fact that the back of the organism is of the same (optical) thickness as the *cilea*.

Our last demonstration is concerned with the 4-D capability of DIHM. By taking a sequence of holograms we can follow the swimming motion using three approaches. We can simply show the reconstructed intensity or phase images at successive times as fast as the frame rate of the CCD chip which can be shown as a movie or as a collection of still pictures. An example is given in Figure 8 for a Copepod.

A second approach is to construct a composite hologram in which many holograms taken at different times are "added". Technically this is possible because holograms are linear in the scattering amplitude and thus the superposition principle holds. *Note that classical scattering is quadratic in the amplitude and superposition of exposures leads to a mess.* The "addition" works as follows: Take holograms h_1, h_2, h_3, h_4,.... at times t_1, t_2, t_3, t_4,.... and then combine these holograms as $(h_1-h_2)+(h_3-h_4)$ +....This packages n holograms into one of the same size. In addition to reducing the size this procedure also has the advantage of eliminating any static features that are not involved in the motion. Standard reconstruction will then produce the 3-D trajectories of many objects simultaneously. The trajectory of a rotifer is shown in Figure 9.

The actual measurement can be done in the Octopus software as shown in Figure 10.

From the length scale known in the reconstruction and the frame rate of the CCD camera we can estimate the swimming speed to be of the order of $v \sim 300$ μm/s. Because we are in the low Reynolds number regime where frictional forces dominate we can go further and equate the kinetic energy to the frictional energy $mv^2/2 = F_{fr}d$ where $F_{fr} = 3\pi\eta dv\lambda$ is Stokes' drag force. If we approximate the cell by a sphere of equal volume we have $\lambda = 1$ and if we approximate it by two touching

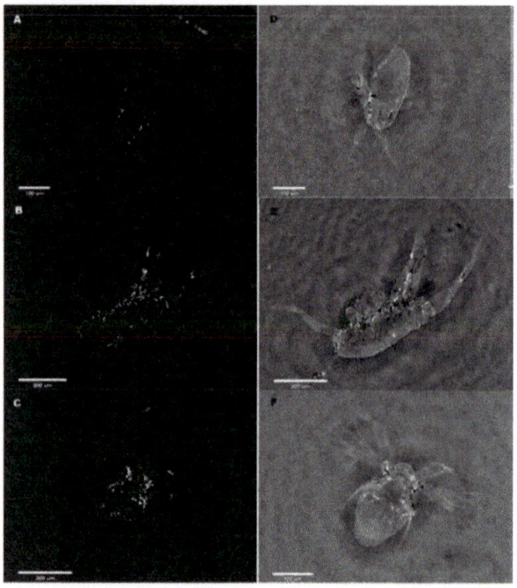

Figure 6: A) Nauplius of Copepod B) Amphipod C) Nauplius of Copepod. A-C is intensity reconstructions while D-F is their respective phase reconstructions.

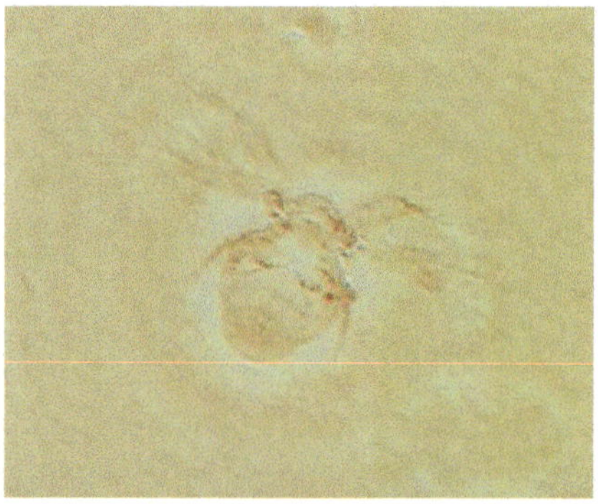

Figure 7: 3-D rendering of the phase image of Copepod Nauplius. The height is a measure of the optical thickness of the organism.

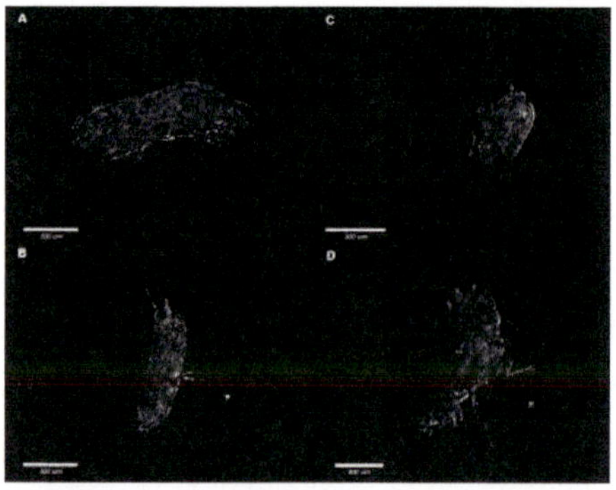

Figure 8: Snapshots of intensity reconstructions of a single Copepod (order: Harpacticoida) at time intervals of 0.12 sec.

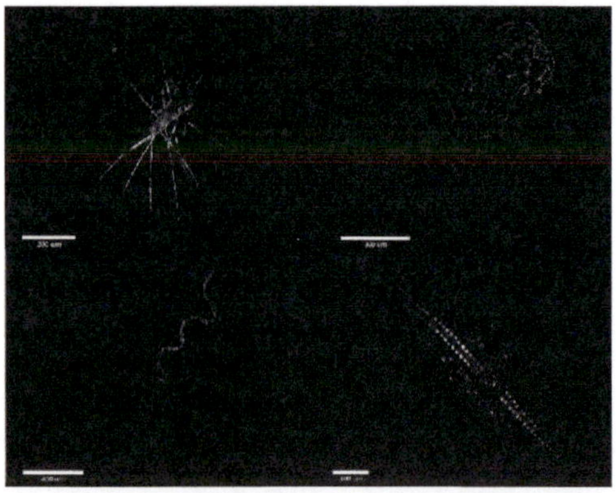

Figure 5: Intensity reconstructions of 4 different species of *Chaetoceros*.

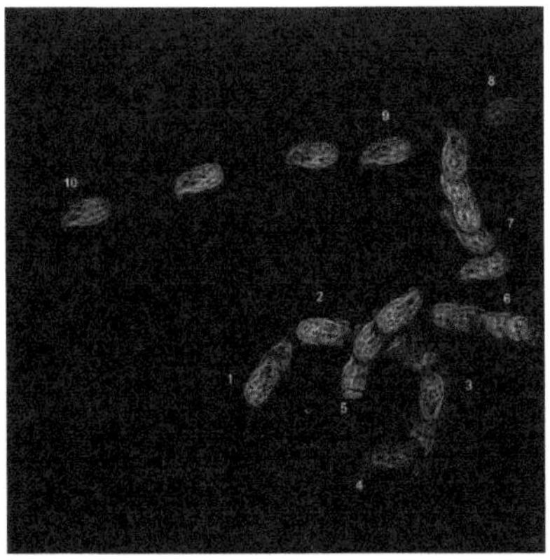

Figure 9: Snapshots, 1/15 sec apart, for a swimming rotifer.

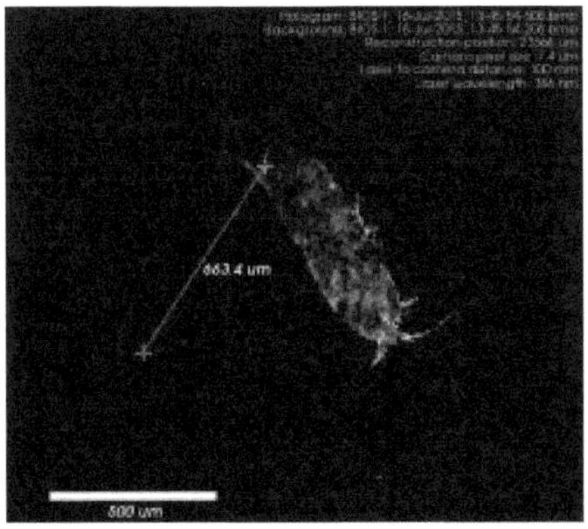

Figure 10: Measuring the distance between two positions of the Copepod of Figure 8 0.3 sec apart.

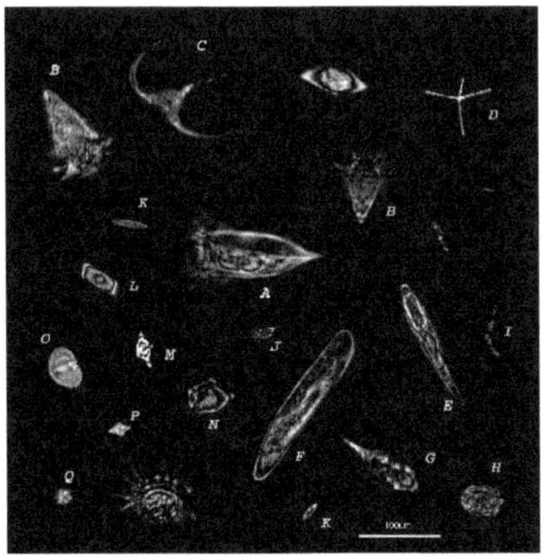

Figure 11: A gallery of marine and some freshwater organisms taken with the submersible DIHM and identified from the image: A: Marine ciliate Favella; B: Marine ciliate *Strombidium conicum*; C: *Dinoflagellate Ceratium* sp.; D: Diatom; E: Rotifer; F: Paramecium; G: Rotifer; H: Ciliate Didinium; I: Swimming ciliate; J: Ciliate; K: Marine pennate diatom; L and M: Pleurosigma/ Gyrosigma sp.; N: Marine silicoflagellate Dictyocha; O: Centric diatom; P: Dinoflagellate (unknown Peridiniale).

spheres it is $\lambda = 1.3$. η is the viscosity of water. Likewise, the power needed to maintain a constant swim speed is this energy divided by the time it takes to swim the distance d. More on such analyses can be found elsewhere for the swimming behaviour of dinoflagellates [13].

Conclusions

The purpose of this paper is to demonstrate that the submersible holographic microscope is the ideal tool to not only image micro-organisms in-situ at any depth of water in oceans, rivers and lakes but to collect in-situ microscopic structural data on the organisms by also generating, from the same set of holographic data, phase images that by applying cross sectional cuts (in the data and NOT of the organism!) from which internal structure is revealed that previously could only be obtained by sophisticated, i.e. expensive and time-consuming methods like x-ray or electron microscopy. As we hope to have demonstrated the resolution is of the order of microns. To do better than that a high resolution desktop holographic microscope can be used to image organisms still alive and swimming in a water sample. Finally we show a gallery of micro-organisms in Figure 11.

Seawater samples were looked at in the lab and in situ measurements were taken. Up to 15 types of microorganisms were tentatively identified, however some were left unknown. Many more organisms (up to a few hundred) within the raw data were left out due to space and time limitations. While individual identification may be in error, it is impressive that each image has superb resolution showing minute features such as antennae, setae, ciliae and tails. Furthermore, there are 3 videos in the supplementary information showing time sequences of A) scallop larvae swimming, B) a Copepod and C) a Ceratium in the tidal current. In the Appendix we show a few more examples of micro-organisms found in our survey. Supplementary information accompanies the manuscript on the Light: Science & Applications website (http://www.nature.com/lsa/).

Acknowledgements

This work was supported by grants (HJK) from the Natural Sciences Research Council of Canada and from the Office of Naval Research, Washington DC. LMR would like to thank the Larouche lab at Dalhousie University for allowing John and I to use their submersible DIHM.

References

1. http://4-deep.com/
2. Bochdansky AB, Jericho MH, Herndl GJ (2013) Development and deployment of a point-source digital inline holographic microscope for the study of plankton and particles to a depth of 6000 m. Limnol Oceanogr Methods 11: 28-40.
3. Jericho SK, Klages P, Nadeau J, Dumas EM, Jericho MH, et al. (2010) In-line digital holographic microscopy for terrestrial and exobiological research. Planet Space Sci 58: 701-705.
4. Xu W, Jericho MH, Meinertzhagen IA, Kreuzer HJ (2001) Digital in-line holography for biological applications. Proc Natl Acad Sci 98: 11301-11305.
5. Garcia-Sucerquia J, Xu W, Jericho SK, Klages P, Jericho MH, et al. (2006) Digital in-line holographic microscopy. Appl Optics 45: 836-850.
6. Jericho MH, Kreuzer HJ (2011) In: Ferraro P, Wax A, Zalevsky Z (eds.) Point Source Digital In-line Holographic Microscopy. Coherent Light Microscopy. Berlin Heidelberg: Springer-Verlag 46: 3-30.
7. Jericho SK, Jericho MH, Kreuzer HJ (2012) In: Emmanuel Reynaud (ed.) Holographic Microscopy of Marine Organisms. Imaging Marine Life: Modern Imaging Techniques in Marine Biology. Weinheim: Wiley-VCH 48-66.
8. Jericho M, Kreuzer HJ, Kanka K, Riesenberg R (2012) Quantitative Phase and Refractive Index Measurements with Point Source Digital In-line Holographic Microscopy. Appl Optics 51: 1503-1515.
9. Garcia-Sucerquia J, Xu W, Jericho MH, Kreuzer HJ (2006) Immersion digital in-line holographic microscopy. Opt Lett 31: 1211-1213.
10. Nickerson BS, Kreuzer HJ (2013) Deconvolution for digital in-line holographic microscopy. Adv Opt Techn 2: 337-344.
11. Missan S, Hrytsenko O (2015) Quantitative phase contrast imaging with digital inline holographic microscopy for viability assessment of mammalian cells in culture. Proc SPIE 9336: 93361-933615.

12. Leertouwer HL, Wilts BD, Stavenga DG (2011) Refractive index and dispersion of butterfly chitin and bird keratin measured by polarizing interference microscopy. Opt Express 19: 24061-24066.

13. Lewis NI, Cembella AD, Xu W, Jericho MH, Kreuzer HJ (2006) Swimming Speed of three species of the marine dinoflagellate Alexandrium as determined by Digital In-line Holography. Phycologia 45: 61-70.

17

Establishment and Characterization of a Novel Kidney-cell Line from Orange-spotted Grouper, *Epinephelus coioides*, and its Susceptibility to Grouper Iridovirus

Sue-Min Huang[1,2], Chien Tu[1], Shu-Ting Kuo[1], Hung-Chih Kuo[3], Chi-Chung Chou[4] and Shao-Kuang Chang[2]*

[1]*Animal Health Research Institute, Council of Agriculture, Taipei, Taiwan*
[2]*Institute of Veterinary Medicine, School of Veterinary Medicine, National Taiwan University, Taipei, Taiwan*
[3]*Department of Veterinary Medicine, National Chiayi University, Chiayi, Taiwan*
[4]*Department of Veterinary Medicine, National Chung-Hsing University, Taichung, Taiwan*

Abstract

A new continuous cell line, designated as GK-7, was developed from the kidney tissue of the marine grouper, *Epinephelus coioides*. The cell line grew well in Leibovitz's L-15 medium supplemented with 15% fetal bovine serum at a range of temperatures from 20 to 32°C, with optimal growth at 25°C. Morphologically, the GK-7 cell line is spindle-shaped epithelial-like cell comfirmed by immunophenotyping with cytokeratin antibody. Chromosome number analysis showed that GK-7 cells of the 50th and 150th cell passages had a modal diploid chromosome number of 48 and 66, respectively. Replication of GIV with the cell line showed that the maximum virus yield reached up $10^{8.4}$ $TCID_{50}$ mL^{-1} with the cells of the 50th passage. Electron micrographs showed abundant cytoplasmic icosahedral virions with a mean diameter of 200 nm in virus-infected cells. Negative staining of ultrathin sections of infected cells showed three-layered membrane enveloped mature viral particles with a diameter about 240 nm. Green fluorescent protein can be expressed in both cell lines at 48 hr after the cell lines were transfected with a green fluorescent reporter gene driven by a cytomegalovirus promoter. Our results showed that the GK-7 cell line provided valuable tools for the isolation and investigation of fish iridovirus and for vaccine production.

Keywords: Fish cell line; *Epinephelus coioides*; iridovirus correspondence

Introduction

Grouper, *Epinephelus spp.*, is one of the most important maricultured fish with great economic value, especially in south-east Asian countries. In Taiwan, the annual cultured grouper production is around 4,500 tons [1-6]. In the last 10 years, iridovirus and nervous necrosis virus (NNV) impacted the grouper industry severely and caused heavy economic losses. Developing an ideal cell line for *in vitro* studies of cell-virus interactions, virus propagation, isolation and vaccine development is critical to disease control in aquaculture. Taiwanese iridovirus isolates consist of three major genotypes : grouper iridovirus (GIV) which belong to *Ranavirus*, red sea bream iridovirus (RSIV), and infectious spleen and kidney necrosis virus (ISKNV) that are classified to *Megalocystivirus* genuses [7]. Hence, development a continuous cell line from tropical fish species for identifying pathogenesis of viral diseases and for vaccine production is imperative. To date, more than 22 cell lines have been developed from grouper for iridovirus and NNV isolation, such as seven cell lines established from eye, fin, heart, swimbladder, kidney, liver and brain of *E. awoara*; four from brain, fin, eye and spleen of *E. coioides* [3,8-23]; five from snout, spleen, live, kidney and swimbladder of *E. akaara* [4,9,15,24,25]; three from brain, gill, heart of *E. quoyanus* [10]; one from heart of *E. lanceolatus* [5]; one from embryo cells of *E. tauvina* [17]; and one from fin of *E. fuscoguttatus* [22]. Although four cell lines derived from *E. coioides*, have been documented, no cell line has been derived from kidney of the species. Fish spleen and kidney are important organs for virus infection, particularly for grouper iridovirus infection [6]. Availability of the cell lines from these tissues would be helpful for providing essential information on virus pathology and vaccine development studies. In this study, we established and characterized a new cell line from kidney tissue of orange spotted grouper. Susceptibility of a new cell line to GIV was confirmed by cytopathic effects (CPE) observation, virus titer determination, electron microscopy and comparison the infection kinetics of GIV replication in earlier and later cell passages in the routine culture system.

Materials and Methods

Primary cell culture

Healthy orange spotted grouper weighing 250 g were obtained from fish farm and maintained in laboratory aquaria equipped with seawater recirculation for 7 days. The fish were anesthetized with an overdose of MS-222 (Sigma-Aldrich, St. Louis, MO) and swabbed with 70% ethanol. The spleen, gill, kidney and fin were immediately removed and washed sequentially in phosphate-buffered saline (PBS) containing 5% chlorex for 10 secs, in 0.1% iodine solution for 2 min, and in antibiotic-antimycotic solution (Gibco, Grand Island, NY) for 1 hour. After washing, the tissues were minced with scissors and transferred to a 60 mm diameter tissue culture dish containing 10 ml of 0.25% trypsin-EDTA solution (Gibco) with a magnetic stirrer for gently digesting and then filtered through a 100 µm nylon mesh. After centrifugation at 1200 rpm for 15 min at 4°C, the filtered cell pellet was resuspended in Leibovitz's L-15 (Life Technologics, Carlsbad, CA) cell

***Corresponding author:** Chang SK, Institute of Veterinary Medicine, School of Veterinary Medicine, National Taiwan University, Taipei 10617, Taiwan
E-mail: changsk@ntu.edu.tw

culture medium containing 20% fetal bovine serum (FBS, Gibco), 200 IU mL^{-1} penicillin and 200 μg mL^{-1} streptomycin. The cells were seeded into 25 cm^2 tissue culture flasks and incubated at 25°C.

Subculture maintenance and cryopreservation

When the primary cell cultures were grown to a complete monolayer, cells were washed with 0.25% trypsin solution and subcultured in fresh L-15 medium containing 20% FBS, 200 IU ml^{-1} penicillin and 200 μg mL^{-1} streptomycin. For the first 10 subcultures of the primary cells, the 50% culture medium was replaced with fresh medium. After the 10 passages, the concentration of FBS was decreased from 20% to 10% for the following 50 subcultures and the antibiotics were reduced to normal concentrations (100 IU ml^{-1} penicillin and 100 μg mL^{-1} streptomycin, Gibco). The subcultures were cultured in the growth medium at a split ration of 1:3 until 150 passages for the experiments. The subcultures were stored in liquid nitrogen after every 10th passage in the freezing medium, which consisted of FBS containing 10% dimethyl sulphoxide. Briefly, 2-day-old subcultures of the cells were harvested by centrifugation and suspended in the freezing medium. The cell suspension were transferred to a cryovial (Nunc) and kept at -80°C for 48 hrs, and then stored in liquid nitrogen (-196°C). For revival, the cryovial was thawed quickly in a water bath at 25°C and removal of the freezing medium by centrifugation, then the thawed cells were transferred to a 25-cm^2 cell culture flasks with 5 ml culture medium. The viability was tested by a haemocytometer counting after trypan blue staining.

Effect of temperature and FBS concentration on cell growth

Optimal temperature and FBS concentrations for the *E. coioides* kidney (designated as GK-7) cell line was studied at the 50 passages. Briefly, cells were seeded onto 25-cm^2 cell culture flasks with 1×10^5 cells per flask and incubated at 25°C for more than 2 h to allow the cell to attach. The flasks with the attached cells were incubated at 20, 25, 28 and 32°C and observed on a daily basis. Similar procedures were performed for determining the effects of various concentrations of FBS (2%, 5%, 10% and 15% FBS) at 25°C. According to the above results, we also investigated the cell growth in L-15 containing 10% FBS at 25°C in 50th and 150th different passages. Cells from each flasks were trypsinized, and counted by three independent hemocytometer counts every three days. The experiments were carried out for 15 days. The mean number of cells and standard deviations (SD) were calculated.

Chromosome number analysis

Chromosome numbers of the GK-7 cells were determined at 50th and 150th passages. The cells were incubated in a 25-cm^2 tissue culture flask until 70% confluent and then treated with 0.2 μg ml^{-1} colcemid (Sigma) for 5 h at 25°C, allowing the cells arrested at metaphase. After gentle pipetting, the cells were collected by centrifugation at 200 *g* for 5 min at 4°C and resuspended in hypotonic 0.75% KCl solution for 30 min. The resuspended cells were then fixed in a fixed solution (3:1 mixture of methanol : acetic acid) for 20 min at room temperature. Following centrifugation at 800 g for 10 min at 4°C, the supernatant was discarded and the cells were fixed in fresh fixed solution and centrifuged. The step was repeated the cell suspension and dropped onto a clean slide glass. After air-dry, the dried cells were stained with 5% Giemsa solution for 20 min at room temperature. One hundred cells at metaphase were counted under a light microscope (Nikon).

Virus

The GIV/90/grouper strain was used in this research. The virus strain was isolated from sick juveniles of grouper (*Epinephelus coioides*) in the field. Our data of viral major capsid protein (MCP) nucleotide phylogenetic analysis showed that the GIV/90/grouper isolates (GenBank accession number KU510328) belong to the GIV (Accession number AAV91066) group of *Ranavirus*. The MCP nucleotide sequence of GIV/90/grouper strain was 100% identical to that of GIV.

Confirmation of the origin of the cell line

To authenticate the origin of the developed GK-7 cell line, the complete sequence of the 16S ribosomal RNA (rRNA) was determined. Total genomic DNA was extracted from the 50th passage for GK-7 cells with a QIAamp DNA Mini kit (QIAGEN) according to the manufacturer's instructions. Primers 16s-f (5'-taccgactcgctagccca-3') and 16s-r (5'-tagtcgttttctcagaca-3') were used to amplify a fragment of 16S rRNA gene that referred to the complete genome of mitochondrion derived from *E. coioides* (Accession number KM377093) to design primers. The polymerase chain reaction (PCR) was carried out in a 25 μl of mixtures containing 1x PCR buffer for Blend Taq (Toyobo), 0.2 mM dNTPs, 10 pmol of each primer, 1.25 unit of Blend Taq DNA polymerase, 100 ng of the extracted genomic DNA. The PCR mixture was incubated in a PTC-200 DNA thermal cycler (MJ Research, Waltham, MA) at 95°C for 5 min of 1 cycle; 35 cycles of 94°C for 30 s, 58°C for 50 s, and 72°C for 1 min ; finally, extension at 72°C for 7 min. PCR products were sequenced using an ABI PRISM 377 DNA sequencer with a BigDye Terminator Kit (Applied Biosystems, Foster city, CA, USA). Sequences of the mitochondrial genes amplified from the GK-7 cell line were examined for similarity with existing sequences submitted in GeneBank database.

Morphological confirmation with immunocytochemistry

Immunophenotyping of the GK-7 cell line was carried out at passage 48 following the method described by Gong et al. Briefly, the cells were grown on coverslips in 6-well plates for 24 h and then fixed with 4% paraformaldehyde at 4°C for 1 h and permeabilized with 0.2% Triton-X 100 at room temperature for 15 min. After blocking with 2% BSA in PBS at room temperature for 2 h, the cells were incubated with four different primary antibodies against mouse anticytokeratin, mouse antivimentin, mouse antifibronectin and goat anit rabbit desmin, respectively (Sigma-Aldrich; catalogue number: C2562, 180-0052, F0791 and D8281), at a dilution of 1:200 at room temperature for 2 h to characterize the cell line. Subsequently, the cells were rinsed three times with PBS and were incubated with goat-anti-mouse IgG FITC-conjugated or goat-anti-rabbit IgG FITC-conjugated (1:320 in PBS containing 1% BSA). In control coverslips, only PBS with 1% BSA was used in place of the primary antibodies. DAPI (diaminophenylindole) at a final concentration of 1 μgmL^{-1} was used to stain DNA in nuclei. All samples were observed with a fluorescence microscope.

GIV infection, propagation and titration

For the virus isolation, the tissues were thawed and homogenized with 10-fold volume of sterile PBS (pH 7.4). The homogenate was centrifuged at 1500 g for 15 min at 4°C and filtered through 0.45 μm membrane filter. The GK-7 cells were seeded onto 6-well plates at about 80% confluence and cultivated for 18 h; then, the medium was discarded and the GK-7 cells in the 6-well plates were inoculated with 0.1 mL of the filtrate. These inoculated cells were incubated at 25°C for 7-14 days and observed daily by inverted microscope. After the 7-14 days, when CPE was not observed, supernatant were treated by the second to fifth passages with the same procedures. Virus titres were determinated by titrating a series of 10-fold dilutions onto each cells in 96-well plates and calculated as $TCID_{50}$ mL^{-1} per flask [20].

Comparison of viral replication efficiency in earlier and later passages

The GIV replication in the GK-7 cell line was evaluated by virus inoculation in the 50th and 150th passages of the GK-7 cell line. Each cell passages were incubated in 25-cm^2 tissue culture flasks at a density of 2 × 10^5 cells for 24 h at 25°C to 80% confluence. After removal of the medium, the cell culture was inoculated with 0.1 mL of virus suspension at a multiplicity of infection (MOI) of 0.01, then 5 mL of maintenance medium containing 2% FBS was added. The cells were examined and their culture supernatants were sampled daily for 7 days after the inoculation. The sampled supernatants were stored at -80°C before determining virus titers and titrated in triplicate.

Electron microscopy

The monolayers of GK-7 cells were infected with 10^6 $TCID_{50}$ mL^{-1} GIV/90/grouper. After appearance of advanced CPE, the virus-infected cells were harvested, pelleted by centrifugation at 3000 rpm for 10 min, and fixed with 2.5% glutaraldehyde in cacodylate buffer (0.1 M, pH 7.2) for 2 h at 4°C. The fixed cells were washed with fresh cacodylate buffer and rinsed in PBS buffer (0.1 M, pH 7.2) for 10 min three times,then post-fixed in 1% osmium tetroxide for 1 h. After being rinsed four times in PBS for 15 min, the fixed cells were dehydrated in graded ethyl alcohol (50, 75, 90, 95, 100%) and embedded in epoxy resin. Ultrathin sections were cut with a ultramicrotome (Reicher-Jung) and were then stained with saturated aqueous uranyl acetate and lead citrate. Preparation of GIV for negative staining were performed as described previously [18]. A drop of the virus infected cells which were treated by three cycles of rapid freezing/thawing were applied onto a carbon coated 300-mesh copper grids and stained with 2% phosphotungstic acid (pH 6.8) for 1 min. Specimens were examined using an electron microscope (Jeol JEM-1000CX II, Japan) and photomicrographs were taken.

Cell transfection

To evaluate transfection efficiency and gene expression in GK-7 cells, the vector pEGFP-N1expressing green fluorescent protein by human cytomegalovirus promoter was prepared according to supplier's instructions (Clontech Laboratories, CA, USA). The cell line was seeded onto 24-well plates at a density of 2 × 10^4 cells per well. After 24 h, subconfluent monolayers were formed and the monolayers were washed once with serum-free mediun. Two micrograms of the plasmid pEGFP-N1 and 12 µL of lipofectamine 2000 (Invitrogen) were dissolved separately in 375 µL of L-15 medium without serum and antibiotics, and then they were mixed and incubated for 20 min to prepare the transfection reagent. Following the incubation, 500 µL of this mixture was added to each well. After 24-72 h, the green fluorescence signals were observed under a fluorescent microscope.

Results

Primary cell culture and subculture

Primary cell cultures of *E. coioides* were carried out in November 2007 using tissues from spleen, gill, kidney and fin. The primary cells attached to the culture flask at day 2 and a monolayer of the cells was formed after 30 days. The cells were subcultured in L-15 medium with 20% FBS at a ratio of 1:2 every 5-7 days for the initial 10 passages, as allowed by the cell growth rates. After the 10 passages, these cells propagated rapidly and thereafter the cultures were adapted to culture medium with 10% FBS at 1:3-1:5 ratio at intervals of 3 to 5 days. After the 50th passage, the cells were split at a ratio of 1:3 every 7 days. Only the cells derived from kidney tissue formed a continuous cell line designated as GK-7. The passage numbers of the cells used in the present study were all around the 50th passage to 150th passage. Morphologically, the initial subcultures of GK-7 cell lines consisted of spindle-shape epithelial-like cells. To date, the cells have undergone 150 passages and showed no apparent morphologyical changes (Figure 1).

Cryopreservation

The GK-7 cells were cryopreserved every ten passages from 5th to 150 th passages. The cryopreserved cells at 15, 25, 45, 55, 65, 100 and 150 passages recovered from liquid nitrogen storage after 6 mo, 1 year, 2 years, 3 years showed 70-90% viability and grew to confluency within 3-5 days in L-15 medium with 10% FBS at 25°C.

Growth studies

To optimize the conditions for cultivation of the GK-7 cell, the cells at the 50th passage were used to investigate the kinetics of cell growth. The GK-7 cells were able to grow at temperatures between 20°C and 32°C, and the highest growth occurred at 25°C (Figure 2a). Growth rates were remarkably reduced at 32°C, indicating that the GK-7 cells were quite sensitive to high temperature. Growth of GK-7 cells at varying levels of FBS concentrations is shown in Figure 2b. Results show that serum is required for the cells to survive and grow and the growth rates of the cell increased with the FBS concentration. With regard the growth rates of GK-7 cell with different passage history, cells of the 50th passage proliferated significantly faster than those of the 150th passage during the observation period of 15 days. The cells of the 50th passage peaked at approximately 1.8 × 10^6 cells/ml after 9 days of seeding and the cells of the 150th passage gavel.2 × 10^6 cells/ml at the same day (Figure 2).

Chromosome number analysis

The results of chromosomes counts of 100 metaphase plates at the 50th passage showed diploid numbers ranging from 20 to 68 with a modal peak at 48 chromosomes and the metaphates at the 50th passage with the normal diploid number, additionally displayed the normal karyotype morphology of GK-7 cells consisted of a pair of subtelocentrics (st) and 23 pairs of telocentrics. The chromosome number distribution at the 150th passage based on 100 metaphase plates, displayed 2n values ranging between 30 and 76 with the modal value of 66 chromosomes and had a karyotype composed of two pairs st and 31 pairs telocentrics (Figure 3).

Characterization of the GK-7 cell line

Four antibodies against cytokeratin, vimentin, fibronectin and desmin, respectively, were used for immunostaining to identify the

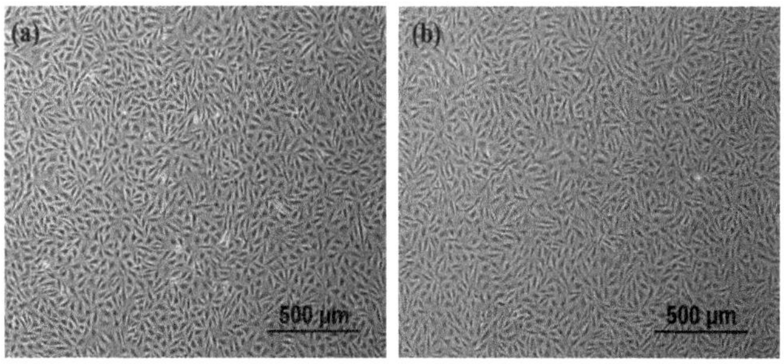

Figure 1: Morphology of GK-7 50 (a) and 150 (b) passages (bar = 500 µm).

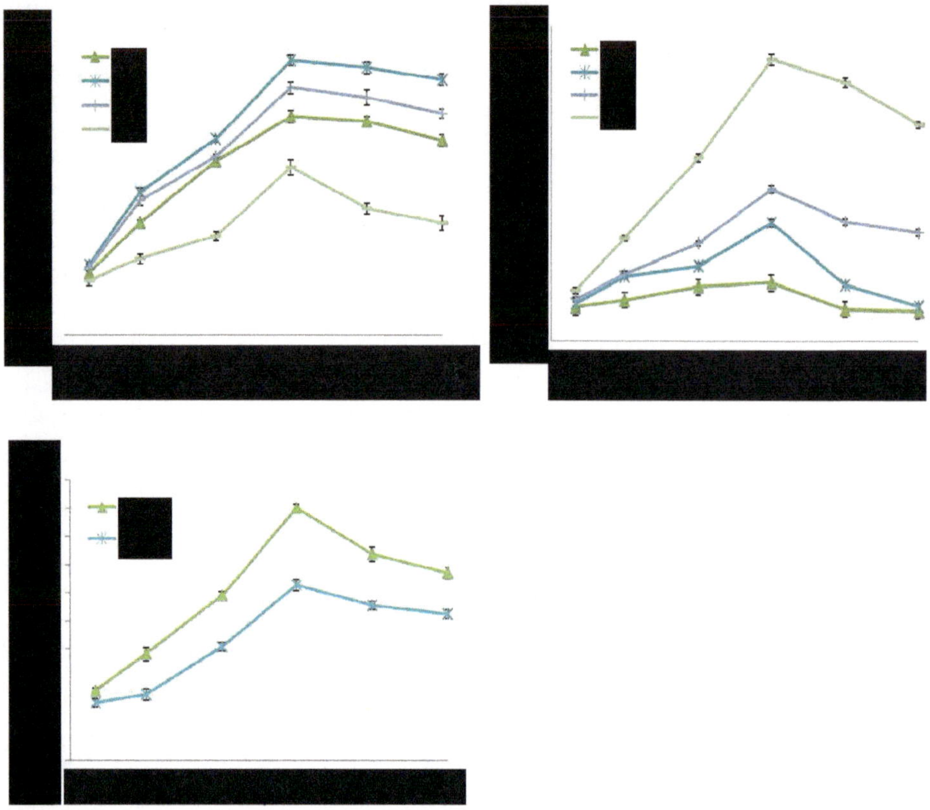

Figure 2: Effect of incubation temperature (a) and fetal bovine serum concentrations (b) on the growth curves of GK-7 cell lines. Growth kinetics of GK-7 cell lines at the passages 50 and 150 (c). Data are means ± S.D. of three measurments.

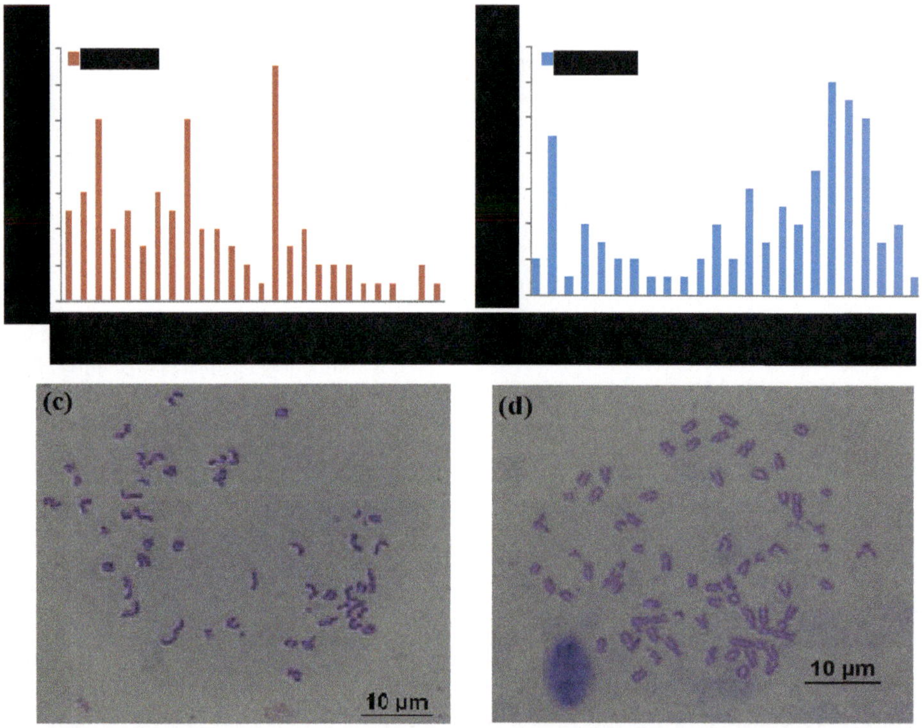

Figure 3: Chromosome number analysis of GK-7 cells. Distribution of chromosome numbers in the 100 examined GK-7 cells at passage 50th (a) and 150th (b). Giemsa-stained metaphase chromosomes of GK-7 cells at passage 50th (400X) (c) and 150th (d) (400X).

feature of the GK-7 cell. Strong fluorescence signals were observed in the cells reacting with antibody against cytokeratin, an epithelial cell marker (Figure 4). No reactivity was observed in control and with anti-vimentin, anti- fibronectin and anti- desmin antibodies. These results confirmed that the GK-7 cells are epithelial cells.

Identification of cell origin

The origin of the GK-7 cell line was examined by the PCR designed for amplifying the conserved region of the 16S rRNA gene. The PCR product of 1706 bp demonstrated a similsr size to that from *E. coioides* mitochondrial DNA (Figure 5). The nucleotide sequence of the PCR product (GenBank accession number KU510327) was 99.9% similar to the known *E. coioides* mitochondrial DNA sequences available in GenBank (Accession number KM377093 and EU043376).

Cytopathic effect and virus replication

Viral infection was detected in both cell passages at 50th and 150th subcultures of GK-7 cells within 3 days after GIV inoculation. Initially, the specific CPE developed as foci areas of rounded, granular and refractive cells, which spread through the cell sheet over 3-4 days to form a network of degenerated cells and completely disintegrated by 5 days post-inoculation (d.p.i) in the cells of the 50th passage. By contrast, the CPE in the cells of the 150 passage appeared slower than those of the 50 passage (Figure 6). Over 7 days of observation, the virus titers of GK-7 cells of the 50th and 150th passages reached up $10^{8.4}$ and $10^{7.2}$ $TCID_{50}$ mL^{-1}, respectively (Figure 7).

Electron microscopy

Ultrathin sections of the GIV-infected GK-7 cells at 2 d.p.i. showed a large number of membrane-bound aggregates and crystalline arrays of GIV particles in the cytoplasm (Figure 8). The nuclear membranes remained largely intact, and chromatin margination and cytoplasmic changes were evident in the GK-7 cells (Figure 8a). The virions were hexagonal and crystalline aggregation in infected cells and the mean diameter of the virions were about 200 nm (Figure 8b). Negatively stained enveloped virus showed a typical hexagonal shape, and the envelope structure of virus particle was three-layered with an inner electron-dense core surrounded by a lighter coat (Figure 8c). The

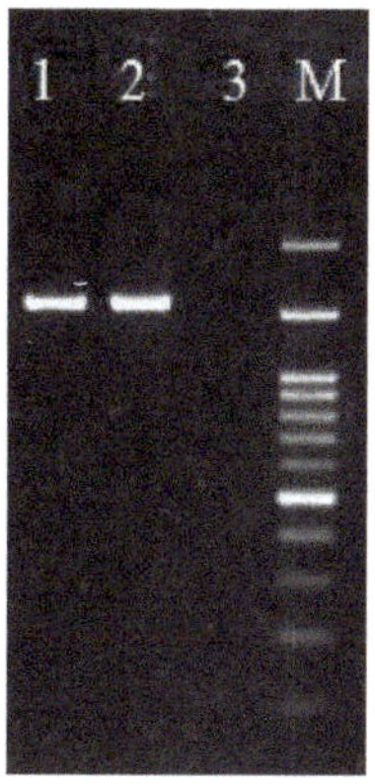

Figure 5: PCR amplification of 1706 bp sequence of the 16S rRNA from E. coioides mitochondrion (M: 100 bp Ladder; Lane1: GK-7 cell Lane 2: E. coioides muscle tissue ; Lane 3: negative control.

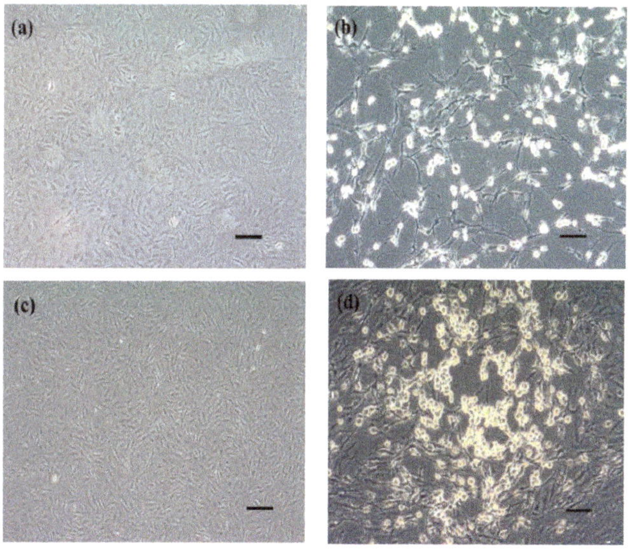

Figure 6: The cytopathic effect (CPE) of GK-7 cells infected by grouper iridovirus (GIV) (bar = 100 μm). (a) GK-7 mock at cell passages50. (b) GK-7 infected by GIV after 5 days at cell passages 50. (c) GK-7 mock at cell passages 150. (d) GK-7 infected by GIV after 5 days at cell passages 150.

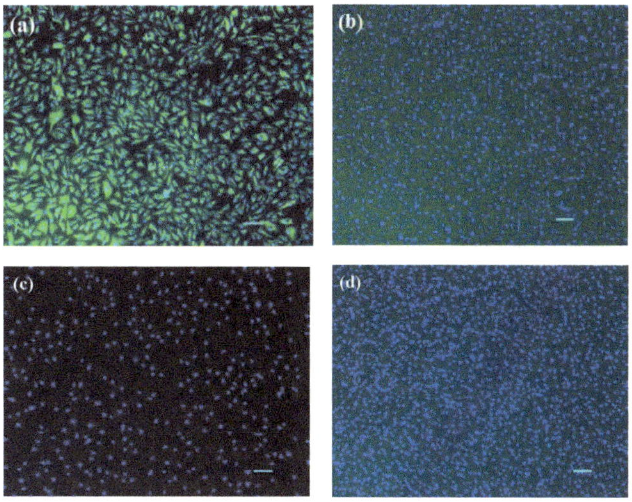

Figure 4: Immumofluorescent assay to detect cytokeratin (a) desmin (b) fibronectin (c) and viementin (d) in GK-7 cells. The cell nuclei were stained with DAPI and presented a blue fluorescence (bar = 100 μ m).

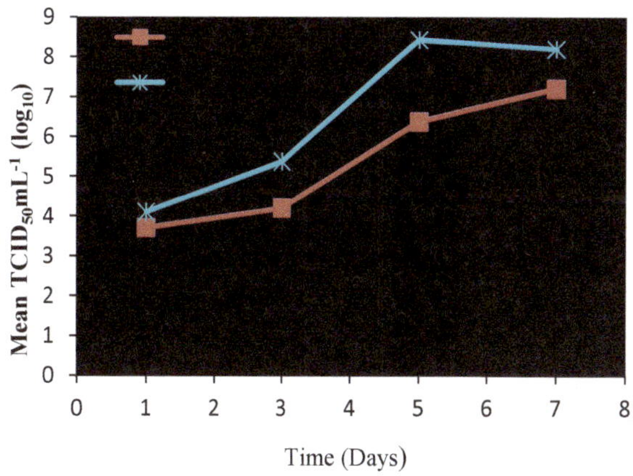

Figure 7: The replication virus titers of GIV infected 50th and 150th passages of GK-7 cells. Data represented the mean of triplicate measurements.

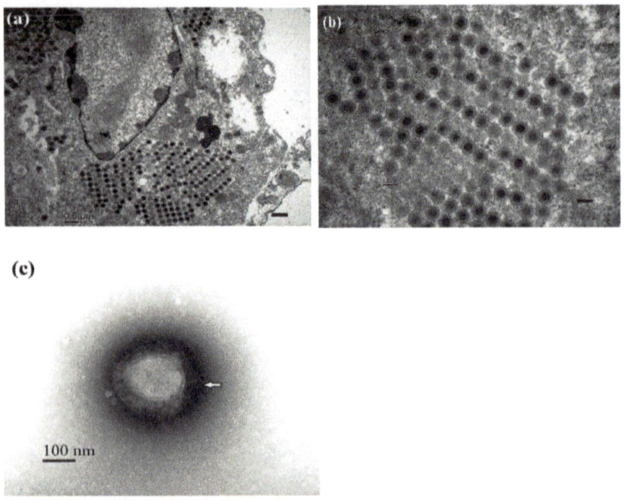

Figure 8: Electron micrographs of ultrathin sections and negative staining in GK-7 cells. (a) Ultrathin sections of grouper iridovirus (GIV)-infeted GK-7 cells, a large amount of virus particles locates in the cytoplasma (bar = 0.5 μm); (b) Virus particles aggregate and crystalline array in GIV-infeted GK-7 cells (bar = 0.2 μm); (c) Negatively stained GIV particle was exhibited a three-layered membrane structure (indicated by arrow) with an inner electron-dense core. (bar = 100 nm).

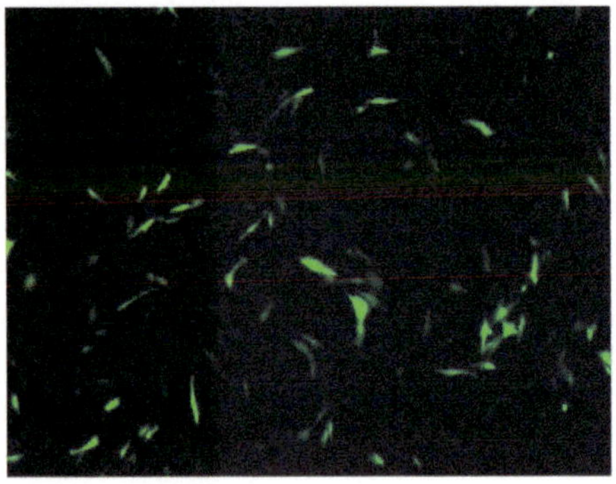

Figure 9: Expression of green fluorescent protein in GK-7 cells at passages 50 with pEGFP-N1 (100X).

particles were larger than those found in ultrathin sections, with a mean diameter about 240 nm.

Cell transfection

The GK-7 cell line was successfully transfected with pEGFP-N1 vecter by lipofectamine 2000. The green fluorescence signals were observed in GK-7 cell line as early as 24 h post-transfection, and showed clear and strong fluorescence signals after 48 h (Figure 9). The transfection efficiency was around 50%, calculated by microscopy.

Discussion

In this study, we established and characterized a novel cell line from the kidney of *E. coioides*, and the cell line was designated as GK-7. Morphologically, the GK-7 cells were composed of spindle-shape epithelial-like cells which was distinct from the GK cells derived from the *E. awoara* composed of a heterogeneous mixture of fibroblasts and epithelial-like cells [13]. Furthermore, the immunocytochemistry demonstrated strong positivity to cytokeratin which are used as specific markers for epithelial cells. These results indicated that the GK-7 cells were of epithelial origin. Similar results were reported in GBC4 clonal cell line from the brain of *E. coioides* [23] and EAGL drived from the liver of *E.akaara* [15].

The GK-7 cells grew efficiently at 20-32°C and showed the optimum growth at 25°C, which was consistent with the conditions for the other established *Epinephelus* cell lines [9]. The advantages of cell lines that grow over a wide temperature range is their potential suitability for isolating both warm-water and cold-water fish viruses [14]. Base on the results show that serum is the higher concentration of FBS, the faster growth rate of cells. However, 2% FBS can support maintenance growth and at least required serum proportion for the GK-7 cell line at low cost.

Growth curves between the GK-7 cells of the 50th and 150th were similar. The results of growth curves performed in GK-7 cell line were more stable. In addition, the cells of the 50th passage displayed higher densitiy than that of the cells of the 150th pssages, showing that the GK-7 cell line with shorter passage history could be propagated more rapidly for performing the culture. In addition, our virus replication data showed that the maximum virus yield in the cell of the 50th passage was faster and higher than that in the cells of 150th passage. Maximum virus yield was produced in the GK-7 cells of the 50th and 150th passage at 5 and 7 d.p.i., respectively. The higher production could result from higher cell density present at the time of infection. These results demonstrated the GK-7 cells of the 50th passage was suitable for both research and product manufacture.

In this study, the GK-7 cell line at the 50th and 150th passages had different chromosome numbers. At the 50th passage, the GK-7 cell contained a modal value of 48 chromosomes; in contrast, at the 150th passage, chromosome numbers varied in a range from 30 and 76 with the modal value at 66 chromosomes. The similar chromsome variability is also observed in *Epinephelus akaara* grouper brain cell line [8]. The modal number of the grouper brain cell line is 48 at the 35th passage and becomes 60 at the 70th passage. A clonal cell line, GBC4, derived from grouper (*Epinephelus coioides*) brain also demonstrates a shift of chromosome number from 22 to 58 [24]. These results indicated that GK-7 cell line similar to most of the established fish cell line, had undergone some degree of cell transformation.

The cell line used to propagate a virus has to be rapid and efficient propagate the virus in large quantities and should be suitable for a wide variety of virus strains. Our study indicated that the GK-7 cell line was susceptible to GIV and was a statisfactory tool to propagate GIV, as showing that viral titers reached $10^{8.4}$ $TCID_{50}$ mL^{-1}. Similar findings have been also reported in the other grouper kidney cell line [13]. These observations indicated that the GK-7 cell line exhibited higher susceptibility to piscine ranaviruses. Enveloped GIV particles were found in the infected cells. Previous ultrastructural studies on ranaviruses reveal that the enveloped virus particles can be seen in the early stages of frog virus 3 (FV3) and SGIV infection [1,17-18]. We first discovered that the GIV also possessed an envelope structure (Figure 8c). In addition, most matured virus particles appeared in the cytoplasm of GIV-infected GK-7 cells (Figure 8a). Similar findings were also reported in a giant grouper (*Epinephelus lanceolatus*) cell derived from heart tissue [5]. In the cells infected, numerous virus particles were observed in the viral factory and in the cytoplasm near the metamorphic nucleus [5]. The ultrastructure and size of GIV are very similar to those pscine and amphibian iridovirus belonging to genus Ranavirus.

The 16S rRNA has been used to identify the origin of various established cells in previous studies [21]. Our data showed that the sequence of GK-7 16S rRNA matched perfectly with the sequence of orange spotted grouper *E. coioides*, 16S rRNA [25,26]. In the absence of 16S rRNA data for *E.coioides*, this is a strong evidence to support that the GK-7 cell line was derived from orange spotted grouper.

The GK-7 cell line we developed allowed foreign genetic materials expressed. Exogenous DNA delivery of a cultured cell is very useful for both basic research and biotechnological applications. In this study, the GK-7 cell line was successfully transfected with pEGFP-N1 plasmid by means of lipofectamine 2000 and transfection efficiency was up to 50% (Figure 9). These results illustrated that this cell line may be able to be genetically manipulated for host-pathogen interactions study.

The present study indicated that the GK-7 cell line is a suitable tools for studying epidemiology and pathology but also can further be applied to develop vaccine to control the GIV-associated diseases.

Acknowledgment

The authors thanks Dr. Fan Lee for his comments on this manuscript. This study was granted by projects 99AS-1.1.5-HI-H2, 100AS-1.1.5-HI-H2, 101AS-10.2.4-HI-H1, 102AS-10.2.4-HI-H1, 103AS-10.1.3-HI-H1 of the Council of Agriculture, Executive Yuan, Taiwan.

References

1. Barunwald J, Nonnenmacher H, Tripier-Darcy F (1985) Ultrastructural and biochemical study of frog virus 3 uptake by BHK-21 cells. J of General Virology 66: 283-293.
2. Cheng TC, Lai YS, Lin IY, Wu CP, Chang SL, et al. (2010) Establishment, characterization, virus susceptibility and transfection of cell lines from cobia, Rachycentron canadum (L), brain and fin. J Fish Dis 33: 161-169.
3. Chi SC, Hu WW, Lo BJ (1999) Establishment and characterization of a continuous cell line (GF-1) derived from grouper, *Epinephelus coioides* (Hamilton): a cell line susceptible to grouper nervous necrosis virus (GNNV). Journal of Fish Diseases 22: 173-182.
4. Gong J, Huang Y, Huang X, Ouyang Z, Guo M, et al. (2011) Establishment and characterization of a new cell line derived from kidney of grouper, Epinephelus akaara (Temminck & Schlegel), susceptible to Singapore grouper iridovirus (SGIV). J Fish Dis 34: 677-686.
5. Guo CY, Huang YH, Wei SN, Ouyang ZL, Yan Y, et al. (2015) Establishment of a new cell line from the heart of giant grouper, Epinephelus lanceolatus (Bloch), and its application in toxicology and virus susceptibility. Journal of Fish Diseases 38: 175-186.
6. Huang C, Zhang X, Gin KYH, Qin QW (2004) In situ hybridization of a marine fish virus, Singapore grouper iridovirus with a nucleic acid probe of major capsid protein. J of Virological Methods 117: 123-128.
7. Huang SM, Tu C, Tseng CH, Huang CC, Chou CC, et al. (2011) Genetic analysis of fish iridoviruses isolated in Taiwan during 2001–2009. Archives of Virology 156: 1505-1515.
8. Huang X, Huang Y, Ouyang Z, Qin Q (2011) Establishment of a cell line from the brain of grouper (Epinephelus akaara) for cytotoxicity testing and virus pathogenesis. Aquaculture 311: 65-73.
9. Huang XH, Huang YH, Sun JJ, Han X, Qin Q (2009) Characterization of two grouper *Epinephelus akaara* cell lines: application to studies of Singapore grouper iridovirus (SGIV) propagation and virus-host interaction. Aquaculture 292: 172-179.
10. Ku CC, Teng YC, Wang CS, Lu CH (2009) Establishment and characterization of three cell lines derived from the rockfish grouper Epinephelus quoyanus: Use for transgenic studies and cytotoxicity testing. Aquaculture 294: 147-151.
11. Lai YS, John JA, Lin CH, Guo IC, Chen SC, et al. (2003) Establishment of cell lines from a tropical grouper, Epinephelus awoara (Temminck & Schlegel), and their susceptibility to grouper irido- and nodaviruses. J Fish Dis 26: 31-42.
12. Lai YS, Murali S, Chiu HC, Ju HY, Lin YS, et al. (2001) Propagation of yellow grouper nervous necrosis virus (YGNNV) in a new nodavirus-susceptible cell line from yellow grouper, *Epinephelus awoara* (Temminck & Schlegel), brain tissue. Journal of Fish Diseases 24: 299-309.
13. Lai YS, Murali S, Ju HY, Wu MF, Guo IC, et al. (2000) Two iridovirus-susceptible cell lines established from kidney and liver of grouper, *Epinephelus awoara* (Temminck & Schlegel), and partial characterization of grouper iridovirus. J of Fish Diseases 23: 379-38.
14. Nicholson BL, Danner DJ, Wu JL (1987) Three New Continuous Cell Lines from Marine Fishes of Asia. In Vitro Cellular & Developmental Biology 23: 199-204.
15. Ou-Yang ZL, Huang XH, Huang EY, Huang YH, Gong J, et al. (2010) Establishment and characterization of a new marine fish cell line derived from red-spotted grouper *Epinephelus akaara*. Journal of fish biology 77: 1083-1095.
16. Parameswaran V, Shukla R, Bhonde RR, Hameed AS (2007) Development and characterization of two new cell lines from milkfish (*Chanos chanos*) and grouper (*Epinephelus coioides*) for virus isolation. Marine Biotechnology 9: 281-291.
17. Qin QW, Chang SF, Ngoh-Lim GH, Gibson-Kueh S, Shi C, et al. (2003) Characterization of a novel ranavirus isolated from grouper *Epinephelus tauvina*. Diseases of Aquatic Organisms 53: 1-9.
18. Qin QW, Lam TJ, Sin YM, Shen H, Chang SF, et al. (2001) Electron microscopic observations of a marine fish iridovirus isolated from brown-spotted grouper, *Epinephelus tauvina*. Journal of Virological Methods 98: 17-24.
19. Qin QW, Wu TH, Jia TL, Hegde A, Zhang RQ (2006) Development and characterization of a new tropical marine fish cell line from grouper, *Epinephelus coioides* susceptible to iridovirus and nodavirus. J of Virological Methods 131: 58-64.
20. Reed LJ, Muench H (1938) A Simple Method of Estimating Fifty Percent Endpoints. American J of Epidemiology 27: 493-497.
21. Wang G, Lapatra S, Zeng L, Zhao Z, Lu Y (2003) Establishment, growth, cryopreservation and species of origin identification of three cell lines from white sturgeon, Acipenser transmontanus. Methods in Cell Science 25: 211-220.
22. Wei Y, Fan T, Jiang G, Sun A, Xu X, et al. (2009) Establishment of a novel fin cell line from Brown-marbled grouper, Epinephelus fuscoguttatus (Forsskål), and evaluation of its viral susceptibility. Aquaculture Research 40: 1523-1531.
23. Wen CM, Huang JY, Ciou JH, Kao YL, Cheng YH (2009) Immunochemical and molecular characterization of GBC4 as a tanycyte-like cell line derived from grouper brain. Comparative biochemistry and physiology 153: 191-201.
24. Wen CM, Lee CW, Wang CS, Cheng YH, Huang HY (2008) Development of two cell lines from Epinephelus coioides brain tissue for characterization of betanodavirus and megalocytivirus infectivity and propagation. Aquaculture 278: 14-21.
25. Zhou GZ, Li ZQ, Yuan XP, Zhang QY (2007) Establishment, characterization, and virus susceptibility of a new marine cell line from red spotted grouper (*Epinephelus akaara*). Marine Biotechnology 9: 370-376.
26. Zhuang X, Ding S, Wang JUN, Wang Y, Su Y (2009) A set of 16 consensus primer pairs amplifying the complete mitochondrial genomes of orange-spotted grouper (Epinephelus coioides) and Hong Kong grouper (Epinephelus akaara). Molecular Ecology Resources 9: 1551-1553.

18

Portunid Crab Fishery Resources From Nagapattinam Coast, South East Coast of India

D Varadharajan* and P Soundarapandian

Faculty of Marine Science, Centre of Advanced Study in Marine Biology, Annamalai University, Parangipettai-608 502, Tamil Nadu, India

Abstract

Marine crabs are plays a vital role, depending upon the directly and indirectly in livelihoods the millions of people around the world. In the last two decades, the employ in the principal fisheries and farming parts has grown faster than the world's population and employment in traditional culture management. The food from land is day by day decreased, as consumption of populations are high, since the people looking forward in seafood. The marine resource study is very important and the fisheries data are require for any coast, the true data depicting the state for future problems and solving the demand of food.

Keywords: Commercial crabs; Resource; Economic; Future plan

Introduction

Crustaceans, such as crabs are very important for human beings, it has a significant quantity of protein, vitamins and other nutrients. They are used for feed and feed additive and extremely used in farm animal and making a medically valuable products. In recent years, the importance of crabs meat and other parts to the health and populations longevity and good price of world market. The demand for seafood currently exceeds, day by day increasing in populations, and occurs in overexploitations [1-3]. In the present study, the landing of crabs data very much useful for further research and welfare of fisherman. In Indian coast commonly available in fifteen edible crabs, since the twelve crabs are commercially valuable of local and export trade. In Indian coastal region of Nagapattinam coast the landing of fisheries data are very important, because these coastal regions a significant quantity for seafood supply for the nation. It is economically very important for the state or country, they are authorised to export, developing and managements of future research. The crab fisheries potential have been conducted by many workers in Indian waters [4]. The same potential has been conducted from Goa, and Palk Bay in Gulf of Mannar [5,6]. Tuticorin from Kakinada and the south Kanara coast Parangipettai coast,from Palk bay region [7]. In the present investigation an attempt has been made to study the fishery potential of commercially important portunid crab along Nagapattinam coast. From the previous crab landing reports the crab population is declining every year. So crab fishery resource study is essential to improve the population of the respective areas.

Materials and Methods

The study was conduct in different landing centers, viz, (station1) Pazhaiyar (latit 11°. 21 N long 79° 49 E), (station2) Poombukar (Lati, 11.13 N, Long, 79.849 E), (station 3) Thirumullaivasal (lati 12030' N, long 71074' E) and (station 4) Nagapattinam (lati 10° 15'-11° 12'46" N, long 79° 37' 30"-79° 50'12"E). Data of catch and effort of crab landings were collected and recorded from four different landing centers. The study was made from April 2009 to March 2010. The collect data for watching the landing of crabs and then fisherman giving more information were collected by interviewing them personally. The average of daily commercial crab landing was worked out from the data thus obtained and raised to the number of fishing days to assess the monthly total crab landings of the centre. The data were, collected for all the months except the second of half April-16 to May-30. The methodology for the present study was followed by standard method [2,3].

Results and Discussion

Pazhaiyar (Station-1)

The (station-1) Pazhaiyar landing centre is positioned third place for male, female, berried female. The total crab landings in Saamiyarpettai are in the following pattern; *P.pelagicus>P. sanguinolentus>S. tranquebarica>S. Serrata>P. vigil>C. feriata>C.variegata>C. natator>C. lucifera>C. Granulata>C.truncata>C. Gladiator* (Figures 1-4).

Poombukar (Station-2)

The (station-1) Poombukar landing centre standing fourth place in total male, female, berried female. The total crab landings are in the following order; *P.pelagicus>P. sanguinolentus>P.vigil> C. feriata>C. lucifera>S. tranquebarica>S. serrata>C. natator>C.variegata>C. granulata>C.truncata>P. gladiator* (Figures 1-4).

Thirumullaivasal (Station-3)

The (station 3) Thirumullaivasal landing centre standing second place in total male, female, berried female and total crabs landed. The total crab landings in Annankovil are in the following order;*P. pelagicus>P. sanguinolentus>S. tranquebarica>C. feriata>P. vigil> C. natator>S. serrata>C.variegata>C. Lucifera>C. granulata>C.truncata> C. gladiator>*(Figures 1-4).

Nagapattinam (Station-4)

The total crab landings in Nagapattinam (station 4) are in the following pattern; *P.pelagicus>P. sanguinolentus>S. tranquebarica>C. feriata>P. vigil>C. lucifera> C. natator>C.variegata>C.truncata>S. serrata>C. granulate>P. gladiator* (Figures 1-4).

***Corresponding author:** D Varadharajan, Faculty of Marine Science, Centre of Advanced Study in Marine Biology, Annamalai University, Parangipettai-608 502, Tamil Nadu, India, E-mail: heartvaradhan@gmail.com

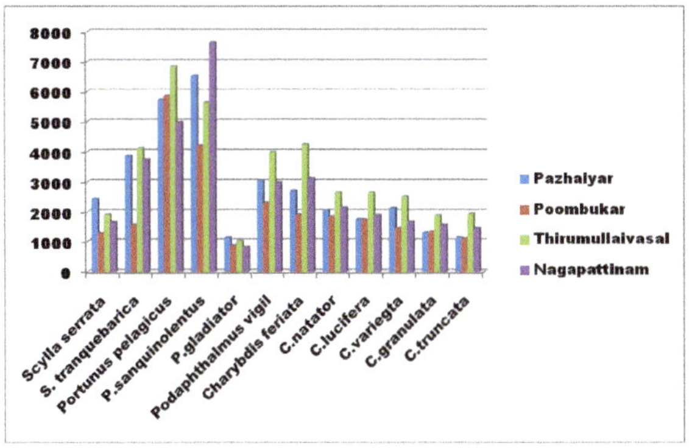

Figure 1: Total male crab landed (kg) from four different landing centre during April-2009 to March- 2010.

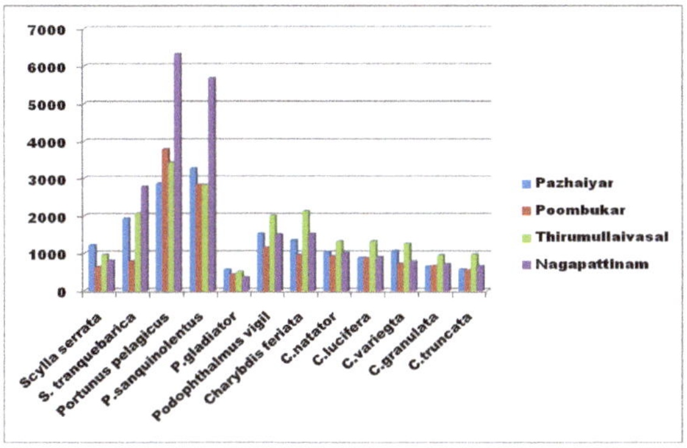

Figure 2: Total female crab landed (kg) from four different landing centre during April-2009 to March-2010.

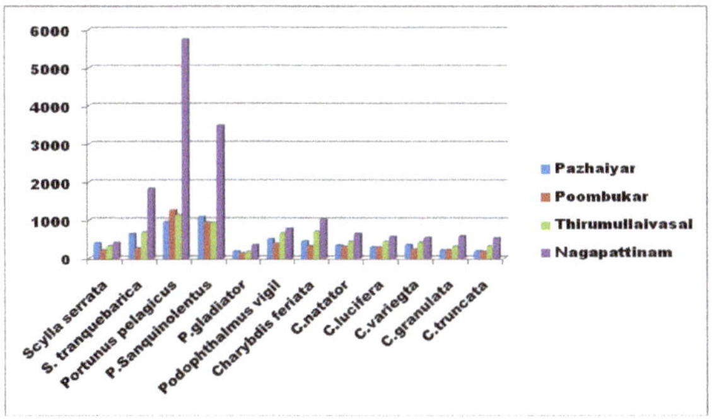

Figure 3: Total berried female crab landed (kg) from four different landing centre during April-2009 to March-2010.

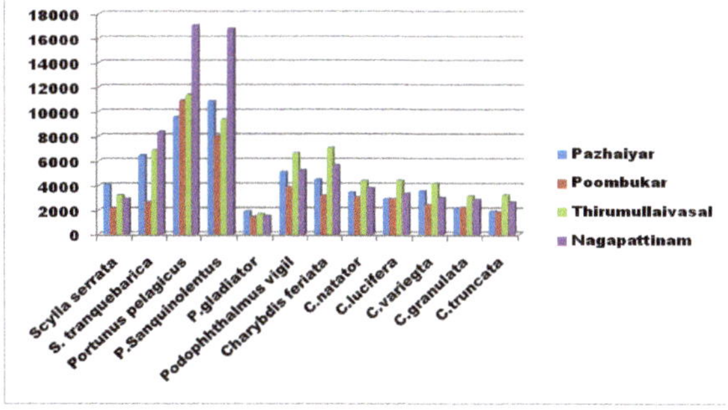

Figure 4: Total crabs landed (kg) from four different landing centre during April-2009 to March-2010.

The landings of total crab in the present study is 240559.8 kgs or (241 t) reported in different four stations. The maximum landing of crabs was recorded in (station-2) Nagapattinam (73,385 kg) and minimum was recorded in (station-2) Poombukar (44,705 kg) followed by other stations. This term agreed with previous studies, the crabs landings in Parangipettai coast 79,445.6 kg, its reported by Soundarapandian [2] and similarly, Arukkattuthurai to Aiyyampattinam coast the landing of crabs 3, 88,196.032 kg, its reported by Varadharajan et al. [3]. The differences in the crab landings in the past two decades are due to many reasons, including development of improved fishing craft and gear, importance of crab meat in the national and international markets and changes in the environmental parameters. The crab meat is not only used for the human consumption but recently most of the crabs are being fished as the source of fishmeal. The present investigations the *P. pelagicus and P. sanguinolentus* constituted around 39% of the total landings forming a major crab fishery along Nagapattinam coast. However, in this study the commercial crabs viz., *S.serrata, S.tranquebarica and P.pelagicus, P. sanguinolentus, P.vigil and Charybdis sp in C. feriata, C. natator, C. lucifera* are available throughout the study periods in almost all stations [8]. The *P. gladiator, C. granulata, C. truncata and C. variegata*, its fishery is purely seasonal and contributes very less as far as total crab fishery is concerned [9]. The quantity of minimum was landed in *Portunus gladiator, Charybdis* sp., are recorded of seasonally [2,3].

The crab fishery before and after tsunami did not show any marked variation except some minor differences in the landings has been reported that the increased landings of crabs again suffered a setback after Tsunami, resulting in a significant fall in crab landings along Indian coast [10]. Being a commodity of lesser export value, no commercial operations have been directed towards crab resource in a big way in the country. However, substantial quantities of crabs are landed every year as by-catches of fisheries trawlers and of indigenous fishing units throughout the country. The commercially crabs are important fisheries resources in marine environment have been collected extensively in part of their distribution range in Indian coastal waters. Since the crabs became important as exportable fisheries resources, overfishing have occurred more severely in some coastal regions [11,12]. As a result of overfishing of local populations, both the size and number of crabs have decreased. The crabs including the male species with strong muscle meat, the female species with nutritious eggs, the molting species with a new soft shell grown beneath the old hard one, and the juvenile species. Owing to their great demand in the markets, marine crabs are extensively cultured in many provinces of southern India. Another anticipated problem is the demand for small crabs by consumers because of their low price in the market. Despite all these problems, it might still be important to consider the application of some guidelines for commercial crab fishing in the coastal [13]. Therefore, the study presents an account of some case studies on the approach of fishermen towards the sustenance of the crab fishery resources they exploit and hence, of their livelihood. The crab fisheries are a highly dynamic system, which is constantly undergoing change due to natural and anthropogenic factors. There is having more or less negative effects on the resources of the coastal regions. In the present study, the different coastal fishery reports are significantly important for the state and the Indian government. Since, the crab landings data are essential to make out the contribution of crabs to the Indian economy. They are from different coast is supplying a significant quantity of seafood for the state, still there is no detailed information on crab catches from this region. In the present study an attempt has been made to identify the fishery potential of commercially important crabs of both marine and estuarine biotopes of Nagapattinam coast. In these coastal provides significant quantity of seafood for the state, so definitely solve the problem in future the demand of food and also the need for conversation strategies.

References

1. Lalitha DS (1985) The fishery and biology of crabs of Kakinada region. Indian J Fish 32: 18-32.

2. Soundarapandian P (2004) Fishery Potential of Commercially Important Crab Portunssanguinolentus (Herbst) along Parangipettai Coast, South East Cost of India. International Journal of Animal and Veterinary Advances 165-173.

3. Varadharajan D, Soundarapandian P, Dinakaran GK, Vijakumar G (2009) Crab Fishery Resources from Arukkattuthurai to Aiyammpattinam, South East Coast of India. Curr Res J BiolSci1: 118-122.

4. Anzari ZA, Harkantra SN (1976) Crab resources of Goa. Sea Exp J 8: 21-24.

5. Ameerhamsa KMS (1978) Fishery of the swimming crab PortunuspelagicusLinnaeus from Palk bay and Gulf of Mannar. Indian J Fish 25: 229-232.

6. Shanmugam S, Bensam P (1980) The fishery for the crab Scylla serrata (Forskal) at Tuticorin during 1974-75. Indian J Fish 27: 102-110.

7. Sukumaran KK, Telang KY, Thippeswamy O (1986) On the fishery and biology of the crab Portunuspelagicus (Herbst) along the south Kanara coast. Indian J Fish 33:188-200.

8. Radhakrishnan CK (1979) Studies on portunid crabs of Porto Novo (Crustacea: Decapoda: Brachyura) Ph.D Thesis, Annamalai University India 1-129.

9. Sanil Kumar S (2000) New horizons in sea crab meat processing. Seafood Export J 31: 41-43.

10. Babu DE (2005) Fishery of Scylla serrata. In: Seminar on sustainable utilizataion of Brackish water crab resources of Godavari Mangroves. Fishing Chimes 25.

11. Rajamani M, Manickaraja M (1998) A note on the fishery of the swimming crab Portunuspelagicus(Linnaeus) from Tuticorin bay. Journal of the Marine Biological Association of India 39: 185-187.

12. Saradha PT (1998) Crab fishery of t h e calicut coast with some aspects of t he population characteristics of portunus sanguinolentus, p. pelagicus and charybdis cruciata. Indian J Fish 45: 375-386.

13. Sukumaran KK, Telang KY, Thippeswamy D (1986) On the fishery and biology of the crab Portunuspelagicus(Herbst) along the south Kanara coast. Indian J Fish 33: 188-200.

In Vitro Test and Molecular Docking of Alkaloid Compound in Marine Sponge *Cinachyrella anomala* against T47D Cell Cycle

Awik Puji Dyah Nurhayati[1]*, Rarastoeti Pratiwi[2], Subagus Wahyuono[3], Istriyati[2], Hari Purnomo[4] and Syamsudin Abdillah[5]

[1]*Biology Department, Mathematic and Natural Science Faculty, Sepuluh November Institute of Technology Surabaya, Indonesia*
[2]*Biology Faculty, GadjahMada University, Yogyakarta, Indonesia*
[3]*Department of Biological Pharmacy, Faculty of Pharmacy, Gadjah Mada University, Yogyakarta, Indonesia*
[4]*Department of Chemistry Pharmacy, Faculty of Pharmacy, Gadjah Mada University, Yogyakarta, Indonesia*
[5]*Department of Pharmacology, Faculty of Pharmacy, Pancasila University, South Jakarta, Indonesia*

Abstract

The compound 1,4,9-triazatricyclo[7,3,1,0]trideca-3,5(13),10-trien-8-ol (SA2014) was isolated from the marine sponge *Cinachyrella anomala*. *In vitro* assay for SA2014 compound was found to be able to induce cell-cycle arrest at the sub-G_1 and G_2/M phases of T47D cancerous cell. A combined dosage between of SA2014 compound and of doxorubicin was able to induce cell-cycle arrest at sub-G_1 and G_2/M phases. Molecular docking approach showed that SA2014 compound inhibited cdk2 enzyme. The strength of interaction between SA2014 and cdk2 (docking score = -65,43) was more stable than the interaction between doxorubicin and cdk2 (-36,59).

Keywords: *C. anomala*; SA2014; cdk2 inhibitor; T47D cell

Introduction

Sponge is a member of Metazoa, which has been successfully through millions of years of evolution. This is evident from the fact that it is widely distributed all over the world, both in the salt water and the fresh water [1]. Sponges that live around the coral reefs are able to produce high-level toxic secondary metabolite as the consequence of extreme water pressure, competition, self-defense mechanism against nudibranch, gastropods, carnivorous fishes [2]. They are also found to have pharmacological potentials [3].

Around 20,000 types of active compounds are produced by sponges. They have wide-ranging chemical class diversity, including sterol, terpenoid, isoprenoid, nonisoprenoid, quinone, brominates, nitrogen heterocyclic, and heterocyclic nitrogen sulfur [4], amino acids, porphyrin and peroxide [5]. There are varied groups of functional compounds from the sponge (OH, OCH_3, OAc, OSO_3, Na^+) [6]. Every shift in the functional groups can potentially shift polarity of each component in a dramatic way. An oncological study on marine sponges found that they could interact with essential components in the cell cycle, enzymes, and other targets [7]. Research conducted by Holland et al showed steroid compounds derived from *Cinachyrella sp* for anticancer activity through inhibition of aromatase for specific targets. In this study the cytotoxic activity against T47D cells of alkaloids *Cinachyrella sp.* The first natural 6-hydroximino-4-en-3-one steroids were isolated from *Cinachyrella* spp and are examples of molecules that can be deployed against a specific type of cancer. They displayed high affinity to aromatase, which is the rate-limiting enzyme that catalyzes the conversion of androgens to estrogens [8].

The development of modern medicine is oriented for specific targets. They are mostly macromolecules, particularly proteins directly involved in the biological activities. Molecular docking is a technique, which provides high accuracy for drug interaction with specific receptors. This way, the technique can be used to predict activities of a molecule (compounds) [9]. The problem discussed in this research is how 1,4,9-triazatricyclo[7,3,1,0]trideca-3,5(13),10-trien-8-ol compound in the marine sponge *C. anomala* interacts with specific receptors participate in the cell cycle.

Materials and Methods

Sponge *C. anomala* was taken from the intertidal area of Kukup Beach, Kemadang Village, Tanjungsari Sub-District, Gunung Kidul Regency, DIY using direct collection technique [6]. The sponge samples were put into plastic bags and stored in an Icebox under a temperature of 5°C until the extraction time. Isolation and identification of alkaloid compound of marine sponge *C anomala* was conducted by Nurhayati [10]. The alkaloid compound from *C. anomala* is cinachyramine derivative, using molecular formula $C_{10}H_{13}N_3O$ 1,4,9-triazatricyclo[7,3,1,0]trideca-3,5(13),10-trien-8-ol [10].

Analysis of cell cycle using flow cytometry method

Cell cycle analysis was conducted using flow cytometry method [11], by distributing 10^6 cell/well, which is distributed into a 6-well plate. T47D cell was treated using 4,9-triazatricyclo[7,3,1,0]trideca-3,5(13),10-trien-8-ol and incubated for 24 hours. The cells were harvested using trypsin-EDTA, then centrifuged (at 2000 rpm, 30 seconds at 4°C). The cell suspension was homogenated and transferred to flowcyto-tube. Flow Cytometry data were analyzed using Modfit LT 3,0 program to find out the cell distribution in each phase of G1, S and G2/M.

Molecular docking of 1,4,9-triazatricyclo[7,3,1,0]trideca-3,5(13),10-trien-8-ol compound with cdk2 inhibitor receptor

Molecular docking was conducted using the software

***Corresponding author:** Awik Puji Dyah Nurhayati, Biology Department, Mathematic and Natural Science Faculty, Sepuluh November Institute of Technology Surabaya, Indonesia, E-mail: awiknurhayati@gmail.com

PLANTS (Protein-Ligand ANT-System). The compound 1,4,9-triazatricyclo[7,3,1,0]trideca-3,5(13),10-trien-8-ol was given SA2014 code. Ligand native of the receptor was NS9 (2s,3s)-3-{[7-(benzylamino)-3-(1-methylethyl))(pyrazolo[1,5-a]pyrimidin-5-yl] amino}butane-1,2,4-triol) - an inhibitor of cdk2. Cdk2 inhibitor is a checkpoint component of cell cycle at G_1 phase [12]. The comparative compound was Doxorubicin, with code receptor 3NS9.PDB in the Protein Data Bank (PDB). Protein and reference ligand (ref ligand) were prepared using YASARA it is computer programe, while ligand preparation used marvinSketch.

Results and Discussion

The 1,4,9-triazatricyclo[7,3,1,0]trideca-3,5(13),10-trien-8-ol test against T47D cell cycle

Flow Cytometry analysis for T47D cell showed that 1,4,9-triazatricyclo[7,3,1,0]trideca-3,5(13),10-trien-8-ol compound could arrest cell cycle in various phases. At a dosage of 31,25 µg/mL, the compound could induce cell cycle arrest in sub-G_1 5,87% and G_2/M 50,50% phases. A combined dosage between 12,25 µg/mL of SA2014 compound and 5 µg/mL of doxorubicin could induce cell cycle arrest in sub-G_1 5,75 and G_2/M 36,89% phases (Table 1). Doxorubicin is a potent anticancer agent for the treatment of breast cancer.

Interaction of 1,4,9-triazatricyclo[7,3,1,0]trideca-3,5(13),10-trien-8-ol compound in marine sponge *C. anomala* and cdk2 enzyme

The results of in silico molecular docking test for the interaction between 1,4,9-triazatricyclo[7,3,1,0]trideca-3,5(13),10-trien-8-ol compound of marine sponge *C. anomala* and cdk2 protein showed a more stable docking score, compared to that of doxorubicin (Table 2 and Figure 1).

Molecular docking approach showed that 1,4,9-triazatricyclo[7,3,1,0]trideca-3,5(13),10-trien-8-ol compound of maringe sponge *C. anomala* (SA2014) against cdk2 enzyme showed that SA2014 was related to amino acids of leucine[128], lysine[129], Prolin[130], asparagin[132], leucin[133], and Isoleucin[192], while

No	Treatment	Sub-G_1 (%)	G_0/G_1 (%)	S (%)	G_2/M (%)
1	T47D cell as control	3,04	43,83	25,60	30,83
2	12,25 µg/mL of SA2014	3,99	47,51	23,20	29,44
3	31,25 µg/mL of SA2014	5,87	30,53	18,90	50,50
4	5 µg/mL of doxorubicin	5,35	44,07	25	30,87
5	12,25 µg/ml of doxorubicin	6,60	42,71	26,30	30,80
6	Combination of 12,25 µg/ mL of SA2014 and 5 µg/ mL of doxorubicin	5,75	43,14	19,90	36,89

Table 1: 1,4,9-triazatricyclo[7,3,1,0]trideca-3,5(13),10-trien-8-ol(SA2014) test against T47D cell cycle.

Compounds	*Docking* score against cdk2
Ligand Native (NS9)	-91.31
doxorubicin	-36.59
SA2014	-65.43

Note: NS9=compound of cdk2 inhibitor, SA2014=structure of,4,9-triazatricyclo[7,3,1,0]trideca-3,5(13),10-trien-8-ol compound of maringe sponge *C. anomala*

Table 2: Docking score of 1,4,9-triazatricyclo[7,3,1,0]trideca-3,5(13),10-trien-8-ol compound of marine sponge *C. anomala* against Cell Devision Protein kinase2 (cdk2).

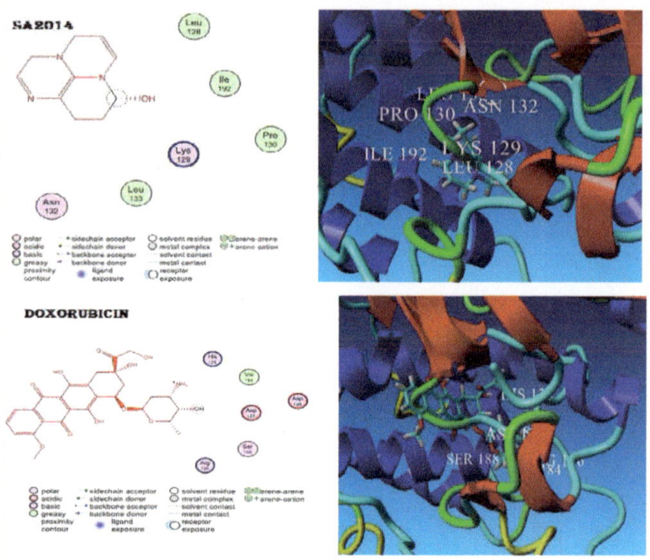

Figure 1: Visualization of *docking* for SA2014 [1,4,9triazatricyclo[7,3,1,0] trideca-3,5(13), 10-trien-8-ol] compound of maringe sponge *C. anomala* and doxorubicin against cdk2 enzyme.

doxorubicin was related to histidin[125], arginin[126], asparagin[127], valine[184], asparagin[185], and serin[188]. In terms of molecular size, 1,4,9-triazatricyclo[7,3,1,0]trideca-3,5(13),10-trien-8-ol compound is smaller than doxorubicin, and its binding to cdk2 enzyme is more stable than doxorubicin. Stability of compound binding with negative docking score showed that its binding to ligand is more stable [13]. In silico assay of molecular docking supported the result of in vivo assay that 1,4,9-triazatricyclo[7,3,1,0]trideca-3,5(13),10-trien-8-ol compound of marine sponge *C. anomala* could induce cell cycle arrest in G_0/G_1 phase of T47D cell.

Inhibition of 1,4,9-triazatricyclo[7,3,1,0]trideca-3,5(13),10-trien-8-ol compound in terms of proliferation and T47D cell cycle in sub-G_1 and G_2/M phases was assumed to play a role as an inhibitor of cdk2. Cdk2 inhibitor could lead to phosphorylation of Thr-14 and Tyr-15 and degenerate cyclin/cdk2 and cyclin A/Cdk2 complexes, thus interfering with the progress of G_1 to S [14].

Cells that are interrupted during the first to the mid-G_1 phase would be postponed in the checkpoint G_1. Checkpoint G_1 depended upon the increasing expression and activation of gene p53 [15]. Gene p53 plays an important role in maintaining genome stability, since p53 serves as a conductor [16]. Gene p53 promoted expression of downstream effector genes, such as p21, gadd45, mdm2 and Bcl-2 associated X protein (Bax) [17] to arrest the cell cycle and improve DNA or apoptosis [18]. The response was due to the fact that effector gene has a certain place in the regulator to recognize P53. After induction, P53 would activate transcription of a number of genes, such as P21 [19]. Gene p21 is related to the inhibitor cyclin A/Cdk2 activity [20]. Over-expression of P21 led to cell arrest at G_1 phase [21].

Conclusion

Molecular docking of 1,4,9-triazatricyclo[7,3,1,0]trideca-3,5(13),10-trien-8-ol compound of maringe sponge *C. anomala* (SA2014) is smaller than doxorubicin, and its binding to cdk2 enzyme is more stable than doxorubicin.

Acknowledgment

Financial support from the Sepuluh November Institute of Technology Surabaya, Indonesia and Directory of Higher Education.

Conflicts of Interest

The authors declare no conflict of interest.

References

1. Utit MJ, Turon X, Becerro MA, Galero J (1996) Feedeing deterrence in sponge. the role of toxicity, fisical defenses, energetic contents on life-history stage. Journal of Experiment Marine Biology and Ecology 205: 187-204.
2. Perdicaris S, Vlachogianni T, Valavanidis A (2013) Bioactive natural substances from marine sponges: new developments and prospects for future pharmaceuticals. Natural Products & Chemistry Research 1: 114.
3. Faulkner DJ (2001) Marine natural Product.Nat Prod Rep 18: 1- 49.
4. Bhakuni D, Rawat DS (2005) Bioactive marine natural products. Springer, New York.
5. Tilvi S, Rodrigues C, Naik CG, Parameswaran PS, Wahidullah S (2004) new bromotyrosine alkaloids from the marine sponge psammaplysilla purpurea. Tetrahedron 60: 10207-10215.
6. Wright AE (1998) Isolation of Marine Natural Product. Human Press.
7. Tohme R, Darwiche N, Gali-Muhtasib H (2011) Review. A Journey Under the Sea: The Quest for Marine Anti-Cancer Alkaloids. Molecules 16: 9665-9696.
8. Holland HL, Kumaresan S, Tan L, Njar VCO (1992) Synthesis of 6-hydroximino-3-oxo steroids, A new class of aromatase inhibitor. J Chem Soc Perkin Trans 1: 585-587
9. Purnomo H (2011) *Kimia komputasi*: molecular docking plants, penambatan molekul plants [protein-ligand-ant-system]. Penerbit Pustaka Pelajar. Yogyakarta.
10. Dulic V, Stein G, Far DF, Reed SI (1998) Nuclear accumulation of p21 cip at the onset of mitosis a role at the G2/Mphase transition. Mol Cell Biol 18: 546-557.
11. Nurhayati APD, Pratiwi S, Wahyuono S, Istriyati Abdillah S (2014) isolation and identification of alkaloid compound of marine sponge *Cinachyrella* sp. (Family Tetillidae). Journal of Advanced Botany and Zoology 2: 1-4.
12. Vermes J, Haanen C, Sperger R (2000) Flow cytometry of apoptotic cell death. Journal of Immunological Methods 243: 167-190.
13. Morgan DO (1995) Principles of CDK regulation. Nature 374: 131-134.
14. Kasta MB, Onyekwere O, Sidransky D (1991) Participation of p53 protein in the cellular response to DNA damage. Cancer Res 51: 6304-6311.
15. Nylander K, Dabelsteen E, Hal PA (2000) The p53 Molecule and its prognosis role in squamous cell carcinomas of the head and neck. Journal of Oral. Pathology & Medicine 29: 413-425.
16. Miyashita T, Reed JC (1995) tumor suppressor p53 is a direct transcriptional activator of the human bax gen. Cell 80: 293-299.
17. Malanga M, Pleschke J (1998) Poly (ADP-ribose) binds to specific domains of p53 and alters its DNA binding function. Journal of Bioogical Chemistry 273: 11839-11843.
18. Zhao H, Antinore S, Lung MJ, Fan F, Blanck P (2000) The central region of GADD45 is required for its interaction with p21/WAF. Experimental Cell Research 258: 92-100.
19. Harper JW, Adami GR, Wei N, Keyomarsi K, Elledge SJ (1993) The p21 Cdk-interacting protein Cip 1 is a potent inhibitor of G1 cyclin-dependent kinases. Cell 75: 805-816.
20. Rudoltz MS, Gary K, Kenneth R, Ruth JM, Gillies M (1996) Molecular biology of the cell cycle: potential for therapeutic applications in radiation oncology. Seminar in Radiation Oncology 6: 284-294.
21. Rodriguez J, Nunez L, Peixinho S, Jimenez C (1997) Isolation and synthesis of the first natural 6-hydroximino 4-en-3-one-steroids from the sponges Cinachyrella spp. Tetrahedron Lett 38: 1833-1836

Mercury Accumulation in Food Chain of Fish, Crab and Sea Bird from Arvand River

Yaghoob Parsa[1], Seyed Seyed Mohammad Bagher Nabavi[2], Seyedeh Narges Nabavi[2] and Mehdi Hosseini[2]*

[1]Department of Marine Biology, Faculty of Biological Science, Shahid Beheshti University, Tehran, Iran
[2]Department of Marine Biology, Faculty of Marine Science, Khoramshahr University of Marine Science and Technology, Iran

Abstract

In this study, concentration of mercury (Hg) was determined in the trophic levels of fish, blue crab and one sea birds species from Arvand river, located in the Khuzestan province in the lowlands of southwestern Iran at the head of the Persian Gulf. The order of mercury concentrations in tissues of the fish species was as follows: liver>gill>muscle, in tissues of the crab was as follows: hepatopancreas>gill>muscle and in tissues of the birds species was as follows: feather>liver>kidney. Therefore, Liver in fish, hepatopancreas in crab and feather in birds exhibited higher mercury concentration than the other tissues. There was a positive correlation between mercury concentrations in fish, crab and birds species with size of its food items. We expected to see higher mercury levels in tissues of female species because they are larger and can eat larger food items. The results of this study show that highest mean mercury levels were found in the birds (*A. crecca*), followed by crab (*P. pelagicus*), pelagic fish (*S. strongylura*) and benthic fish (*E. diacanthus*). Mean value of mercury in fish species, *S. strongylura* were (0.54 μg g^{-1} dry weight), *E. diacanthus* (0.27 μg g^{-1} dry weight), crab *P. pelagicus* (0.74 μg g^{-1} dry weight) and bird species *A. crecca* was (3.4 μg g^{-1} dry weight).

Keywords: Accumulation; Mercury; Food chain; Sea bird; Arvand River

Introduction

Mercury is a highly toxic metal, which, because of its volatility, can be rapidly spread all over the world from its natural and anthropogenic sources. These properties combined with its huge biomagnification in the food chain, make it the metal of the most environmental concern [1,2]. The anthropogenic mercury sources are both diffuse and local. The diffuse sources are mainly the burning of fossil fuel, especially coal, the incineration of municipal solid waste, and the cement production; the local sources consist mainly of effluents from industrial activities, especially in developing countries [3,4].

Most mercury pollution residues in aquatic environments, where mercury is converted to methyl mercury (MeHg) by aquatic biota. Because of the high affinity of MeHg to sulphydryl groups of proteins, this heavy metal is rapidly incorporated into the food chain, bioaccumulating in aquatic organisms, and biomagnifying from one trophic level to the next [5]. Other forms of mercury do not magnify in concentration up the food chain. Methyl mercury is created by bacteria in highly organic portions of aquatic systems, such as the sediment of river and wetlands. The zooplanktons pick up the methyl mercury as they filter the water and feed on algae. When small fish eat zooplankton, the methyl mercury builds up in their bodies as the fish grow bigger and older. Small fish are eaten by larger fish, and the concentration of methyl mercury increases at each step in the aquatic food chain. It is highest in large walleye, northern pike, and other predatory fish [3]. It's the methyl mercury in these fish that poses the greatest threat to human health. Therefore, the people who rely on fish for much of their diet are most at risk from methyl mercury, which can hamper normal development of the central nervous system. In adults, exposure to methyl mercury can result in damage to the nervous system and organs [4].

It is known that certain forms of mercury can readily accumulate within organisms tissues at much higher levels than those in the water column and in sediment [5]. Mercury is non-biodegradable contaminant that could enter aquatic food chains and consequently accumulates in organisms positioned in various trophic levels [6]. Fish, which usually occupies the last levels of aquatic food chains, are considered as the main aquatic pathway for metals to be transferred into human body [6]. Biological and ecological factors such as ecological needs, habitat, feeding habits have significant influences on metals bioaccumulation, bioavailability and therefore on their transference. However, feeding habit plays a significant role in the accumulation of pollutants in organism's tissues [7].

The blue swimming crab *Portunus pelagicus* belongs to the phylum; arthropoda which make up about three quarters of living animal species. For the fauna, the species most studied are the benthic macroinvertebrates, particularly those with low mobility, which accumulate larger concentrations of metals compared to animals that live in open water. Among the most-studied are the crab Portunidae which has important characteristics that allow the study of bioaccumulation: they feed mainly on a wide variety of fish, bivalves, plant, crustaceans and benthic animals, as well as the sediment itself and they have a slow growth rate and long life cycle. Therefore, this crab species is especially appropriate for use in studies of environmental impact by metals pollution.

As birds are ordinarily at the top of the food web, they are valuable for environmental monitoring [8]. Fish-eating birds, as top level carnivores, often have high levels of contaminants [9]. Mercury accumulation in many sea birds has been frequently reported [9-15]

***Corresponding author:** Mehdi Hosseini, Department of Marine Biology, Faculty of Biological Science, Shahid Beheshti University, Tehran, Iran
E-mail: smhbio@yahoo.com

but determination of mercury in birds such as kingfishers is rare and less common [16]. However, since kingfishers have a wide range of distribution, are avid eaters (consuming more than 50% of their body weight each day), and obtain virtually all their food from aquatic systems [16] they make for excellent bio monitoring species.

This study was to determine the levels of mercury in the trophic levels of benthic fish, pelagic fish, crab and sea bird residing in Arvand River, Iran. The second aim was to determine the ability of each species and tissue in mercury accumulation.

Materials and Methods

Study area

Sampling sites were selected along the Arvand River, northwest coasts of the Persian Gulf (Figure 1). The Arvand River, the border between Iraq and Iran, is the biggest river in the Persian Gulf. It passes three main cities including Al-Basre in Iraq, Abadan and Khoramshahr in Iran. For people of these cities, the Arvand River is considered as a main resource of seafood and drinking water. This river is formed by the confluence of Shatt al-Arab in Iraq and Karoon River in Iran. In addition to receiving effluents of more than seven big and small Iranian and Iraqi cities, there are many non-pointed and pointed metals sources along its course [1,17]. This river is surrounded by many petrochemical units such as Abadan petrochemical complex and petroleum refinery. Also, metals concentration may be due to discharge of sewage and urban effluents and related to the oil tankers traffic in the river. In addition, the Arvand River carries about 48 tons of oil residues to the northwest Persian Gulf annually [17]. Other sources of pollution in this area, including wars and invasions, are yet to be methodically investigated [1].

Two fish species (*Epinephelus diacanthus* and *Strongylura strongylura*) were caught in October of 2011. The samples placed on ice, immediately transported to the laboratory on the same day and stored at -20°C until analysis [18].

For analysis, muscle and liver of each fish were dissected, freeze-dried and crushed to uniform particle size [19]. It was then drained under folds of filter, weighed, wrapped in aluminum foil and then frozen at 10°C prior to analysis. The tissues were placed in clean watch glasses and were oven dried at 105°C for 1 hour and later cooled in the desiccators. Each sample of fish was homogenized in an acid-cleaned mortar and 2 g were digested in triplicate in a water bath at 60°C for 6 h after adding 2.5 mL each of concentrated HNO_3 and H_2SO_4 [18,19].

Each species was properly cleaned by rinsing with distilled water to remove debris, planktons and other external adherent. The samples were dissected with sterilized scissors and tweezers to remove samples

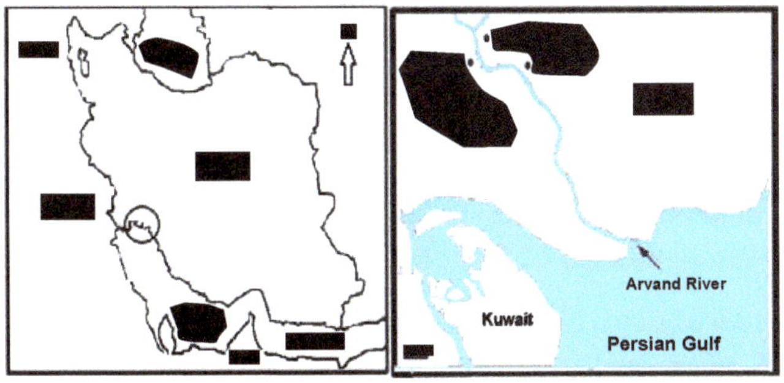

Figure 1: Map of Persian Gulf and study area.

Organism	Scientific name	Feeding habitat	Sex	n	Weight
Fish benthic	*Epinephelus diacanthus*	Benthic predator : consumes mainly benthic invertebrates, detritus and plant	Female	35	72 ± 7.2
			male	40	70 ± 1.3
Fish pelagic	*Strongylura strongylura*	Consumes mainly benthic fish, also invertebrates, plant and crustacean	Female	42	89 ± 4.9
			male	39	82 ± 1.4
Blue crab	*Portunus pelagicus*	Fish predator: consumes fish pelagic, detritus and plant	Female	45	98 ± 1.7
			male	51	83 ± 2.3
Sea bird	*Anas crecca*	Omnivorous: consumes mostly crab, fish, shrimp and marine invertebrates	Female	9	85 ± 3.5
			male	7	69 ± 1.4

Table 1: Scientific name, feeding habitat, sex and weight ((Mean ± SE g) of the specimens.

of three tissues (hepatopancreas, muscle and gill), in standardized locations: (i) muscle of the chelar propodus, due to higher metal accumulation was verified (ii) hepatopancreas tissue, which has a particularly high metabolic rate and (iii) gills, because of their osmoregulatory function. It was then drained under folds of filter, weighed, wrapped in aluminum foil and then frozen at 10°C prior to analysis. The tissues were placed in clean watch glasses and were oven dried at 105°C for 1 hour and later cooled in the desiccators.

The birds *Anas crecca* was collected from Arvand River. These birds were collected from hunters who had shot them during October of 2011. The specimens were weighed, stored in polyethylene bags, and kept at -20°C until they were dissected and analyzed. Samples were thawed and liver; feather and kidney were separately dissected from the bodies of the specimens. Samples were freeze-dried and homogenized [20]. Finally they were changed into the powder. Powdered sample and feather sample (finely cut) were directly weighed (50–100 ± 0.1 mg) into the precleaned combustion boats.

Each sample was analyzed for mercury by the mineralization method with HNO_3 at 65 percent, according to Ruelas-Inzunza J et al. [20]. Analyses were optimized by hollow cathode lamps (LCO), according to the metallic element analyzed, and samples were read using a GBC-932 AA atomic absorption spectrophotometer [21]. The equipment was calibrated using metal stock solutions (1000 ppm). The recovery means for mercury was 102% respectively.

The data were tested for normality using a Shapiro–Wilk's test. The data were not normally distributed. Mercury concentrations between tissues and species were tested for mean differences among species using One-Way analysis of variance (ANOVA) fallowed by Duncan post hoc test [22].

Results and Discussion

Table 1 shows scientific name, trophic level and mean body weight for the species of fish, crab and bird samples. Mercury concentrations were calculated in microgram per gram wet basis (μg g^{-1} dry weight). In order to check the validity of the measurements, reference material (Multi-4, Merck) was used.

During current study, mercury concentrations in tested tissues of all the fishes decreased in order liver>gill>muscle and in tissues of crab followed the hierarchical pattern hepatopancreas>gill>muscle (Table 2). Metals are taken up by fishes and crab from food and water, distributed throughout fish body by blood and eventually accumulated in target organs. Some tissues such as liver in fish and hepatopancreas in crab are considered as target organs for mercury accumulation [23]. The

Species		Hepatopancreas	Gill	Muscle	Liver	Feather	Kidney
E. diacanthus	Mean ± SE		0.21 ± 0.6	0.12 ± 0.4	0.49 ± 0.4		
	Range		0.09–0.71	0.07–0.73	0.07–0.73		
S. strongylura	Mean ± SE		0.54 ± 0.3	0.32 ± 0.7	0.78 ± 0.7		
	Range		0.28–2.1	0.11–1.2	0.07–0.73		
P. pelagicus	Mean ± SE	1.13 ± 0.1	0.68 ± 0.3	0.41 ± 0.5			
	Range	0.25–1.8	0.15–1.1	0.11–0.91			
A. crecca	Mean ± SE			1.1 ± 0.6	3.6 ± 0.1	5.2 ± 0.4	1.6 ± 0.2
	Range			0.75–4.8	1.1–6.5	1.3–8.6	0.92–4.1

Table 2: Mercury concentration (μg g^{-1}) in tissues of fish species, blue crab and sea bird species from Arvand River.

very high levels mercury in the liver and hepatopancreas in comparison to other tissues may be related to the content of metallothionein protein in liver tissues. Metallothionein protein that plays a significant role in the regulation and detoxification of mercury is produced in high levels in liver or hepatopancreas tissue [24]. This protein contains a high percentage of amino group, nitrogen and sulphur that sequester metals in stable complexes [23,24]. In general, the accumulation of mercury in the liver or hepatopancreas could be resulted from the abundance of metallothioneins proteins in these tissues in comparison to other tissues. The research same, the comparison on mercury accumulation between all tissues fish show that bioaccumulation of mercury was more in liver than other tissues [18]. Some other researchers also reported that in crustacean bioaccumulation of mercury was more in hepatopancreas than other tissues [25,26].

Gills usually reflect the concentrations of metals in surrounding water [7]. This organ is directly in contact with water and suspended materials thus could absorb different substances from the surrounding environment. They also serve a variety of physiological functions such as osmoregulation and gas exchange. Due to these functions, gills have remarkable influences on the exchange of toxic metals between a fish and its environment [27]. However, the muscle tended to accumulate less mercury in comparison to the liver and gills. This finding may reflect the low concentration of metallothioneins in the muscle tissue [7].

Mercury concentrations in tested tissues of the bird decreased in order feather>liver>kidney (Table 2). Mercury concentrations are usually found at the highest levels in feathers followed by liver, kidney and muscle tissues [11,28]. In birds, feathers play a major role in mercury excretion [13]. Mercury accumulates especially well in bird feather because it has high affinity for the sulfhydryl groups in keratin [28].

Once metals enter a bird, they can be stored in internal tissues such as the kidneys and the liver [28]. Many seabirds would demethylate organic Hg in tissues such as the liver and kidneys, and store a large portion of their Hg burdens in inorganic form [13]. Reports indicate that in a variety of wild caught birds, liver mercury levels are consistently higher than kidney levels which in turn exceed those in muscle [8,11,15]. This suggested markedly with studies which have reported levels of mercury in the kidney was higher than in the liver tissue [13,14,29].

Lewis and Furness [29] have reported mercury levels in the kidney that were most elevated in relation to the levels in the liver among black headed gull (*Larus ridibundus*) which were given the highest dose of mercury. The ratio of levels in the kidney and liver may, therefore, be an indicator of mercury poisoning. Higher kidney to liver ratios may indicate elevated mercury levels [29].

The result also showed that there are differences in mercury concentration in the different species. The indicated variability of metal concentration in the different species depends on their habitats [30] and feeding habits and food sources [31].

In present study, we considered two groups of fish including, benthic fish (*E. diacanthus*) and pelagic species (*S. strongylura*) as candidate biological indicators for evaluating the effects of trophic levels and habitats on metal accumulation. *E. diacanthus* lives in close association with sediment and feeds mainly upon benthic organism, detritus and tiny invertebrates. *S. strongylura* is a pelagic species that feeds on fish and mollusca. The blue swimming crab *P. pelagicus* feed mainly on a wide variety of fish, bivalves and crustaceans.

Also, we considered one species of bird *Anas crecca* that eat mainly aquatic crustacean and also fish. Therefore, *E. diacanthus* feeds on sediment and benthic organism, S. strongylura species feeds on *E. diacanthus*, blue crab *P. pelagicus* feed on both fish and bird *Anas crecca* feed on fishes and crab species.

Despite of being benthic and non-migratory fish, *E. diacanthus* indicated high mercury accumulation in their tissues. Because *E. diacanthus* feed on sediment and benthic organism and is close to bottom sediment and receive more sediment-associated mercury. Ratkowsky et al. [32] studied mercury contamination of Derwent Estuary in Australia. They found that there is a relationship between the frequency of high concentrations of mercury in fish tissues and feeding habits of the fish. According to Yi et al. [22], heavy metals concentrations in food chain increase in the following order benthic species followed by other species.

The high concentration of mercury in the blue crab species, *P. pelagicus* may be related to fish eating habits of the species. The diet of *P. pelagicus* consists of benthic fish. Fishes have been reported as a vector of the transfer of mercury element to top marine predators of the food chains [7,22].

The highest concentration of mercury was detected in tissues of bird *A. crecca*. Because organisms that are high on the trophic level might be expected to accumulate higher levels of bioaccumalative metals [7,22].Thus, in terms of mercury accumulation, the expected ranking in our study is pelagic fish<benthic fish<blue crab<bird. Therefore, this finding could confirm that mercury concentration is heavily controlled by habitat and feeding habits [33].

Since larger organisms generally exhibit higher contaminant level in their bodies [18,33] and organisms that eat higher organisms also accumulate more contaminants when comparing to organisms that eat a range of different foods or eat smaller organisms. We expected to see higher mercury levels in tissues of female crab (Figure 2) because they are larger and can eat larger food items. In general, mercury levels have been shown to increase with size and age of the ingested crab and it tends to be higher in species that occupy higher trophic levels [34], based on this logic we predicted that there should be higher levels of

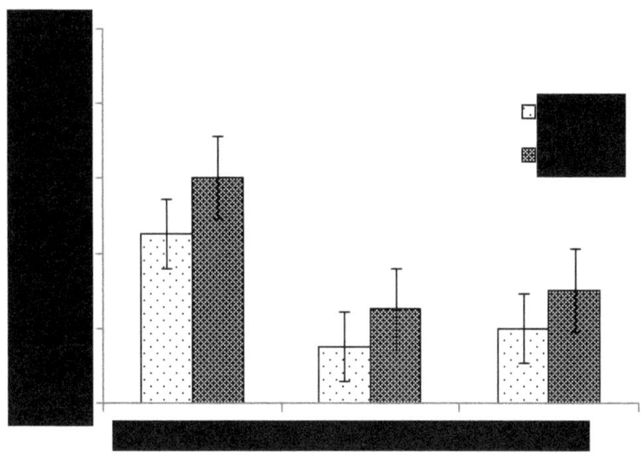

Figure 2: Mercury concentration (μg g^{-1}) in the tissues of male and female of blue crab *P. pelagicus*.

mercury in the larger predators. Gewurtz et al. [26] have shown that higher mercury levels in female crustacean were due to the increased consumption of food.

Several studies indicate a difference in metal body burden between male and female birds [35,36] while other studies report negligible differences in mercury between sexes [13,14,28]. We found that mercury values were larger in tissue of male of the species than the females (Figure 3). This finding merits future investigation since a larger sample size will allow a more accurate analysis and detection of significant differences if they indeed exist. Other reports indicate that although female birds can get rid of mercury in their eggs, the amount they shed in this way is usually small compared to the amount put into feathers during molt [35,36]. Therefore the small difference that has been reported in Hg body burdens in male and female is consistent with our current data. Zamani et al. [28] have shown that higher mercury levels were in female bird because they are larger and can eat larger food items.

Overall, these four species (fish, crab and bird) feed at comparable trophic levels and exhibit similar foraging behavior. The fish and crab species are (residents of Arvand River) only have access to local food, but bird *A. crecca*, which is migratory, can obtain food from other regions. We conclude that mercury concentrations in fish and crab species reflect mercury contaminant in Arvand river, but mercury burden of bird *A. crecca* is a reflection of diversity in food items from wider geographical locations with perhaps much higher mercury pollution than water of Iran and Iraq.

Consequently, mercury can be transferred to higher trophic level by biomagnification. This finding confirms that mercury have the ability to biomagnified through the aquatic food chains. In addition, their concentrations in high trophic level depend on the organisms of lowest trophic level. The results of this study show that highest mean mercury level were found in the bird *A. crecca*, followed by blue crab, benthic fish and pelagic fish. Barbosa et al. [24] studied the biomagnification of mercury in a marine food web in Rio Negro, Brazil. They found that mercury concentrations varied widely in all species; however, they showed a trend that depended on fish feeding strategies. The highest mean level was found in the piscivorous benthic, followed by piscivorous pelagic and herbivorous pelagic. They concluded that mercury is biomagnified through the food web. Cheng et al. [23] investigated mercury biomagnification in some food webs in the aquaculture pond ecosystem of the Pearl River Delta, China. They reported that the concentrations of mercury in the high trophic levels of the food web depend on those concentrations in lower trophic levels.

Correlation

There was significant correlation between mercury concentrations in fish and crab tissues (r=0.81, P<0.01, Figure 3a). It is suggested that significant correlation between mercury concentration in fish and crab tissues may be related to high variability of mercury levels in the fish.

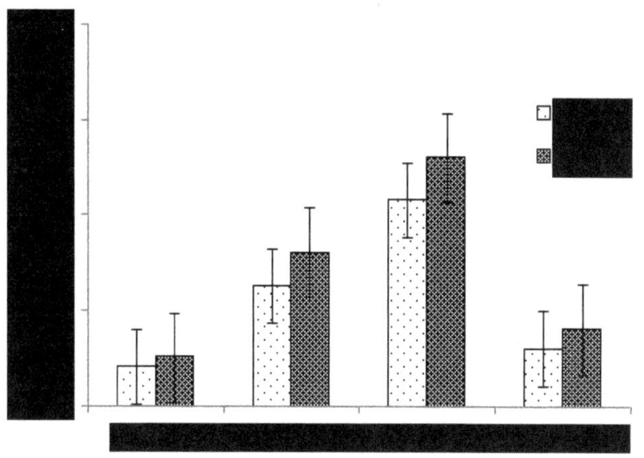

Figure 3: Mercury concentration (μg g^{-1}) in the tissues of male and female of migratory common teal *A. crecca*.

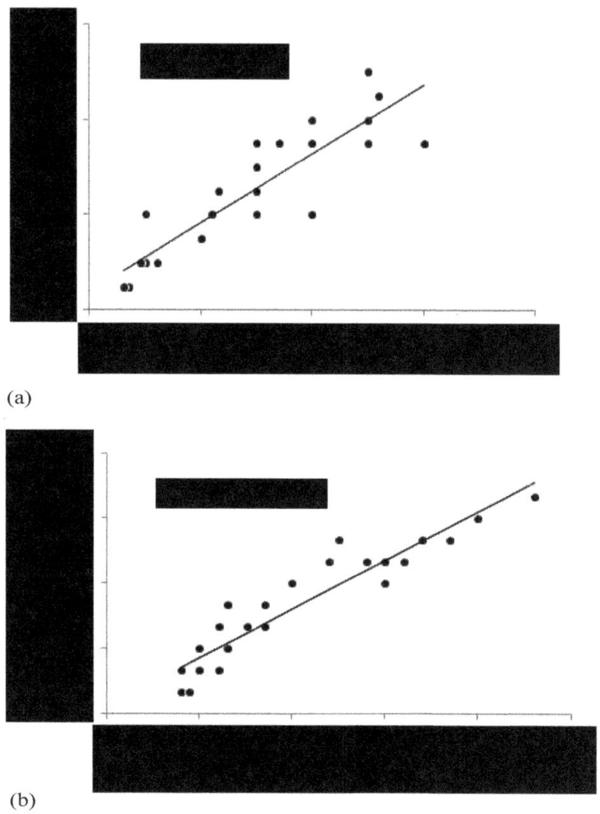

Figure 3a, b: Relationship of mercury concentrations (μg g^{-1}) between fish and crab (a), between crab and bird (b).

E. diacanthus is benthic species and is more common in the bottom sediment and receive more sediment associated metals. Also, there was significant correlation between mercury concentrations in crab and bird tissues (r=0.86, P<0.001, Figure 3b). Therefore, this finding could be due to the differences in their ecological niches and feeding habitats. Therefore, *Anas crecca* could be considered as suitable biomonitor agents for mercury contamination in the study area.

Conclusion

Our results indicated that the levels of mercury varied among species and tissues. Mercury can be transferred to higher trophic level by biomagnification. Also, mercury concentrations in high trophic level depend on the organisms of lowest trophic level. Therefore, the results of this study show that highest mean mercury level were found in the bird (*A. crecca*), followed by blue crab (*P. pelagicus*), benthic fish (*E. diacanthus*) and pelagic fish (*S. strongylura*). There was a positive correlation between mercury concentrations in organisms with size of food items. Therefore, we expected to see higher mercury levels in tissues of female species because they are larger and can eat larger food items. The results confirmed that the concentration of mercury in organisms strongly affected by habitat and feeding habit, however, the influences of habitat appears to be more than feeding habit. The fish species are traditional food for fishermen and traditional coastal communities, where it is an important source of protein. From a public health standpoint, measurement of mercury levels in the tissues of fishes is highly dangerous for human health. Therefore, studies of mercury concentrations in coastal areas are relevant and useful for monitoring the health of environmental compartments, maintenance of biodiversity, and for assuring the quality of life, mainly for humans. Finally, there is a need to develop and refine bird model to serve as a sentinel of ecosystem health, to help provide early warning indications for possible human exposure.

Acknowledgement

The author would like to thank Prof. Bahram Kiabi for field assistance. This work was funded by Environmental Protection of Tehran, Iran.

References

1. Sheppard C, Al-Husiani M, Al-Jamali F, Al-Yamani F, Baldwin R, et al. (2010) The Gulf: A young sea in decline. Mar Pollut Bull 60: 13-38.
2. Al-Saleh I, Shinwari N (2002) Preliminary report on the levels of elements in four fish species from the Arabian Gulf of Saudi Arabia. Chemosphere 48: 749-755.
3. Fitzgerald WF, Lamborg CH, Hammerschmidt CR (2007) Marine Biogeochemical Cycling of Mercury. Chem Rev 107: 641-662.
4. Pirrone N, Costa P, Pacyna JM, Ferrara R (2001) Mercury emissions to the atmosphere from natural and anthropogenic sources in the Mediterranean region. Atmos Environ 35: 2997-3006.
5. Beltrame MO, Marco SGD (2010) Influences of sex, habitat, and seasonality on heavy-metal concentrations in the burrowing crab (Neohelice granulate) from a coastal lagoon in Argentina. Arch Environ Contam Toxicol 58: 746-756.
6. Svensson BG, Schutz A, Nilsson A, Akesson I, Akesson B, et al. (1992) Fish as a source of exposure to mercury and selenium. Sci Total Environ 126: 61-74.
7. Bustamante P, Bocher P, Cherel Y, Miramand P, Caurant F (2003) Distribution of trace elements in the tissues of benthic and pelagic fish from the Kerguelen Islands. Sci Total Environ 313: 25-39.
8. Horai S, Watanabe I, Takada H, Iwamizu Y, Hayashi T, et al. (2007) Trace element accumulations in 13 avian species collected from the Kanto area, Japan. Sci Total Environ 373: 512-525.
9. Burger J, Gochfeld M (1991) Lead, mercury, and cadmium in feathers of tropical terns in Puerto Rico and Australia. Arch Environ Contam Toxicol 21: 311-315.
10. Thompson DR, Furness RW (1989) The chemical form of mercury stored in South Atlantic seabirds. Environ Pollut 60: 305-317.
11. Kim EY, Murakami T, Saeki K, Tatsukawa R (1996) Mercury levels and its chemical form in tissues and organs of seabirds. Arch Environ Contam Toxicol 30: 259-266.
12. Monteiro LR, Granadeiro JP, Furness RW, Oliveira P (1999) Contemporary patterns of mercury contamination in the Portuguese Atlantic inferred from mercury concentrations in seabird tissues. Mar Environ Res 47: 137-156.
13. Saeki K, Okabe Y, Kim EY, Tanabe S, Fukuda M, et al. (2000) Mercury and cadmium in common cormorants (Phalacrocorax carbo). Environ Pollut 108: 249-255.
14. Nam DH, Anan YI, kemoto T, Okabe Y, Kim EY, et al. (2005) Specific accumulation of 20 trace elements in great cormorants (Phalacrocorax carbo) from Japan. Environ Pollut 134: 503-514.
15. Houserova P, Kuban V, Kracmar S, Sitko J (2007) Total mercury and mercury species in birds and fish in an aquatic ecosystem in the Czech Republic. Environ Pollut 145: 185-194.
16. Baron LA, Ashwood TL, Sample BE, Welsh C (1997) Monitoring bioaccumulation of contaminants in the belted kingfisher (Ceryle alcyon). Environ Monit Assess 47: 153-165.
17. Al-Hello AA, Al-Obaidy AM (1997) The chemistry of Shatt Al-Arab Water from Qurna to Al-Fao. Mar Mesopotamica 12: 190-201.
18. Abdolahpur Monikh F, Safahieh AR, Savari A, Doraghi A (2012) Heavy metal concentration in sediment, benthic, benthopelagic, and pelagic fish species from Musa Estuary (Persian Gulf). Environ Monit Assess 185: 215-222.
19. Basset J, Denney RC, Jeffery GH, Mendhan J (1981) Vogel:Analise Inorganica Quantitativa, fourthed. Guanabara S. A, Rio de Janeiro.
20. Ruelas-Inzunza J, Paez-Osuna F (2004) Trace metals in tissues of resident and migratory birds from a lagoon associated with an agricultural drainage basin (SE Gulf of California). Arch Environ Contam Toxicol 47: 117-125.
21. Athanasopoulos N (1993) Flame methods manual for atomic absorption. GBC Scientific Equipment PTY Ltd, Victoria
22. YI Y, Wang Z, Zhang K, Yu G, & Duan X (2008) Sediment pollution and its effect on fish through food chain in the Yangtze River. International Journal of Sediment Research 23: 338-347.
23. Cheng Z, Liang P, Shao D-D, Wu S-C, Nie X-P, et al. (2011) Mercury biomagnification in the aquaculture pond ecosystem in the Pearl River Delta. Arch Environ Contam Toxicol 61: 491-499.
24. Barbosa AC, de Souza J, Dorea JG, Jardim WF, Fadini PS (2003) Mercury biomagnification in a Tropical Black Water, Rio Negro, Brazil. Arch Environ Contam Toxicol 45: 235-246.
25. Pourang N, Nikouyan A, Dennis JH (2005) Trace element concentrations in fish, surficial sediments and water from northern part of the Persian Gulf. Environ Monit Assess 109: 293-316.
26. Gewurtz SB, Bhavsar SP, Fletcher R (2011) Influence of fish size and sex on mercury/PCB concentration: importance for fish consumption advisories. Environ Int 37: 425-434.
27. Farkas A, Salanki J, Specziar A (2003) Age and size-specific patterns of heavy metals in the organs of freshwater fish Abramis brama L. populating a low-contaminated site. Water Res 37: 959-964.
28. Zamani-Ahmadmahmoodi R, Esmaili-Sari A, Ghasempoury SM, Savabieasfahani M (2008) Mercury levels in selected tissues of three kingfisher species; Ceryle rudis, Alcedo atthis, and Halcyon smyrnensi, from Shadegan Marshes of Iran. Ecotoxicology 18: 319-324.
29. Lewis SA, Furness RW (1991) Mercury accumulation and excretion in laboratory reared black-headed gull Larus ridibundus chicks. Arch Environ Contam Toxicol 21: 316-320.
30. Yılmaz AB, Yılmaz L (2007) Influences of sex and seasons on levels of heavy metals in tissues of green tiger shrimp (Penaeus semisulcatus de Hann, 1844). Food Chemistry 101: 1664-1669.
31. Caussy D, Gochfeld M, Gurzau E, Neagu C, Ruedel H (2003) Lessons from case studies of metals: investigating exposure, bioavailability, and risk. Ecotoxicol Environ Saf 56: 45-51.
32. Ratkowsky DA, Dix TG, Wilson KC (1975) Mercury in fish in the Derwent

Estuary, Tasmania, and its relation to the position of the fish in the food chain. Australian Journal of Marine and Freshwater Research 26: 223-231.

33. Agah H, Leermakers M, Elskens M, Fatemi SMR, Baeyens W (2009) Accumulation of trace metals in the muscle and liver tissues of five fish species from the Persian Gulf. Environ Monit Assess 157: 499-514.

34. Phillips GR, Lenhart TE, Gregory RW (1980) Relation between trophic position and mercury accumulation among fishes from the Tongue River reservoir. Environ Res 22: 73-80.

35. Furness RW (1993) Birds as monitors of pollutants. In: Furness RW, Greenwood JJD, (eds). Birds as monitors of environmental change. Chapman & Hall, London, UK, 103.

36. Gochfeld M, Burger J (1987) Heavy metal concentrations in the liver of three duck species: influence of species and sex. Environ Pollut 45: 1-15.

Mercury in Different Tissues of Grey Herons (*Ardea cinerea*) from the Volta Lake, Ghana

Kwaansa-Ansah EE*, Agyemang D and Opoku F

Department of Chemistry, Kwame Nkrumah University of Science and Technology, Kumasi, Ghana

Abstract

Birds as high consumers and account for accumulative processes in marine pollution accumulate contaminants such as mercury (Hg). Therefore, in this study, the concentration of total mercury (THg) in ten (10) different tissues of *Ardea cinerea* from the Volta Lake, Ghana was measured. The THg concentrations in *Ardea cinerea* tissues follow the order: lungs<heart<brain<gizzard<intestines <flesh<blood <kidney<liver<feather. The lowest THg concentration occurred in lung tissue and the highest in feathers tissues (1.23 ± 0.57 and 2.45 ± 0.29 μgg^{-1}, respectively). Correlational analyses showed that total mercury concentrations in tissues are significantly related to age of the birds ($p<0.05$).The results showed that the accumulation of total mercury varied among the various tissues. However, *Ardea cinerea* species showed relatively high THg concentration in their feather tissues.

Keywords: Total mercury; *Ardea cinerea*; Tissues; Feather; Volta lake

Introduction

Mercury as a highly toxic non-essential metal strongly influences humans and wildlife with direct effects on behaviour, neurology, endocrinology and development [1,2]. The source of Hg is from both natural and anthropogenic sources as well as human activities [3]. Hg contaminates oceanic islands and polar and sub-polar regions due to its high volatility and long atmospheric residence time [4,5]. Exposure to mercury results in impaired vision and hearing, dizziness, vomiting, headache, muscular weakness, allergies, depressed immune system and brain damage [6].

The Volta Lake is a unique and an extraordinary natural resource that provide drinking water, food, recreation, employment and transportation to millions of Ghanaians as well as tourism. Volta Lake is the World's second largest and Africa's largest artificial inland water body which occupy 3.6% of the total land area of Ghana. The Volta Lake supply about 90% of the nation's inland fish supply.

It is known that certain forms of mercury can readily accumulate within organisms tissues at much higher levels than those in the water column and in sediment [7]. Mercury as a non-biodegradable contaminant can enter aquatic food chains and consequently accumulates organisms at the various trophic levels [8]. Biological and ecological factors have significant influences on metals bioaccumulation and bioavailability [9].

Birds are valuable tool for environmental monitoring because they are always at the top of the food chain and often have high levels of Hg contamination [10-14]. Feathers are among the most attractive tissue to monitor, because they are considered as the main route for Hg elimination [14]. Mercury accumulation in piscivorous birds has extensively being studied in the past [15-18], but to the best of our knowledge data on mercury contamination in different tissues of grey herons from the Volta Lake, Ghana is lacking. This work is aiming to evaluate the residual accumulation of mercury in different tissues of grey heron's in Ghana.

Materials and Methods

Study area

The Volta Lake was formed in 1965 as a result of the construction of the Akosombo dam which generates hydroelectric power for Ghana. It has a surface area of 8500 km^2 and covers about 4% of Ghana's total land area. It covers 1,232 communities with a population of over 300,000 people living along the shoreline whose livelihood depend on the Volta Lake [19].

Sampling

Our overall protocol, under appropriate state and federal permits was to collect grey herons from shorebirds netted during their migratory stopover on Volta Lake in Ghana. 10 grey herons were collected as by catch of long line fishing vessels or found recently dead on land killed following collision with the Scientific considered for this study. The grey herons were weighed in the field, immediately frozen and kept at -20°C until analyses. Birds were thawed overnight and dissected for kidney, intestines, lungs, heart, gizzard, blood, brain, liver, flesh and feathers that give the best measure of the Hg contamination of a bird. The birds were transported to the Department Of Wildlife, Faculty of Renewable Natural Resources, Kwame Nkrumah University of Science and Technology, Kumasi for measurement of their biometrical features. After measurement of biometrical features, the birds were kept frozen at -4°C until analysis.

Analysis

During the analysis, the birds were dissected and the various tissues separated into kidney, intestines, lungs, heart, gizzard, blood, brain, liver, flesh and feathers. Each tissue was washed with deionized water minced into smaller pieces and homogenized. The samples were digested for total mercury determination by an open flask

***Corresponding author:** Kwaansa-Ansah EE, Department of Chemistry, Kwame Nkrumah University of Science and Technology, Kumasi, Ghana
E-mail: eekwaansaansah@yahoo.com

procedure developed by Akagi and Nishimura [20]. In the procedure, approximately 0.5 g of each tissue was weighed into a 50 mL digestion tube and 1mL deionized water, 2 mL of a mixture of HNO_3 and $HClO_4$ (1:1) and 5 mL H_2SO_4 was added. The samples were digested at 200°C ± 5 for 30 minutes, cooled to room temperature and topped to the 50mL mark with deionized water. A blank and standard solution was digested using 25, 50 and 100 µL of1 µg/mL standard Hg solution. Total mercury concentrations were determined using an automatic Mercury Analyser (Model HG-5000). The reducing reagent used for the analysis was 0.5 mL of 10% (w/v) $SnCl_2.2H_2O$ in 1 MHCl. Quality assurance samples analysed included procedural blanks, replicate samples and post-digestion spikes. The accuracy of the procedure was determined by analysis of standard reference materials (SRMs). The SRMs included National Research Council of Canada (NRCC) DOLT-4 (dogfish liver) and TORT-2 (lobster hepatopancreas). All CRMs were recovered at ± 10% of the certified value; 0.27 ± 0.06 for Tort-2 and 2.58 ± 0.22 for Dolt-4. Statistical analysis was conducted using IBM Statistical Package for the Social Sciences '20. A significance level of $\alpha<0.05$ was used for all tests.

Results and Discussion

The mean concentration of mercury in the lungs, heart, brain, gizzard, intestines, fresh, blood, kidney, liver and feathers tissues were 1.23 ± 0.57, 1.30 ± 0.58, 1.45 ± 0.46, 1.49 ± 0.62, 1.68 ± 0.56, 1.94 ± 0.55, 1.86 ± 0.58, 2.26 ± 0.31, 2.32 ± 0.30 and 2.45 ± 0.29, respectively (Table 1).The lowest Hg concentration occurred in lung tissue and the highest in feathers tissues (1.23 ± 0.57 and 2.45 ± 0.29 μgg^{-1}, respectively).The concentration of THg in the tissues studied was found to increase in the following order: lungs<heart<brain<gizzard<intestines<flesh<blood<kidney<liver<feathers.

During current study, mercury concentrations in tested tissues of all the fishes increased in order lungs < heart <brain<gizzard<intestines<flesh<blood<kidney<liver<feathers (Figure 1). Mercury concentrations are usually found at the highest levels in feathers followed by liver, kidney, and muscle tissues [17,21]. In birds, feathers play a major role in mercury excretion [22]. As a result of high affinity for the sulfhydryl groups in keratin, mercury can accumulates in bird feather [21,23]. Previous reports indicate that Hg accumulates more in liver than kidney follow by muscle [11,17,18]. Lewis and Furness [24] also reported lower mercury levels in kidney than liver in black-headed gull (*Larus ridibundus*).

Frodello et al. [25] and Storelliet al. [26] reported that the liver appeared to be the preferential organ for mercury accumulation, followed by the kidney. The THg concentration in the liver is accounted for the biotransformation of demethylates organic mercury into less

Tissue	Mean THg concentration	Range of THg Concentration
Lung	1.23 ± 0.57	0.77 - 2.70
Heart	1.30 ± 0.58	0.79 - 2.76
Brain	1.45 ± 0.46	1.18 - 2.57
Gizzard	1.49 ± 0.62	0.96 - 2.79
Intestines	1.68 ± 0.56	1.15 - 2.79
Flesh	1.94 ± 0.55	1.91 - 2.80
Blood	1.86 ± 0.58	1.20 - 2.85
Kidney	2.26 ± 0.31	1.97 - 2.85
Liver	2.32 ± 0.30	2.02 - 2.86
Feather	2.45 ± 0.29	2.05 - 2.84

Table 1: Mean THg concentration (μgg^{-1}) in tissues of *Ardea cinerea*.

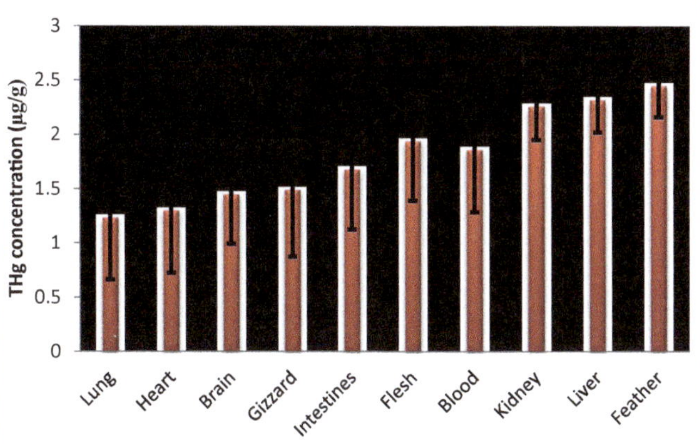

Figure 1: Total Mercury concentration (µg/g) in tissues of the *Ardea cinerea*.

toxic [27]. The levels of mercury found in the kidneys are explained on the basis that the kidney is the major reservoir of inorganic mercury in birds as well as mammals [28]. The kidney contains a thiol-rich protein called metallothionein and exposure to toxic metals triggers the production of this protein and binds tightly to the metal [29]. Spalding et al. [30], found that kidney disease and gout were present in great white herons are greater than 25 µg/g. Though elevated mercury levels were found in the kidney tissues of the birds in this study, the highest concentration of THg (2.85 µg/g) found in the kidney tissues of *Ardea cinerea* was about nine times less than the concentration reported to cause diseases and adverse effects in birds.

Higher concentration of mercury found in the feathers may be due to the fact that mercury is excreted mainly via the feathers [31]. Feathers which represent the major pathway for elimination of mercury in birds are useful indicators for assessment of whole bird mercury burdens [15,32,33]. Studies on the Great skua (*Catharcta skua*) shows that mercury was more contaminated in the feathers [34]. Mercury in the food eaten during feather growth considerably affects the mercury levels of the feathers [35].

Boening [36] and Ochoa-Acuna et al. [37] found some differences in Hg concentrations in feathers across taxonomic bird groups. Mercury concentrations in birds also depend on body size, moult strategy, migration patterns and physiology [38]. Fish-eating birds are at risk of higher contents of Hg because its circulation is associated mainly with water basins [37,39].

Pearson correlation coefficient between the age and THg concentration in all the tissues studied showed a very strong linear relationship. Total mercury concentrations in the tissues of *Ardea cinerea* were positively correlated with age ($p<0.05$). This relationship suggests that as grey herons mature, mercury is accumulated in all the tissues studied as such they can be useful bio-indicators for monitoring Hg pollution in the Volta Lake and other aquatic ecosystems.

Conclusion

The results showed that the accumulation of total mercury varied among the various tissues. However, *Ardea cinerea* species showed relatively high THg concentration in feather tissues. From a public health perspective, levels of mercury in the tissues of birds are very alarming. Therefore, studies on mercury contamination in birds especially around the coastal areas are relevant and useful for environmental monitoring.

Competing Interests

The authors declare that they have no competing interests.

Acknowledgement

The authors are very grateful to the National Council for tertiary Education (NTCE), Ghana for a research grant under the Teaching and Learning Innovation Fund (TALIF-KNUSTR/3/005/2005). We are also grateful to the Department Of Wildlife, Faculty of Renewable Natural Resources, Kwame Nkrumah University of Science and Technology (KNUST), Kumasi, Ghana for the use of their facilities for this work.

References

1. Scheuhammer AM, Meyer MW, Sandheinrich MB, Murray MW (2007) Effects of environmental methylmercury on the health of wild birds, mammals, and fish. Ambio 36: 12-18.
2. Grandjean P, Satoh H, Murata K, Eto K (2010) Adverse effects of methyl mercury: environmental health research implications. Environ Health Persp 118: 1137-1145.
3. Selin NE (2009) Global biogeochemical cycling of mercury: a review. Annu Rev Environ Resourc 34: 43-63.
4. Fitzgerald WF, Engstrom DR, Mason RP, Nater EA (1998) The case for atmospheric mercury contamination in remote areas. Environ Sci Technol 32: 1-7.
5. Ebinghaus R, Kock HH, Temme C, Einax JW, Löwe AG, et al. (2002) Antarctic springtime depletion of atmospheric mercury. Environ Sci Technol 36: 1238-1244.
6. Risher JF (2003) Elemental mercury and inorganic mercury compounds: Human health aspects. World Health Organization, Geneva.
7. Beltrame MO, Marco SGD (2010) Influences of sex, habitat, and seasonality on heavy-metal concentrations in the burrowing crab (Neohelice granulate) from a coastal lagoon in Argentina. Arch Environ Contam Toxicol 58: 746-756.
8. Hosseini M, Nabavi SMB, Parsa Y (2013) Bioaccumulation of Trace Mercury in Trophic Levels of Benthic, Benthopelagic, Pelagic Fish Species, and Sea Birds from Arvand River, Iran. Biological Trace Element Research 156: 175-180.
9. Bustamante P, Bocher P, Cherel Y, Miramand P, Caurant F (2003) Distribution of trace elements in the tissues of benthic and pelagic fish from the Kerguelen Islands. Sci Total Environ 313: 25-39.
10. Furness RW, Greenwood JD (1993) Birds as Monitors of Environmental Changes British Trust for Ornithology. 1stEdn UK: Thetford Norfolk 342.
11. Horai S, Watanabe I, Takada H, Iwamizu Y, Hayashi T, et al. (2007) Trace element accumulations in 13 avian species collected from the Kanto area, Japan. Sci Total Environ 373: 512-525.
12. Burger J, Gochfeld M (2004) Marine birds as sentinels of environmental pollution. Eco Health 1: 263-274.
13. Burger J, Gochfeld M (1991) Cadmium and lead in common terns (Aves, Sterna hirundo): Relationship between levels in parents and eggs. Environ Monit Assess 16: 253-258.
14. Monteiro LR, Furness RW (1995) Seabirds as monitors of mercury in the marine environment. Water Air Soil Pollut 80: 851-870.
15. Thompson DR, Furness RW (1989) The chemical form of mercury stored in South Atlantic seabirds. Environ Pollut 60: 305-317.
16. Monteiro LR, Granadeiro JP, Furness RW, Oliveira P (1999) Contemporary patterns of mercury contamination in the Portuguese Atlantic inferred from mercury concentrations in seabird tissues. Mar Environ Res 47: 137-156.
17. Kim EY, Murakami T, Saeki K, Tatsukawa R (1996) Mercury levels and its chemical form in tissues and organs of seabirds. Arch Environ Contam Toxicol 30: 259-266.
18. Houserova P, Kuban V, Kracmar S, Sitko J (2007) Total mercury and mercury species in birds and fish in an aquatic ecosystem in the Czech Republic. Environ Pollut 145: 185-194.
19. Braimah LI (2001) Volta lake fisheries management plan: Fisheries sub-sector capacity building project. 18-32.
20. Akagi H, Nishimura H (1991) In: Suzuki T, Imura N, Clarkson TW (eds) Speciation of mercury in the environment. Advances in mercury toxicology New York: Plenum Press 53-76.
21. Zamani-Ahmadmahmoodi R, Esmaili-Sari A, Ghasempoury SM, Savabieasfahani M (2008) Mercury levels in selected tissues of three kingfisher species; Cerylerudis, Alcedoatthis, and Halcyon smyrnensi, from Shadegan Marshes of Iran. Ecotoxicology 18: 319-324.
22. Al-Hello AA, Al-Obaidy AM (1997) The chemistry of Shatt Al-Arab water from Qurna to Al-Fao. Mar Mesopotamica 12: 190-201.
23. Saeki K, Okabe Y, Kim EY, Tanabe S, Fukuda M, et al. (2000) Mercury and cadmium in common cormorants (Phalacrocoraxcarbo). Environ Pollut 108: 249-255.
24. Lewis SA, Furness RW (1991) Mercury accumulation and excretion in laboratory reared black-headed gull Larus ridibundus chicks. Arch Environ Contam Toxicol 21: 316-320.
25. Frodello JP, Romeo M, Viale D (2000) Distribution of mercury in the organs and tissues of five toothed-whale species of the Mediterranean. Environ Pollut 108: 447-452.
26. Storelli A, Marcotrigiano GO (2005) Accumulation of mercury, cadmium, lead and arsenic in Swordfish and Bluefin tuna from the Mediterranean Sea: A comparative study. Mar Pollut Bull 50: 1004-1007.
27. Kinabo C, Lyatuu H (2009) Mercury contamination in domestic ducks in Geita Northwest Tanzania. Tanz J Sci 35: 38-45.
28. United States Department of the Interior (1998) Guidelines for Interpretation of the Biological Effects of Selected Constituents in Biota, Water, and Sediment: Mercury. Report No. 3. National Irrigation Water Quality Program Information, United States Department of the Interior.
29. Ware RA, Burkholder PM, Chang LW (1975) Ultra structural changes in renal proximal tubule after chronic organic and inorganic mercury intoxication. Environ Res 121-140.
30. Spalding MG, Bjork RDG, Powell VN, Sundlof SF (1994) Mercury and cause of death in great white herons (Ardeaherodiasoccidentalis). J Wildlife Manage 58: 735-739.
31. Parslow JLF, Jeffries DJ, Hason HM (1973) Gannet mortality incidents in 1972. Marine Poll Bull 4: 41-43.
32. Liu QG, Chen Y, Li JL, Chen, LQ (2007) The food web structure and ecosystem properties of a filter-feeding carps dominated deep reservoir ecosystem. Ecol Model 203: 279-289.
33. Furness RW, Muirhead SJ, Woodbum M (1986) Using bird feathers to measure mercury in the environment: relationships between mercury content and moult. Marine Pollution Bulletin 17: 27-30.
34. Bearhop S, Waldron S, Thompson D, Furness R (2000) Bio-amplification of mercury in great skua Catharactaskua chicks: The influence of trophic status as determined by stable isotope signatures of blood and feathers. Mar Pollut Bull 40: 181-185.
35. Mautinom M, Bell JU (1986) Experimental Lead Toxicity in the Ring-necked Duck. Environmental Research 41: 538-545.
36. Boening DW (2000) Ecological effects, transport, and fate of mercury: A general review. Chemosphere 40: 1335-1351.
37. Ochoa-Acuna H, Sepulveda MS, Gross TS (2002) Mercury in feathers from Chilean birds: influence of location, feeding strategy, and taxonomic affiliation. Mar Pollut Bull 44: 340-349.
38. Dauwe T, Bervoets L, Pinxten R, Blust R, Eens M (2003) Variation of heavy metals within and among feathers of birds of prey: effects of molt and external contamination. Environ Pollut 124: 429-436.
39. Stewart FM, Phillips RA, Catry P, Fumess RW (1997) Influence of species, age and diet on mercury concentrations in Shetland seabirds. Marine Ecology Progress Series 151: 237-244.

Fatty Acids Profile, Atherogenic (IA) and Thrombogenic (IT) Health Lipid Indices in *Leiognathusbindus* and *Upeneussulphureus*

Mansoreh Ghaeni[1], Khadijeh Nabi Ghahfarokhi[2]* and Ladan Zaheri[2]

[1]*Department of Fisheries, Ahvaz Branch, Islamic Azad University, Ahvaz, Iran*
[2]*Fisheries Scientific Association, Islamic Azad University, Ahvaz, Iran*

Abstract

In this study, the fatty acids composition and the related health lipid indices (IA, atherogenic and IT thrombogenic) of Orange fin pony fish (Leiognathusbindus) and sulphur goatfishes (*Upeneussulphureus)* were studied in fall and spring in Mahshahr port in south of Iran. Fatty acids were analyzed by Gas Chromatograph (FID). The fatty acid composition of *Leiognathusbindus* and *Upeneussulphureus* showed a relevant proportion (26.95% and 25.34% respectively) of poly-unsaturated fatty acids (PUFAs) with a prevalence of the n – 3 series. The IA and IT indices resulted comparable in spring and fall. The ratio of W3/W6 was different in fall and spring for both species.

Keywords: *Leiognathusbindus; Upeneussulphureus;* Atherogenic; Thrombogenic

Introduction

Recently great interest has been devoted to the study of the lipid composition of fish and fish products which are recognized as important sources of *n*-3 fatty acids. In fact a wide class of health beneficial fatty acids characterizes their nutritional value and their therapeutic effects are raising the commercial interest of this food mainly for its high content of healthy fatty acids [1,2].

The lipids of marine fish have gained attention because they are rich in fatty acids, especially docosahexaenoic acid (DHA) and eicasapentaenoic acid (EPA), which are predominant. Among the essential fatty acids are important for the human health. Most fatty acids can be synthesized in the body, however a human body lacks the enzymes required to produce the two essential fatty acids EPA and DHA. These fatty acids must be taken from the diet [3].

The lipid content of fish is highly variable between and within species. Many factors appear to contribute to this variability, including food availability; catch location, fish size, and maturity stage, biological variations, sampled tissue, ration size and starvation [4].

In this study we selected two marine fish; goat fish and pony fish because they are trash fish in our trawlers and the world total catch of them is 31757 and 217506 ton respectively [5]. They were distributed in many country and those species commonly used as Surimi raw materials, their production trends and their abundance in the Southeast Asian waters were identified [6] Even though these species have less demand in fresh condition, there is considerable market for dry fish and also as fishmeal especially in poultry industry. *Upeneussulphureus* (sulphur goatfishes) is from Mullidae family in the order Perciformes [7] that distributed from Red Sea, Persian Gulf, Madagascar, Seychelles, Réunion, Pakistan, India, Sri Lanka, Andaman Sea, Indonesia, New Guinea, Fiji, New Caledonia, Philippines and southern Japan [8].

Leiognathusbindus (Orange fin pony fish) known as slip mouths or slimy are a small family, Leiognathidae, of fishes in the order Perciformes [9]. They reported in the Red Sea (Port Sudan), Persian Gulf, coasts of India and Sri Lanka, and elsewhere in the eastern Indian Ocean; eastward to the western Central Pacific [5].

Fishery, distribution, biology and population dynamics of them have been studied in detail [10-20]. Since no detailed information is available on fatty acid profile from these species so the objective of this study was to determine fatty acid profiles goatfish and pony fish and to compare their nutritional content in fall and spring season in Mahshahr port.

Materials and Methods

Samples

Goat fiah and pony fish were obtained from Fishermen at Mahshahr port randomly in spring and fall season. Number of fish was 30 from each species. The fish were caught the night before the procedure, kept in ice and transferred to the laboratory for analyses. The fishes were weighed, deheaded, eviscerated and cleaned prior to freezing. In an attempt to obtain a homogeneous sample from each species, their flesh were removed from their backbones, minced, blended and immediately extracted. The mean weights and total length of the fishes were: 78.25 g and 17 cm for goat fish and 13.25 g and 6.5 cm for pony fish.

Lipid extraction

Lipid extractions were performed on minced fish samples (25 g each) using the extraction methods and chloroform-methanol. Methylene chloride (100 μL) and 1 mL 0.5 M NaOH in methanol were added to oil extracts in a test-tube and heated in a water bath at 90°C for 10 min. The test tubes were removed from the water bath and allowed to cool before addition of 1 mL 14% BF3 in methanol. The test tubes are heated again in a water bath for 90°C for 10 min, and cooled to room temperature. One mL distilled water and 200-500 μL hexane was added to the test tubes and then FAME was extracted by vigorous shaking for about 1 min. Following centrifugation, the top layer was transferred into a sample bottle for GC analysis [21].

***Corresponding author:** Khadijeh Nabi Ghahfarokhi, Fisheries Scientific Association, Islamic Azad University, Ahvaz, Iran, E-mail: gonomwo@yahoo.com

Fatty acids analysis

Fatty acid analyses were carried out using the IUPAC II.D.19 method. Fatty acids were analyzed using a HP Agilent 5890 system Gas Chromatograph equipped with SP-2330 and a flame ionization detector (FID). Separation of fatty acid methyl esters was achieved by using fused silica capillary column (30 m×0.25 mm×0.20 μm film thickness). The oven temperature was set at 120°C for 2 min then reached to 220°C with a ramp rate of 5°C/min, and then held for 15 min. The injector and detector temperatures were maintained at 155°C and 260°C, respectively. The carrier gas was helium 10 psi with a split ratio of 1/50. The air and hydrogen of pressure were 338 ml/min and 45 ml/min respectively. Results were expressed as the percentage of each fatty acid with respect to the total fatty acids [22,23].

Indexes of lipid quality

The saturated/unsaturated fatty acids (SFA/UFA) ratios were calculated including trans fatty acids in the UFA group. The atherogenicity (AI) and thrombogenicity (TI) indices were also calculated according to the following equations [24].

From the data on the fatty-acid composition, the following were calculated:

1) Index of atherogenicity (IA): indicating the relationship between the sum of the main saturated fatty ac-ids and that of the main classes of unsaturated, the former being considered pro-atherogenic (favoring the adhesion of lipids to cells of the immunological and circulatory system), and the latter anti atherogenic (inhibiting the aggregation of plaque and diminishing the levels of esterified fatty acid, cholesterol, and phospholipids, there- by preventing the appearance of micro- and macro- coronary diseases).

$$IA = [(4 \times C14:0) + C16:0 + C18:0] / \left[\sum MUFA + \sum PUFA - n6 + \sum PUFA - n3\right]$$

2) Index of thrombogenicity (IT): showing the tendency to form clots in the blood vessels. This is defined as the relationship between the pro-thrombogenetic (saturated) and the anti-thrombogenetic fatty acids (*MU-FAs*, PUFAs – *n6* and *PUFAs – n3*) [1,23].

The following equation was applied:

$IT=(C14:0+C16:0+C18:0)/(0.5MUFA+0.5PUFA\text{-}n6+3PUFA\text{-}n3+PUFA\text{-}n3/PUFA\text{-}n6)$

Statistical Analysis

Data were subjected to analysis of variance (ANOVA) and mean comparison was carried out using Duncan's test. Statistical analysis was performed with SPSS. Significance was established at P<0.05.

Results

Fatty acid profile

The fatty acid composition of goatfish and pony fish are shown in Table 1. The amount and number of fatty acid were different in fall and spring for each species (Figure 1). A total of 24 fatty acids in goatfish and 26 fatty acids for pony fish were identified in this study in fall. The determination of the fatty acid compositions of goatfish and pony fish were performed by GC. The composition of saturated (SFA), monounsaturated (MUFA), polyunsaturated fatty acid (PUFA) and high unsaturated fatty acid (HUFA) of goatfish were found to be 52.4%, 23.5%, 18.9% and 15.47% in fall, while those of pony fish were found to be 50.004%, 24.919%, 17.27% and 14.62%, respectively. Accordingly, the SFA level in gold band goatfish and the MUFA level in striped red mullet were the highest among the other fatty acid groups. The PUFA/SFA ratios for goatfish and pony fish were 0.36 and 0.34 respectively in fall.

Fatty acid	*Leiognathus bindus*		*Upeneus sulphureus*	
	fall	**spring**	**fall**	**spring**
(12:0)	0.1	--	0.25	-
(14:0)	5.49	5.95	2.9	2.45
(15:0)	0.47	---	0.34	-
(16:0)	29.57	33.97	27.49	39.7
(17:0)	1.04	1.36	0.92	1.76
(18:0)	10.75	10.25	16.41	11.97
(20:0)	0.15	1.78	1.97	---
(21:0)	0.43	----	0.53	---
(22:0)	0.6	1.06	---	1.8
(23:0)	0.084	---	---	----
(24:0)	1.32	---	1.59	---
(14:1)	1.57	1.08	1.36	0.89
(16:1)	9.59	13	5.27	6.97
(18:1)	12.09	18.4	15.40	24.45
(20:1)	0.71	1.11	0.83	---
(22:1)	0.080	---	0.35	---
(24:1) ω9	0.87	---	0.29	---
(18:2) ω6	0.66	0.8	1.12	0.85
(20:2) ω6	0.29		0.25	---
(18:3) ω3	0.36	0.61	0.29	0.33
(18:3) ω3	1.07	0.47	1.54	0.5
(20:3) ω6	0.27	---	0.23	---
(20:4) ω6	2.99	---	4.58	---
(22:5) ω3	-	1.66	1.74	0.46
EPA ω3	5.7	4.63	4.65	2.43
DHA ω3	3.35	3.78	4.5	4.33
SFA	50.004	54.37	52.4	57.68
MUFA	24.919	33.59	23.5	33.31
PUFA	17.27	11.95	18.9	8.9
HUFA	14.62	10.07	15.47	7.22

Table 1: Fatty acid profile (%) of *Leiognathus bindus* and *Upeneus sulphureus* in fall and spiring.

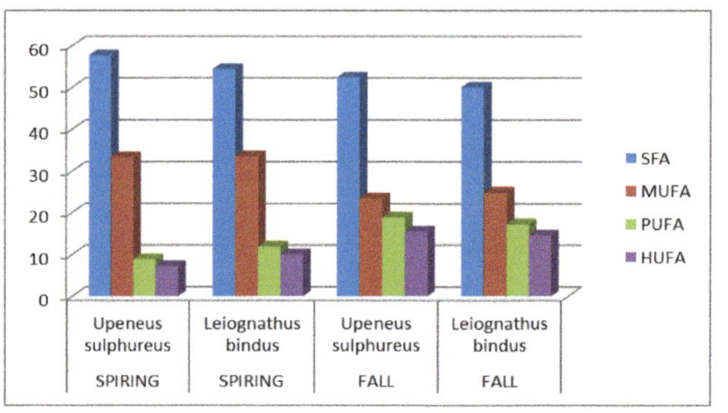

Figure 1: Fatty acid profile (%) of *Leiognathus bindus* and *Upeneus sulphureus* in fall and spiring.

Lipid quality indices

Table 2 and Figure 2 show total lipids IA and IT values for *Leiognathus bindus* and *Upeneus sulphureus* fat. IT values were significantly higher in *Upeneus sulphureus* than *Leiognathus bindus* in spring. The ratio of

Species/indices	spring	fall	spring	fall	spring	fall
	IT		IA		W3/W6	
Leiognathus bindus	0.886	0.810	1.493	1.476	7.47	3.01
Upeneus sulphureus	1.154	0.852	1.445	1.308	6.54	2.02

Table 2: IA and IT indices of *Leiognathus bindus* and *Upeneus sulphureus* in fall and spring.

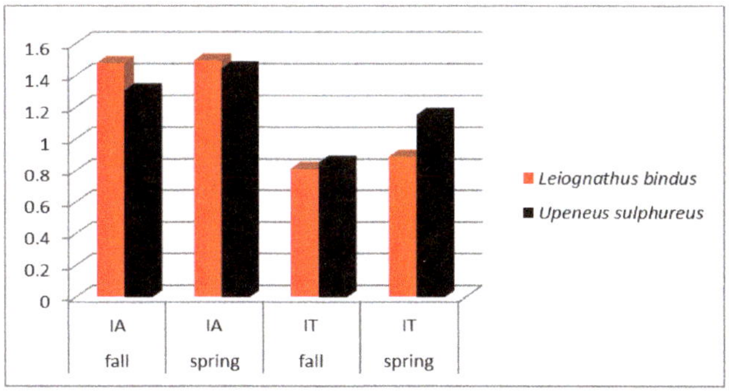

Figure 2: IA and IT indices of *Leiognathus bindus* and *Upeneus sulphureus* in fall and spring.

W3/W6 was different in fall and spring for both species. IA value for *Upeneus sulphureus* was the lowest.

IA=Index of Athero-genicity; IT=Index of Thrombogenicity

Discussion

Fish consumption is increasingly recommended by health authorities, not only for its high-quality protein content, but also for being a source of fatty acids considered highly beneficial for human health (n3 and n6). Therefore, it is not surprising that there is higher demand for fish with a growing concern for the health aspects of the diet [25].

In conclusion *Leiognathus bindus* and *Upeneussulphureus* showed a considerable amount of saturated and unsaturated fatty acids in both fall and spring. Polyunsaturated fatty acids (PUFA) are found to be one of the highest compounds found in these fishes and n-3 PUFA is studied extensively in both types of fishes. The n-3 polyunsaturated fatty acid which is primarily docosahexaenoic acid (DHA) and eicosapentaenoic acid (EPA) was found in appreciable amounts and would be suitable for inclusion in the formulation of highly unsaturated low-fat diets and to help prevent diseases.

The AI and IT values obtained are higher than those reported for reared and wild sharp snout sea bream, *Diploduspuntazzo* [4].

In view of the tissue fatty acid composition, depending on the time of the year or the season, reared fish could be more suitable than wild fish for human consumption.

References

1. Garaffo MA, Vassallo-Agius R, Nengas Y, Lembo E, Rando R, et al. (2011) Fatty Acids Profile, Atherogenic (IA) and Thrombogenic (IT) Health Lipid Indices, of Raw Roe of Blue Fin Tuna (*Thunnus thynnus* L.) and Their Salted Product "Bottarga". Food Nutr Sci 2: 736-743.
2. Sanfilippo M, Reale A, Ziino M, Romeo V, Lembo E, et al. (2011) Chemical Composition and Nutritional Value of Engraulis encrasicolus (Linnaeus, 1758) Caught by Driftnet "Menaide" along Sicilian Coast: a Natural Food for Mediterranean Diet. World Journal of Fish and Marine Sciences 3: 44-50.
3. Öksüz A, Özyılmaz A, Küver Ş (2011) Fatty Acid Composition and Mineral Content of *Upeneus moluccensis* and *Mullus surmuletus*. Turkish Journal of Fisheries and Aquatic Sciences 11: 69-75.
4. Rueda FM, Herna´ndez MD, Egea MA, Aguado F, Garcı´a B, et al. (2001) Differences in tissue fatty acid composition between reared and w ild sharpsnout sea bream, Diplodus puntazzo (Cetti, 1777). Br J Nutr 86: 617-622.
5. Woodland DJ, Premcharoen S, Cabanban AS (2001) Bony fish: leiognathidae. www.FAO.org
6. (2009) End-of-Project Meeting of the Working Party on Information Collection for Economically Important Species as Surimi Raw Materials in the Southeast Asian Region. SEAFDEC.
7. Randall JE, Kulbicki M (2005) A Review of the Goatfishes of the Genus *Upeneus* (Perciformes: Mullidae) from New Caledonia and the Chesterfield Bank, with a New Species and Four New Records. Zoological Studies 45: 298-307.
8. Uiblein F, Heemstra PC (2010) A taxonomic review of the Western Indian Ocean goatfishes of the genus *Upeneus* (Family Mullidae), with descriptions of four new species. Smithiana Bulletin 11: 35-71.
9. Froese R, Pauly D (2006) "Leiognathidae" in FishBase.
10. Rajkumar U (2006) Fishery and population dynamics of silverbellies off Visakhapatnam. J Mar Biol Ass India 48: 213-219.
11. Shadi A, Dehghan Mediseh S, kouchanian P, Gandomi Y (2011) Length-Weight Relationships for 6 Fish Species from Khuzestan (North of Persian Gulf), Iran. World J Fish Marine Sci 3: 129-131.
12. Sparks JS (2006) A New Species of Ponyfish (Teleostei: Leiognathidae:Photoplagios) from Madagascar, with a Phylogeny for Photoplagios and Comments on the Status of Equula lineolata Valenciennes; The American Museum of Natural History Central Park West at 79th Street, New York, NY 10024 Number 3526: 20.
13. Kwak SN, Klumpp DW (2004) Temporal variation in species composition and abundance of fish and decapods of a tropical seagrass bed in Cockle Bay, North Queensland, Australia. Aquatic Botany 78: 119-134.
14. Ra HK, Choi Y, Lim H (2005) Taxonomic review of the genus Liognathus from Korea. Korean Journal of Ichthyology 17: 91-97.
15. Golani D, Fricke R, Golani BA (2011) First record of the Indo-Pacific slender ponyfish *Equulites elongatus* (Günther, 1874) (Perciformes: Leiognathidae) in the Mediterranean. Aquatic Invasions 5: S75-S77.
16. Murty VS (1986) Studies on the growth and population dynamics of silverbelly *LEIOGNATHUS BINDUS* (VALENCIENNES) in the trawling grounds off Kakinada. Indian Journal of Fisheries 33: 3
17. Abraham KJ, Joshi KK, Murty VSR (2011) Taxonomy of the fishes of the family Leiognathidae (Pisces, Teleostei) from the West coast of India. Zootaxa 2886: 1-18.
18. Chakrabarty P, Amarasinghe T, Sparks JS (2008) Rediscription of ponyfishes of Sri Lanka and the status of *Aurigequla* fowler 1918. Cey J Sci 37: 143-161.
19. Gholami M, Zoriasatein N (2005) Identification and evaluation of Leioganathidae family in Persian Gulf. 6th conference of science and marine technology, Tehran.
20. Chukwuemeka U, Ndukwe GI, Audu TO (2008) Comparison of Fatty Acids Profile of Some Freshwater and Marine Fishes. Internet Journal of Food Safety 10: 9-17.
21. Turan H, Sonmez G, Kaya Y (2007) Fatty acid profile and proximate composition of the thornback ray (*Raja clavata*, L. 1758) from the Sinop coast in the Black Sea. J Fisheries Sci 1: 97-103.
22. Kaya Y, Turan H (2008) FATTY ACIDS COMPOSITION OF ANCHOVY (*Engraulis encrasicolus* L. 1758) OIL PRODUCED IN SINOP-TURKEY. Journal of Fisheries Sciences 2: 693-697.
23. Larqué E, Garaulet M, Pérez-Llamas F, Zamora S, Tebar FJ, (2003) Fatty acid composition and nutritional relevance of most widely consumed margarines in Spain. Grasas y Aceites 54: 65-70.
24. Senso L, Sua´rez MD, Ruiz-Cara T, Garcı´a-Gallego M (2007) On the possible effects of harvesting season and chilled storage on the fatty acid profile of the fillet of farmed gilthead sea bream (*Sparus aurata)*. Food Chemistry 101: 298-307.
25. Muhamad NA, Mohamad J (2012) Fatty Acids Composition of Selected Malaysian Fishes. Sains Malaysiana 41: 81-94.

Reproductive Biology of the Commercially Important Portunid Crab, *Portunus sanguinolentus* (Herbst)

Soundarapandian P*, Varadharajan D and Boopathi A

Faculty of Marine Sciences, Centre of Advanced Study in Marine Biology, Annamalai University, Tamil Nadu, India

Abstract

The male reproductive system of *P. sanguinolentus* is bilaterally symmetrical creamy to whitish in colour, composed of a pair of testes, a pair of vas differentia, and a pair of ejaculatory ducts internally, and a pair of pleopods externally as accessory reproductive organs, present on the inner side of the abdominal flab. The vas differentia has been divided into three distinct regions, based on the morphological and functional criteria: Anterior (AVD), Median (MVD) and Posterior (PVD) vas deferens. The gonads of the male measuring below 8.5 cm carapace width (CW) has immature stage, above 10.5 cm CW has mature stage, whereas the CW in between 14.5-10.5 cm was maturing stage. The female reproductive system composed of a pair of ovaries, a pair of seminal receptacles (or) spermatheca, and a pair of oviducts open to the exterior through the female genital opening situated on the left and right sternites of sixth thoracic segment. The ovaries of *P. sanguinolentus* are categorized into five stages, according to the size, colour and external morphology of the ovaries. They are immature, early maturing, late maturing, ripe and spent. Fifty percent of the female crabs attained sexual maturity when they reached the size of 9.1- 9.5 cm CW. Fifty percent of the male crabs attained sexual maturity when they reached the size of 9.5-10.0 cm CW. However, the smallest berried female attained sexual maturity in the present study was measured at 8.6 cm CW. Based on the change in colour, increase in the size and change in the shape of the berry eggs, three different stages of egg development was observed. Stage-I has pale yellow to deep yellow coloured egg mass was 260.16 ± 0.24 μm. Stage-II has yellow to gray coloured egg mass was 290.2 ± 0.31 μm. Stage-III has deep gray to black coloured egg mass was 340.32 ± 0.34 μm.

Keywords: Fluids; Hormones; Pheromones; Genital system; Three spot crab

Introduction

The crab fishery in India is fast developing, and there is a vast scope for the crab meat, both national and international markets. Crabs rank third next to the shrimps and lobsters for their esteemed seafood delicacy and also the value of fishery they support [1]. The main species supporting the crab fisheries in India are *Scylla serrata*, *S. tranguebarica*, *Portunus pelagicus*, *P. sanguinolentus*, *Charybdis feriata*, *C. lucifera* and *C. truncata*. Reproduction is the main mechanism to maintain species proliferation and continuity, and in brachyuran crabs, is extremely diversified ultimately shaped to maximize egg production and offspring survivorship [2,3]. The reproductive cycle of crustaceans has been widely studied, mainly of those species that have commercial value or ecological potential [4-6]. Many works have been conducted on various aspects of the reproductive biology of many marine animals [7-9]. Study of gonads, size at sexual maturity and fecundity are important aspects of reproductive biology of crab. This information is often required to manage crab fisheries on sustainable basis. Usually two types of criteria are applied to determine the size at which sexual maturity is attained. One is reproductive criteria or physiological maturity in which developmental stages of male and female gonads and smallest size of berried female is taken as an indication of being sexually mature [10]. Another one criteria is the functional maturity in which the relative growth of chela in males and abdominal width in females, with respect to the body size (Carapace width) is to find out the size at sexual maturity. Size at which male and female brachyuran crabs attain sexual maturity has been worked out by several researchers in portunid crabs; *P. sanguinolentus*, *P. pelagicus*, *Ovalipes punctatus*, *O. stephonsoni* and *Liocarcinus depurator* and *Scylla* species [11-18].

The study of breeding season and fecundity is important to have a full understanding of the population dynamics. Fecundity or reproductive potential can be defined as the number of ova shed during particular spawning season or the number of eggs produced in a single batch of spawning [19,20]. The later one is also known as "batch fecundity". Those who have studied batch fecundity in *Portunus* species was given by Reeby et al. [21] [22-25]. The recognition of biological data about reproduction is required for judicious management and exploitation of this resource. Hence in the present study, an attempt has been made on the various aspects of reproductive biology of the male and female crabs of *P.sanguinolentus*, such as morphology of the reproductive tract, gonad development, Gonado Somatic Index (GSI), size at maturity and fecundity.

Materials and Methods

Description of the experimental animal

Systematic position

Phylum: Arthropoda

Order: Decapoda Latreille

Family: Portunidae Rafinesque

***Corresponding author:** Soundarapandian P, Faculty of Marine Sciences, Centre of Advanced Study in Marine Biology, Annamalai University, Parangipettai-608 502, Tamil Nadu, India, E-mail: soundsuma@gmail.com

Genus: *Portunus* Weber

Species: *P. sanguinolentus* Herbst

Nomenclature: Common and vernacular names of *P.sanguinolentus* in India are given below:

Common name-Three spotted crab (or) Blood spotted crab

Gujarathi-Karachla

Marathi-Khekhada

Kannada-Denji

Malayalam-Kavalan njandu

Tamil-Mukkannu nandu

Telugu-Chukkala peeta

Oriya-Kankda/Cherala petta

Bengali-Lajjaboti kankra

The three spotted swimming crab, *P. sanguinolentus,* is one of the most common and commercially important crab found along the east cost of India [26,27]. It is commonly known as three-spot swimming crab [28]. In Tamil Nadu, the local fisherman calls as "Mukkannu nandu". This crab got its common name from three prominent red to maroon spots on the posterior part of the carapace. They are widely distributed in the Indo-pacific region from the east coast of South Africa to Hawaician waters [29]. Their size and sex composition varies with depths as the juveniles, and adult males typically inhabit estuaries and inshore waters about 10-30 m deep [30,31]. In contrast, females are abundant in 40-80 m depths [31]. The berried crab mostly burried in the sand. It carries the eggs until complete their development and the larvae hatch as zoea, which drift about in the water current. After hatching, zoea transforms into a megalopa, which acquires benthic habits and grows into juvenile crab.

Distinguishing characters (Plate 1)

- The carapace is much broader than its length.
- Anterolateral border bears nine spines, of which the posterior most is the longest.
- There is no spine on the posterior border of the arm of the cheliped.
- Behind the chelipeds, the three pairs of legs are normal, and the last pair is oar-like for swimming.
- The anterior pair of abdominal appendages is slender and straight with marginal spines at the tip.
- The colouration of the carapace is characteristically in a dark gray ground colouration, the posterior half of the carapace has three large reddish round spots, of which one is median and two laterals, each spot encircled by a whitish ring.

Distribution: Geographical distribution of *P. sanguinolentus* ranges from Indo-Pacific, Southern Africa, Red sea, Persian Gulf, India, Pakistan, China Sea, Philippines, Japan, Australia and Tahiti.

Fishery information: Caught all along the Indian Coasts. Common size is 10-12 cm in commercial catches; growing to a maximum size of 15 cm across the carapace.

Areas of abundance/landings: Along the coasts of Tamil Nadu (Chennai), Andhra Pradesh (Kakinada), Kerala and Karnataka.

Peak season: March–June and October to December. Caught in trawl, shore-seine and crab traps.

The crabs for the present study was collected from Parangipattai landing centre (Lat.11°29' N; Long.79°46' E) and brought to the laboratory by using plastic container (Plate 2). The crabs were segregated as male and females. They were weighed individually and the size of the carapace width was measured. Sexes were determined by examining the abdominal morphology.

The pleopods and gonopores were analyzed to find out the mating and extrusion of eggs. Presence of the eggs or egg remnants or their absence on pleopods, colour of the egg mass was also noted. The crabs were dissected to study the morphology of the reproductive tract and stages of maturation, size, weight, colour of the gonads and Gonadosomatic Index (GSI). The dissected gonads were measured to the nearest 0.001 g by using the electronic balance, and subsequently, gonadosomatic index was calculated by the following formula.

$$GSI = \frac{\text{Wet weight of the gonad}}{\text{Wet weight of the animal}} \times 100$$

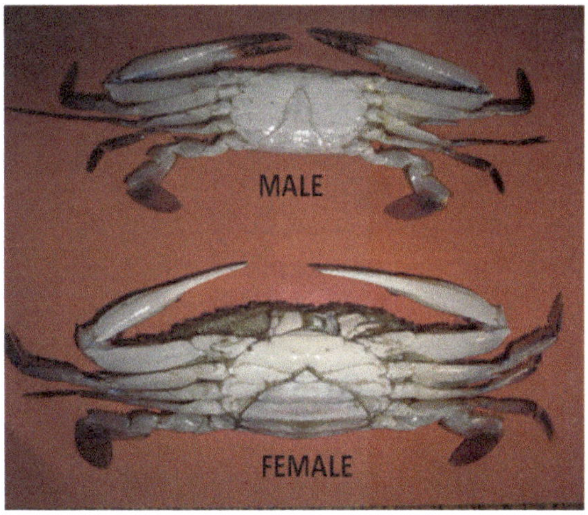

Plate 1: Male and female crabs of *P. sanguinolentus.*

Plate 2: Different size crabs of *P. sanguinolentus.*

Recently spawned females with bright yellow or orange coloured eggs were selected for egg counting (Fecundity). To estimate the fecundity, pleopods bearing eggs were removed carefully and the weight of the whole egg mass (berry) was taken to the nearest 0.01 g, then four subsamples of berried eggs (0.1 g) were cut from different locations of the egg mass. Number of eggs in each four sub samples was counted with the help of counting tray under stereomicroscope. Average no of eggs present in the samples was then calculated. In order to find out the total weight of the egg mass, the eggs were removed from the pleopods, and weight of the pleopods was taken. This was subtracted from the weight of the whole egg mass. Fecundity was then calculated by the following formula [23].

$$N=W\times n/w$$

Where W=total weight of the egg mass only; w=average weight of the four egg samples; n=average number of eggs in the samples; N=total number of eggs (Fecundity).

Results

Morphology of the male reproductive system

The male reproductive system of *P. sanguinolentus* is bilaterally symmetrical creamy to whitish in colour, composed of a pair of testes, a pair of vas differentia, and a pair of ejaculatory ducts internally, and a pair of pleopods externally as accessory reproductive organs, present on the inner side of the abdominal flab. The testes are flat and highly coiled, roughly in the form of "H" located on the dorsal portion of the hepatopancreas sandwiched between the hepatopancreas and the hypodermis of the carapace, and continuing laterally to the stomach upto vas deferens. Just before the posterior stomach and anterior to the heart, the left and right testis are bridged by a commissure.

The testes are connected with the vas deferens through a short small duct called vas efferens. Vas differentia is a pair of elongated and coiled tubules, which extend longitudinally from the posterior testes upto the posterior region of the body. The vas differentia has been divided into three distinct regions, based on the morphological and functional criteria: Anterior (AVD), Median (MVD) and Posterior (PVD) vas deferens. The AVD are white, tightly coiled, and lying on either side of the median line of cephalothorax, posterior to the dorsal part of the stomach. The coils of the AVD increase in size posterio-ventrally and lead into the middle vas deferens, which are milky-white meandering tubules of a higher caliber than both the testes and AVD. The posterior vas deferens are whitish in colour arises from the posterior end of the median vas deferens, which are massive for its proximal part, but gradually narrow before opening to the ejaculatory duct. Each posterior VD is connected with an ejaculatory duct, which is a smooth narrow duct extending between the musculature of the swimming peddles. The ejaculatory duct leads into the slender weak tube like genital papilla (or) penis, which are located at the base of the swimming legs. Each penis passes into the two pairs of abdominal appendages called pleopods (or) gonopods situated in the inner side of the abdominal flab. The first pleopod is made up of two segments, the basal one is broad to the sternal wall and the terminal one is long tube-like and tapering towards the tip, which is actually inserted into the seminal receptacle of the female during copulation. The second pleopod helps in passing the seminal fluids from the penis into the funnel like portion of the first pleopod (Plate 3).

Developmental stages of male gonads

Immature: The gonads of the immature crabs are small and creamy in colour, on either side of the stomach. Testes and vas differentia are not clearly differentiated. Gonads of males measuring below 85 mm Carapace Width are in immature stage.

Maturing: Testes and vas differentia are well developed and clearly differentiated, and creamy white in colour. Testes are large coiled tube. which spreading laterally and posteriorly to the stomach. Anterior vas differentia has enlarged middle and posterior vas differentia is straight and opaque extending to both the side of the heart.

Mature: Testes showed further enlargement, as vas differentia are coiled and very much swollen occupying full body cavity. The AVD and MVD are enlarged and milky white in colour, PVD enlarged and convoluted, but still opaque. Gonads of males measuring above 105 mm CW are in mature stage.

Gonadosomatic Index (GSI) in male crabs

The Gonadosomatic index of the male and female crabs was studied according to the size and weight of the crab. The GSI of the male crab (0.48) was increased from immature crabs to matured crabs (0.79). The size of the animals was ranging between 8.0-8.5 cm CW to 14.1-14.5 cm CW (Table 1).

Morphology of the female reproductive tract

The female reproductive system composed of a pair of ovaries, a pair of seminal receptacles (or) spermatheca, and a pair of oviducts (open to the exterior through the female genital opening situated on the left and right sternites of sixth thoracic segment). The oviducts pass ventrally

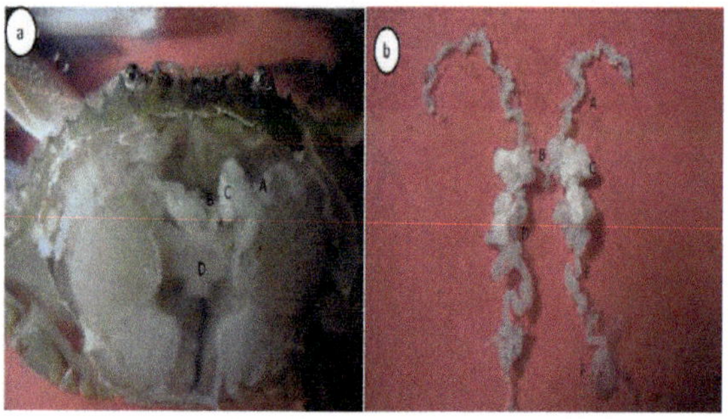

Plate 3: Figure Ia. Disssected male crab with gonad; A-Testis, B-commissure, C-Anterior vasdeferens, D-Heart; b.Male reproductive system; A-Testis, B-Commissure, C-Anterior vasdeferens, D-Median vas deferens, E-Posterior vas deferens, F-Ejaculatory duct.

S.No.	Carapace width (Cm)	Mean body weight (g)	Mean gonad weight (g)	GSI
1	8.0-8.5	29.93 ± 0.61	0.145 ± 0.28	0.48 ± 0.04
2	8.6-9.0	34.50 ± 2.35	0.178 ± 0.04	0.51 ± 0.02
3	9.1-9.5	41.35 ± 1.97	0.226 ± 0.07	0.54 ± 0.04
4	9.6-10.0	49.18 ± 2.42	0.313 ± 0.09	0.63 ± 0.18
5	10.1-10.5	57.73 ± 2.92	0.37 ± 0.11	0.64 ± 0.17
6	10.6-11.0	65.90 ± 5.37	0.431 ± 0.17	0.65 ± 0.24
7	11.1-11.5	70.0 ± 2.51	0.46 ± 0.08	0.65 ± 0.09
8	11.6 -12.0	79.33 ± 4.36	0.523 ± 0.29	0.65 ± 0.12
9	12.1-12.5	102.76 ± 2.67	0.69 ± 0.47	0.67 ± 0.62
10	12.6-13.0	118.83 ± 1.22	0.834 ± 0.05	0.70 ± 0.04
11	13.1-13.5	121.43 ± 4.97	0.88 ± 0.46	0.73 ± 0.08
12	13.6-14.0	131.1 ± 1.01	1.01 ± 0.13	0.77 ± 0.29
13	14.1-14.5	140.2 ± 2.17	1.10 ± 0.47	0.79 ± 0.46

Table 1: Mean body weight, Mean gonad weight, Gonadosomatic Index (GSI) of the male crab (Values are Mean ± SD).

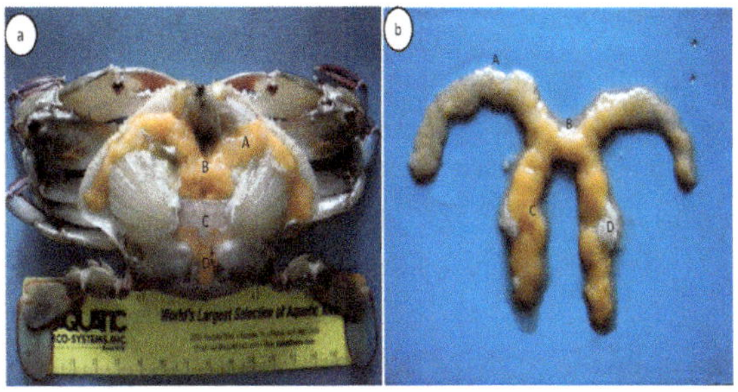

Plate 4: Figure IIa: Dissected female crab with ovary, A-Anterior horn, B-Commisssure, C-Heart, D-Posterior horn; b.Dissected ovary, A–Anterior horn, B–Commissure, C-Posterior horn, D–Spermatheca.

from the seminal receptacle. The ovary is surrounded by a fibrous connective tissue, which separates the ovary from the surrounding hemocoel. The ovary is roughly in the form of a 'H' and lies on top of the hepatopancreas, extending on both sides along the anterior margin of the cephalothorax are called anterior horns. The ovaries run in posterior cardial direction to the cardiac stomach and just posterior to the stomach, anterior horns are joined by a commissure. The posterior horns pass posteriodorsoventrally and laterally, and applied dorsally to the seminal receptacles which are sandwitched on either side between the lateral wall of the body cavity and lateral part of the pericardium and the posterior horns extend up to posterior margin occupies all spaces of the body cavity. The posterior horns are equal in size at the posterior end (Plate 4).

Stages of ovarian development: The ovaries of *P. sanguinolentus* are categorized into five stages according to the size, colour and external morphology of the ovaries. They are immature, early maturing, late maturing, ripe and spent.

Immature: Small, flattened ovaries, white (or) translucent in colour, occupy small areas of the body cavity.

Early maturing: Ovary is larger than previous one, light orange in colour, slightly nodulated, not extending into hepatopancreatic region.

Late matured: Larger ovaries yellowish or orange in colour, nodulated, extending into hepatopancreatic region and occupies almost all the space of the body cavity.

Ripe: Very large ovaries, dark yellow or orange in colour, highly nodulated, hepatopancreas is completely hidden or immersed, ovary fills up all the spaces of the body cavity.

Spent: Small, translucent and greatly reduced ovaries unspawned ova are visible throughout the fibrous connective tissue.

Gonadosomatic Index in females: The GSI in females were increased from1.95 ± 0.37 in immature crabs to ripened crabs 3.97 ± 0.31; however, in spent crab has low value of GSI as similar to immature crabs (Table 2).

Size at first maturity: Fifty percent of the female crabs attained sexual maturity when they reached the size of 9.1-9.5 cm carapace width. Fifty percent of the male crabs attained sexual maturity when they reached the size of 9.5-10.0 cm carapace width. However, the smallest berried female attained sexual maturity in the present study was measured at 8.6 cm carapace width.

Ovigerous females: At the time of spawning, the eggs from the ovary of female crabs were extruded through their gonopores and get attached to the cluster of long and very smooth setae on endopodite of the pleopods. Based on the change in colour, increase in the size and change in the shape of the berry eggs, three different stages of egg development was observed.

- **Stage-I**

Pale yellow to deep yellow coloured egg mass, no eyespots was visible in the eggs, the mean size of the eggs was 260.16 ± 0.24 μm (Table 3)

- **Stage-II**

Yellow to gray coloured egg mass, eyespots was observed, mean size of the egg was 290.2 ± 0.31μm (Table 3).

- **Stage-III**

Deep gray to black coloured egg mass, eyespots and chromatophores were discernible, the mean size of the egg was 340.32 ± 0.34 μm (Table 3).

Fecundity: Seventeen ovigerous females were used for the calculation of fecundity. The fecundity was found to be from 283963 to 967293 eggs, in the animals had the carapace width of 10.6-14.1 cm.

Discussion

The focus of the present day aquaculture was moving towards the development of hatchery technology and crab forming to meet the human protein demand. Berried females are important to start a commercial hatchery, but berried females are not available throughout the year from the wild. So, the production of berried females in a controlled condition is essential. Before that one should know the morphology of the reproductive system and reproductive biology of a particular crab is important.

Morphology of the male reproductive system

The morphology of the male reproductive system of *P. sanguinolentus* consisting of an "H" shaped structure with a pair of testes, pair of vas deferentia and ejaculatory ducts was similar to the reproductive systems described for other decapod crustaceans in general and Portunid crabs, in particular [32-35]. The descriptions of the testes and commissure of *P. sanguinolentus* matched the descriptions of *P. marmaratus*, *P. sanguinolentus*, *C. sapidus*, *Thenus orentalis*, *Panulirus laevicauda* and *Goniopsi cruentata* [36-40]. However, did not report the existence of a commissure between the testes of *C. opilio* and *M. branchyodactyla* respectively [41,42]. Testes of *P. sanguinolentus* are tubular organs composed of numerous microscopically visible lobules connected to the semniferous duct; consequently, they have been classified as lobular testes according to the categories (lobular and tubular), and are similar to the previous reports of some other brachyurans [34,40,43,44]. Whereas in few species of Brachyura have tubular testes, such as *Menippe mercenaria,Eriocheir sinensis*, *Chionoecetes opilio*, *Pachygrapsus crassipes* and *C. bairdi* [45-49]. However, they seem to be distributed along different groups of Brachyura (Majoidea, Grapsoidea and Xanthoidea). Chiba studied that lobular testes are very common in portunid crabs. This is the reasons why some brachyurans have a tubular testicular arrangement, whilst others have a lobular one, is unknown [42,50]. The testes connected to the vas deferens by means of a small duct known as vas deferens has been described in *Callinectus sapidus* and *P. pelagicus* [35,43]. Vas deferens (VD) is a pair of elongated and convoluted tubules which extend longitudinally in the posterior region of the body [34,51]. In *P. sanguinolentus*, the vas deferens was divided into three distinct regions: the anterior, middle

S.No.	Carapace width (Cm)	Mean body weight (g)	Mean gonad weight (g)	GSI
1	8.0-8.5	29.60 ± 1.3	0.88 ± 0.09	2.97 ± 0.37
2	8.6-9.0	33.8 ± 1.38	1.21 ± 0.05	3.57 ± 0.12
3	9.1-9.5	37.3 ± 3.08	1.48 ± 0.13	3.98 ± 0.10
4	9.6-10.0	40.7 ± 3.47	1.78 ± 0.23	4.37 ± 0.06
5	10.1-10.5	47.5 ± 4.60	2.1 ± 0.27	4.42 ± 0.09
6	10.6-11.0	60.2 ± 2.54	2.95 ± 0.19	4.90 ± 0.17
7	11.1-11.5	76.7 ± 7.61	3.27 ± 0.39	4.26 ± 0.24
8	11.6 -12.0	87.15 ± 4.94	3.46 ± 0.17	3.97 ± 0.31
9	12.1-12.5	99.67 ± 2.67	3.64 ± 0.47	3.65 ± 0.36
10	12.6-13.0	114.96 ± 8.30	3.97 ± 0.27	3.45 ± 0.18
11	13.1-13.5	129.8 ± 8.43	4.45 ± 0.13	3.42 ± 0.27
12	13.6-14.0	137.85 ± 7.46	1.74 ± 0.18	1.26 ± 0.21

Table 2: Mean Body weight, Mean gonad weight, Gonadosomatic index (GSI) of the female crab (Values are Mean ± SD).

Stages of development in berried females	No of specimens observed	Colour of the berry	Shape of the eggs	Mean size of the eggs (μm)
Stage I	20	Pale to dark yellow	Spherical	260.16 ± 0.24
Stage II	15	Yellow to Gray	Spherical	290.2 ± 0.31
Stage III	10	Gray to Black	Elliptical	340.32 ± 0.34

Table 3: The colour, shape and size of the berried egg (The values are in Mean ± SE).

and posterior vas deferens as has been reported in other crabs based on their morphological and functional criteria in *P. sanguinolentus*, *C. sapidus*, *T. orentalis* and *M. branchydactyla*. However, other studies characterized two in *G. cruentata*, four in *S. chacei* and *D. puligator* into as many as eight regions [40,52,53]. Different criteria and different microscopic and macroscopic foci may account for this diversity [40]. The presence of diverticula in the VD of Branchyura has been widely described particularly in spider crabs, which present numerous ramified diverticula in the posterior region of the VD [34,36,37,40,43,53]. Diverticula play an important role in increasing the secretion, absorption and storage of spermatophores and seminal fluids [53,54]. The terminal portion of the reproductive system is the ejaculatory duct, which is a smooth duct extending between the musculature of the swimming pedals, as already described in Brachyura [55].

Developmental stages of male gonads

Very few works have been described for the maturity stages in male crabs. Diesel [56] has described six maturity stages in the males of *C. irroratus*. Campbell observed five stages of maturation in *C. iroratus,* based on a modified version of who indicated six stages of gonad development [56]. Krol et al. [57] indicated five stages of maturation in male crab *C. amnicola*. Haefner [58] reported four different morphometric categories of male gonads in *M. branchydactyla*. Lawal-Are [59] recognized three stages of maturation based on the testis development in *P. pelagicus* and *P. sanguinolentus*. In *P. pelagicus,* three stages of maturation was reported based on the development of vas deferens by [22].

The morphology of female reproductive system

The female reproductive system of *P.sanguinolentus* is more or less similar to that of other Portunid crabs. It is closely related with the structure observed by in *P. sanguinolentus* [34,35]. Almeida and Buckup [32] and Johnson [37] observed similar structure in *S. serrata* and in *C. sapidus*. George [35] noticed that the posterior prolongation of the right side ovary is shorter and narrower than the left side in *P. sanguinolentus*. This condition is also observed in *P. sanguinolentus* at the present study. Corgo and Freire [60] in *Cancer pagurus* and Estampador [33] in *S. serrata* observed that the posterior prolongations of the ovary are connected at the posterior end. In *P. sanguinolentus* differ from such species by posterior prolongation of the ovaries are permanently separated. In many species of branchyurans, fresh mating female crabs are indicated externally by hardened mass of spermatozoa called sperm plug, together with associated secretions producing from the vulva. Similar observation was made in the present study, as well as the study of Sukumaran and Neelakantan [61]. The mated females inhibit further copulation with other males, by prevent the loss of sperm or keep deletrous materials from entering the female reproductive tract [61,62].

Stages of ovarian development

Stages of ovarian development have been determined based on the morphological and histological analysis of the ovary. In the present study, the morphological analysis of the ovaries of *P. sanguinolentus* demonstrated that a gradual development of the ovaries confirmed the maturation into five maturation stages. Among different workers who have studied the maturation of ovaries in brachyuran's crabs, there is a little consistency as to the number of maturity stages recognized. Six stages of maturation were recorded by Diesel [56] in rock crab *C. irroratus* and five stages in *Chaceon quinquedens*. Hartnoll [63], Parker [64] recorded five maturity stages in *P. pelagicus;* similarly classified five ovarian developmental stages. Dhas et al. [65] reported four maturity stages, except the spent stage in *P. sanguinolentus*. In the deep sea golden crab, *C. fenneri*, the developmental stages were classified into six stages by Sumpton et al. [66]. The microscopical analysis of the ovaries of *P. sanguinolentus* in the present study demonstrated that a gradual development of the ovaries confirmed the maturation into five maturity stages. The examination of gonad is one of the most accurate techniques, but precise estimates rely on some previous knowledge about specific patterns of growth and reproduction to aid the interpretation of the results, as well as knowledge of short-term breeding cycle to avoid confusion between spent and immature specimens [67].

Gonadosomatic index

The GSI in females showed clear cyclic pattern, with high values during January-February in *P. sanguinolentus* and during September and January–February in *P. pelagicus* [59]. In the present study, the GSI was studied only based on the size and weight of the crab, but not in seasonal wise. The GSI has the great relationship with the size and weight of the crabs.

Size at first maturity

The reproductive capacity of certain crustacean species can be assessed by the study of sexual maturity [68]. The size at onset of sexual maturity in brachyurans can be evaluated considering different criteria, including growth allometry, gonad development stages, presence of spermatic bags or sperm in spermathecas, vestige eggs on the ovigerous areas, or the presence of eggs in the abdomen as was analyzed by several authors [31,69-72]. Sexual maturity in brachyurans has been determined in various ways, based on the analyses of morphological maturity, relative growth criteria and physiological and functional maturity [73]. According to Muino [74] the consecutive changes observed during gonadal development are important because besides enabling and estimate of physiological sexual maturity [74-77]. In brachyuran crabs, chela in males and abdomen in females are considered as secondary sexual characters because of their functions in reproduction [78]. The male crab uses its chela for territorial defense, combat, mating and courtship, as well as in carrying and holding the female during copulation. The abdomen in adult females forms an incubation chamber for the developing eggs, which are attached to the setose pleopods. The increase in relative growth of the male chela and female abdomen at the puberty moult brings these structures to fully functional size at sexual maturity. Hence the relative growth of chela in males and abdomen in females has been used to determine size at which puberty moult occurs or functional maturity attained. Cobo and Fransozo [79] from Cochin, west cost of India, reported that the size of the smallest berried *P. sanguinolentus* as 8.5 cm CW and the size at maturity range was 8.1-9.6 cm CW. Sumpton et al. [12] who worked out physiological maturity in both male and female *P. sanguinolentus* from Queensland, Australia, reported that the males and females attain sexual maturity at 8.3 and 7.4 cm CW, respectively. Reeby et al. [13] reported that male *P. sanguinolentus* attains full sexual maturity (meaning functional as well as physiological) at 8.1-8.5 cm CW from Karwar, west cost of India. They did not work out sexual maturity in *P. sanguinolentus*.

In south west coast of India, the male and female may undergo a pupertal moult, at a size ranging between 8.0-8.5 cm and 8.0-9.0 cm CW, respectively. Hartnoll [80] reported that male and female attain full maturity at 8.3-8.9 mm CW and 8.1-9.3 mm CW, respectively. In the present study, the smallest berried female attained sexual maturity at 86 mm CW, while the male and female crab attained sexual maturity when those reached the size of 95-100 mm CW and 9.0-9.5 cm CW, respectively. According to the present investigation, it was observed that the functional and physiological maturities occur almost at the same size.

The size at which maturity occurs can vary with latitude or location, and within individuals at any location [59]. Pillai and Nair [81] compared geographic variation in the size of sexually mature females in five species of crabs along the east and west coast of North America. Four of the five species show significant geographic variation in size at onset of maturity. The differences in size at maturity among population of the same species of crab may be attributed to variation in moult increment, and in the number of moults [81]. Environmental factors such as temperature and salinity can also affect size at sexual maturity in crabs. Rhasheed and Mustaquim [82] who investigated the effect of temperature and salinity on size at sexual maturity of female blue crab *C. sapidus* from nine Texas bay systems, stated that size at maturity can vary along the Texas coast, as temperature and salinity vary from bay to bay. He also mentioned that seasonal and annual variation in temperature and salinity in the bay could also affect size at onset of maturity.

Berried females

Egg bearing females occurs throughout the year, with proportion of females bearing egg masses being peak during three different months August, January and March along Parangipettai coast, whereas November to March in South west coast, whereas December to May and July to August in Calicut coast [83-85]. Peaks of higher breeding intensity may be associated with variation in temperature, salinity, food availability, rainfall and photoperiod [86]. In brachyuran crabs inhabiting tropical waters usually breed throught the year, whereas those found in temperate waters breeds only in certain months [87]. It is generally suggested that near the tropics, reproduction occurs year round because environmental conditions are generally favorable for gonad development [88]. However, both continuous and seasonal reproductive patterns are found in subtropical and tropical regions [89].

Fecundity

Fecundity is defined as the number of eggs produced by an individual female [90]. Fecundity is calculated as the number of eggs carried externally by the female [23]. In general, the fecundity of decapod crustaceans is evaluated as the number of eggs or weight of eggs produced by a female in a single egg batch (Batch fecundity), and is positively correlated with body size (or) weight of females [68]. Fecundity is significantly related to the female crab's carapace width [68]. The larger females have potential to produce more eggs than smaller females [91]. Fecundity is a species specific character, not only regarding the number of eggs extruded in a single batch, but also the frequency of brood production during the life cycle of crustaceans [92].

In general, the fecundity of the similar species was varied between different locations. Ryan [34] reported that the fecundity in *P. sanguinolentus* was ranged between 961000 eggs to 2250000 eggs. Dhas et al. [65] has reported that feacundity was ranged between 268400 to 668300 eggs from Mangalore. Reeby et al. [21] has been reported that the eggs ranged between 1, 58,608-7,12,526 eggs from Karwar. Kumar et al. [23] reported that the fecundity from Karachi was between 2,25,649 to 5, 24,456 eggs. Kailola et al. [93] have been recorded from Northern Taiwan that the eggs was ranged between 4,10,000 to 24,40,000. In the present study, the fecundity was ranged between 2,83,963 to 9,67,293 eggs.

The fecundity was varied in relation to latitudinal range, habitat structure and food availability [94]. Food availability is the most important factor as feeding factor in paramount for yolk formation [94]. Fecundity determines the reproductive potential of a species and the stock size of its population [95]. Information on fecundity is crucial for the management of crab fisheries [95]. A clear knowledge of the fecundity plays a significant role to evaluate the commercial potentialities of crab stock, and also be used to assess the abundance and reproductive potential of the spawning stock. The discrepancy may be affected by environmental factors, including predation, parasitization and temperature, which may affect the balance between the optimal number and the size of eggs, and also by the loss of eggs during incubation period, and/or during handling as the crabs were obtained from commercial landings [24,96].

The male reproductive system of *P. sanguinolentus* is bilaterally symmetrical creamy to whitish in colour, composed of a pair of testes, a pair of vas differentia, and a pair of ejaculatory ducts internally, and a pair of pleopods externally as accessory reproductive organs, present on the inner side of the abdominal flab. The vas differentia has been divided into three distinct regions, based on the morphological and functional criteria: Anterior (AVD), Median (MVD) and Posterior (PVD) vas deferens [97,98]. The gonads of the male measuring below 8.5 cm carapace width (CW) has immature stage, above 10.5 cm CW has mature stage, whereas the CW in between 14.5-10.5 cm was maturing stage. The female reproductive system composed of a pair of ovaries, a pair of seminal receptacles (or) spermatheca, and a pair of oviducts open to the exterior through the female genital opening situated on the left and right sternites of sixth thoracic segment. The ovaries of *P. sanguinolentus* are categorized into five stages, according to the size, colour, and external morphology of the ovaries. They are immature, early maturing, late maturing, ripe and spent.

Fifty percent of the female crabs attained sexual maturity when they reached the size of 9.1-9.5 cm CW. Fifty percent of the male crab's attained sexual maturity when they reached the size of 9.5-10.0 cm CW. However, the smallest berried female attained sexual maturity in the present study was measured at 8.6 cm CW. Based on the change in colour, increase in the size and change in the shape of the berry eggs, three different stages of egg development was observed. Stage-I has pale yellow to deep yellow coloured egg mass, no eyespots was visible in the eggs, the mean size of the eggs was 260.16 ± 0.24 µm. Stage-II has yellow to gray coloured egg mass, eyespots was observed, mean size of the egg was 290.2 ± 0.31µm. Stage-III has deep gray to black coloured egg mass, eyespots and chromatophores were discernible, the mean size of the egg was 340.32 ± 0.34 µm. Seventeen ovigerous females were used for the calculation of fecundity. The fecundity was found to be from 283963 to 967293 eggs in the animal's had the carapace width of 10.6-14.1 cm.

References

1. Mohammed Saved A, Rajeev Rahavan P (2001) Mud Crab-culture and fattening techniques, status and prospects. Sea Exp J 32: 25-29.
2. Hartnoll RG, Gould P (1989) Brachyuran life history strategies and the optimization of egg production. Proc Zool Symp 59: 1-9.
3. LopezGreco LS, Hemandez JE, Bolanos JE, Rodriguez EM, Hemandez G (2000) Population features of *Microphrys bicomutus* Latreille, 1825 (Brachyura, Maiidae) from Isla Margarita Verezuela. Hydrobiol 439: 151-159.
4. Reigada ALD, Negreiros-Fransozo ML (1999) Maturidade sexual em *Hepatus pudibundus* (Decapoda, Branchyura, Calappidae). Iheringia, Ser Zool 86: 159-164.
5. Pinheiro MAA, Fransozo A (2002) Reproductive dynamics of the speckled swimming crab *Arenaeus cribrarius* (Lamarck, 1818) (Brachyura, Portunidae), on the north coast of Ssao Paulo State. Brazil J Crust Biol 22: 416-428.
6. Castiglioni DS, Negreiros-Fransozo ML (2006) Ciclo reproductivo do caranguejo violinista *Uca rapax* (Smith) (Crustacea, Branchyura, ocypodidae) abitante de umestuario degradado em Paraty Rio de Janeiro Brasil Revi. Brasil de zool 23: 331-339.
7. O' Brien L (1999) Factors influencing the rate of sexual maturity and the effect on spawning stock for Georges Bank and Gulf of Maine Atlantic cod *Gadus morhua* stocks. J Northw Atl Fish Sci 25: 179-203.
8. Duran P, Junquera S, Alvarez MS (1999) Scientific Council Meeting–JUNE 1999, Yellowtail Flounder length at maturity in the Grand Bank (1995-1998). Northwest Atlantic Fisheries Organization.
9. Morita K, Morita SH (2002) Rule of age and age at maturity: Individual variation in the maturation history of resident white spotted charr. J Fish Biol 61: 1230-1238.
10. Ahmed M, Mustaquim J (1974) Population structure of four species of Porcellanid crabs (Decapoda: Anomura) occurring on the coast of Karachi. Mar Biol 26: 173-182.
11. Campbell GR, Fielder DR (1986) Size at sexual maturity and occurrence of ovigerous females in three species of commercially exploited portunid crabs in SE Queensland. Proc Royal Soc Queensland, Australia 97: 79-87.
12. Sumpton WD, Smith GS, Potter MA (1989) Notes on the biology of the portunid crabs, *Portunus sanguinolentus* (Herbst) in subtropical Queensland waters. Australian Journal of Marine and Freshwater Research 40: 711-717.
13. Reeby J, Prasad PN, Kusuma MS (1990) Size at sexual maturity in the male crabs of *Portunus sanguinolentus* and *P. pelagicus*. Fishery Technology 27: 115-119.
14. Xiao Y, Kumar M (2004) Sex ratio and probability of sexual maturity of females at size, of the blue swimmer crab, *Portunus pelagicus* (Linnaeus) off southern Australia. Fisheries Research 68: 271–282.
15. Du Preez HH, McLachalan A (1984) Biology of the three-spot swimming crab, *Ovalipes punctatus*, I. Morphometrics and relative growth (Decapoda: Portunidae). Crusta 47: 72-82.
16. Haefner PA (1985) Morphometry, reproduction, diet, and epizoites of *Ovalipes stephensoni* Williams, 1976 (Decapoda, Brachyura). J Crustacean Biol 5: 658-672.
17. Muino R, Fernandez L, Gonzalez-Gurriaran E, Freire J, Vilar JA (1999) Size at maturity of *Liocarcinus depurator* (Brachyura: Portunidae): A reproductive and morphometric study. J Mar Biol Assoc UK 79: 295-303.
18. Tongdee N (2001) Size distribution, sex ratio and size at maturity of Mudcrab (*Scylla* spp.) in Ranong Province, Thailand. Asian Fish Sci 14: 113-120.
19. Pillay TR (1964) An aid to identification of the Mysid cat fishes of Bangladesh. Bangladesh J Zool 2: 1-12.
20. Potter IC, Chrystal PJ, Loneragan NR (1983) The biology of the blue manna crab *P. pelagicus* in an Australian estuary. Mar Biol 78: 75-85.
21. Reeby J, Prasad PN, Kusuma N (1990) Fecundity of *Portunus* species from Karwar Waters. J Fish Sci Technol 27: 153-154.
22. Lestang SD, Hall NG, Potter IC (2003) Reproductive biology of the blue swimming crab (*Portunus pelagicus*, Decapods: Portunidae) in five bodies of water on the west coast of Australia. Fish Bull 101: 745-757.
23. Kumar MS, Xiao Y, Venema S, Hooper G (2003) Reproductive cycle of the blue swimmer crab, *Portunus pelagicus*, off Southern Australia. J Mar Biol Assoc UK 83: 983-994.
24. Takween W, Qureshi NA (2005) Population structure and reproductive biology of four species of swimming crabs (Crustacea: Brachyura: Portunidae) from coastal area of Karachi, Pakistan. Pakistan Journal of Marine Sciences 14: 107-121.
25. Arshad A, Kamarudin MS, Saad CR (2006) Study on fecundity embryology and larval development of blue swimming crab *Portunus pelagicus* (Linnaeus, 1758) under laboratory conditions. Res J Fish Hydrobiol 1: 35-44.
26. John Samuel N, Thirunavukkarasu N, Soundarapandian P, Shanmugam A, Kannupandi T (2004) Fishery potential of commercially important portunid crabs along Parangipettai coast. In*:* Pro Ocean Life Food Med Expo 165-173.
27. Varadharajan D, Soundarapandian P, Dinakaran GK, Vijakumar G (2009) Crab fishery resources from Arukkattuthurai to Aiyammpattinam, South East Coast of India. Curr Res J Biol Sci 1: 118-122.
28. Carpenter KE, Niem VH (1998) FAO species identification guide for fishing purpose. The living marine resources of the Western Central Pacific, vol. 2, Cephalopods, Crustaceans, Holothurian and Sharks. FAO Report 687: 1-139.
29. Apel M, Spiridonov VA (1998) Taxonomy and zoogeography of the portunid crabs (Crustacea: Decapoda: Brachyura: Portunidae) of the Arabian Gulf and adjacent waters. Fauna of Arabia 17: 159-331.
30. Chapgar BR (1957) On the marine crabs (Decapoda: Brachyura) of Bombay State. J Bombay Nat Hist Soc 54: 399-439.
31. Campbell A, Eagles MD (1982) Size at maturity and fecundity of rock crabs, *Cancer irroratus,* from the Bay of Fundy and southwestern Nova Scotia. Fish Bull 81: 357-362.
32. Almeida AO, Buckup L (2000) Occurrence of protandric hermphroditism in a population of the neotropical freshwater crayfish *Parastacus brasiliensis* (Parastacidae). J Crust Biol 20: 224-230.

33. Estampador EP (1949) *Scylla* II. Comparitive studies on spermatogenesis and oogenesis. Philippine Journal of Science 78: 301-353.

34. Ryan EP (1967) Structure and function of the reproductive system of the crab, *Portunus sanguinolentus* (Herbst) (Brachyura: Portunidae). Mar Biol Asso India 2: 522-544.

35. George MJ (1963) Biology of the edible crab *Neptunus sanguinolentus* Herbst. University of Madras, India 199.

36. Mouchet S (1931) Spermatophores des Crustaces Decapods Anomures et Brachyoures et castration parasitaire chez quelques Pagures. Annales de Station Oceanographique de Salammbo 6: 1-203.

37. Johnson PT (1980) History of the blue crab, *Callinectes sapidus*, a model for the decapoda. Praeger Publishers, New York, USA 440.

38. Burton TE (1995) The spermatid pathway and associated reproductive structures of squat lobster *Thenus orientalis* (Lund, 1973). Invertebr Reprod Dev 28: 53-61.

39. Lima AVP (1995) Estodo do Sistema Reprodutivo de machos da Lagosta verde *Panulirus laevicauda* (Latreille, 1871) (Decapoda: Palinuridae). Dissertaçao, Universidade Federal da Paraiba, Brazil 200.

40. Garcia TM, Silva JF (2006) Testis and vas deferens morphology of the red-clawed mangrove tree crab (*Goniopsis cruentata*) (Latreille, 1803). Braz Arch Biol Technol 49: 339-345.

41. Beninger PG, Elner RW, Foyleand TP, Odense PH (1988) Functional anatomy of the male reproductive systemand the female spermatheca in the snowcrab *Chionoecetes opilio* (*O. Fabricius*) (Decapoda: Majidae), and a hypotheses for fertilization. J Crustacean Biol 8: 322-332.

42. Simeo CG, Ribes E, Rotllant G (2009) Internal anatomy and ultra structure of the male reproductive system of the spider crab *Maja brachydactyla* (Decapoda: Brachyura). Tissue Cell 45: 345-361.

43. Cronin LE (1947) Anatomy and histology of the male reproductive system of *Callinectes sapidus*. J Morphol 81: 209-239.

44. Castilho GG, Ostrensky A, Pie MR, Boeger WA (2007) Morphology and histology of the male reproductive system of the mangrove land crab *Ucides cordatus* (L.) (Crustacea, Brachyura, Ocypodidae). Acta Zool 89: 157-161.

45. Binford R (1913) The germ-cells and the process of fertilization in the crab, *Menippe mercenaria*. J Morphol 24: 147-201.

46. Hoestlandt H (1948) Recherches surla biologie de l'*Eriocheir sinensis* en France. Ann Ins Oceanogr 24: 1-116.

47. Kon T, Honma Y (1970) Studies on the maturity of the gonad in some marine invertebrates—IV: Seasonal changes in the testes of the Tanner crab. Nippon Suisan Gakkai Shi 36: 1028-1031.

48. Sapelkin AA, Fedoseev VY (1981) Structure of male reproductive system of Tanner crabs. Sov J Mar Biol 7: 37-43.

49. Chiba A, Honma Y (1972) Studies on the maturity of the gonad in some marine invertebrates—IV: Seasonal changes in the testes of the lined shored crab. Nippon Suisan Gakkai Shi 38: 317-329.

50. Beninger PG, Lanteigne C, Elner RW (1993) Reproductive process revealed by spermatophore dehiscence experiments and by histology, ultrastructure, and histochemistry of the female reproductive system in the snow crab *Chionoecetes opilio* (O. Fabricius). J Crust Biol 13: 1-16.

51. McLaughlin PA (1993) Internal anatomy. In: Mantel LH (Ed), The Biology of Crustacea, Vol 5 Academic Press, New York, USA 1-52.

52. Hinsch GW, McKnight CE (1988) The vas deferens of the Spanish lobster, *Scyllarides chacei*. Int J Invertebr Reprod Dev 13: 267-280.

53. Manjon-Cabeza ME, Raso JEG (2000) Morphological reproductive aspects of males of *Diogenes pugilator* (Roux, 1829) (Crustacea, Decapoda, Anomura) from Southern Spain. Sarsia 85: 195-202.

54. Fasten N (1915) The male reproductive organs of some common crabs of Puget Sound. Puget Sound Mar Sta Pub 1: 35-45.

55. Adiyodi KG, Anilkumar RG (1988) Accessory sex glands. In: Adiyodi KG, Adiyodi RG (Eds.), Rep Biol Invert, Vol 3, John Wiley and Sons, Kerala, India 261-318.

56. Diesel R (1991) Sperm competition and the evolution of mating behaviour in *Brachyura,* with special reference to spider crabs (Decapoda, Majidae). In: Bauer RT, Martin JW (Eds.), Crust Sex Biol Columbia University Press, New York, USA 145-163.

57. Krol RM, Hawkins WE, Overstreet RM (1992) Reproductive components. In: Microscopic Anatomy of Invertebrates. Decapod Crustacea 10: 295-343.

58. Haefner PA (1976) Distribution, reproduction and molting of the rock crab, *Cancer irrigates* say 1917 in the mid Atlantic Bight. J Nat Hist 10: 377-397.

59. Lawal-Are AO (2010) Reproductive biology of the blue crab, *Callinectes amnicola* (De Rocheburne) in the Lagos Lagoon, Nigeria. Turkish J Fish Aquat Sci 10: 1-7.

60. Corgo A, Freire J (2006) Morphometric and gonad maturity in the spider crab *Maja brachydactyla*: A comparison of methods for estimating size at maturity in species with determinate growth. ICES Mar Sci 63: 851-859.

61. Sukumaran KK, Neelakantan B (1996) Relative growth and sexual maturity in the marine crabs, *Portunus* (*Portunus*) *sanguinolentus* (Herbst) and *Portunus* (*Portunus*) *pelagicus* (Linnaeus), along the Southwest coast of India. Indian J Fish 43: 215-223.

62. Pearson J (1909) *Cancer pagurus* (The edible crab). Memories of L. pool Marine Biological Commission No. 16: 262.

63. Hartnoll RG (1969) Mating in Brachyura. Crusta 16: 161-181.

64. Parker GA (1970) Sperm competition and its evolutionary consequences in the insects. Biol Rev 45: 525-567.

65. Dhas MA, Subramoniam T, Varadarajan S, Govindharajulu P (1980) Germinal zone activity and oocyte differentiation in the marine crab *Portunus pelagicus*. Proc Indian Natl Sci Acad B Biol Sci 46: 287-292.

66. Sumpton WD, Potter MA, Smith GS (1994) Reproduction and growth of the commercial sand crab, *Portunus pelagicus* (L.) in Moreton Bay, Queensland. Asian Fish Sci 7: 103-113.

67. Sukumaran KK, Teland KR, Thippeswamy D (1986) On the fishery and biology of the crab *Portunus sanguinolentus* (Herbst) along the south Kanara coast. Indian J Fish 33: 188-200.

68. Erdman RB, Blake NJ (1988) Reproductive ecology of female goldan crabs, *Geryon fenneri* manning and holthuis from Southeastern Florida. J Crust Biol 8: 392-400.

69. Flores AAV, Paula J (2002) Sexual maturity, larval release and reproductive output of two brachyuran crabs from a rocky intertidal area in central Portugal. Invert Reprod Develop 42: 21-34.

70. Hines AH (1982) Allometric constraints and variables of reproductive effort in brachyuran crabs. Mar Biol 69: 309-320.

71. Brown RB, Powell GC (1972) Size at maturity in male Alaskan Tanner crab, *Chionoecetes bairdii*, as determined by chela allometry, reproductive tract weights, and size of precopulatory males. J Fish Res Bd Can 29: 423-427.

72. Lopez Greco LS, Rodriguez EM (1998) Size at the onset of sexual maturity in *Chasmagnathus granulata* Dana, 1851 (Grapsidae, Sesarminae): A critical overall view about the usual criteria for its determination. Proc Fourth Int Crust Cong 675-689.

73. Pinheiro MAA, Fransozo A (1998) Sexual maturity of speckled swimming crab *Arenaeus cribrarius* (Lamark, 1818) (Decapoda, Brachyura, Portunidae), in the Ubatuba litoral, Silo Paulo State, Brazil. Crusta 71: 434-452.

74. Muino R (2002) Fecundity of *Liocarcinus depurator* (Brachyura; Portunidae) in the Ria de Arousa (Galicia, North-west Spain). J Mar Biol Ass UK 82: 109-116.

75. Sampedro MP, Gonzalez-Guirrian E, Freire J, Muino R (1999) Morphometry and sexual maturity in the spider crab *Maja squinado* (Decapoda: Majidae) in Galicia, Spain. J Crust Biol 19: 578-592.

76. Hartnoll RG (1974) Variation in growth pattern between some secondary sexual characters in crabs (Decapoda, Brachyura). Crusta 27: 131-136.

77. Oshiro LMY, Silva R, Silveira CM (1999) Rendimento de carne nos caranguejos Guaia, *Menippe nodifrons* Stimpson, 1859 e Guaiamum, *Cardisoma guanhumi* Latreille, 1825 (Crustacea, Decapoda, Brachyura) da Baia de Sepetiba/Rj. Acta Biologica Leopoldensia 21: 83-88.

78. Silva ZS, Oshiro LMY (2002) Aspectos reprodutivos de *Goniopsis cruentata* (Latreille) (Crustacea, Branchyura, Grapsidae) na Baia de Sepetiba, Rio de Janeiro, Brazil. Revist Brasil De Zool 193: 907-914.

79. Cobo VJ, Fransozo A (2003) External factors determining breeding season in

the red mangrove crab *Goniopsis cruentata* (Latreille) (Crustacea, Branchyura, Grapsidae) on the Sao Paulo State Nothern coast, Brazil. Revi Brasil De zool 20: 213-217.

80. Hartnoll RG (1978) The determination of relative growth in crustacea. Crusta 34: 281-293.

81. Pillai KK, Nair NB (1976) Observations on the breeding cycles of certain estuarine crustaceans of the southwest coast of India. J Mar Biol Ass India 15: 754-770.

82. Rhasheed S, Mustaquim J (2010) Size at sexual maturity, breeding season and fecundity of three spot swimming crab *Portunus sanguinolentus* (Herbst) (Decapoda, Brachyura, Portunidae) occurring in the coastal waters of Karachi, Pakistan. Fish Res 103: 56-62.

83. Hines AH (1989) Geographic variation in size at maturity in brachyuran crabs. Bull Mar Sci 45: 356-368.

84. Fisher MR (1999) Effect of temperature and salinity on size at maturity of female blue crabs. Trans Am Fish Soc 128: 499-506.

85. Radhakrishnan CK (1979) Studies on Portunid crabs of Porto Novo (Crustacea: Decapoda: Brachyura). PhD, Thesis, Annamalai University, India.

86. Pillai KK, Nair NB (1971) The annual reproductive cycles of *Uca annulipes*, *Portunus pelagicus* and *Metapenaeus affinis* (Decapoda, Crustacea) from the southwest coast of India. Mar Biol 11:152-166.

87. Saradha PT (1998) Crab fishery of the Calicut coast with some aspects of the population characteristics of *Portunus sanguinolentus*, *P. pelagicus* and *Charybdis cruciata*. Indian J Fish 45: 375-386.

88. Emmerson WD (1994) Seasonal breeding cycles and sex ratios of eight species of crabs from Mgazana, a mangrove estuary in Transkei, South Africa. J Crust Biol 14: 158-168.

89. Warner GF (1977) The Biology of Crabs. Paul Elek (Scientific Books) Ltd., London, UK 202.

90. Meusy J, Payen GG (1988) Female reproduction in malacostracan. Crustacea Zool Sci 5: 217-265.

91. Litulo C (2005) Population biology of the fiddler crab *Uca annulipes* (Brachyura: Ocypodidae) in a tropical East African mangrove (Mozambique). Est Coast Shelf Sci 62: 283-290.

92. Lagler KF (1949) Studies of freshwater biology. Ann Arbor Michigan, USA 119.

93. Kailola PJ, Williams MJ, Stewart PC, Riechelt RE, McNee A, et al. (1993) Australian Fisheries Resources, Bureau of Resource Sciences, Canberra, Australia 422.

94. Sastry AN (1983) Ecological aspects of reproduction. In: Vernberg FJ, Vernberg W (Eds), The Biology of Crustacea, Vol 7, Environ Adap Acad Press, New York, USA 179-254.

95. Lee HH, Hsu CC (2003) Population biology of the swimming crab *Portunus sanguinolentus* in the waters off Northern Taiwan. J Crust Biol 23: 691-699.

96. Litulo C (2004) Fecundity of the pantropical fiddler crab *Uca annulipes* (H. ilne Edwards, 1837) (Brachyura: Ocypodidae) at costa do sol mangrove, map up to bay, southern Mozambique, Western Indian Ocean. J Mar Sci 3: 87-91.

97. Mantelatto FLM, Fransozo A (1997) Fecundity of the crab *Callinectus ornatus* Ordway, 1863 (Decapoda, Brachyura, Portunidae) from the Ubatuba region, Sao Paulo, Brazil. Crusta 70: 214-224.

98. Smith CC, Fretwell SD (1974) The optimal balance between size and number of offspring. Am Nat 108: 499-506.

Mercury Accumulation and the Mercury-PCB-Sex Interaction in Summer Flounder

Madenjian CP[1]*, Jensen OP[2], Krabbenhoft DP[3], DeWild JF[3], Ogorek JM[3] and Vastano AR[2]

[1]*US Geological Survey, Great Lakes Science Center, Ann Arbor, Michigan, USA*
[2]*Department of Marine and Coastal Sciences, Rutgers University, New Brunswick, New Jersey, USA*
[3]*US Geological Survey, Wisconsin Water Science Center, Middleton, Wisconsin, USA*

Abstract

Patterns in the relative differences in contaminant concentrations between the sexes of mature fish may reveal important behavioral and physiological differences between the sexes. We determined whole-fish total mercury (Hg) concentrations in 23 female summer flounder (*Paralichthys dentatus*) and 27 male summer flounder from New Jersey coastal waters. To estimate the change in Hg concentration due to release of eggs at spawning, Hg concentration in the somatic tissue and ovaries of 5 of the 23 female summer flounder were also determined. To ascertain whether most of the Hg in the summer flounder was methylmercury (MeHg), whole-fish MeHg concentrations were determined in all 50 summer flounder. Whole-fish Hg concentrations averaged 113 ng/g for females and 111 ng/g for males. Thus, females were 2% higher in Hg concentration than males, on average, but the difference was not statistically significant. Based on Hg determinations in the somatic tissue and ovaries, we predicted that Hg concentration of females would increase by 3.7%, on average, immediately after spawning due to release of eggs. On average, 92% of the Hg in the summer flounder was MeHg. To determine whether the effect of sex on Hg concentration was significantly different from the effect of sex on polychlorinated biphenyl (PCB) concentration, we paired our Hg determinations with PCB determinations from a previous study, and applied regression analysis. Sex significantly interacted with contaminant type (Hg or PCBs), as males were 43% higher in PCB concentration than females, whereas females were 2% higher in Hg concentration than males. Males eliminating Hg from their bodies at a faster rate than females was a likely explanation for this discrepancy between the two contaminant types. Overall, the Hg and PCB concentrations in the summer flounder were relatively low, and therefore our findings also had implications for continued operation of the summer flounder fishery.

Keywords: Androgens; Fish consumption advisories; Hg-elimination rates; Methylmercury; PCBs; Sex differences; Summer flounder

Introduction

Three species of freshwater teleost fishes (lake trout (*Salvelinus namaycush*), burbot (*Lota lota*), and lake whitefish (*Coregonus clupeaformis*)) share the characteristic of the ratio of whole-fish polychlorinated biphenyl (PCB) concentration in males to whole-fish PCB concentration in females substantially exceeding the ratio of whole-fish total mercury (Hg) concentration in males to whole-fish Hg concentration in females [1]. This apparent pattern has been attributed to males eliminating Hg from their bodies at a faster rate than females, whereas long-term elimination of PCBs by fish is negligible for both sexes [1-4]. The faster rate of Hg elimination in males has been linked to certain androgens [2]. For these three fish species, mature males were between 20% and 35% higher in whole-fish PCB concentration than similarly aged mature females, and this difference appeared to be primarily due to a greater rate of energy expenditure in mature males stemming from higher activity and a higher resting metabolic rate (or standard metabolic rate [SMR]) [5-7]. A consequence of this greater rate of energy expenditure is a higher rate of food consumption, which, in turn, results in a higher rate of PCB accumulation. In contrast to the relative differences in PCB concentrations between the sexes, male lake trout were only 8% higher in Hg concentration than female lake trout, and females were actually higher than males in Hg concentration for both burbot and lake whitefish.

Documentation of a substantially greater ratio of PCB concentration in males to PCB concentration in females compared with the ratio of Hg concentration in males to Hg concentration in females in a species of a marine teleost fish would mark an important milestone in assessing the pervasiveness of this characteristic throughout all teleost fish populations. To date, this characteristic has been documented in populations of freshwater teleost fishes only [1]. Marine fish differ from freshwater fish in their processes for osmoregulation, which is the maintenance of water and salt balance within the bodies of fish [8]. Perhaps these physiological differences between freshwater and marine fishes exert an influence on the difference between the two abovementioned ratios, which would then suggest that factors other than the sex-related difference in Hg-elimination rates may be contributing to the relative difference in Hg concentrations between the sexes. On the other hand, if this apparent relationship between the two abovementioned ratios holds true for marine teleost fishes, then evidence to support the contention that males eliminate Hg at a faster rate than females is further strengthened. Likewise, the implication that certain androgens are linked to enhanced Hg-elimination rate in males is further corroborated.

Of the flatfish fisheries operating along the U. S. Atlantic coast, the summer flounder or fluke (*Paralichthys dentatus*) fishery is most important [9,10]. Both the summer flounder population and the

***Corresponding author:** Madenjian CP, US Geological Survey, Great Lakes Science Center, Ann Arbor, Michigan, USA, E-mail: cmadenjian@usgs.gov

commercial fishery extend from Massachusetts to North Carolina. The summer flounder recreational fishery is most concentrated in New York and New Jersey waters of the Atlantic Ocean [9-11]. Although the spawning season for summer flounder occurs during September through March, peak spawning typically occurs during October and November. Summer flounder is one of the important predators inhabiting the western Atlantic coastal ecosystem, and the diet of adults primarily consists of fish and squid [12]. Female summer flounder grow considerably faster than male summer flounder [11]. To the best of our knowledge, the difference in whole-fish Hg concentrations between the sexes of summer flounder has yet to be quantified. Adult male summer flounder exceeded adult female summer flounder in whole-fish PCB concentration by 43%, and this difference has been attributed to a higher energy expenditure rate in males and the growth dilution effect [13].

The overall goal of our study was to characterize Hg accumulation in mature summer flounder from the coastal waters of New Jersey. Specific objectives included: (1) quantify the difference in whole-fish Hg concentrations between the sexes of summer flounder caught from a spawning aggregation off the New Jersey coast, (2) quantify the difference between somatic tissue Hg concentration and ovary Hg concentration in female summer flounder, (3) estimate the change in whole-fish Hg concentration of female summer flounder associated with release of eggs at spawning, (4) determine whether the bulk of Hg in summer flounder is MeHg, (5) determine whether the effect of sex on Hg concentration was significantly different from the effect of sex on PCB concentration in summer flounder, and (6) discuss the implications of our findings with regard to the continued operation of the summer flounder fishery. Of special interest was whether the ratio of Hg concentration in males to Hg concentration in females was less than the ratio of PCB concentration in males to PCB concentration in females for summer flounder.

Methods

Field methods

The same 50 adult summer flounder used by Madenjian et al. [13] in their study on the difference in PCB concentrations between the sexes were used in our study. In brief, these fish were captured via trawling by a commercial fishing boat operating in coastal waters about 30 km east of Barnegat Light, New Jersey on 12 November 2013. We purchased 85 summer flounder from Viking Village, the commercial fish producer at Barnegat Light, New Jersey, and then transported the fish on ice in coolers to the Rutgers University Marine Field Station (RUMFS) in Tuckerton, New Jersey for further processing. Total length (TL) of each summer flounder was measured to the nearest mm, and each summer flounder was weighed to the nearest g. Sex and maturity of each summer flounder was determined by visual inspection of the gonads. We bagged each fish individually, placed a cardboard tag with a unique identification number in the bag, and then kept the fish frozen at -20°C until further processing. Refer to Madenjian et al. [13] for more details.

Hg, MeHg, and age determinations

To determine Hg concentration in somatic tissue and ovaries, five female summer flounder were randomly selected and partially thawed. Otoliths were then removed for aging purposes, ovaries were removed and weighed to the nearest g, and the remaining somatic tissue was also weighed to the nearest g. We homogenized ovaries and somatic tissue separately in appropriately sized blenders. About 100 g of each somatic tissue homogenate was placed in a contaminant-free glass jar, sealed with a lid, and then stored at -20°C. All of the available homogenate of each pair of ovaries (between 50 and 85 g) was placed in a contaminant-free glass jar, sealed with a lid, and then stored at -20°C. For the 18 remaining female summer flounder, we partially thawed each fish and removed the otoliths for aging. Then, each whole fish was homogenized using appropriately sized blenders, and approximately 100 g of the homogenate was then placed in a contaminant-free glass jar, sealed with a lid, and stored at -20°C. Of the remaining 62 males, 27 were randomly selected and then processed in the same manner that the 18 females used for whole-fish Hg determinations were processed. Frozen homogenates were shipped to the US Geological Survey (USGS) Mercury Research Laboratory in Middleton, WI for Hg and MeHg determinations. All samples were lyophilized prior to mercury analysis. At the National Oceanic and Atmospheric Administration (NOAA) Northeast Fisheries Science Center at Woods Hole, MA, the summer flounder otoliths were aged via thin sectioning and enumeration of annuli, under supervision of NOAA fishery scientists. Otoliths for one of the summer flounder were damaged during their removal, and therefore an age was not assigned to this fish. Hg concentrations in fish homogenates were measured using a Nippon MA-2000 direct combustion analyzer (Tokyo, Japan), following USEPA Method 7473 [14]. Details are provided by Madenjian et al. [1,4]. An aliquot of fish homogenate is combusted at 850°C, resulting in the reduction of all species of mercury to gaseous elemental mercury, which is detected in the sample stream by atomic absorption. We performed quality assurance/quality control (QA/QC) protocols throughout the Hg determination process, including regular checks of instrument calibration, analysis of analytical blanks, sample replicates, standard reference materials (SRMs), and laboratory practices to prevent sample contamination. Calibration of the instrument is checked prior to each run, and is based on a 7-point polynomial calibration curve (R^2>0.995). Triplicate analyses were performed once every 10 samples and the relative standard deviation (RSD) for the triplicate analyses ranged from 0.8 to 16.5% (mean of 5.1%). International Atomic Energy Agency Reference Material 407 (IAEA 407) was used as the SRM, and at a minimum an SRM analysis was included in every set of 10 samples. In total, 17 SRM determinations were made, and the average SRM recovery was 105.3%, which is well within the range of acceptance (75-125% recovery) for the USGS Mercury Research Laboratory. We estimated a detection limit for Hg of 7 ng/g (dry weight basis), based on multiple analyses of a fish homogenate (IAEA 407) and following USEPA protocol [15]. All summer flounder Hg concentrations were reported on a wet weight basis. MeHg concentrations in the fish homogenates were determined with a Brooks Rand MERX automated methylmercury analyzer (Brooks Rand Instruments, Seattle, Washington, USA), following an adaptation of USEPA method 1630 [16]. Samples were extracted in 4.5 M nitric acid for 8 hours at 50°C rather than using the distillation procedure outlined in USEPA method 1630. The aliquot was treated with sodium tetramethylborate, with the resulting ethylated mercury purged onto Tenex traps, thermally desorbed, mass-separated with a gas chromatograph column, and detected by cold vapor atomic fluorescence. Refer to Madenjian et al. [1,4] for more details. In fish, MeHg is typically the predominant form of mercury [17], but inorganic mercury can make a substantial contribution to Hg found in fish in some cases [18]. QA/QC protocols were followed throughout the MeHg determination process. The instrument was calibrated daily with a 6-point calibration curve (R^2>0.995). Triplicate analyses were performed once every 10 samples and RSD for the triplicate analyses ranged from 0.9 to 10.5% (mean of 5.6%). IAEA 407 was used as the SRM, and a total of 11 SRM determinations were made. Mean SRM recovery was 110.2%, which is well within the range of acceptance (75-125%) for the USGS Mercury Research Laboratory. We estimated a detection limit for MeHg of 2 ng/g (dry weight basis), based on multiple analyses of a fish homogenate (IAEA 407) and following USEPA protocol [15]. All summer flounder MeHg concentrations were expressed on a wet weight basis.

Data analyses

The mass balance approach described by Niimi [19] was used to calculate whole-fish Hg and whole-fish MeHg in each of the five females selected for mercury determinations of ovaries and somatic tissue. The body burden of mercury (either Hg or MeHg) in a portion of the fish is the weight of mercury contained in that portion of the fish. Body burdens in the ovaries and somatic tissue were calculated by multiplying the mercury concentration in the tissue by the weight of the tissue. Then, these body burdens were summed to yield the whole-fish body burden, and this sum was then divided by the sum of the weights for the ovaries and somatic tissue to yield the whole-fish mercury concentration. In addition, the expected percent change in mercury concentration due to release of eggs at spawning was estimated for each of the five females by calculating the ratio of mercury concentration in the somatic tissue to the estimated whole-fish mercury concentration, using the procedure described by Niimi [19]. For each of the five females, gonadosomatic index (GSI) was calculated by dividing the weight of the ovaries by the total weight of the fish and then multiplying by 100.

We considered four analyses of covariance (ANCOVA) models to test whether sex had a significant effect on Hg concentration. Likewise, we considered four ANCOVA models to test whether sex had a significant effect on MeHg concentration. For each ANCOVA, mercury (either Hg or MeHg) concentration was the dependent variable, sex was the main effect, and TL, weight, Fulton's condition *K* (equal to weight $\cdot 10^5 \cdot TL^{-3}$), or age was the covariate. We found that the covariate had an insignificant effect on mercury concentration in all eight ANCOVA models. Thus, a two-sample *t* test was used to determine whether sex had a significant effect on Hg concentration and whether sex had a significant effect on MeHg concentration. Assumptions of normality and homogeneity of variances were met for the two-sample *t* test applied to Hg concentration data (Shapiro-Wilk statistic: W=0.97; P=0.1512; Folded *F* statistic: F=1.25; df=22, 26; P=0.5792). For the two-sample *t* test applied to MeHg data, the homogeneity of variances assumption was met (Folded *F* statistic: F=1.05; df=22, 26; P=0.9015), but the normality assumption was violated (Shapiro-Wilk statistic: W=0.93; P=0.0071). However, given the robustness of the two-sample *t* test to departures from the normality assumption [20], its application to the MeHg data was still appropriate. Data for 23 females and 27 males were included in the statistical analyses. The ratio of mean Hg concentration of males to mean Hg concentration of females and the ratio of mean MeHg concentration of males to mean MeHg concentration of females were then calculated. We set α=0.05 for all of our statistical testing.

To determine whether the bulk of the Hg in the summer flounder was MeHg, the fraction of Hg as MeHg was calculated by the ratio of MeHg concentration to Hg concentration for each of the 50 summer flounder. The mean of these ratios was then calculated using data for all 50 summer flounder. In addition, we calculated mean ratios by sex.

To determine whether contaminant type (Hg or PCBs) significantly interacted with sex, we tested whether the regression line of PCB concentration as a linear function of Hg concentration for females was significantly different from the regression line of PCB concentration as a linear function of Hg concentration for males. To perform this analysis, Hg concentration was paired with PCB concentration for each of the 50 summer flounder, using PCB concentration data from Madenjian et al. [13]. A simple linear regression analysis of PCB concentration as a function of Hg concentration was performed for both females and males. We used an *F* test for equality of regression lines to determine whether the two regression lines were significantly different from one another.

Results

Differences in whole-fish Hg and MeHg concentrations between the sexes

Female summer flounder were substantially larger than male summer flounder, even though mean age varied little between the sexes (Table 1). Females ranged from 2 to 6 years in age, while age of males ranged from 3 to 8 years.

Mean age for males (4.2 years) was just slightly higher than mean age for females (3.7 years). Females averaged 497 mm in TL, while males averaged 433 mm in TL. Females averaged 1264 g in weight, while males averaged 863 g in weight (Table 1). Gonad condition of all 50 summer flounder was ripe or nearly ripe.

Hg concentration did not significantly vary between the sexes of summer flounder (*t* test: t=0.25; df=48; P=0.8056). Mean whole-fish Hg concentration averaged 113 ng/g for females and 111 ng/g for males (Table 1). Ratio of mean whole-fish Hg concentration in males to mean whole-fish Hg concentration in females was equal to 0.98. Thus, females exceeded males in Hg concentration by 2%. Hg concentration trended neither upward nor downward with increasing age of summer flounder (Figure 1).

MeHg concentration did not significantly vary between the sexes of summer flounder (*t* test: t=0.43; df=48; P=0.6656). Mean whole-fish MeHg concentration averaged 104 ng/g for females and 101 ng/g for males (Table 1). Ratio of mean whole-fish MeHg concentration in males to mean whole-fish Hg concentration in females was equal to 0.96. Thus, females exceeded males in MeHg concentration by 4%.

Hg and MeHg concentrations in ovaries and somatic tissue

Mean Hg concentration in somatic tissue was about five times higher than mean Hg concentration in the ovaries (Table 2). On average, whole-fish Hg concentration of the females was expected to increase by 3.7% immediately after spawning due to release of eggs. Similarly, mean MeHg concentration in somatic tissue exceeded mean Hg concentration in the ovaries by roughly a factor of five (Table 2). On average, whole-fish MeHg concentration of the females was expected to increase by 3.6% immediately after spawning due to release of eggs. Mean GSI for the females was 4.3% (Table 2).

Characteristic	Females		Males	
	n	Mean	n	Mean
Total length (mm)	23	497 (7)	27	433 (7)
Weight (g)	23	1264 (53)	27	863 (48)
Age (years)	23	3.7 (0.2)	26	4.2 (0.2)
Hg (ng/g)	23	113 (7)	27	111 (6)
MeHg (ng/g)	23	104 (6)	27	101 (6)

Table 1: Mean values for total length, weight, age, Hg concentration, and MeHg concentration, by sex, of the summer flounder from New Jersey coastal waters, November 2013, used in the study. Standard error of the mean enclosed within parentheses. n = number of fish.

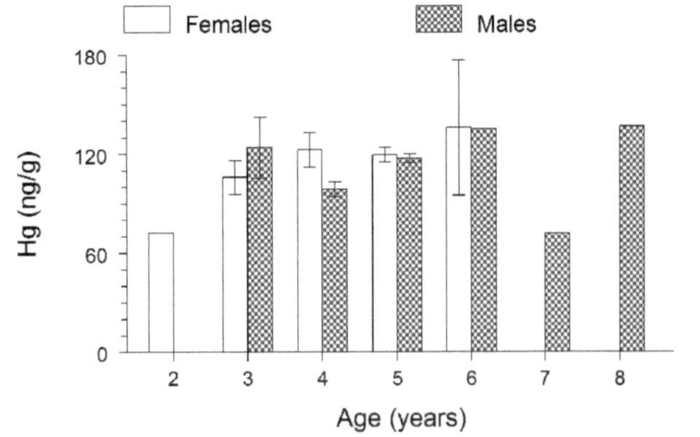

Figure 1: Mean whole-fish Hg concentration, by age and sex, for summer flounder from New Jersey coastal waters, November 2013. Error bars represent one standard error. Sample sizes of females were 1, 12, 6, 2, 2, 0, and 0 for ages 2, 3, 4, 5, 6, 7, and 8, respectively. Sample sizes of males were 0, 7, 12, 4, 1, 1, and 1 for ages 2, 3, 4, 5, 6, 7 and 8, respectively.

Characteristic	Mean
Hg concentration in ovaries (ng/g)	21 (1)
Hg concentration in somatic tissue (ng/g)	119 (11)
Expected percent change in whole-fish Hg concentration immediately after spawning due to release of eggs (%)	+3.7 (0.3)
MeHg concentration in ovaries (ng/g)	21 (2)
MeHg concentration in somatic tissue (ng/g)	109 (11)
Expected percent change in whole-fish MeHg concentration immediately after spawning due to release of eggs (%)	+3.6 (0.3)
GSI (%)	4.3 (0.4)

Table 2: Mean values for Hg concentration in ovaries, Hg concentration in somatic tissue, expected percent change in whole-fish Hg concentration immediately after spawning due to release of eggs, MeHg concentration in ovaries, MeHg concentration in somatic tissue, expected percent change in whole-fish MeHg concentration immediately after spawning due to release of eggs, and gonadosomatic index (GSI) of the five female summer flounder from New Jersey coastal waters, November 2013, used in the study. Standard error of mean enclosed within parentheses.

Proportion of Hg represented by MeHg

Mean ratio of whole-fish MeHg concentration to whole-fish Hg concentration (standard error of mean in parentheses) was 0.92 (0.02), based on data for all 50 summer flounder. Thus, on average, MeHg accounted for 92% of the Hg found in the summer flounder. By sex, ratio of whole-fish MeHg concentration to whole-fish Hg concentration averaged 0.93 (0.02) for females and 0.92 (0.03) for males. Thus, ratio of MeHg concentration to Hg concentration varied little between the sexes.

Sex-contaminant type (Hg or PCBs) interaction

Sex significantly interacted with contaminant type (Hg or PCBs) (F=5.57; df=2, 46; P=0.0068) (Figure 2). Thus, the effect of sex on Hg concentration was significantly different from the effect of sex on PCB concentration. Males were 43% higher in PCB concentration, but females were 2% higher in Hg concentration. The ratio of Hg concentration in males to Hg concentration in females was lower than the ratio of PCB concentration in males to PCB concentration in females in summer flounder (Figure 3). A similar pattern has been observed in lake trout, burbot and lake whitefish.

Discussion

The most likely explanation for the ratio of Hg concentration in males to Hg concentration in females being substantially lower than the ratio of PCB concentration in males to PCB concentration in females in summer flounder was that males eliminated Hg from their bodies at a considerably faster rate than females. In contrast, long-term elimination of PCBs by fish has been shown to be negligible for both sexes [21]. Faster Hg-elimination rates in males compared with females have been reported for northern pike (*Esox lucius*) and laboratory mice (*Mus musculus*). Based on the tracking of isotopically enriched Hg body burdens in six (three males and three females) northern pike over a 3-year period, Hg-elimination rate in mature males was estimated to be more than twice as fast as Hg-elimination rate in mature females [2]. As juveniles, rates of Hg excretion in urine did not vary between the sexes of laboratory mice [22]. However, as the mice approached maturity, males eliminated Hg in their urine at a rate three to four times higher than that of females. As adults, males continued to eliminate Hg in their urine at a substantially faster rate than females. Enhancement of Hg-elimination rate in male laboratory mice has been attributed to testosterone [22-25]. Madenjian et al. [1,2,4] hypothesized that certain androgens, such as testosterone and 11-ketotestosterone, enhance Hg-elimination in male teleost fishes as well. The pattern of the ratio of PCB concentration in males to PCB concentration in females exceeding

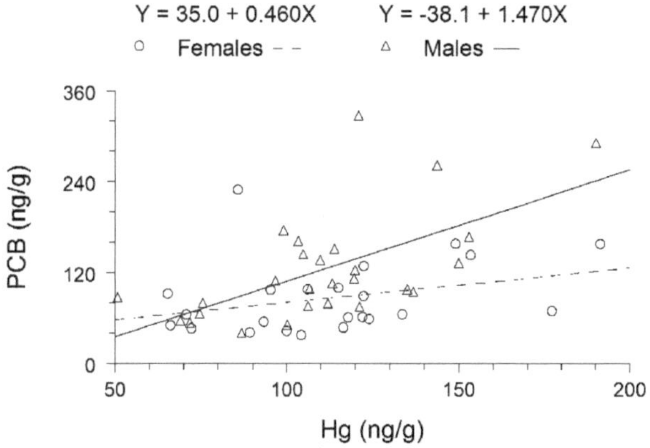

Figure 2: Whole-fish PCB concentration as a function of whole-fish Hg concentration for summer flounder from New Jersey coastal waters, November 2013. Displayed lines are based on simple linear regression analysis. PCB concentrations were from Madenjian et al. [13].

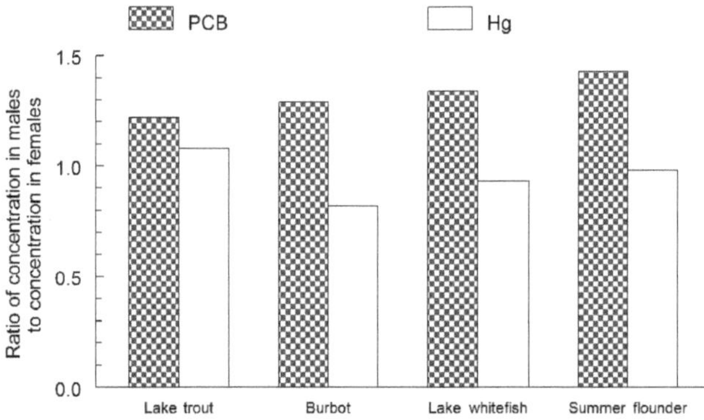

Figure 3: Comparison of the ratio of whole-fish PCB concentration in males to whole-fish PCB concentration in females with the ratio of whole-fish Hg concentration in males to whole-fish Hg concentration in females. Ratio of whole-fish Hg concentration in males to whole-fish Hg concentration in females for summer flounder was based on the results of this study. The other ratios were from previous studies [1,3,4,6,13].

the ratio of Hg concentration in males to Hg concentration in females persists. This phenomenon has now been documented in summer flounder, a marine fish species, as well as in three freshwater fishes, namely lake trout, burbot, and lake whitefish. Because testosterone and 11-ketotestosterone are found in all male teleost fishes [26], we may expect this pattern to hold true for all species of teleost fishes.

The magnitude of the difference between the ratio of PCB concentration in males to PCB concentration in females and the ratio of Hg concentration in males to Hg concentration in females was not solely dependent on the relative size of the testes (GSI). Because androgens have been linked to enhanced Hg-elimination rates, and because androgen concentration in the blood is positively related to GSI of males, Madenjian et al. [4] proposed that the difference between these two ratios should increase with increasing GSI. Of the four species examined to date, burbot exhibited the highest GSI of males, as GSI of males averaged 10.8%. Mean GSIs of males of the other three species were 3.1% for lake trout, 2.1% for lake whitefish, and 1.7% for summer flounder [1,4,13]. The largest difference between these two ratios was for burbot, thereby supporting the hypothesis by Madenjian et al. [4]. However, the differences between the two ratios for lake whitefish and summer flounder were substantially greater than that for lake trout, yet GSIs for lake whitefish and summer flounder were slightly lower than that for lake trout. Clearly, factors other than GSI also contributed to the difference between the two ratios. We must point out that androgen concentration in the blood is not solely dependent on GSI [4]. That is, androgen concentration can be relatively high despite a relatively low GSI. Based on results from experiments on laboratory mice [23-25], concentrations of certain androgens were important determinants of Hg-elimination rate. But, perhaps factors other than androgen concentration also influenced Hg-elimination rate. Research on various strains of laboratory mice has suggested that Hg-elimination rate may be influenced by glutathione content in the liver and kidneys, and that glutathione content in the liver and kidneys varies substantially between strains of laboratory mice [24]. Of course, differences in androgen concentration between strains may be driving these differences in glutathione content in the liver and kidneys between strains. Interestingly, Kostyniak [27] showed that Hg-elimination rate of males of the CFW strain of laboratory mice was five times higher than that of males of the CBA/J strain of laboratory mice. Moreover, estimates of Hg-elimination rates from laboratory experiments indicated that lake whitefish were capable of eliminating Hg from their bodies at a rate nearly three times faster than that for lake trout [1]. These estimates were derived from mixed-sex fish populations held in laboratory tanks. In addition, Hg-elimination rate in fish has been shown to increase with increasing water temperature [28]. In sum, the ratio of PCB concentration in males to PCB concentration in females minus the ratio of Hg concentration in males to Hg concentration in females was positive for all four fish species, suggesting that males eliminated Hg at a faster rate than females in all four species. The magnitude of the difference between the two ratios may be positively related to androgen concentration, but other factors may also be involved. Certainly, Hg-elimination rate can vary considerably from one species to another.

Expected mean percent increase in Hg concentration of females due to release of eggs at spawning was lower in summer flounder than in other fish species that have been studied. We estimated that Hg concentration of summer flounder females would increase by 3.7%, on average, immediately after spawning due to release of eggs. In lake whitefish from northern Lake Huron, Hg concentration of females was expected to increase by 17.9%, on average, immediately after spawning [1]. In burbot from Great Slave Lake (Northwest Territories, Canada), females were expected to increase in Hg concentration by 6.2%, on average, due to shedding of eggs at spawning [4]. Mean percent increases in Hg concentration of females due to release of eggs at spawning were estimated to be 15.3% in rainbow trout (*Oncorhynchus mykiss*), 16.0% in white sucker (*Catostomus commersoni*), 10.9% in white bass (*Morone chrysops*), 5.5% in smallmouth bass (*Micropterus dolomieu*), and 22.4% in yellow perch (*Perca flavescens*) [19]. The relatively modest change in Hg concentration of summer flounder females due to shedding of eggs was attributable, in part, to the relatively low GSI of female summer flounder. Female summer flounder averaged 4.3% in GSI, whereas mean GSI of females of the other species ranged from 7.4% to 22.3% [1,4,19].

We found that 92%, on average, of the Hg in summer flounder from New Jersey coastal waters was MeHg. Similarly, 91% of the Hg in burbot from Great Slave Lake was MeHg [4], and 91% of the Hg in lake whitefish from northern Lake Huron was MeHg [1]. In contrast, 84% of the Hg in fillets of white perch (*Morone americana*) from the Hackensack River (NJ, USA) was in the inorganic mercury form, while only 16% of the Hg was MeHg [18]. Raymond and Rossmann [17] reported that between 60 and 70% of the Hg found in alewives (*Alosa pseudoharengus*) from Lake Michigan was MeHg, whereas all (100%) of the Hg in Lake Michigan lake trout was MeHg. Further, all (100%) of the Hg in Lake Erie burbot was determined to be MeHg [4]. We also found that the percentage of Hg represented by MeHg differed little between the sexes of summer flounder. Similarly, the percentage of Hg represented by MeHg showed a minimal amount of variation between the sexes of burbot from Great Slave Lake [4], as well as lake whitefish from northern Lake Huron [1].

Long-term elimination of Hg by fish is very slow, but measurable. Trudel and Rasmussen [28] reported half-lives of Hg in fish ranging from 130 to 1030 days, based on results taken from the literature for experiments that were greater than 90 days in duration. Using regression modeling, Hg-elimination rate was found to depend on fish weight and water temperature. However, sex of the fish was not considered as an explanatory variable in the regression analyses [28]. Results from more recent research have suggested that the half-life of Hg in fish is greater than previously reported estimates [29,30]. In other words, the rate of elimination of Hg by fish is slower than once thought. Using isotopically enriched Hg in whole-lake experiments, half-lives of Hg in yellow perch and northern pike were estimated to be 489 days and 1193 days, respectively [29,30]. In a follow-up study, half-lives of Hg in male and female northern pike were estimated to be 950 days and 2039 days, respectively; this was the first study, to the best of our knowledge, to examine the difference in Hg-elimination rates between the sexes of fish [2]. Long-term elimination of PCBs by fish is so slow that it is undetectable [21]. In four experiments designed to estimate elimination rates, PCB body burden in fish showed no further decrease after an initial phase (several days in duration), during which PCB body burden did decrease immediately after dosing of the fish.

Results from previous studies have shown that ratio of Hg concentration in males to Hg concentration in females can vary substantially from one species of fish to another [1]. As mentioned above, male lake trout were 8% greater in whole-fish Hg concentration than female lake trout [3], whereas female burbot were 22% greater in whole-fish Hg concentration than male burbot [4]. Based on muscle tissue determinations, males averaged higher Hg concentrations than females in some species of fish, whereas females averaged higher Hg concentrations than males in other species of fish [31-34]. For example, Bastos et al. [34] investigated sex-related differences in muscle Hg concentrations of 41 species of freshwater fishes from the Madeira River in the Amazon, and they reported that males exceeded females in Hg concentration in 23 species, while females exceeded males in Hg concentration in the other 18 species. A significant difference in Hg concentrations between the sexes was only detected in 4 of the 41 species. For 3 species, males averaged a significantly higher Hg concentration than females, whereas females averaged a significantly higher Hg concentration than males in the other species. Bastos et al. [34] acknowledged that low sample sizes contributed to the low proportion of significant differences detected between the sexes. As previously explained, we propose that this interspecific variation in the ratio of Hg concentration in males to Hg concentration in females is primarily driven by interspecific variation in the elimination rate of Hg by males.

Our findings have implications for continued operation of the summer flounder fishery. Thresholds of 300 and 1000 ng/g have been established by the US Environmental Protection Agency (EPA) and US Food and Drug Administration (FDA), respectively, for Hg concentration in fish eaten by citizens, and either restricted consumption or no consumption is advised when these thresholds are exceeded [35]. Mean whole-fish Hg concentration in the 50 summer flounder from our study was 112 ng/g, and the maximum whole-fish Hg concentration was 191 ng/g. Fillet Hg concentration is about 10% greater than whole-fish Hg concentration in most fishes [36]. Applying this conversion factor to our summer flounder Hg data, mean and maximum Hg concentrations in the summer flounder fillets were estimated to be 123 and 210 ng/g, respectively. Thus, all of the summer flounder fillets would be below the EPA and FDA thresholds for Hg concentration. Deshpande et al. [37,38] determined Hg concentration in the white muscle tissue of summer flounder caught along the New Jersey coast in the New York Bight Apex area during 1993, and Hg concentration in these fillets averaged 40 ng/g. The considerably lower Hg concentrations reported by Deshpande et al. [37,38] compared with Hg concentrations from our study were probably due, at least in part, to a difference in summer flounder sizes between the two studies. Summer flounder used in the Deshpande et al. [37,38] study averaged only 361 mm in TL, whereas summer flounder used in our study averaged 463 in TL. Our estimated average fillet Hg concentration of 123 ng/g was similar to the average fillet Hg concentration of 136 ng/g reported by Staudinger [39] for summer flounder caught in the continental shelf waters of the Atlantic Ocean between New Jersey and Massachusetts during 2002-2003 using bottom trawls. Average size of summer flounder used in our study was similar to summer flounder average size from the Staudinger [39] study. Determinations of fillet Hg concentrations of mixed species of flounder (Pleuronectiformes) from New Jersey supermarkets yielded a mean concentration of 50 ng/g [40]. The lower mean Hg concentration from the Burger and Gochfeld [40] study compared with our mean Hg concentration may simply have been due to other species of flounder being lower in Hg concentration than summer flounder. The FDA threshold for PCB concentration in fish eaten by people is 2000 ng/g [41]. Mean and maximum whole-fish PCB concentrations of the summer flounder from our study were 107 and 327 ng/g, respectively [13]. Thus, all of the summer flounder PCB concentrations were well below the FDA guideline. Mean PCB concentration of the summer flounder used in the Deshpande et al. [37,38] study was less than the detection limit of 50 ng/g. The considerably lower PCB concentrations reported by Deshpande et al. [37,38] compared with our summer flounder PCB concentrations were attributable to two factors. First, Deshpande et al. [37,38] determined PCB concentrations in fillets, whereas we determined whole-fish PCB concentrations. Typically, whole-fish PCB concentrations are substantially higher than fillet PCB concentrations [42]. Second, the summer flounder used in the Deshpande et al. [37,38] study were considerably smaller than the summer flounder used in our study.

Although we did not detect a significant increase in Hg concentration with increasing summer flounder TL, Staudinger [39] found that Hg concentration of summer flounder significantly increased with increasing TL. This discrepancy between the two studies could be attributable to differences in TL ranges between the two studies. Our summer flounder ranged from 381 to 560 mm in TL, while the summer flounder used by Staudinger [39] ranged from 280 to 750 mm in TL. We would have likely detected a significant increase in Hg concentration with increasing TL had we sampled from a slightly wider range of TLs. Some of the summer flounder used in the Staudinger [39] study were probably immature fish, because most summer flounder less than 290 mm TL are immature [43].

Conclusion

Mean whole-fish Hg concentration of summer flounder caught in New Jersey coastal waters was 113 ng/g for mature females and 111 ng/g for mature males, but this difference was not statistically significant. However, sex did significantly interact with contaminant type (Hg or PCBs), as males were 43% greater in PCB concentration than females, but females were 2% higher in Hg concentration. This significant interaction effect was attributed to males eliminating Hg from their bodies at a faster rate than females, whereas long-term elimination of PCBs by both sexes was negligible. The ratio of PCB concentration in males to PCB concentration in females exceeding the ratio of Hg concentration in males to Hg concentration in females has now been documented in summer flounder, a marine species, as well as in three freshwater fish species, namely lake trout, burbot, and lake whitefish. MeHg represented 92% of the Hg found in the summer flounder. Both Hg and PCB concentrations in the summer flounder were relatively low, well below the EPA and FDA guidelines, and therefore our findings had implications for continued operation of the summer flounder fishery.

Acknowledgments

Ganzorig Batsaikhan, Maria Berezin, Kaycee Coleman, Julia Criscione, Christopher Filosa, James Fiorendino, Christopher Free, Evan Kwityn, Andrew Lahr, Margaret Shaw, and Thomas Siciliano assisted with homogenization of the fish tissues. Jason Morson obtained the summer flounder and provided further assistance throughout this study. Blanche Jackson and Eric Robillard contributed their expertise in aging summer flounder otoliths. James Hurley reviewed the manuscript and made helpful suggestions for its improvement. Use of trade, product, or firm names does not imply endorsement by the United States Government. This article is Contribution 2029 of the US Geological Survey Great Lakes Science Center.

References

1. Madenjian CP, Ebener MP, Krabbenhoft DP (2016) Mercury accumulation, and the mercury-PCB-sex interaction, in lake whitefish (*Coregonus clupeaformis*). Environments 3: 7.
2. Madenjian CP, Blanchfield PJ, Hrenchuk LE, Van Walleghem JLA (2014) Mercury elimination rates for adult northern pike *Esox lucius*: evidence for a sex effect. Bull Environ Contam Toxicol 93: 144-148.
3. Madenjian CP, Keir MJ, Whittle DM (2011) Sexual difference in mercury concentrations of lake trout (*Salvelinus namaycush*) from Lake Ontario. Chemosphere 83: 903-908.
4. Madenjian CP, Stapanian MA, Cott PA, Krabbenhoft DP, Edwards WH, et al. (2015) Females exceed males in mercury concentrations of burbot *Lota lota*. Arch Environ Contam Toxicol 68: 678-688.

5. Madenjian CP (2011) Sex effect on polychlorinated biphenyl concentrations in fish: a synthesis. Fish and Fisheries 12: 451-460.

6. Madenjian CP, Stapanian MA, Cott PA, Rediske RR, O'Keefe JP (2014) Polychlorinated biphenyl concentrations of burbot *Lota lota* from Great Slave Lake are very low but vary by sex. Arch Environ Contam Toxicol 66: 529-537.

7. Madenjian CP, Ebener MP, Sepúlveda MS (2015) PCB concentrations of lake whitefish (*Coregonus clupeaformis*) vary by sex. Journal of Great Lakes Research 41: 1185-1190.

8. Marshall WS, Grosell M (2006) Ion transport, osmoregulation and acid-base balance. In: Evans DH, Claiborne JB (eds.) The physiology of fishes. Taylor & Francis CRC Press, Boca Raton, FL, USA.

9. Terceiro M (2002) The summer flounder chronicles: Science, politics, and litigation 1975-2000. Reviews in Fish Biology and Fisheries 11: 125-168.

10. Terceiro M (2011) The summer flounder chronicles II: new science, new controversy 2001-2010. Reviews in Fish Biology and Fisheries 21: 681-712.

11. Morson JM, Bochenek EA, Powell EN, Gius JE (2012) Sex at length of summer flounder landed in the New Jersey recreational party boat fishery. North American Journal of Fisheries Management 32: 1201-1210.

12. Wuenschel MJ, Able KW, Vasslides JM, Byrne DM (2013) Habitat and diet overlap of 4 piscivorous fishes: variation on the inner continental shelf off New Jersey. Fish Bull 111: 352-369.

13. Madenjian CP, Jensen OP, Rediske RR, O'Keefe JP, Vastano AR, et al. (2016) Differences in energy expenditures and growth dilution explain higher PCB concentrations in male summer flounder. PLoS ONE 11: e0147223.

14. United States Environmental Protection Agency (2007) Method 7473: mercury in solids and solutions by thermal decomposition, amalgamation, and atomic absorption spectrophotometry. Office of Water, Washington, DC, USA.

15. United States Environmental Protection Agency (1990) Guidelines establishing test procedures for the analysis of pollutants. Office of Water, Washington, DC, USA.

16. United States Environmental Protection Agency (1998) Method 1630: methyl mercury in water by distillation, aqueous ethylation, purge and trap, and cold vapor atomic fluorescence spectrometry. Office of Water, Washington, DC, USA.

17. Raymond B, Rossmann R (2009) Total and methyl mercury accumulation in 1994-1995 Lake Michigan lake trout and forage fish. Journal of Great Lakes Research 35: 438-446.

18. Weis P, Ashley JTF (2007) Contaminants in fish of the Hackensack Meadowlands, New Jersey: size, sex, and seasonal relationships as related to health risks. Arch Environ Contam Toxicol 52: 80-89.

19. Niimi AJ (1983) Biological and toxicological effects of environmental contaminants in fish and their eggs. Canadian Journal of Fisheries and Aquatic Sciences 40: 306-312.

20. Scheffé H (1959) The analysis of variance. John Wiley & Sons, New York, NY, USA.

21. Madenjian CP, Hesselberg RJ, DeSorcie TJ, Schmidt LJ, Stedman RM, et al. (1998) Estimate of net trophic transfer efficiency of PCBs to Lake Michigan lake trout from their prey. Environ Sci Technol 32: 886-891.

22. Hirayama K, Yasutake A (1986) Sex and age differences in mercury distribution and excretion in methylmercury-administered mice. J Toxicol Environ Health 18: 49-60.

23. Yasutake A, Hirayama K, Inoue M (1989) Mechanism of urinary excretion of methylmercury in mice. Arch Toxicol 63: 479-483.

24. Tanaka T, Naganuma A, Kobayashi K, Imura N (1991) An explanation for strain and sex differences in renal uptake of methylmercury in mice. Toxicology 69: 317-329.

25. Tanaka T, Naganuma A, Miura N, Imura N (1992) Role of testosterone in γ-glutamyltranspeptidase-dependent renal methylmercury uptake in mice. Toxicol Appl Pharmacol 112: 58-63.

26. Borg B (1994) Androgens in teleost fishes. Comparative Biochemistry and Physiology 109C: 219-245.

27. Kostyniak PJ (1980) Differences in elimination rates of methylmercury between two genetic variant strains of mice. Toxicol Lett 6: 405-410.

28. Trudel M, Rasmussen JB (1997) Modeling the elimination of mercury by fish. Environ Sci Technol 31: 1716-1722.

29. Van Walleghem JLA, Blanchfield PJ, Hintelmann H (2007) Elimination of mercury by yellow perch in the wild. Environ Sci Technol 41: 5895-5901.

30. Van Walleghem JLA, Blanchfield PJ, Hrenchuk LE, Hintelmann H (2013) Mercury elimination by a top predator, *Esox lucius*. Environ Sci Technol 47: 4147-4154.

31. Monteiro LR, Isidro EJ, Lopes HD (1991) Mercury content in relation to sex, size, age and growth in two scorpionfish (*Helicolenus dactylopterus* and *Pontinus kuhlii*) from Azorean waters. Water, Air, and Soil Pollution 56: 359-367.

32. Stafford CP, Hansen B, Stanford JA (2004) Mercury in fishes and their diet items from Flathead Lake, Montana. Transactions of the American Fisheries Society 133: 349-357.

33. Gewurtz SB, Bhavsar SP, Fletcher R (2011) Influence of fish size and sex on mercury/PCB concentration: Importance for fish consumption advisories. Environ Int 37: 425-434.

34. Bastos WR, Dórea JG, Bernardi JVE, Manzatto AG, Mussy MH, et al. (2016) Sex-related mercury bioaccumulation in fish from the Madeira River, Amazon. Environ Res 144: 73-80.

35. Cai Y, Rooker JR, Gill GA, Turner JP (2007) Bioaccumulation of mercury in pelagic fishes from the northern Gulf of Mexico. Canadian Journal of Fisheries and Aquatic Sciences 64: 458-469.

36. Becker DS, Bigham GN (1995) Distribution of mercury in the aquatic food web of Onondaga Lake, New York. Water, Air, and Soil Pollution 80: 563-571.

37. Deshpande AD, Draxler AFJ, Zdanowicz VS, Schrock ME, Paulson AJ (2002) Contaminant levels in the muscle of four species of fish important to the recreational fishery of the New York Bight Apex. Mar Pollut Bull 44: 164-171.

38. Deshpande AD, Draxler AFJ, Zdanowicz VS, Schrock ME, Paulson AJ, et al. (2000) Contaminant levels in the muscle of four species of fish important to the recreational fishery of the New York Bight Apex. NOAA Technical Memorandum NMFS-NE-157, Northeast Fisheries Science Center, Woods Hole, MA, USA.

39. Staudinger MD (2011) Species and size-specific variability of mercury concentrations in four commercially important finfish and their prey from the northwest Atlantic. Mar Pollut Bull 62: 734-740.

40. Burger J, Gochfeld M (2005) Heavy metals in commercial fish in New Jersey. Environ Res 99: 403-412.

41. United States Food and Drug Administration (2015) Tolerances of polychlorinated biphenyls (PCBs), 21 CFR. Health and Human Services, Public Health Service, Rockville, MD, USA.

42. Amrhein JF, Stow CA, Wible C (1999) Whole-fish versus filet polychlorinated-biphenyl concentrations: An analysis using classification and regression tree models. Environmental Toxicology and Chemistry 18: 1817-1823.

43. Northeast Fisheries Science Center (2013) 57th Northeast Regional Stock Assessment orkshop (57th SAW) Assessment Report. US Department of Commerce, Northeast Fisheries Science Center Reference Document pp: 13-16.

Ribosomal Internal Transcribed Spacer (ITS) DNA Variation in *Millepora*

Craig S Tepper[1]* and Sophia C Gaynor[1,2]

[1]*Department of Biology, Cornell College, Mt. Vernon, IA 52314, USA*
[2]*Interdisciplinary Graduate Program in Genetics, University of Iowa, Iowa City, IA 52242, USA*

Abstract

Two main growth forms of *Millepora* (fire coral) are present around the islands of The Bahamas: one exhibits a strong, blade-like structure, *Millepora complanata*, and the other possesses a delicate branch-like structure, *Millepora alcicornis*. The phylogenetic relationship of these corals has been under considerable debate for over a century due to the existence of a wide-range of intermediate growth forms. Recent genetic analysis using ribosomal DNA (rDNA) suggests the existence of two distinct reproductively isolated cryptic clades that are independent of morphology [1]. However, using repeated rDNA sequences for phylogenetic construction can lead to false phylogenies if repeated sequences have not undergone concerted evolution, a process involving homogenization of individual repeats of a multigene family. We analyzed twenty rDNA clones isolated from a single bladed *Millepora* colony and found that although variant rDNA sequences were present, rDNA appears to be largely homogenized.

Keywords: rDNA ITS (Internal transcribed spacer); *Millepora*; Concerted evolution; Phenotypic plasticity; Intragenomic sequence variation; DGGE (Denaturing gradient gel electrophoresis)

Introduction

For centuries, biologists have attempted to group organisms based on shared characteristics in order to understand the evolutionary relationships of the tremendous diversity of life. Phylogenic relationships are estimated using morphological, behavioral, and other phenotypic characters. However, these characters may not accurately represent evolutionary relationships because evolution is not always divergent [2,3]. Two species can independently evolve the same features due to similar habitats and favored adaptations. Therefore, two species that are not closely related may end up more phenotypically similar to each other than to their closest relatives.

The problem of understanding evolutionary relationships between organisms is particularly acute in reef-building coral assemblages that serve as the foundation of complex reef ecosystems [4]. The evolutionary history and current speciation in this diverse class of animals remain paradoxical. For example, although some corals reproduce during non-synchronous spawning, most corals reproduce in annual, synchronous mass spawning events that provide numerous opportunities for interspecific hybridization [5]. Vollmer and Palumbi [6] examined *Acropora cervicornis*, *A. palmata*, and *A. prolifera*. They demonstrated that *A. prolifera* was a first-generation hybrid descendent of *A. cervicornis* and *A. palmata* and did not deserve a separate species designation. Hence, coral diversity can occur not only by conventional species formation, but also through inter-species hybridization. Further compounding the problem is that morphological traits used to construct phylogenies are not always useful for corals [4]. Coral taxonomic classification (as well as our current understanding of coral evolution) is based upon morphological characters of the calcareous skeleton [7]. Unfortunately, the calcareous skeleton of many marine organisms shows considerable phenotypic plasticity. The architecture of the coral skeleton is affected by environmental factors such as underwater irradiance, water motion, water temperature, and sedimentation [8]. Additionally, calcification rates are affected by lunar, diurnal and seasonal fluctuations [9].

The coral phylogenetic issue is particularly problematic in the calcareous hydrozoan coral, *Millepora*, which is one of the most common skeleton-forming animals on reefs. This group of corals, known as fire coral because of its painful sting, is represented by multiple species and is nearly ubiquitous on reefs in the Atlantic, Indian and Pacific Oceans [10]. Millepores are important reef framework builders, second only to the scleractinia corals [10]. The morphology of the Millepores is highly variable and shows phenotypic plasticity [1,4,11]. Stearn and Riding [4] reported that the various growth forms of *Millepora* in the Caribbean range from thinly encrusting sheets and delicate dendroid branches for *M. alcicornis*, to thicker, rigid bladed forms for *M. complanata* (Figure 1). It is this variation in morphology that has led to constant controversy about Millepore classification. Currently, species designation within the genus is mainly based on growth form, geographical distribution and morphological differences such as surface texture, nematocyst structure, and the size and shape of pores [10].

Tepper et al. [1] examined the evolutionary relationship of the two commonly found Millepores (*M. complanata* and *M. alcicornis*) in the northern Caribbean off the coast of San Salvador, The Bahamas. In addition to these two recognizable morphologies, they reported the existence of numerous intermediate forms, of which the specific species status is questionable. Because of the wide range of growth forms (Figure 2), the question arises whether the blade (*M. complanata*) and branching (*M. alcicornis*) forms are separate species or represent phenotypic variations of one highly variable species complex.

The inherent uncertainty in using morphological characters as a way to establish phylogenetic relationships can be aided by using genetic markers to distinguish between closely related growth forms. Among the most widely used genetic markers are the internal transcribed spacer (ITS) regions of ribosomal DNA (rDNA). Ribosomal RNA genes are organized in clusters of hundreds to thousands of tandem repeats per

***Corresponding author:** Craig S. Tepper, Department of Biology, Cornell College, Mt. Vernon, IA, USA, 52314, E-mail: ctepper@cornellcollege.edu

Figure 1: Photographs depicting the typical growth forms of the two reported *Millepora* species found in the Bahamas. 1A. *Millepora alcicornis* (branched) 1B. *Millepora complanata* (bladed).

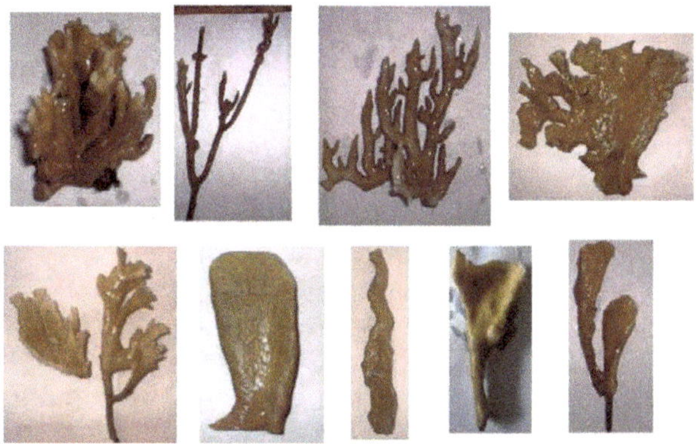

Figure 2: Photographs depicting the wide range of Millepore growth forms found on reefs around San Salvador, The Bahamas.

chromosome, each of which consists of three coding regions (18S, 5.8S and 28S), two internal transcribed spacer regions (ITS-1 and ITS-2), one external transcribed spacer (ETS), and one non-transcribed spacer (NTS) region [12].

The two ITS regions have been used for detecting differences between conspecifics and are useful in studying closely related individuals due to their higher mutation rates [13]. Takabayashi et al. [14] reported that the ITS region varies from 2 to 31% in different coral species, making this region ideal for comparative analyses between populations. Meroz-Fine et al. [15] utilized DNA sequence information from the ITS regions of the Red Sea Fire Coral, *Millepora dichotoma,* to show that the currently recognized single species with two growth forms (blade and branching) was instead two distinct species. Forsman et al. [16] reported that they were unable to use morphological characters to identify species diversity in the coral genus *Porites*. Using rDNA ITS regions and mitochondria gene markers, they revealed numerous cryptic patterns of species diversity in *Porites*.

Tepper et al. [1] used rDNA ITS regions and identified the existence of two Millepore clades. Each clade contained members of all three morphologies (bladed, branched, and intermediate growth forms). Their analysis suggests the existence of two cryptic clades that are independent of morphology and reproductively isolated.

Although phylogenetic analysis based on rDNA has helped untangle evolutionary relationships, the use of rDNA sequences has proven to be problematic because of the existence of variability among the repeated rDNA units, which may cause extensive differentiation even within single individuals [17]. The problem concerns the assumption that these repeated sequences have homogenized via concerted evolution, a process involving homogenization of individual repeats of a multigene family [18-22]. Concerted evolution results in the production of uniform sequences in all repeats in a given species. Two mechanisms, unequal crossover and gene conversion, have been proposed to contribute to the process of concerted evolution [18]. However, evidence exists that concerted evolution may not be complete in all species. Vollmer and Palumbi [17] showed that concerted evolution did not completely homogenize rDNA arrays, thereby accounting for the intragenomic variation they observed in rDNA repeats for the scleractinian coral *Acropora* leading to the construction of false phylogenies. LaJeunesse and Pinzón [23], using a single colony of *Acropora valida,* reported the existence of 23 unique ITS-2 sequences out of 29 that were sequenced, some differing by up to 28%. Dorado and Sánchez [24], using the gorgonian coral *Pseudopterogorgia bipinnata*, reported an ITS-2 sequence variation of 14.4% among 37 samples. However, phylogenetic analyses of the ITS regions for corals in the genera *Pavona*, *Platygyra*, *Porites* and *Siderastrea* [25-27] demonstrated clear phylogenetic relationships due to the lack of intragenomic variability within ITS regions. These results indicate that sequences of rDNA ITS regions are useful for phylogenetic analysis as long as the level of intragenomic variability is assessed.

Sequencing cloned rDNA ITS regions to generate phylogenetic trees, as was done by Tepper et al. [1] for the Millepores, can lead to to the detection of variants in repeated sequences resulting in construction of false phylogenies unless repeated rDNA arrays have been homogenized [28]. In order to determine if the two cryptic Millepore clades identified by Tepper et al. [1] are valid, we are assessing intragenomic rDNA sequence variation.

Here we report that although minor rDNA sequence variation is present, Millepore intraspecific rDNA sequence divergence is consistently lower than interspecific sequence divergence. Since other investigators have reported high levels of rDNA intragenomic variation using single colonies of coral [17,23,24], our results imply that the rDNA regions appear to be homogenized in Millepores residing in patch reefs around San Salvador, The Bahamas.

Materials and Methods

Field-site description

Millepore colonies used in this study were collected in May 2010 from a patch reef at French Bay (23°56'59"N, 74°32'50"W), located on the southern end of San Salvador, The Bahamas (Figure 3). San Salvador is located on the eastern edge of the Bahamian island chain and is characterized by its karst and hyper saline lakes. Colonies were held in a flow-through seawater tank for no more than two days before coral DNA was isolated.

DNA extraction, PCR amplification and DNA sequencing

A single bladed form of *Millepora* was broken into four pieces (Figure 4) and DNA was extracted, PCR amplified, cloned, and sequenced as described below. Five clones were isolated and sequenced from each coral fragment.

Genomic DNA was isolated from colonies of *Millepora* using a procedure modified from Rowan and Powers [29] and Lopez et al. [30] and described in Tepper et al. [1].

ITS rDNA PCR amplification was performed as described in Tepper et al. [1] using the coral specific primer A18S

Figure 3: Satellite image of San Salvador, The Bahamas showing the French Bay collection site (modified from NASA Earth Observatory).

Figure 4: A single bladed *Millepora* colony was broken into four pieces (A-D). Five rDNA ITS clones were isolated from each piece and intragenomic sequence variation was assessed.

(5'GATCGAAC-GGTTTAGTGAGG3'), and the universal primer ITS 4 (5'TCCTCCGCTTATTGATATGC3') [14]. The primers amplify approximately 781 nucleotides [1].

Amplified PCR products were run on 1.2% (w/v) low melting agarose gels, excised, ligated into pGEM-T vectors (Promega) and transformed into competent JM109 *E. coli* host cells [1]. Following blue-white selection, positive colonies were harvested and plasmid DNA was isolated using the Zyppy Plasmid Miniprep Kit (Zymo Research). Plasmids containing ITS rDNA inserts were sequenced in both directions with fluorescently-labeled M13 forward and reverse primers [1] using a LI-COR 4300 DNA Analyzer (LI-COR, Lincoln, NE). Sequence reaction products were analyzed using e-Seq V3.0 (LI-COR). A BLAST query of the National Center of Biological Information's (NCBI) sequence database confirmed that the sequences were most similar to other *Millepora* rDNA samples.

Intragenomic sequence variation analysis

Sequences were manually aligned to correspond with published alignments. Percent divergence and unrooted phylogenetic reconstructions (based on Maximum Parsimony) were implemented in ClustalW and ClustalX version 2.0 [31]. Percent divergence calculations were based on pair-wise treatments of sequences. Ribosomal DNA sequence from *M. exaesa* was used as the outgroup [32].

Denaturing gradient gel electrophoresis (DGGE)

Amplification and primers for PCR-DGGE analysis were as described above except the primer ITS 4 was modified with a 39 bp GC clamp [33] (5'CGCCCGCCGCGCCCCGCGCCCGTC CCGCCGCCCCCGC CCTCCTCCGCTTATT GTATG3'). In order to determine if the PCR amplified rDNA products were heterogeneous or homogeneous, PCR products from one bladed and one branched form of *Millepora* were run on 6% polyacrylamide denaturing gels containing a 40-80% gradient (2.8M urea/16% formamide to 5.6M urea/32% formamide). Gels were pre-run at 90 volts and 55°C for 30 minutes. Once PCR products were loaded, the gels were pre-heated at 60°C for 10 minutes without voltage followed by 150 volts at 60°C for 20-24 hrs. The gels were stained for 1-2 hrs in TAE (40 mM tris, 10 mM sodium acetate, 1 mM EDTA, pH 8) containing ethidium bromide (5 μg/mL) and destained in fresh TAE buffer for 30 minutes. Following staining, the gels were photographed under UV light.

Results

Ribosomal DNA variation within individuals and between species

Twenty different clones were isolated and sequenced from a single colony of bladed *Millepora* (Figure 3). Of the 20 rDNA clones, 11 (55%) were identical in sequence. The amplified rDNA fragment (781 bp, not including primers) contained partial fragments of the 18S (127 bp) and 28S (39 bp) rDNA genes, the entire sequence of the 5.8S (158 bp) gene, and the entire sequence of the ITS rDNA, ITS-1 (242 bp) and ITS-2 (215 bp). No length variation was observed in any of the rDNA sequences. The average pairwise sequence percent divergence (the number of non-nucleotide matches between two sequences divided by the length of the sequence) among clones within a single bladed *Millepora* colony was 0.025% with a range of 0-0.087% (Table 1). The expected Taq polymerase cloning error rate is approximately 0.01% [34]. Within individual rDNA variation was observed across four of five rDNA regions. The lone exception was the partial 18S rDNA sequence, which displayed no variation in the twenty clones. The levels of within-individual variation were highest for the ITS regions, reaching 0.017% (average of 0.012%) in ITS-1 and 0.172% (average of 0.047%) in ITS-2. Among the three rDNA genes, the variation ranged from 0, for all three rDNA genes, to a high of 0.026% for the partial 28S rDNA fragment.

Average percent interspecific sequence divergence between the bladed *Millepora* colony and the Mediterranean *M. exaesa* (Table 1) was 0.653% overall (ranging from 0.649-0.655%), 0.696% in ITS-1 (ranging from 0.694-0.703%) and 0.691% for ITS-2 (ranging from

	Overall Avg	18s Avg	ITS-1 Avg	5.8s Avg	ITS-2	28s Avg
	(Range)	(Range)	(Range)	(Range)	(Range)	(Range)
Bladed *Millepora*	0.025	0	0.012	0.003	0.047	0.005
	(0-0.087)	--	(0-0.017)	(0-0.013)	(0-0.172)	(0-0.026)
Millepora exaesa	0.653	0.126	0.696	0.356	0.691	0.054
	(0.649-0.655)	--	(0.694-0.703)	(0.354-0.361)	(0.689-0.703)	(0.051-0.077)

Table 1: Ribosomal DNA variation. Average pairwise sequence divergence (%) an among 20 sequenced bladed *Millepora* clones showing levels of rDNA variation observed within an individual coral colony and between species. The between species comparison is represented by the 20 sequenced *Millepora* clones compared to a single *M. exaesa* sequence [31]. Ranges are also provided.

0.689-0.703%). As observed for within colony variation, interspecific rDNA variation was highest in the ITS regions and lowest in the rDNA genes.

Phylogenetic reconstructions based on Maximum Parsimony were used to analyze the level and pattern of diversity among sequence variants generated from the 20 bladed *Millepora* clones and unrooted phylogenies based on ITS-1 and ITS-2 rDNA sequences were constructed (data not shown). All 20 sequences from the bladed *Millepora* colony were very tightly clustered together and difficult to differentiate. The sequences showed little divergence which supports the percent divergence data. The published outgroup sequence of *M. exaesa* [33] occupied a distant phylogenetic position in the reconstruction, indicating its divergence from the bladed *Millepora* colony.

Assessing intragenomic variation using denaturing gradient gel electrophoresis (DGGE)

DGGE analysis offers a method for visualizing intragenomic variation when direct (non-plasmid based) sequencing is difficult as in the millepores [1]. We chose to amplified rDNA sequences of a single colony of both bladed and branched *Millepora* morphologies because these colonies were previously identified to belong to different clades that were separated by five SNPs [1]. Interestingly, the results revealed the different morphologies contained a different dominant and a rare sequence variant (Figure 5). Sequence variation within rDNA is still common despite concerted evolution and is prevalent in ITS regions [23]. A rare rDNA variant can arise in a single generation as a result of a single point mutation. However, there is a low probability that the variant will spread through the rDNA array and become dominant [23].

The DGGE gel results for *M. complanata* consistently showed the presence of only two variant rDNA forms although only 11 out 20 (55%) rDNA sequences were identical. A comparison of the sequence data showed that of the nine rDNA that did not match, three were identical and the remaining 6 sequences contained single unique nucleotide substitutions. A possible explanation for the appearance of only two rDNA variants observed on the DGGE gels might be the lack of sensitivity of DGGE to detect low copy number single nucleotide rDNA variants. Although all corals examined to date contain some level of rDNA intragenomic variation, LaJeunesse and Pinzón [23] have shown that DGGE can be used to identify dominant sequences which are best suited for phylogenetic reconstructions that are relevant to the evolutionary history of the species. The appearance of only two rDNA variants in the DGGE gels (Figure 5) and the tight clustering of the 20 cloned rDNA fragments observed in the unrooted tree (data not shown) indicate there are only minor nucleotide differences in the rDNA repeats.

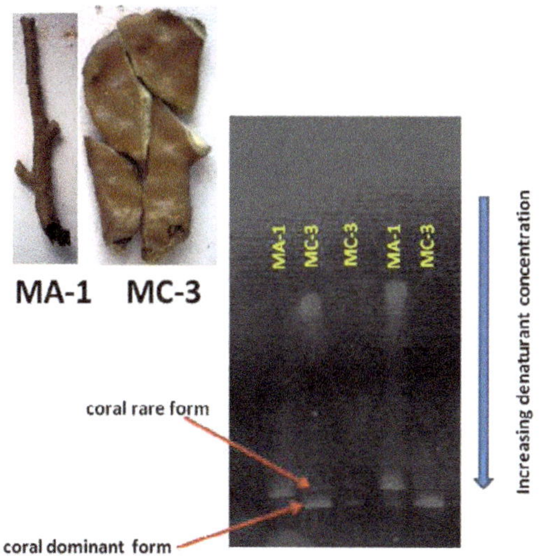

Figure 5: PCR-DGGE analysis of the amplified rDNA region from both branched (MA-1) and bladed (MC-3) *Millepora* morphologies. Arrows indicate the appearance of prominent and variant rDNA sequences.

Discussion

Eukaryotic genomes typically contain hundreds to thousands of tandemly repeated copies of the 18S, 5.8S and 28S genes, and the two internal transcribed spacers, ITS-1 and ITS-2 [12]. As part of a multigene family, the numerous copies of rDNA are expected to become homogenized through mechanisms of concerted evolution [18]. Homogenization leads to greater similarity among members of a repeated sequence family within a species than between species. This is the reason why rDNA sequences have been extensively used for species-level phylogenetic analysis [13]. However, basing species-level phylogenetic reconstructions on ITS regions of rDNA can be problematic because of incomplete homogenization due to the existence of polymorphisms, unequal crossover, gene conversion and interspecific hybridization among repeated units which may cause extensive differentiation even within single individuals [35].

Phylogenetic studies based on rDNA have been used extensively to provide numerous insights into scleractinian coral evolution [25-27,32,36,37]. However, Vollmer and Palumbi [17] demonstrated that individual coral colonies can contain high levels of intragenomic variation, and rDNA sequences may not be suitable for species-level phylogenetic analysis. Although most of the intragenomic variation was uncovered in the genus *Acropora*, their results implied that the rate of concerted evolution, and hence homogenization of repeated sequences, is quite variable in corals. This high rate of intragenomic variation may be due to interspecific hybridization as well as incomplete concerted evolution [17].

The level of intragenomic variation present in *Millepora* is low as demonstrated in the pairwise percent divergence data (Table 1). In addition to the low levels of average genetic divergence in the three conserved rDNA genes, both ITS regions were also extremely low, indicating that the rDNA array may be homogenized in the population of Millepores inhabiting patch reefs surrounding San Salvador, The Bahamas. If enough time has passed for concerted evolution to take place, low levels of intragenomic variation makes sense because the stem-loop secondary structures of both ITS-1 and ITS-2 rRNA play a major role in the excision and maturation of rRNA and reduced tolerance of nucleotide changes are expected as a result of the ITS splicing machinery [38]. Hence, only variants carrying minor changes outside of secondary stem formation in DNA sequence are tolerated. This assumption is supported by the two closely related rDNA variants observed for both bladed and branched forms of *Millepora* on the DGGE gel (Figure 5). Since DGGE analysis is based on the melting characteristics of the double-stranded PCR products, variant DNA fragments that migrate close together are expected to have small changes in DNA sequence.

In contrast to the low sequence divergence within an individual, the overall pairwise sequence percent divergence average for the bladed form of *Millepora* compared to *M. exaesa* is 0.653% (a range of 0.649-0.655%). When individual rDNA regions are compared, the conserved genes had the lowest sequence divergence (range of 0.051% for the partial 28S gene to 0.361% for the complete 5.8S gene) and the ITS regions had the highest (range of 0.694% for ITS-1 to 0.703% for both ITS-1 and ITS-2; Table 1). The results indicate that the rDNA sequences appear to be homogenized in this population of *Millepora*. Hence, PCR amplification followed by rDNA cloning and sequencing may be a reliable method for understanding the phylogenetic relationship of the Millepores surrounding San Salvador, The Bahamas.

Our study validates the use of rDNA as a potential phylogenetic tool to determine species-level distinctions in the Millepores and confirms the work of Chen et al. [36] who showed that high intragenomic rDNA sequence diversity seems to be unique to *Acropora* due to its ability to cross hybridize and then have one of the parent species backcross to the hybrids. Tepper et al. [1] showed that the two clades of Millepores surrounding San Salvador, The Bahamas are not hybridizing because they do not share any SNPs at the five ITS locations and appear to be reproductively isolated. The percent sequence divergence data clearly show that there are low levels of intragenomic rDNA variation, suggesting that the phylogenetic signals are informative for species-level comparisons. Similar patterns of low intragenomic variation have been reported for other corals [25-27].

Although the Millepores may be an exception, other lines of evidence indicate that traditional cloning methods overestimate the intragenomic diversity [23]. In order to avoid using phylogenetic markers that can lead to false phylogenies, excising bands from DGGE gels for sequence analysis allows for the detection of prevalent and rare intragenomic variants that may be scored as distinct phylogenetic signals by traditional cloning methods [24].

Most phylogenetic studies use primary DNA sequence information; however, Chen et al. [36] have shown that rRNA secondary structures are useful in phylogenetic reconstructions because they contain characters not found in the primary structure. The transcript folding structure of the ITS-2 region provides the proper orientation for the ribosomal coding regions when they are processed into small and large rRNA. Functional ITS sequences fold themselves to form secondary structures that are conserved and can be used as diagnostic indicators of taxonomic difference. ITS predicted secondary structure has been useful in providing reliable phylogenetic information in many corals including scleractinians (with the exception of *Acropora*) [36], the hexacoral *Zoanthus*, [39], and octocorals [24,40]. Grajales et al. [40] reported that predicted ITS-2 secondary structure led to the construction of new phylogenetic relationships for the octocoral species, *Eunicea*. We are beginning to examine whether predicted rRNA ITS-2 secondary structure can aid in untangling the phylogenetic relationships of the Millepores.

Acknowledgments

This work was conducted in the Bahamas under a permit granted by The Bahamas Environment, Science and Technology (BEST) Commission. We would like to thank Cornell College researchers Aye Mon, Elise Mead, and Helen Pope for their assistance with the molecular data. We would also like to thank Dr. Donald T. Gerace, Chief Executive Officer, and Dr. Tom Rothfus, Executive Director of the Gerace Research Center, San Salvador, Bahamas. Completion of this work was made possible by Cornell College Faculty Development Grants (CST) and a Ringer Endowed Fellowship (CST), McElroy Research Grants (CST), a GRC Student Grant (SCG) and the LI-COR Biosciences Genomics Education Matching Funds Program (CST).

References

1. Tepper CS, Squiers L, Hay C, Gorbach D, Friend D, et al. (2012) Cryptic species: A mismatch between genetics and morphology in Millepora. Marine Science 2: 57-65.
2. Hendry AP (2009) Speciation. Nature 458: 162-164.
3. Schluter D (2009) Evidence for ecological speciation and its alternative. Science 323: 737-741.
4. Stearn CW, Riding R (197) Forms of the Hydrozoan Millepora on a recent coral reef. Lethaia 6:187-200.
5. Harrison PL, Babcock RC, Bull GD, Oliver JK, Wallace CC, et al. (1984) Mass spawning in tropical reef corals. Science 223: 1186-1189.
6. Vollmer SV, Palumbi SR (2002) Hybridization and the evolution of coral reef diversity. Science 296: 2023-2025.
7. Wallace CC, Willis BL (1994) Systematics of the coral genus Acropora: Implications of new biological findings for species concept. Annual Review of Ecology and Systematics 25: 237-262.
8. Barnes DJ, Chalker BE (1990) In: Dubinsky Z (Ed.) Calcification and photosynthesis in reef building corals and algae. Coral Reef Ecosystems of the World. Elsevier, Philadelphia, PA 109-131.
9. Barnes DJ, Lough JM (1989) The nature of skeletal density banding in scleractinian corals: Fine banding and seasonal patterns. Journal of Experimental Marine Biology and Ecology 126: 119-134.
10. Lewis JB (2006) Biology and ecology of the hydrocoral Millepora on coral reefs. Adv Mar Biol 50: 1-55.
11. Ruiz-Ramos DV, Weil E, Schizas NV (2014) Morphological and genetic evaluation of the hydrocoral Millepora species complex in the Caribbean. Zoological Studies 53: 1-15.
12. Prokopowich CD, Gregory TR, Crease TJ (2003) The correlation between rDNA copy number and genome size in eukaryotes. Genome 46: 48-50.
13. Hillis DM, Dixon MT (1991) Ribosomal DNA: Molecular evolution and phylogenetic inference. Quarterly Review of Biology 66: 411-453.
14. Takabayashi M, Carter DA, Loh WKW, Hoegh-Guldberg O (1998) A coral specific primer for PCR amplification of the internal transcribed spacer region in ribosomal DNA. Molecular Ecology 7: 925-931.
15. Meroz-Fine E, Brickner I, Loya Y, Ilan M (2003) The hydrozoan coral Millepora dichotoma: Speciation or phenotypic plasticity? Marine Biology 143: 1175-1183.
16. Forsman ZH, Barshis DJ, Hunter CL, Toonen RJ (2009) Shape-shifting corals: Molecular markers show morphology is evolutionarily plastic in Porites. BMB Evolutionary Biology 9: 45-53.
17. Vollmer SV, Palumbi SR (2004) Testing the utility of internally transcribed spacer sequences in coral phylogeny. Molecular Ecology 13: 2763-2772.
18. Elder JF, Turner BJ (1995) Concerted evolution of repetitive DNA sequences in eukaryotes. Quarterly Review of Biology 70: 297-320.
19. Harpke D, Peterson A (2006) Non-concerted ITS evolution in Mammillaria (Cactaceae). Molecular Phylogenetics and Evolution 41: 579-593.
20. Harris DJ, Crandall KA (2000) Intragenomic variation within ITS1 and ITS2 of fresh-water crayfishes (Decapoda: Cambaridae): Implications for phylogenetic and microsatellite studies. Molecular Biology and Evolution 17: 284-291.
21. Sang T, Crawford D, Stuessy TF (1995) Documentation of reticulate evolution in peonies (Paeonia) using internal transcribed spacer sequences of nuclear ribosomal DNA: Implications for biogeography and concerted evolution. Proceedings of the National Academy of Sciences 92: 6813-6817.
22. Zimmer EA, Martin SL, Beverley SM, Kan YW, Wilson AC (1980) Rapid duplication and loss of genes coding for the α chains of hemoglobin. Proceedings of the National Academy of Sciences 77: 2158-2162.
23. LaJeunesse TC, Pinzón JH (2007) Screening intragenomic rDNA for dominant variants can provide a consistent retrieval of evolutionary persistent ITS (rDNA) sequences. Molecular Phylogenetics and Evolution 45: 417-422.
24. Dorado D, Sánchez JA (2009) Internal transcribed spacer 2 (ITS2) variation in Gorgonian coral Pseudopterogorgia bipinnata in Belize and Panama. Smithsonian Contributions to the Marine Sciences 38: 173-179.

25. Forsman ZH, Guzman HM, Chen CA, Fox GE, Wellington GM (2005) An ITS region phylogeny of Siderastrea (Cnidaria: Anthozoa): Is S. glynni endangered or introduced. Coral Reefs 24: 343-347.

26. Lam K, Morton B (2003) Morphological and ITS1, 5.8S, and partial ITS2 ribosomal DNA sequence distinctions between two species Playtygyra (Cndaria: Scleractinia) from Hong Kong. Marine Biotechnology 5: 555-567.

27. Moothien Pillay KRM, Asahida T, Chen CA, Terashima H, Ida H (2005) ITS ribo-somal DNA distinctions and genetic structure of populations in two sympatric species of Pavona (Cnidaria: Scleractinia) from Mauritius. Zoological Studies 45: 132-144.

28. http://earthobservatory.nasa.gov/IOTD/view.php?id=76134.

29. Rowan R, Powers DA (1991) Molecular genetic identification of symbiotic dinoflagelletes (Zooxanthellae). Marine Ecology Progress Series 71: 65-73.

30. Lopez JV, Kersanach R, Rehner SA, Knowlton, N (1999) Molecular determination of species boundaries in corals: Genetic analysis of the Montastraea annularis complex using amplified fragment length polymorphisms and a microsatellite marker. Biology Bulletin 196: 80-93.

31. Larkin MA, Blackshields G, Brown NP, Chenna R, McGettigan PA, et al. (2007) ClustalW and Clus-talX version 2.0. Bioinformatics 23: 2947-2948.

32. Odorico DM, Miller DJ (1997) Internal and external relationships of the Cnidaria: implications of primary and predicted secondary structure of the 5'-end of the 23S-like rDNA. Proceedings of the Royal Society, London B Biological Sciences 264: 77-82.

33. Sheffield VC, Cox DR, Lerman LS, Meyers RM (1989) Attachment of a 40-base pair G+C rich sequence (GC-clamp) to genomic fragments by polymerase chain reaction results in improved detection of single-base changes. Proceedings of the National Academy of Sciences 86: 232-236.

34. Eckert KA, Kunkel TA (1990) High fidelity DNA-synthesis by the Thermus aquaticus DNA polymerase. Nucleic Acids Res 18: 3739-3744.

35. Suh Y, Thien LB, Reeve HE, Zimmer EA (1993) Molecular Evolution and phylogenetic implications of the internal transcribed spacer sequences of ribosomal DNA in Winteraceae. American Journal of Botany 80: 1042-1055.

36. Chen CA, Chang C, Wei NV, Chen C, Lein Y, et al. (2004) Secondary structure and phylogenetic utility of the ribosomal internal transcribed spacer 2 (ITS2) in scleractinian corals. Zoological Studies 43: 759-771.

37. Lopez JV, Knowlton N (1997) Discrimination of species in the Montastraea annularis complex using multiple genetic loci. Proceedings of the 8th International Coral Reef Symposium 2: 1613-1618.

38. Cote CA, Peculis BA (2001) Role of ITS2-proximal stem and evidence of indirect recognition of processing sites in pre-rRNA processing in yeast. Nucleic Acids Research 29: 2106-2116.

39. Aguilar C, Reimer JD (2010) Molecular phylogenetic hypotheses on Zoanthus species (Anthozoa: Hexacorallia) using RNA secondary structure of the internal transcribed spacer 2 (ITS2). Marine Biodiversity 40: 195-204.

40. Grajales A, Aguilar C, Sanchez JA (2007) Phylogenetic reconstruction using secondary structures on internal transcribed spacer 2 (ITS2, rDNA): Finding the molecular and morphological gap in the Caribbean gorgonian corals. BMC Evolutionary Biology 7: 90-98.

Is Fortification of Methionine Necessary in Soya Bean (Glycine Max) Based Feeds for *Oreochromis andersonii* (Castelnau, 1861) Raised in Semi – Concrete Ponds?

Kefi AS*, Chungu NP, Mupenda N and Mumba C

National Aquaculture Research and Development Centre, P. O. Box 22797, Kitwe, Zambia

Abstract

An experiment was conducted to determine the optimal amount of methionine that can be incorporated in the soya bean based *Oreochromis andersonii* feeds. Three levels (2%, 4% and 6%) of methionine included in soya bean based feed (30% and 10% crude protein and crude lipid respectively) were tested on *O. andersonii* for seventy eight (78) days in hapas erected in a semi–concrete pond (250 m^2) arranged in a complete randomized design. Specific growth rate (SGR% day^{-1}), mean body weight gain (MBWG) (g) and apparent feed conversion efficiency (AFCE %) differed ($P<0.05$) among the treatments although the 4% and 6% gave similar ($P>0.05$) SGR (%), MBWG (g) and AFCE (%) but higher than the control and the lowest methionine level used in the experiment. Trend analysis showed a linear relationship (F=3.358, $P=0.02$) between Gonadosomatic Index (GSI) and methionine level with the highest methionine level producing fish with the highest GSI. Furthermore, the highest methionine level produced a significantly higher ($P<0.05$) GSI than in other treatments. However, females and larger sized fish appeared to mature earlier than males and smaller fish respectively. Fish body moisture and ash were similar ($P>0.05$) in all the treatments. It is, therefore, recommended that soya bean based feed can be fortified with 4% methionine to improve the *O.andersonii* growth. An economic analysis is required to ascertain the cost effectiveness of using the synthetic amino acid in soya bean fish feeds for *O. andersonii*.

Keywords: Methionine; *Oreochromis andersonii*; Soya bean; Growth and Reproduction

Introduction

The use of soya bean has now generally been accepted as the best alternative to the use of fish meal in the livestock feeds including fish. This is because it has similar balanced amino acid (AA) profile although its methionine level is lower than that of fish meal from as low as 0.55% in the full fat soya beans [1]. It is considered to meet requirements for some fish species [2,3] such as common carp, tilapia, channel catfish, salmonids, red drum, striped bass and marine shrimp [4]. It has also relatively high amounts of vitamins (thiamine, niacin, B–Complex and carotene) [5,6]. Furthermore, soya bean is readily available on the market at a relatively lower price [7]. If properly cooked, its low levels of cellulose and complex carbohydrates provide an additional advantage of using the plant in aquaculture nutrition. It has also low fibre content coupled with high digestibility [8].

Osman et al. [9] speculated low levels of both methionine and lysine in commercial fish feeds. Some authors [10] have, therefore, recommended the fortification of plant based diets with synthetic methionine in order to balance the AAs. Methionine is a sulfur containing and an essential AA required by fish as well as terrestrial vertebrates for normal growth and metabolic functions. This is because methionine is used in the protein synthesis and as a sulphur source which is needed for synthesis of other sulphur - containing biochemicals [11]. It is in the form of S–adenosyl–methionine which serves as a principal donor of methyl group, a major contributor to the whole body pool of carbon units that are required for trans–methylations, the biosynthesis of choline, thymidine, polyamines and creatine [12-14].

Oreochromis andersonii is indigenous to Zambia and has been adopted as the fish species of choice for aquaculture. It is naturally in lagoons of the Upper and Middle Zambezi River, Kafue systems [15] and Lake Kariba [16]. Adults occupy deep water while juveniles remain inshore. It feeds on detritus and zooplankton although bigger individuals also take insects and other invertebrates. In aquaculture the requirement for soya bean protein and lipid has been studied. Kefi et al. [17] found 30% crude protein to be ideal economically. Furthermore, Kefi et al. [18] observed that the minimal lipid requirement was lower than 10% although the optimal lipid requirement was observed to be 15%.

Although there are similar studies on methionine that have been conducted in other fish species, there has been no published work on *O. andersonii.* For instance Osman et al. [9] found 3.05% methionine of the crude protein to be optimal for *O. niloticus*. The experiment was conducted to determine the optimal methionine inclusion levels in soya bean based diet fed to *O. andersonii* juveniles.

Materials and Methods

The study to investigate the optimal methionine levels in soya bean based feed was carried out at National Aquaculture Research and Development Centre (NARDC), Kitwe for seventy eight days in hapas placed in a 250 m^2 semi-concrete pond. *Oreochromis andersonii* was the experimental fish species used in this study with varying levels of methionine (2%, 4% and 6%) as treatments. The percentages of methionine used were based on what Osman et al. [9] reported in *O.*

***Corresponding author:** Kefi AS National Aquaculture Research and Development Centre, P. O. Box 22797, Kitwe, Zambia
E-mail: askefi@yahoo.com

niloticus and a slight lower methionine of 2% was used as the start level. The experimental pond was prepared by draining it, setting up of hapas (1 × 0.9 × 1 m) and refilling with water to 60 cm level. The treatments were allocated to hapas in a randomized block design in triplicates. Two hundred and fifty *O. andersonii* (19.58 ± 0.103; mean ± SD) were seined from 250 m^2 semi-concrete pond and were taken to concrete tank (50 m^2) for conditioning for 48 hours. Thereafter, 20 fish were randomly sampled and their morphometric measures (weight, standard length and total length) taken as described by Skelton [15] before being placed in each hapa. The remaining fish were kept in the same concrete tank for possible replacements upon mortality that occurred within the first fourteen days of the experiment.

Isonitrogenous (30%) and isocaloric (10%) soya bean based feed (soya bean 54.6%, maize meal 43.4%, vitamin starter 1% and di - calcium phosphate 1%) fortified with DL-methionine feed grade 99% (Sumitomo Chemical Limited, Japan) levels at 2%, 4% and 6% was prepared and provided to fish at 3% live body weight twice a day with an exception of Saturdays and Sundays.

Selected water quality parameters (dissolved oxygen, water temperature, conductivity and Chlorophyll *a*) were taken once daily three times in a week (Monday, Wednesday and Friday) at 10 hours. Mortality was also monitored every day throughout the study.

At the end of the experiment, all the fish were weighed and the length measured. Three fish were randomly selected from each treatment and analyzed for dry matter and ash content according to the procedures of AOAC [19]. The ash was determined as total inorganic matter by incineration of the sample in an Advantech electric furnace at 550°C for 5 hours. The remaining inorganic material was reduced to their stable form, oxides or sulphates were considered as ash. Then the ash was calculated as follows:

$$Ash\ (\%) = \frac{weight\ of\ ash}{weight\ of\ original\ sample}\ x\ 100$$

Moisture was determined by drying samples in an Advantech electric furnace maintained at 105°C for 5 hours. The difference between the initial weight of the sample and that after drying was recorded and moisture calculated as follows:

$$Moisture\ (\%) = \frac{Difference\ between\ before\ and\ after\ drying\ (g)}{Weight\ of\ sample\ before\ drying}\ x\ 100$$

The remainder of the fish was then placed in buckets filled with 10% formalin for 10 days. After this period, the fish were dissected and the gonads recovered. Upon recovery, the maturity stage of the gonad was identified as described by Balarin [20]. The gonads were then weighed on sensitive analytical balance.

Statistical analysis

Growth of the fish was calculated following the equations; Weight gain (g)=Final fish weight (g)–Initial fish weight (g); Specific growth rate =100(LnWT1 - LnWT0)/t where, Ln is natural logarithm, WT0 is the initial weight (g), WT1 is the final fish weight (g) and t is time in days. Feed convention efficiency (FCE%) was calculated as follows: 100(body weight gain/feed taken). Condition factor was calculated as follows: 100,000W/L^3 where W is weight (g) while L is standard length (mm) of fish [21]. The gonad somatic index (GSI) was calculated as follows: 100(weight (g) of gonads/weight (g) of fish).

Analysis of variance (ANOVA) planned contrasts were performed to determine the differences between use and non-use of methionine and among methionine levels means which were deemed significant at *P*<0.05 for parametric data. Before analysis, parametric data were tested for normality using Shapiro–Wilk test and the homogeneity of variance using Levene's test for Equality of Variances. Contrasts tests not assuming equal variances were used if the assumption of equal variance was violated.

The stages of maturity of fish were analyzed using the multinomial logistic regression models where size (weight), sex and methionine entered as factors and stages of gonads as dependent variable. In order to eliminate cells with zero frequencies, the gonadal stages where grouped into three categories. These were immature, inactive (inactive and inactive-active) and ripe (active, ripe and ripe running) only. The ripe gonads were used as a reference group. The size (weight) was collapsed into 10 g intervals. Interactions between and among the factors were tested after the main effects were maintained in the model if significant (*P*<0.05), and similarly factors not having significant (*P*>0.05) effect were taken out from the model.

The model used was follows:

$$Y=a + b_1X_1 + b_2X_2 + b_3X_3 + b4X4 + \varepsilon$$

Where:

Y=Dependent variable (stages of gonads)

X_1 to X_4=Independent variables

X_1=Sex of fish

X_2=Weight of fish (g)

X_3=Methionine level (%)

X_4=interaction effect

b_1 to b_4=regression coefficients

a=Constant

ε=Residual

Cross tabulations and Chi-square (χ^2) were used to examine the associations between the reproductive status of fish and sex, and methionine levels.

Statistical Package for Social Scientist (SPSS) 15.0 (SPSS Inc) softwares were used in analyzing the data. Microsoft excel was used in the production of figures and graphs. Untransformed data are presented to facilitate interpretation.

Results

Planned comparisons revealed a significant effect (*P*<0.05) of methionine on SGR. Trend analysis revealed that SGR (%) increased proportionally (F =30.504, *P*=0.0001) with methionine level. Similar trend was observed in FCE (%) (F=13.725, *P* =0.0001). The final mean weight, condition factor (K) and standard length (mm) did not differ significantly (*P*>0.05) among the treatments. However, the highest final mean weight was recorded at 4% methionine level. Specific growth rate (SGR) differed significantly (*P* <0.05) among methionine levels. However, the control did not differ (*P*>0.05) from that of 2% while the 4% methionine level was similar (*P*>0.05) to that of 6%. The mean body weight gain differed significantly (*P*<0.05) with the higher levels of methionine recording the highest body weight gain. The 4% and 6% methionine did not differ significantly (*P*>0.05). Similarly the control and the lowest methionine level (2%) used did not differ significantly (*P*> 0.05). The same trend was observed in the feed conversion efficiency

Parameter	Treatment (methionine level) (%)			
	0	2	4	6
Initial mean weight (g)	19.833 ± 0.218	19.733 ± 0.203	19.500 ± 0.201	19.267 ± 0.199
Final mean weight (g)	25.259 ± 0.807	24.898 ± 0.678	26.319 ± 0.735	26.136 ± 0.737
Standard length (mm)	92.298 ± 1.034	91.286 ± 0.840	93.604 ± 0.886	92.870 ± 0.904
SGR (%day^{-1})	0.336 ± 0.015^a	0.323 ± 0.014^a	0.414 ± 0.108^b	0.422 ± 0.015^b
Mean body weight gain (g)	5.426 ± 0.218^a	5.165 ± 0.203^a	6.819 ± 0.201^b	6.869 ± 0.199^b
Condition factor (K)	3.099 ± 0.046	3.092 ± 0.033	3.034 ± 0.038	3.051 ± 0.035
GSI	0.126 ± 0.027^a	0.162 ± 0.035^a	0.270 ± 0.092^a	0.356 ± 0.065^b
FCE (%)	12.905 ± 0.638^a	12.304 ± 0.583^a	16.342 ± 0.616^b	16.648 ± 0.605^b
Ash (%)	15.857 ± 0.996	13.485 ± 5.248	10.217 ± 0.743	19.274 ± 7.761
Dry matter (%)	37.861 ± 0.650	41.750 ± 1.983	63.026 ± 13.685	44.536 ± 4.606
*Mortality (%)	5.00	6.67	6.17	5.67

Means ± SE with different superscripts differ significantly (*P*<0.05) across the rows.

*No statistical significance determined since it was calculated on pooled sample.

Table 1: Growth, morphometric, feed utilisation, organ, reproductive and parameters of *O.andersonii* subjected to different levels of methionine.

(%) (Table 1).

There were no significant differences (*P*>0.05) observed in the ash and dry matter of the fish among the methionine levels used in the experiment. However, the 4% methionine level recorded the lowest ash and highest dry matter. Dry matter was lowest in the control group (Table 1).

Gonadosomatic index (GSI) increased linearly with methionine levels and this was significant (F=3.358, *P*=0.02). The highest methionine level used produced fish with the highest GSI and this was significant different (*P*<0.05) from the other treatments including the control. There was a relationship (*P*<0.05) between the independent variables (sex, fish size and methionine levels) and the stages of maturity of the gonads. However, only size and sex had a significant relationship to the maturity stage of the fish and therefore, it was excluded from the final model. The females were less likely to be immature (Odds Ratio=0.069) or inactive Odds Ratio=0.271) compared to males. Additionally small fish were more likely to be immature (Odds Ratio=42.99) or inactive Odds Ratio=4.959) than the bigger fish. However, the medium sized fish did not differ (*P*>0.05) from the reference group which is large fish. This shows that the females are more likely to be in advanced stage of maturity than males indicating that females matured earlier than males. Consequently, large fish seemed to be advanced in maturity status (Table 2).

Significant differences (*P*<0.05) were observed in the stages of maturity with ripe stage recording the highest reproductive stage although this was not significant different (*P*>0.05) from the immature group. Inactive group was the lowest but was not significant different (*P*>0.05) from the immature group (Figure 1).

According to sex, there were more immature males (79.4%) than females and this was significant (*P*<0.05). Although the inactive group was not significant (*P*>0.05) between the sexes, the females (46.8%) were fewer than the males. There number of females in the ripe stage were significantly (*P<0.05*) more than the females (Figure 2).

There were no significant differences (*P*>0.05) observed in the selected water quality parameters determined (Table 3).

Discussion

Protein requirements are defined as the minimum amount needed to meet requirements for amino acids and to achieve maximum growth [22]. There have several varying results reported by authors probably due to variances in experimental conditions. Wilson [23] estimated 1.8 to 4.0% dietary methionine of the dietary protein as the requirement for commonly cultured species of fish. Jackson and Capper [23] reported that *O.mossambicus* appears to have a lower than 0.53% dietary methionine that may provide a satisfactory growth. In the current experiment, a practical diet fortified with methionine did not produce fish with different final mean weight. However, SGR (%) day^{-1}, weight gain and FCE (%) differed significantly. However, the 4% and 6% methionine produced similar but superior to the control and 2% methionine level. The insignificant final (*P*>0.05) weight exhibited by the fish subjected to all the treatments may be due to shorter experimental period. The rate at which fish grow is dependent on a number of factors including species, age, genetic potential, water temperature, health, and quantity and quality of food. Specific growth rate has been used as a suitable index for determining fish growth. In the current experiment SGR (%) day^{-1} was found to be highest at the

Stage of maturity	Sex of fish	B (Coefficient)	S. E	Wald	Exp (B)
Immature	Female	2.672	0.459	33.843*	0.069
	Small fish	3.761	1.151	10.681*	42.990
	Medium fish	2.205	1.140	3.737	9.066
Inactive	Female	-1.305	0.404	10.413*	0.271
	Small fish	1.601	0.744	4.637*	4.959
	Medium size	0.821	0.720	1.300	2.272

*Significant different (*P*<0.05)

Table 2: Logistic regression on likelihood of maturity stages, sex and size of fish

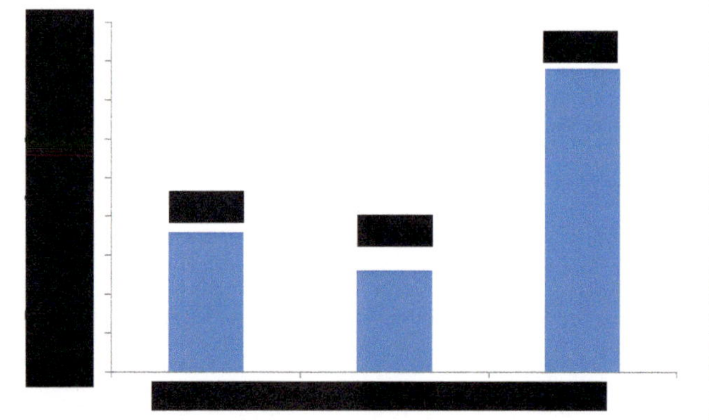

Figure 1: Percentage of gonad stages observed in *O. andersonii*

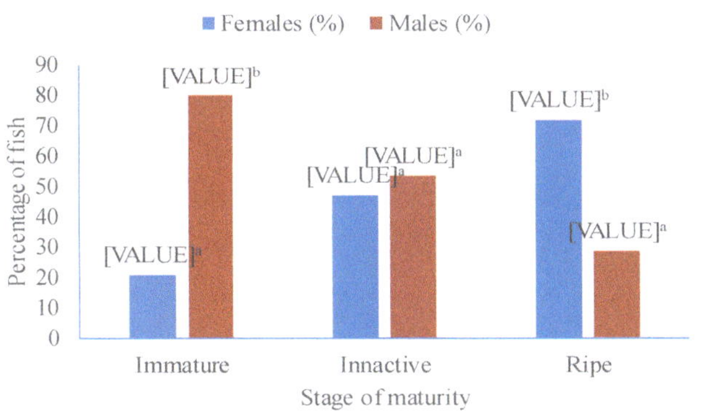

Figure 2: Percentage of gonad stages observed in *O. andersonii* according

Parameter	Treatment (methionine level) (%)			
	0	2	4	6
Water Temperature (°C)	17.292 ± 0.101	17.353 ± 0.104	17.344 ± 0.103	17.4 ± 0.1
Electrical conductivity (μS/cm)	91.341 ± 1.416	91.111 ± 1.435	91.211 ± 1.422	91.791 ± 1.462
Dissolved Oxygen (mg/L)	5.048 ± 0.257	5.407 ± 0.224	4.935 ± 0.267	5.012 ± 0.275
Chlorophyll *a* (mg/L)	0.02 ± 0.007	0.048 ± 0.056	0.051 ± 0.038	0.098 ± 0.111

Table 3: Water quality parameters determined in hapas with *O. andersonii*

methionine level (6%) used although this was not significant different (P>0.05) from that of 4%. This was similar to the weight gain (g) and FCE (%). However, the final mean weight was not significant (P>0.05) although the 4% methionine level produced the fish with the highest final mean weight. Similar results were found by Nguyeni and Davis [24] who did not find any positive effect of methionine addition to soya bean based feed on the final weight on Red tilapia and *O. niloticus*. However, they did not report on SGR (%) day^{-1}. Osman et al. [9] found that the dietary requirements of methionine for *O. niloticus* fingerlings to be 3.05%of the dietary crude protein. Our results are, therefore, consistent with the finding of Wilson [23] Osman et al. [9].

To the best of our knowledge there have been no studies conducted on the effect of methionine on the reproduction of *O. andersonii*. Therefore, our results cannot be referred to any study but can be a benchmark for future studies. Although, the level of methionine did not affect the rate of maturity of *O. andersonii,* there was a positive linear relationship between the GSI and methionine level. The increase in GSI as the methionine increases indicates an elevated gonad mass due to supplementation of the soya bean feed with methionine. This shows a high investment in reproduction as the methionine increases. We have not come across any study that has directly implicated the methionine to reproduction efficiency in fish. Therefore, the increase in gonad mass of the fish with the highest methionine may be attributed to the satisfying the amino acid composition of the feed as high protein level has been associated with high GSI [25,26]. However, sex and size of fish appears to affect the maturity of the *O. andersonii*. The females were found to mature earlier than males and larger fish were more at an advanced maturity stage than small fish. Similar results were observed by Kefi [26]. This observation emphasizes the need to culture all males only if growth before maturity is to be maximized.

The study reveals that 4% and 6% methionine level give similar but superior results on the growth rate of *O. andersonii*. However, methionine seemed to have increased GSI of *O. andersonii*. Therefore, 4% methionine level inclusion in soybean based fish feed would be recommended in the growth enhancement of *O. andersonii*. However, there is need to conduct an economic analysis of utilization of methionine in the soya bean based fish feeds. Furthermore, there is need to study the fortification of the soya bean based fish feeds with methionine for a longer period in aquaculture facility being used in the production of fish for commercial purposes.

Acknowledgement

Authors wish to thank Indian Council of medical research (ICMR), New Delhi for financial support. We also want to thank Shilpa Virdi, Gaurav Sharma and Lekhraj in collection and processing of samples.

References

1. NRC (2012) Nutrient Requirements of Swine. (11th rev edn) Natl. Acad. Press, Washington DC.
2. NRC (1993) Nutrient requirements of fish. National Research Council, National Academy Press, Washington, D.C.
3. Deguara S (1998) Effect of supplementary enzymes on the growth and feed utilisation of gilthead sea bream, *Sparus aurata*. PhD Thesis, University of Stirling, Stirling, Scotland.
4. Cremer MC, Enhua Z, Jian Z, O'Keefe T (2008) Soy protein concentrate replaces fishmeal in fingerling test diets in China. American Soybean Association, St. Louis, Missouri 63141 USA.
5. Martin R, Ruberte R (1980) Legume. In technique and plants for tropical subsistence farm. United States Department of Agriculture 19-20.
6. Fafioye OO, Fagade SO, Adebisi AA, Oni J, Omoyinmi GAK (2005) Effect of dietary soybean (*Glycine max (L) Merr*) on growth and body composition of African Catfish (*Clarias gariepinus*, Burchell) fingerlings. Turk J Fish Aquat Sc 5: 31-35.
7. Guillaume J, Kaushik S, Bergot P, Metailler R (2001) Nutrition and feeding of fish and crustaceans, Praxis, Cornwall, United Kingdom.
8. Gatlin DM. III, Barrows FT, Braown P, Dabrowski K, Gaylord TG et al. (2007) Expanding the utilization of sustainable plant products in aquafeeds: a review. Aquacult Res 38: 551-579.
9. Osman FM, Amer MA, Mahfouz SA, Wasly KM (2008) Reviewing the optimal level of methionine and lysine for tilapia fingerlings (*Oreochromis niloticus*). Proceedings of 8th International Symposium on Tilapia in Aquaculture, 903-919.
10. Davies OA, Ezenwa NC (2004) Groundnut cake as alternative protein source in the diet of *Clarias gariepinus* fry. International Journal of Science and Nature 1: 73-76.
11. Wu G, Davis DD (2005) Interrelationship among methionine, choline and betaine in Channel Catfish *Ictalurus punctatus*. J. World Aquac. Soc 36: 337-345.
12. Alam MD, Teshima SI, Ishikawa M, Koshio, S (2000) Methionine requirement of juvenile Japanese flounder, *Paralichthys olivaceus*. J. World Aquac. Soc 31: 618-626.
13. Mai K, Wan J, Ai Q, Xu W, Liufu Z et al. (2006) Dietary methionine requirement of large yellow croaker, *Pseudoscia enacrocea*. Aquaculture 253: 564-572.
14. Bae JY, Ok IH, Lee S, Hung SSO, Min TS et al. (2011) Re-evaluation of Dietary Methionine Requirement by Plasma Methionine and Ammonia Concentrations in Surgically Modified Rainbow Trout, *Oncorhynchus mykiss*. Asian Australas J Anim Sci 7: 974-981.
15. Skelton P (2001) A Complete Guide to the Freshwater Fishes of Southern Africa, Struik, Cape Town, South Africa.
16. http://www.fishbase.org/summary/Oreochromis-andersonii.html.
17. Kefi AS, Kang'ombe J, Kassam D, Katongo C (2013) Effect of Dietary Soyabean (*Glycine max* (L.) Merr.) Protein Level on Growth and Feed Utilization of *Oreochromis andersonii* (Castelnau, 1861). Pak J Nutr 12: 990–995.

18. Kefi AS, Kang'ombe J, Kassam D, Katongo C (2013) Optimal Dietary Plant Based Lipid on Growth of *Oreochromis andersonii* (Castelnau, 1861). Turk J Fish Aquat Sc 13: 503–508.

19. AOAC (Association of Official Analytical Chemists) (2002) Methods of Analysis of the Association *of Official Analytical Chemists.* Association of Official Analytical Chemists, Inc., Arlington, U. S. A.

20. Balarin JD (1983) A guide to Tilapia breeding: the Baobab Hatchery Technique. Tilapia culture section, Baobab Farm Limited, Mombasa, Kenya.

21. Williams JE (2000) The Coefficient of Condition of Fish. Chapter 13 *in* Schneider, James C. Manual of fisheries survey methods II: with periodic updates. Michigan Department of Natural Resources, Fisheries Special Report 25, Ann Arbor.

22. Jackson A, Capper J (1982) Investigation into the requirements of the tilapia (*Sarotherodon mossambicus)* for dietary methionine, lysine and argnine in semi – synthetic diets. Aquaculture, 29: 289-287.

23. Wilson RP (2002) Amino acids and proteins. In: Fish Nutrition, (3rd Ed.) Ed. J. E. Halver and R. W. Hardy. Academic Press, Inc., New York, USA 143-179.

24. Nguyen TN, Davis DA (2009) Evaluation of Alternative Protein Sources to Replace Fish Meal in Practical Diets for Juvenile Tilapia, *Oreochromis* spp. J World Aquac Soc 40:113–121.

25. Kithsiri HM, Sharma P, Zaidi SGS, Pal AK, Venkateshwarlu G (2010) Growth and reproductive performance of female guppy, *Poecilia reticulate* (Peters) fed diets with different nutrient levels. Indian J Fish, **57**: 65–71.

26. Kefi AS (2014) Influence of dietary soya beans (*glycine max* (l.) merr.) protein and lipid combination, and androgen (17 α – methyl testosterone) levels on the growth and reproduction of *Oreochromis andersonii* (Castelnau, 1861). PhD thesis, Bunda College of Agriculture, University of Malawi, Lilongwe, Malawi.

Exploring Sheraoh Island at South-Eastern Qatar: First Distributional Records of Some Inland and Offshore Biota with Annotated Checklist

Mahmoud M Kardousha*, Abdulrahman Al-Muftah and Jassim A Al-Khayat

Department of Biological and Environmental Sciences, Qatar University, Doha, Qatar

Abstract

Sheraoh island is the most remote among Qatari islands, nearly 73.5 km south-eastern of the mainland of the peninsula. In March 2010, a preliminary general survey of the species was conducted. Observations were made partly by wading and snorkeling in the intertidal zone and partly by surveying the inland habitats. In total the 31 species were recorded, 4 species of halophyte plants: *Halopeplis perfoliata, Suaeda vermiculata, Salsola baryosoma and Zygophyllum qatarense,* one unexpected terrestrial mammal: the cape hare *Lepus capensis,* 2 species of visiting birds: *Motacilla flava* and *Anthus campestus* and one species of sea bird.

Phalacrocorax nigrogularis, one dolphin species: *Sousa chinensis.* The largest richness was found in the marine habitats, 3 species of sea weeds: *Colpomenia sinusa, Padina boergesenii and Sargassum boveanum,* 3 species of corals: *Anomastraea irregularis, Platygyra lamellina* and *Porites harrisoni,* 2 species of nudibranchs: *Chromodoris cazae* and *Chromodoris sp. (unidentified).,* 4 crab species: Hermit crab (unidentified), *Etisus laevimanus, Grapsus albolineatus* and *Lambrus prensor,* 4 species of echinoderms: *Pentacentral mammillatus, Ophiothela venusta, Diadema setosum* and *Echinometra mathaei,* 5 species of bony fish: *Doryrhamphus sp.* (larva), *Cypserulus oligolepis, Lethrinus nebulosus, Cephalopholis miniatus* and *Hemiramphus marginatus,* one species of sea snakes: *Hydrophis sp.* The most striking result is to collect an unknown nudibranch species and to report the cape hare which is unexpected inhabitant. All figures of this study are published for the first time in Qatar.

Keywords: Arabian Gulf; Qatari Islands; Sheraoh island; Marine Biota; Nudibranchs; Cape hare

Introduction

Although there is a growing number of marine ecosystem studies of Qatari waters, the islands have not received much attention. There have been a few publications on island' biodiversity in the region, but with the exception of recent studies on the Qatari lizards [1] and vegetation [2], there are no (to our knowledge) first hand records of biodiversity in Sheraoh island except few observations have been recorded by some visitors [3]. Islands are important priorities for biodiversity conservation that hold more than one-fifth of the world's plant species and about one-quarter of the world's threatened mammals and one-third of the world's threatened birds [4]. In Qatar, the most prominent islands are located in the north and eastern side of Qatari peninsula. They include Rakan island, Halul island, Al-Aaliya island, Al-Safliah island, Al-Ashat island and Sheraoh island. They play a crucial role as resting and breeding sites for different types of seabirds, nesting sites for the green and hawksbill turtles and some of them harbor coral reefs around which host a highly rich diversity of marine species. There is a growing number of studies focusing on islands biodiversity in Qatar. Vegetation (Halul island [2], lizards (three islands were investigated: Halul, Al-Alia and Al-Safliyah [1] and the coral reefs and associated biota of Halul and Al-Safliyah islands [5,6]. In addition, Halul island is a very important island for nesting hawksbill turtles and different sea birds [3]. The biodiversity of Sheraoh island had never been studied and the only information available is visitor observations related to nesting activities of hawksbill turtles [3]. Thus, the current study is the first biodiversity survey of the island. There is a need of further research to help in setting a conservation agenda for the wild life of Arabian Gulf islands.

Materials and Methods

Qatari islands

The archipelago of Qatar peninsula, south of Arabian Gulf, is located between the latitude 24° 27' and 26° 10' North and longitudes 50° 45' and 51° 40' East with an area of 11,521 km^2. The most prominent islands are located north, north-eastern and south-eastern sides of the peninsula. summarizes the Qatari islands with its latitudes and longitudes. The Sheraoh island is the most remote island in the southeastern side (Figure 1). It is located at the far margin of Qatari waters. The island has different biotypes including rocky shores, intertidal and subtidal sand banks, shallow lagoons and coastal sand dunes (Figures 2 and 3) (Table 1).

Two days' cruise to sheraoh island

The observations were made during 2 days' field work in March 2010. The survey was conducted partly by wading and snorkeling in the intertidal zone and by surveying the inland habitats. Two benthic samples were collected offshore. Samples were preserved in plastic collection jars containing 5% seawater-diluted formalin, brought to the laboratory, sorted, and identified. Some specimens were photographed in *situ*. All specimens were investigated by using Stereo-microscope and identified based on the available atlases, field guides and publications such as marine biota [7], sea shore organisms [8], fishes [9], halophytes [10] and corals [11,12].

***Corresponding author:** Mahmoud M Kardousha, Department of Biological and Environmental Sciences, Qatar University, Doha, PB Box 2713, Qatar
E-mail: mkardousha@qu.edu.qa

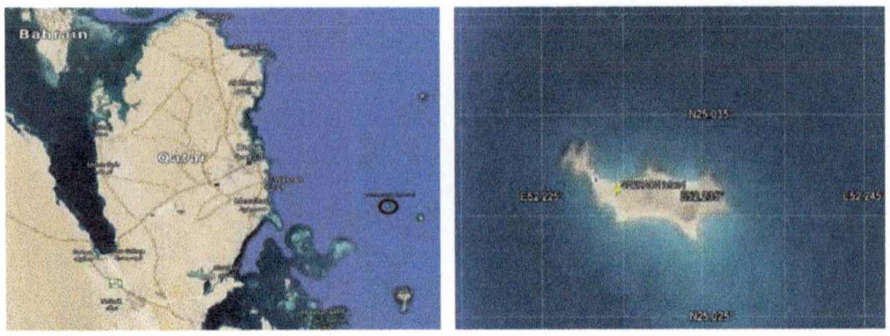

Figure 1: A satellite google map showing State of Qatar with the remote Sheraoh island: (A) encircled and (B) enlarged.

Figure 2: A shallow lagoon in Sheraoh island.

Figure 3: A rocky shore in Sheraoh island.

Results

A total of 32 species were collected/ encountered along the inland, coastal, and tidal areas of the island during the expedition with different groups and species (Table 2).

Inland assemblage

The vegetation of the island was found to be dominated by halophyte communities with *Halopeplis perfoliata* (Figure 4), *Suaeda vermiculata* (Figure 5), *Salsola baryosoma* and *Zygophyllum qatarense* (Figure 6).

- **Birds**

Two types of visitor birds were observed, the western yellow wagtail *Motacilla flava* (Figure 7), and Tawny Pipit *Anthus campestus* (Figure 8). A dead body of Socotra cormorant *Phalacrocorax nigrogularis* was also found (Figure 9).

- **Mammals**

One unexpected mammalian species, the cape hare *Lepus capensis* was seen running very fast and camouflaged under shrubs. A discovered skeleton with stiff tufts of hairs between the pads of the feet obviously indicating that the Cape hare is a common inhabitant on the island (Figure 10). A dolphin skeleton was also found with well-defined long beak and prominent teeth, the upper jaw is shorter than the lower and backbone is humped. These features characterize the endo-pacific humped back dolphin *Sousa chinensis* (Figure 11).

Coastal area assemblage

Along the coastal area, many pieces of coral skeletons were observed scattered everywhere which indicated the dominancy of three species, *Anomastraea irregularis, Platygyra lamellina* and *Porites harrisoni*. The brown algae were plentiful along the coastal area and three species were dominated, *Colpomenia sinusa* (Figure 12), *Padina boergesenii* (Figure 13) and *Sargassum boveanum* (Figure 14). Rocks in the upper intertidal zone were covered by green algae (Figure 15).

Intertidal and subtidal biotopes

The assemblage of intertidal and subtidal diversity was found highly enriched by different types of marine biota.

- **Echinoderms**

Two sea stars were commonly observed, the large sea star *Pentacentral mammillatus* (Figure 16) and the brittle star *Ophiothela venusta* (Figure 17). Similarly, skeletons of the sea urchin *Diadema setosum* (Figure 18) and *Echinometra mathaei* were also common (Figure 19).

- **Nudibranchs**

Two species of nudibranchs were collected from subtidal zone, *Chromodoris cazae* (Figure 20) and un-specified nudibranch related to genus *Chromodoris* which is characterized by prominent red spots and yellow margin (Figure 21).

- **Crustacea (Decapods)**

The decapods *Etisus laevimanus* (Figure 22), *Grapsus albolineatus* (Figure 23) *and Lambrus prensor* (Figure 24) were common in the tidal and intertidal zones and were observed invading sandy and rocky coastal areas. Sand towers indicated presence of ghost crab (Figure 25). In addition, Hermit crabs were observed among the subtidal zone (Figure 26).

Offshore biotype

Five species of bony fish was encountered, a larval stage of pipe

Name	Latitude in Decimal Degrees	Longitude in Decimal Degrees
Rakan Island	26.175723°	51.220295°
Halul Island	25.673681°	52.410462°
Al-Aaliya Island	25.403958°	51.568908°
Al-Saflia Island	25.346331°	51.576800°
Palm Tree Island	25.339780°	51.533572°
Old Palm Tree	25.305338°	51.527821°
Banana Island	25.297635°	51.643587°
Pearl Island	25.371761°	51.559459°
Al-Ashat Island	24.751651°	51.609290°
Sheraoh Island	25.030992°	52.229545°

Table 1: The different prominent Qatari islands with latitudes and longitudes.

Collected and observed Biotypes	Related Biotypes of Arabian Gulf coasts
Phaeophyte Algae: **Scytosiphonaceae** - *Colpomenia sinusa* (Figure 12) **Dictyotaceae** - *Padina boergesenii* (Figure 13) - *Sargassum boveanum*	The 3 species of these Phaeophytes are commonly distributed in all Arabian Gulf coasts including Qatar
Halophyte plants: **Chenopodiaceae** - *Halopeplis perfoliata* (Figure 4) - *Suaeda vermiculata,* (Figure 5) - *Salsola baryosoma* - *Zygophyllum qatarense* (Figure 6)	The 4 species of these halophytes are dominated in all Arabian Gulf countries including Qatar
Corals: **Siderastreidae** - *Anomastraea irregularis* (Figure 5) **Faviidae** - *Platygyra lamellina* **Poritidae** - *Porites harrisoni*	Commonly distributed and recorded from all AG waters including Qatar
Nudibranch Molluscans: **Chromodorididae** - *Chromodoris cazae* (Figure 20) - *Chromodoris sp.* (Figure 21)	*Chromodoris cazae* is commonly distributed only in waters of United Arab Emirates and Qatar. the second un-identified species is observed for the first time from Arabian Gulf waters.
Decapod Crustacea: **Grapsidae** - *Grapsus albolineatus* (Figure 23) **Xanthidae** - *Etisus laevimanus* (Figure 22) **Parthenopidae** - *Lambrus prensor* (Figure 24) - Hermit crab (Figure 26)	Commonly distributed in Arabian Gulf
Echinoderms: **Oriasteridae** - *Pentacentral mammillatus* (Figure 16) **Ophiotrichidae** - *Ophiothela venusta* (Figure 17) **Diadematidae** - *Diadema setosum* (Figure 18) **Echinometridae** - *Echinometra mathaei* (Figure 19)	Commonly distributed in Arabian Gulf waters
Bony Fish: **Exocoetidae** - *Cypserulus oligolepis* **Syngnathidae** - *Doryrhamphus sp.* (larva) **Lethrinidae** - *Lethrinus nebulosus* **Serranidae** - *Cepahlopholis miniata* **Hemiramphidae** - *Hemiramphus marginatus*	Commonly distributed in Arabian Gulf waters
Reptiles: ***Hydrophis sp.***	It common inhabitant in Arabian Gulf and Indian ocean
Birds: **Phalacrocoracidae** - *Phalacrocorax nigrogularis* (Figure 9) **Motacillidae** - *Motacilla flava* (Figure 7) - *Anthus campestus* (Figure 8)	Frequently seen
Mammals: **Delphinide** - *Sousa chinensis* (Figure 11) **Leporidae** - *Lepus capensis* (Figure 10)	*Sousa chinensis* is a common member of marine mammals fauna in the Arabian Gulf. The desert cape hare *Lepus capensis* is an unexpected inhabitant on this remote island.

Table 2: An annotated list of observed and collected Biota from Sheraoh Island, southeastern of Qatar in March 2010.

fish *Doryrhamphus sp.*, flying fish *Cypserulus oligolepis*, *Lethrinus nebulosus*, *Cephalopholis miniatus* and *Hemiramphus marginatus*. One marine reptile was also observed from the vessel, the common Arabian Gulf sea snakes, *Hydrophis sp.*

Discussion

Qatari islands are numerous, and most of them are located nearshore of the main land of the peninsula. The most remote ones are Halul and Sheraoh islands which are located at northeastern and southeastern of Qatar respectively. The islands are usually flat, made of calcareous rock and covered by sand. There is very limited information available on Sheraoh island, although, it is described as a heaven for sea birds, nesting turtles specially the hawked bill types and small fringed reefs [3]. Due to its isolation and long distance from the mainland, Sheraoh island has not been visited by many people. Only fishermen and a few others visit the islands and they typically do not stay long. It is not frequently disturbed, the island still maintains very large breeding

Figure 4: *Halopeplis perfoliata.*

Figure 5: *Suaeda vermiculata.*

Figure 6: *Zygophyllum qatarense.*

Figure 7: *Motacilla flava.*

Figure 8: *Anthus campestus.*

Figure 9: Dead body of Socotra cormorant *Phalacrocorax nigrogularis.*

Figure 10: Skeleton of the cape hare *Lepus capensis.*

Figure 11: Skeleton of the humped back dolphin *Sousa chinensis.*

Figure 12: *Colpomenia sinusa.*

Figure 13: *Padina boergesenii.*

Figure 14: *Sargassum boveanum.*

Figure 15: Upper intertidal zone covered by green algae.

Figure 16: The large sea star *Pentacentral mammillatus*.

Figure 17: Brittle star *Ophiothela venusta*.

aggregations of seabirds and sea turtles that have been lost from the mainland and islands near the mainland.

The Arabian Gulf and Qatari islands are extremely important as nesting sites for green and hawksbill turtles. In addition, they are of international importance for seabirds that either breed on the islands or rest there during their annual migrations. The islands also host other species from different biota. The reefs around the islands host a rich diversity of marine species including algae, seagrass, coral, invertebrates, fishes, turtles, and occasionally sea snakes and marine mammals [12].

One of the most interesting species which was collected from Sheraoh' biota was one unidentified species of nudibranchs that is characterized by prominent red wide spots and yellow margin. It is probably related to Australian species *Chromodoris splendida* [13]. No organized study of nudibranchs in Qatar has yet been undertaken, however, some collections have been reported. As far as references have revealed, 6 species have been collected from Qatari waters, *Chelidonura livida, Flabellina rubrolineatus, Chromodoris annulata, C. obsolete, C. cazae and C. sp16* [14-16]. *Chromodoris cazae* has only reported from Arabian Gulf, specifically from Qatar and United Arab Emirates [17].

The observations and collecting of coral skeletons and algal remains from intertidal zone indicated that there is a high diversity of corals and algae in the surrounding waters. The skeletal material of coral reefs was found covering a wide area of coastal region. The scattered skeletons

Figure 18: Skeletons of the sea urchin *Diadema setosum*.

Figure 19: *Echinometra mathaei*.

Figure 20: The nudibranch *Chromodoris cazae* which recorded only from Qatari and Emirati waters, Arabian Gulf.

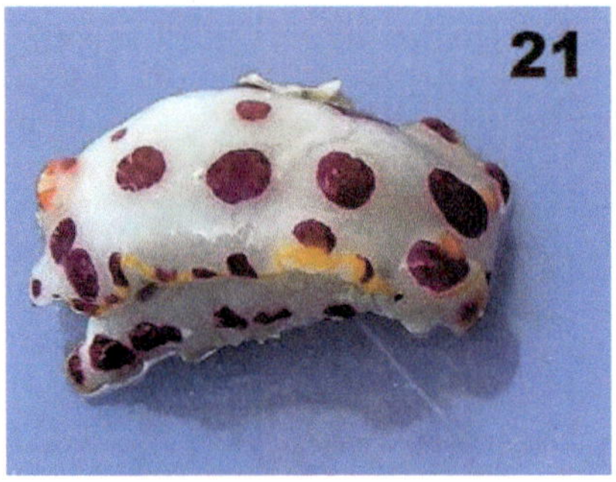

Figure 21: *Chromodoris sp.* a nudibranch which have not ever seen before in Arabian Gulf waters.

Figure 22: The smooth spooner crab *Etisus laevimanus*.

Figure 23: The mottled crab *Grapsus albolineatus*.

Figure 24: The parthenopid crab *Lambrus prensor*.

Figure 25: Sand towers indicated presence of ghost crab.

Figure 26: Hermit crabs were observed among the subtidal zone.

were dominated by the brain coral *Platygyra lamellina* and *Porites harrisoni* which are the most common Arabian Gulf corals [12,18,19].

Among the more active offshore animal types are the crabs which mostly with high distribution near sand and rocky coastal areas. The rock crab *Etisus laevimanus*, the mottled crab *Grapsus albolineatus* and the parthenopid crab *Lambrus prensor* were common along the coastal area. The prominent erected sand towers of Ghost crabs also common along the coastal area [20].

We were unexpected to find a population of Cape hare *Lepus capensis* at the island. The cape hare is a common mammalian member of Arabian Peninsula including Qatar. However, its presence in a remote isolated island like Sheraoh was unexpected. The Cape hare has been recorded also from Al-Saflia island [3]. It could be interesting to study this population using DNA techniques to trace its phylogeny. This will empathize if this population was introduced to the island or if it has been isolated for a long period. As islands near the mainland are commonly disturbed by human presence, the remote and isolated Sheraoh island could be used for studies on natural ecosystem on undisturbed islands [20].

Acknowledgement

The authors would like to thank Qatar University for supporting this study. Special thanks go to Dr. Juha Alatalo, Department of Biological and Environmental sciences, Qatar University for his kind revision of the manuscript. Thanks are also extended to Dr. Perumal Balakrishnan, lecturer of GIS in the Department for preparing the location table and the map. We are also grateful to senior students of marine sciences minor for their invaluable help in collecting specimens especially by snorkeling.

Competing Interest

The authors declare that they have no competing interests.

References

1. Castilla AM, Al-Kubaisi DJ, Davies C, Mushtak A, Al-Mam WJ, et al. (2014) The lizards living in Qatar. Eds Al-Hemaidi AAM, Al-Hajari SA, Al-Subai K, Mohtar RH, Pelegri JM, Castella AM. Ministry of Environment, Doha, Qatar p: 349.
2. Abulfatih HA, Bari EMA, Alsubaey A, Ibrahim YM (2001) Vegetation of Qatar. Scientific and applied research center (SARC), Qatar University, Al-Ahleia Printing Press, Doha, Qatar p: 360.
3. Gillespie F (2008) Discovering Qatar. Creative writing and photography. Rimons, France pp: 148.
4. Hofmann M (2016) IUCN-strategies for the conservation of Island biodiversity, in international conference on island evolution, ecology, and conservation, 18-22 July 2016, University of Azores at Angra do Heroismo, Terceira island, Azores, Portugal.
5. Al-Ansi MA, Al- Khayat JA (1999) Preliminary study on coral reef and its associated biota in Qatari waters, Arabian Gulf. Qatar Univ. Sci J 19: 294-311.
6. Al-Khayat JA (2005) Some Macrobenthic Invertebrates in the Qatari Waters, Arabian Gulf. Qatar Univ. Sci J 25: 126-136.
7. Carpenter KE, Krupp F, Jones DA, Zajonz U (1997) FAO species identification guide for fishery purposes. The living marine resources of Kuwait, Eastern Saudi Arabia, Bahrain, Qatar, and the United Arab Emirates. FAO, Rome p: 293.
8. Jones D (1986) A field guide to the seashores of Kuwait and the Arabian Gulf. University of Kuwait, Kuwait p: 192.
9. Sivasubramaniam K, Ibrahim MA (1982) Common fishes of Qatar (Scientific Atlas of Qatar). Doha modern printing press, Doha, Qatar p: 172.
10. Batanouny KH (1981) Ecology and Flora of Qatar. Aldern press Ltd, Oxford, Great Britain p: 245.
11. Loughland RA, Al-Abdulkader KA (2012) Marine atlas of western Arabian Gulf. Saudi Aramco environmental protection publication, 2nd edition p: 359.
12. Bruckner A, Rowlands G, Riegl B, Purkis S, Williams A, et al. (2012) Atlas of Saudi Arabian Red Sea Marine Habitats. Khaled bin Sultan Living Oceans Foundation p: 273.
13. Rudman WB (1983a) TheChromodorididae(Opisthobranchia:Mollusca) of the Indo- WestPacific: *Chromodoris splendida,* C. as groups. Zoological J of the Linnaean Society 78: 105-173.
14. Qatar Natural History Group (QNHG) (2007), newsletter, No1.
15. Rudman WB (2007) *Chromodoris sp* 16 [In] Se Slug Forum. Australian Museum, Sydney.
16. Macdonald IA (2004) *Chromodoris annulata,* color markings. *[Message in] Sea Slug Forum.* Australian Museum.
17. Rudman WB (2004) *Chromodoris cazae* Gosliner and Behrens. [In] Sea Slug Forum. Australian Museum, Sydney.
18. Basson PW, Burchard JE, Hardy JT, Price ARG (1981) Biotopes of the western Arabian Gulf (Marine life and environments of Saudi Arabia) Shenval 80, Harlow, Great Britain p: 284.
19. Kilbane D, Graham B, Mulcahy R, Onder A, Pratt M (2008) Coral relocation for impact mitigation in Northern Qatar. Proceedings of the 11th International Coral Reef Symposium, Ft. Lauderdale, Florida, 7-11 July 2008, Session number 24.
20. Giraldes BW, Al-Maslamani I, Chatting M, Smyth D (2015) Marine collection in Qatar - basis for biodiversity management, Qatar University Life Science Symposium 2015.

Seaweed Incorporated Diet Improves Astaxanthin Content of Shrimp Muscle Tissue

Kunal Mondal[1]*, Subhra Bikash Bhattacharyya[2] and Abhijit Mitra[2]

[1]*Faculty of Fishery Sciences, West Bengal University of Animal and Fishery Sciences, Kolkata-700094, India*
[2]*Department of Oceanography, Techno India University, Kolkata- 700091, India*

Abstract

Marine shrimp (*Penaeus monodon*) is one of the major candidate species for export oriented aquaculture which dominates the seafood market in regions of European Union, Southeast Asia and USA. Carotenoid rich seafood has now become one of the important criteria in determining the quality of the product to be exported. Recent trends in supplementing fish diets with natural pigment source is an alternative to the utilization of expensive synthetic pigments. In this context, green seaweed *Enteromorpha intestinalis* was selected as a source of carotenoid for inclusion in the formulated diet of *Penaeus monodon*. Astaxanthin being an important category of carotenoid pigment was monitored in shrimp muscle tissue during the feeding trial. Significant variation ($p<0.05$) was observed between the experimental groups as confirmed through ANOVA thus exhibiting higher astaxanthin content of shrimps (18.70 ± 4.48 ppm) fed with seaweed incorporated diet as compared to control (15.80 ± 2.33 ppm). The present experiment therefore emphasizes on the quality improvement of aquaculture product through dietary inclusion of seaweed as a source of astaxanthin.

Keywords: *Penaeus monodon*; Seaweed; Astaxanthin content; Quality improvement

Introduction

Inclusion of algae as dietary supplement in animals have been investigated by researchers for years as a source of pigment [1]. In context to aquaculture, the effects of dietary inclusion of algae have resulted in improved performance including better animal product quality [2-4]. To be more specific the commercial production of shrimps and prawns as an edible item represents one of the fastest growing areas of aquaculture [5] with high consumer appeal and attractive market due to their body colouration or pigmentation which is a direct measure of its body astaxanthin content. Carotenoid utilization by various aquatic species is well documented since it plays a regulatory role in providing antioxidant and pro-vitamin 'A' activity, enhancing immune response, improved reproductive performance, growth, maturation and photo-protection [6]. These natural compounds also help the species to combat against stressful environment [7]. Algae and higher group of plants are the major producers of carotenoid which comprises a family of over 600 natural fat soluble pigments [8]. Studies reveal that alternative utilization of plant pigments in formulated diets have improved the body pigmentation of farmed crustaceans, particularly penaeids in order to achieve better market price [9,10].

Therefore the present work is an attempt to utilize *E. intestinalis* as a natural astaxanthin source for farmed tiger shrimp (*Penaeus monodon*) in relation to its quality improvement.

Materials and Methods

Collection of seaweed and preparation of experimental diets

Live and healthy *E. intestinalis* was collected from Bali Island (22°04'35.17" N latitude and 88°44' 55.70" E longitude) of Sundarbans mangrove eco-region during low tide condition. The collected material was rinsed in ambient water and then with distilled water, oven-dried at 50°C and finally processed to make powder. Experimental diet was formulated through incorporation of the particular seaweed (Diet ENT) at a level of 5%. Simultaneously a control diet (Diet C) was also formulated to study the comparative performance (Table 1).

Feeding trial

A feeding trial was run at experimental aquaria for 90 days of duration. Shrimp juveniles were procured from hatchery and stocked in aquaria at a density of 2 individuals/m^2. Experimental diets were randomly assigned, the shrimps were fed twice daily and the uneaten feed was checked at regular intervals.

Analysis of astaxanthin content in shrimp muscle tissue

The astaxanthin content was analyzed according to the standard spectrophotometric method [11]. Its value in % was converted to ppm level for easy interpretation of data. The body colouration of shrimps after boiling was compared with Roche SalmoFan™ colour score.

Ingredients	Diet C	Diet ENT
Fish meal	35	30
Soybean oil cake	11	11
Mustard oil cake	11	11
Rice polish	23	23
Wheat flour	16	16
Oyster shell dust	2	2
Shark oil	2	2
Enteromorpha intestinalis (as a source of natural astaxanthin)	-	5

Table 1: Formulation of experimental diets.

***Corresponding author:** Kunal Mondal, Faculty of Fishery Sciences, West Bengal University of Animal and Fishery Sciences, Kolkata-700094, India, E-mail: chow@affrc.go.jp

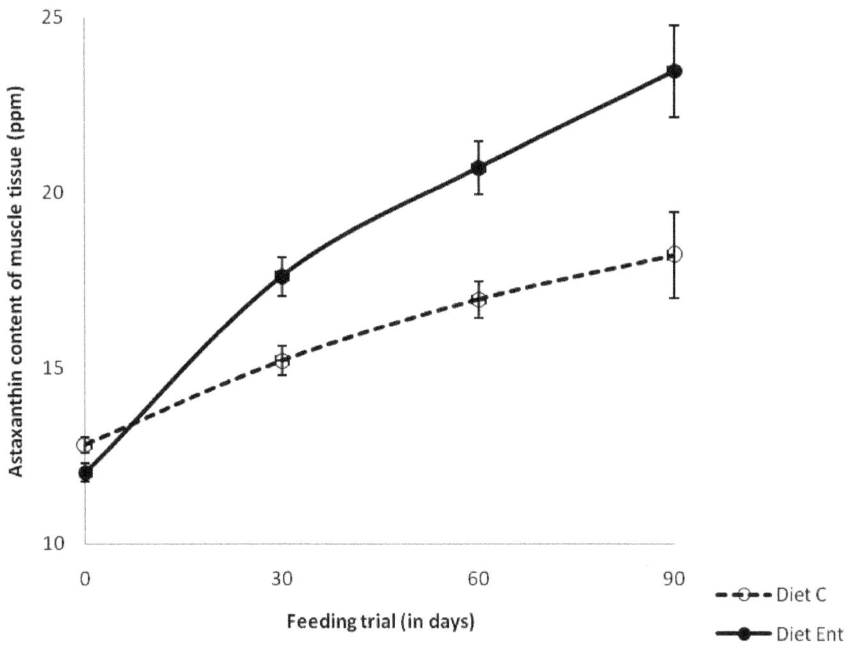

Figure 1: Variations in astaxanthin content of shrimp muscle tissue (ppm) throughout the feeding trial.

Statistical analysis

The collected data were finally subjected to one-way analysis of variance (ANOVA). All statistical calculations were performed with SPSS 9.0 for Windows.

Results and Discussion

The astaxanthin content in muscle tissue was higher in shrimps fed with Diet ENT as compared to Diet C throughout the feeding trial (Figure 1). A darker orange-red colouration was observed in shrimps fed with Diet ENT after boiling them in water for 5 minutes when compared with Roche SalmoFan™ colour score. The colour score was 30 for Diet ENT fed shrimps whereas a score of 27 was recorded from shrimps fed with Diet C. ANOVA results showed significant variation ($p<0.05$) in average astaxanthin contents which may be attributed to the capability of *P. monodon* to easily convert the fraction of seaweed astaxanthin into tissue astaxanthin. The seaweed sample selected for the investigation is found to contain ~120.78 ppm astaxanthin as reported from the present study region [12-14]. Carotenoid, particularly astaxanthin content of feed is one of the major factor influencing the colour development in animals [15] but at the same time scientific knowledge about several factors like dietary pigment source, their dosage level, feeding duration, dietary composition and magnitude of carotenoid esterification is also required to identify these interaction processes [16-20]. The study showed significantly different astaxanthin content of the shrimp species which are in agreement to the observations that crustaceans exhibit strong tendency towards selection of specific carotenoids at a specific rate for their metabolic absorption [16]. Similar work conducted from the present study region reveal that *P. monodon* when fed with diet containing red algae *Catenella repens* at a level of 5% improved the body astaxanthin content [21]. The search for natural astaxanthin was not only limited to the seaweed resources, rather salt-marsh grass *Porteresia coarctata* was also tested as a natural dietary astaxanthin source in *P. monodon* feed with better results from the present geographical locale [12,22]. In continuation, such natural carotenoid supply to the diet of shrimps has been studied for *P. japonicus* and *Litopenaeus vannamei* too from different parts of the globe. The ingredients of natural origin that have been used in the diet are red yeast (*Phaffia rhodozyma*) and microalgae *Dunaliella salina* [23]; *Chnoospora minima* [24]; *Spirulina* sp. [25,26]; *Haematococcus pluvialis* [26] and *Isochrysis galbana* [27]. An usual trend of marked increase in the body carotenoid content has been observed when organisms were fed with plant pigment source diets. For e.g. feed supplemented with 50 ppm algal material (*Dunaliella salina*) improved the body colouration of *P. monodon* [28]. Three types of diet when provided to *P. semisulcatus* containing natural carotenoid sources like red pepper and marigold flower resulted in higher carotenoid accumulation in body tissues [29]. However researchers from Mexico also reported that feed incorporated with cultivated green alga *Ulva clathrata* significantly improved the body pigmentation of farmed shrimp *L. vannamei* [30].

Conclusion

Improved product quality of *P. monodon* clearly reflects the transforming potential of algal astaxanthin into the body tissues by the particular culture species. Thus the present study provides a baseline information about the natural astaxanthin pool of Sundarbans mangrove eco-region which may serve as an alternative to synthetic astaxanthin in animal diets that are much more expensive in terms of costing.

Acknowledgement

The authors would like to acknowledge Department of Science and Technology, Govt. of India for providing financial assistance.

References

1. Strand A, Herstad O, Liaaen-Jensen S (1998) Fucoxanthin metabolites in egg yolks of laying hens. Comparative Biochemistry and Physiology 119: 963-974.
2. Moss SM (1994) Growth rates, nucleic acid concentrations and RNA/DNA ratios of juvenile white shrimp, *Penaeus vannamei* Boone, fed different algal diets. Journal of Experimental Marine Biology and Ecology 182: 193-204.

3. Penaflorida VD, Golez NV (1996) Use of seaweed meals from *Kappaphycus alvarezii* and *Gracilaria heteroclada* as binders in diets of juvenile shrimp *Penaeus monodon*. Aquaculture 143: 393- 401.

4. Cruz-Suarez LE, Ricque-Marie D, Tapia-Salazar M, Guajardo-Barbosa C (2000) Uso de harina de kelp (*Macrocystis pyrifera*) en alimentos para camaro´ n, in Avances en Nutricio´n Acuı´cola V - Memorias del Quinto Simposium Internacional de Nutricio´n Acuı´cola, Me´rida, Me´xico. Universidad Auto´noma de Nuevo Leon, Monterrey, Mexico. p. 227-266.

5. Rosenberry B (2005) World shrimp farming 2005. Shrimp News International p. 276.

6. Howell BK, Matthews AD (1991) The carotenoids of wild and blue disease affected farmed tiger shrimp (*Penaeus monodon*, Fabricus). Comparative Biochemistry and Physiology 98: 375-379.

7. Meyers SP (1994) Developments in world aquaculture, feed formulations and role of carotenoids. Pure and Applied Chemistry 66: 1069-1076.

8. Britton G, Armitt GMS, Lau YMA, Patel K, Shone CC (1981) Carotenoproteins, in Carotenoid Chemistry and Biochemistry (Pergamon press). Oxford: p. 237-251.

9. Lorenz T (1998) A review of the carotenoid astaxanthin as a pigment source and vitamin for cultured Penaeus prawn. NatuRosea Technical Bulletin 51: 1-7.

10. Liao IC, Chien YH (1994) Culture of kuruma prawn *Penaeus japonicus* in Asia. World Aquaculture 25: 18-33.

11. Schuep W, Schierle J (1995) Astaxanthin determination of stabilized, added astaxanthin in fish feeds and pre-mixes, in Carotenoids Isolation and Analysis (Birkhauser Verlag Basel). p. 273-276.

12. Mitra A, Ghosh R, Mallik A, Mondal K, Zaman S, Banerjee K (2013) Study on the role of mangrove based astaxanthin in shrimp nutrition, in Sensitivity of Mangrove Ecosystem to Changing Climate (Springer Publishers, India). Springer: p 297-310.

13. Banerjee K, Ghosh R, Homechaudhuri S, Mitra A (2009) Biochemical composition of marine macroalgae from Gangetic Delta at the Apex of Bay of Bengal. African Journal of Basic and Applied Science 1: 96-104.

14. Chakraborty S, Santra SC (2008) Biochemical composition of eight benthic algae collected from Sunderban. Indian Journal of Marine Sciences 37: 329-332.

15. Moretti VM, Mentasti T, Bellagamba F, Luzzana U, Caprino F, et al. (2006) Determination of astaxanthin stereoisomers and colour attributes in flesh of rainbow trout (*Oncorhynchus mykiss*) as a tool to distinguish the dietary pigmentation source. Journal of Food Additives and Contaminants 23: 1056-1063.

16. Meyers SP, Latscha T (1997) Carotenoids, in Crustacean Nutrition, Advances in World Aquaculture (Baton Rouge, LA, USA). p. 164-193.

17. Bjerkeng B (2000) Carotenoid pigmentation of salmonid fishes - recent progress, in Advances en Nutricio´n Acuı´cola V - Memorias del Quinto Simposium Internacional de Nutricio´n Acuı´cola, Me´rida, Me´xico, Universidad Auto´noma de Nuevo Leo´ n, Monterrey, Mexico. p. 89.

18. Buttle LG, Crampton VO, Williams PD (2001) The effect of feed pigment type on flesh pigment deposition and colour in farmed Atlantic salmon, *Salmo salar* L. Aquaculture Research 32: 103-111.

19. Gomes E, Dias J, Silva P, Valente L, Empis J, et al. (2002) Utilization of natural and synthetic sources of carotenoids in the skin pigmentation of gilthead sea bream (*Sparus aurata*). Journal of European Food Research Technology 214: 287- 293.

20. White DA, Page GI, Swaile J, Moody AJ, Davies SJ (2002) Effect of esterification on the absorption of astaxanthin in rainbow trout, *Oncorhynchus mykiss* (Walbaum). Aquaculture Research 33: 343-350.

21. Banerjee K, Mitra A, Mondal K (2010) Cost-effective and eco-friendly shrimp feed from red seaweed *Catenella repens* (Gigartinales: Rhodophyta). Current Biotica 8: 23-43.

22. Mitra A, Mondal K, Banerjee K (2011) Effect of salt-marsh grass (*Porteresia coarctata*) diet on growth performance of black tiger shrimp, *Penaeus monodon*, in Diversification of Aquaculture (Narendra Publishing House, New Delhi), p. 219-232.

23. Chien YH, Jeng S (1992) Pigmentation of kuruma prawn *Penaeus japonicus* Bate, by various pigment sources and levels and feeding regimes. Aquaculture 102: 333-346.

24. Menasveta P, Worawattanamateekul W, Latscha T, Clark JS (1993) Correction of black tiger prawn (*Penaeus monodon* Fabricus) colouration by astaxanthin. Aquaculture Engineering 12: 203-213.

25. Liao WL, Nur-E-Bohran SA, Okada S, Matsui T, Yamaguchi K (1993) Pigmentation of cultured black tiger prawn by feeding with a *Spirulina* supplemented diet. Nippon Suisan Gakkaishi 59: 165-169.

26. Chien YH, Shiau WC (1998) The effects of *Haematococcus pluvialis*, *Spirulina pacifica* and synthetic astaxanthin on the pigmentation, survival, growth and oxygen consumption of kuruma prawn, *Penaeus japonicus* Bate, in Book of Abstracts of World Aquaculture '98 Baton Rouge (World Aquaculture Society, LA, USA), p. 156.

27. Pan CH, Chen YH, Chen JH (2001) Effects of light regime, algae in water and dietary astaxanthin on pigmentation, growth and survival of black tiger prawn *Penaeus monodon* post-larvae. Zoological Studies 40: 371-382.

28. Boonyaratpalin M, Thongrod S, Supamattaya K, Britton G, Schlipalius LE (2001) Effects of β-carotene source, *Dunaliella salina*, and astaxanthin on pigmentation, growth, survival and health of *Penaeus monodon*. Aquaculture Research 32: 182-190.

29. Gocer M, Yanar M, Kumlu M, Yanar Y (2006) The effects of red pepper, marigold flower and synthetic astaxanthin on pigmentation, growth and proximate composition of *Penaeus semisulcatus*. Turkish Journal of Veterinary and Animal Sciences 30: 359-365.

30. Cruz-Sua´rez LE, Tapia-Salazar M, Nieto-Lo´pez MG, Guajardo-Barbosa C, Ricque-Marie D (2009) Comparison of *Ulva clathrata* and the kelps *Macrocystis pyrifera* and *Ascophyllum nodosum* as ingredients in shrimp feed. AquacultureNutrition 15: 421- 430.

Last "Atypical" Beaked Whales Mass Stranding in the Canary Islands (July, 2004)

A Fernández[1]*, E. Sierra[1], V. Martín[2], M. Méndez[1], S. Sacchinni[1], Y. Bernaldo de Quirós[1], M. Andrada[1], M. Rivero[1], O. Quesada[1], M. Tejedor[2] and Arbelo M[1]

[1]*Veterinary Histology and Pathology, Institute of Animal Health, Veterinary School, University of Las Palmas de Gran Canaria, Canary Islands, Spain*
[2]*Society for the Study of Cetaceans in the Archipelago Canary (SECAC) – Canary Islands, Spain*

Abstract

In July 2004, four *Ziphius cavirostris* stranded in the Canary Islands several days after an international naval exercises were conducted north of the Canary Islands. During the maneuvers high intensity mid-frequency sonar was used. Three of the animals were fully necropsied. Abundant fresh non-digestive aliment was found in all stomachs. Hemorrhages were a constant finding in several organs. Although "*in vivo*" gas embolism could not be established due to decomposition, systemic fat embolism was diagnosed in all three beaked whales. Epidemiological and pathological findings were highly consistent with an "atypical" beaked whale mass stranding that was temporally and spatially associated with sonar. This was the last atypical mass stranding in the Canary Islands once an anti-sonar moratorium was established around the islands, following the EU parliament recommendation and Spanish government resolution in 2004.

Introduction

The possibility that exposure to sonar stimuli could lead to the stranding of cetaceans is an issue of environmental concern because of cetacean stranding and deaths temporally and geographically coincident with naval sonar operations [1,2]. Beaked whales (BWs) (family *Ziphiidae*), one of the deepest-diving whales, are of particular interest because of a notable relationship between military operations employing mid-frequency sonar and BW mass stranding [1,2].

Canary Islands have been a place where more atypical mass stranding associated with naval maneuvers have occurred before 2004 [1,3,4]. Some common epidemiologic factors in this type of mass stranding were: (1) the oceanographic features were characteristic of steep-slope regions; (2) the species involved were predominantly BWs; and (3) these events were temporally associated with naval maneuvers that employed high intensity, low- or mid frequency range sonar signals [3,5].

Fourteen BWs were stranded in the Canary Islands close to the site of an international naval exercise (Neo-Tapon 2002) held on 24 September 2002. A detailed pathological study revealed unusual "bubble" lesions and fat emboli in most of the necropsied beaked whales. These findings constituted a new pathological entity in marine mammals consistent with decompression sickness in human divers [6,7] and was most likely due to a severe changes in the dive behavior of these deep divers [3,6-8]. Epidemiological features, pathological studies, gas analysis and naval exercises information are considered the main diagnostic parameters to approach a diagnosis of this type of atypical mass stranding [3,4,7,9].

Here we report the last beaked whale mass stranding occurred in the Canary Islands in July 2004, associated with naval exercises and antisubmarine sonar activities before an anti-sonar moratorium was recommended by European Union parliament and established around the islands by the Spanish and Canary Islands governments in 2004 [10,11].

Material and Methods

A military international naval exercises (MEDSHARK/Majestic Eagle '04 - MS/ME04) involving many warships, aircraft and submarines from ten allied countries took place north of Canary Islands from 11th through 15th July (2004).

Information about the maneuvers indicated that several warships used mid-frequency sonar (between 2-10 kHz) during the naval maneuvers (Figure 1). Sonar was used in different areas and days to the north of Fuerteventura and Lanzarote islands at 70 nm (on the 11th and 12th July), 85 nm (13th), 60 nm (14th and 15th of July) and at 240 nm on the 16th of July (MEDSHARK/Majestic Eagle '04 - MS/ME04).

Four beaked whales (*Z. cavirostris*) were found either floating close to or stranded on Lanzarote and Fuerteventura (the two closest islands to the exercises area) in a six days period from 21st until 26th July (Figure 1). Apart from the last carcass found, all three stranded whales were necropsied [7] and sampled for histological studies as well for microbiological analyses. Samples consisted of skin and fat, skeletal muscle, lung, liver, spleen, lymph nodes and kidney. Other organs were too decomposed for being analyzed.

Results

A 525 cm long male beaked whale (*Z. cavirostris*) showing good body condition and several postmortem shark bites was the first whale to be found on 21st of July in (Los Ancones, 29° 00' 53" N; 13° 28' 21" W) Lanzarote Island. The animal showed a code 3 (advanced autolysis). Necropsy showed emphysematous lung lobules and hemorrhages in different digestive organs including acoustic fat. Kidney presented parasites (*Crassicauda spp.*), and the stomach contained abundant partially digested squids.

***Corresponding author:** A Fernández, Veterinary Histology and Pathology, Institute of Animal Health, Veterinary School, University of Las Palmas de Gran Canaria, Canary Islands, Spain, E-mail: afernandez@dmor.ulpgc.es

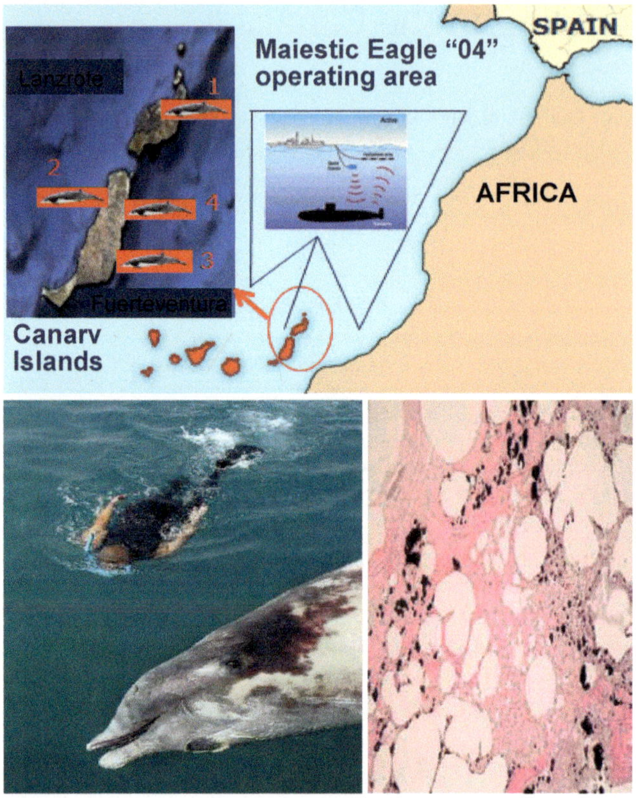

Figure 1: (a) Naval exercises operating area. On the left side, Fuerteventura and Lanzarote islands with the locations where each beaked whale stranded. **(b)** A beaked whale floating near shore and a diver approaching the animal to fix it in order to bring it to the coast. **(c)** A histological picture of a thoracic lymph node showing numerous gas bubble together abundant fat globules of different size (Osmium post-fixation technique x 200). (Anti-Clock wise).

On the 23rd of July, a 500 cm long female beaked whale was found floating near shore in Fuerteventura Island (Nuevo Horizonte, 28º 24' 48" N; 13º 50' 53" W). The cetacean presented a good nutritive condition in an advanced autolytic status (code 4). The animal presented no caudal fin with several shark bites in the abdominal area as well as in the melon. During necropsy, abundant fresh undigested food (cephalopods and crustaceans) was present within the stomach compartments. Numerous parasites (*Crassicauda spp.*) were identified in the kidneys, and multiple calcified and fibrotic nodules were identified in the mesenteric arteries. The liver was very flaccid with enlarged venous vessels. Lung lobules were very emphysematous with gas-filled bullae, very likely due to postmortem autolysis. Hemorrhages were observed in the mandibular acoustic fat.

On the 24th of July, a 575 long male *Z. cavirostris* was found stranded in Fuerteventura Island (Majanicho, 28º 44' 21" N; 13º 56' 06" W) a few kilometers away from the previous stranding. Nutritive condition was similar to first beaked whale, and the carcass also showed severe shark bites especially in the abdominal cavity and head. During necropsy, this whale showed similar gross observations as in the previous described whale, including a large quantity of undigested fresh food in the stomach. Part of the digestive and reproductive tract was not present due to shark predation at sea.

Finally, four days later, on the 26th an extremely putrefactive beaked whale of undetermined sex was found beached in Fuerteventura Island (Puerto Lajas, 28º 32' 25" N; 13º 49' 75" W). The carcass was considered code 5 (advanced decomposition) and therefore too autolysed for necropsy.

Histologically, no inflammatory or neoplastic disease potentially responsible for the cause of death was detected in any organs from the three investigated beaked whales. Emphysematous areas in the lung parenchyma as well as within the lymph nodes were demonstrated histologically. Gas bubbles were microscopically detected within many vessels or tissues, but it was not possible to determine if these were due to putrefaction or some other process. However, using osmium post fixation technique, fat emboli were detected intravascularly in the lungs, kidneys, liver sinusoids and within the lymph nodes in all three stranded whales.

No herpes virus or *Morbillivirus* were detected molecularly by PCR. Isolated enterobacteria from these carcasses were not considered as bacterial postmortem proliferation and no pathogenic ones.

Discussion

Based on the stranding data, this event was defined as an "atypical" beaked whale mass stranding [2], involving *Z. cavirostris*, the most common whale species associated with this kind of stranding. As before it was also temporally and spatially associated with naval activities using high intensity mid-frequency sonar [3,4].

It was suggested that, based on the sea currents during and after the naval exercises, a carcass of that weight would be driven south from the area of the naval exercise to the northern part of the Canary Islands (Lanzarote and Fuerteventura), where these four carcasses eventually beached. This would suggest that beaked whales were originally quite close to the area where naval exercises took place when alive. It would also suggest that the otherwise healthy BWs died relatively quickly at sea before stranding dead on the beach or found floating in seawater. This indicates that beaked whale mass mortalities linked to naval sonars may have directly lethal effects without the need for stranding alive on the beach [5].

These three cases also support the observation (Fernandez, personal communication) that *Z. cavirostris* as a species always appear to float when the body is dead in the water, in contrast to other species from the same *Ziphiidae* family, like *M. densirostris* or *M. bidens* which may sink after death. This observation suggests that other some acoustically-impacted BW species may die at sea and subsequently sink, so the real impact of sonar on this family may be under-represented by stranding.

From the pathological point of view, the high amount of "gas bubble" found in the three necropsied whales could not be considered as an evidence of "gas embolism" since all three animals showed an advanced postmortem autolysis and no gas analysis methodology was then available as it is now [9]. However, the evidence of fat emboli in these beaked whales together with the fact that all three died at sea, are highly suggestive that fat emboli was not due to trauma during stranding but was triggered and developed at sea and probably contributing to the whales relatively rapid death. This also shows that fat embolism is an important pathological finding that can still be investigated even in moderate to advanced autolytic stranded cetaceans [7,12].

The results from mortalities occurred in 2002 and before [3,7] and the results present here provide scientific support for a non-binding EU parliament resolution in 2004 ([http://www.nrdc.org/media/pressreleases/050210b.asp] and [http://www.nrdc.org /media/docs/050210a.pdf]) on the deployment of high intensity sonars until a global assessment of its cumulative environmental impacts on whales, dolphins, fish and other marine life is completed [10].

Following this resolution the Spanish parliament decided to establish a sonar moratorium around the Canary Islands [11] as a preventive measure to avoid these whales to be affected during naval exercises, mainly in Fuerteventura coasts (Canary Islands). After seven years of this moratorium (2004), we can now present here the last beaked whale mass stranding linked to naval exercises with sonar activities in the Canary Islands occurred in July 2004. No military sonar activities has been deployed since then and no new atypical beaked whales mass stranding has occurred in the Canary Islands from 2004 until 2011. This undoubtedly represents the best supportive evidence for the EU parliament resolution [10] and Spanish moratorium [11] in 2004 when the last mass stranding occurred in the Canary Islands.

Acknowledgment

This work was supported by the Spanish Ministry of Science and Innovation with two research projects: (AGL 2005-07947), and (CGL 2009/12663), as well as the Government of Canary Islands, project, SolSub C200801000288. Cabildo de Fuerteventura (Antonio Gallardo) and Reserva de la Biosfera de Fuerteventura.

References

1. Simmonds MP, Lopezjurado LF (1991) Whales and the military. Nature 351: 448.
2. Frantzis A (1998) Does acoustic testing strand whales? Nature 392: 29.
3. Cox TM, Ragen TJ, Read AJ, Vose E, Baird RW, et al. (2006) Understanding the impacts of anthropogenic sound on beaked whales. J Cetacean Res Manage 7: 117-187.
4. Amico AD, Gisiner RC, Ketten DR, Hammock JA, Johnson C, et al. (2009) Beaked whale strandings and naval exercises. Aquatic Mammals 35: 452-472.
5. Evans DL, England GR (2001) Joint Interim Report: Bahamas Marine Mammal Stranding Event of 15-16 March 2000. US Department of Commerce and US Navy: Washington DC.
6. Jepson PD, Arbelo M, Deaville R, Patterson IA, Castro P, et al. (2003) Gas-bubble lesions in stranded cetaceans. Nature 425: 575-576.
7. Fernández A, Edwards JF, Rodríguez F, Espinosa de los Monteros A, Herráez P, et al. (2005) "Gas and fat embolic syndrome" involving a mass stranding of beaked whales (family Ziphiidae) exposed to anthropogenic sonar signals. Vet Pathol 42: 446-457.
8. Tyack PL, Zimmer WM, Moretti D, Southall BL, Claridge DE, et al. (2011) Beaked whales respond to simulated and actual navy sonar. PLoS One 6: e17009.
9. de Quirós YB, González-Díaz O, Saavedra P, Arbelo M, Sierra E, et al. (2011) Methodology for in situ gas sampling, transport and laboratory analysis of gases from stranded cetaceans. Sci Rep 1: 193.
10. European-Parliament (2004) European Parliament resolution on the environment effects of high-intensity naval sonars.
11. Daily sessions of the Congress of Deputies. Cdl Deputies, Editor 2004. File Number 180/000 345.
12. Jepson PD, Deaville R, Patterson IA, Pocknell AM, Ross HM, et al. (2005) Acute and chronic gas bubble lesions in cetaceans stranded in the United Kingdom. Vet Pathol 42: 291-305.

Feeding Ecology and Prey Preference of Grey Mullet, *Mugil cephalus* (Linnaeus, 1758) in Extensive Brackish Water Farming System

Asish Mondal[1,2]*, Deepta Chakravortty[2], Susmita Mandal[1], Bhattacharyya SB[1,2] and Abhijit Mitra[2]

[1]*Kakdwip Research Centre of Central Institute of Brackish water Aquaculture Kakdwip, South 24 Parganas, West Bengal-743 347, India*
[2]*Department of Oceanography, Techno India University, Salt Lake campus, Kolkata, India*

Abstract

Feeding ecology and prey preference of *Mugil cephalus* reared in extensive brackish water farming system was studied based on monthly examination of stomach contents over a period of 10 months (February-November, 2014). Feeding intensity, planktonic constituents in water and food constituents in the stomach were estimated numerically. The percentage compositions of food items in the stomach falling under different groups were then compared with that of fish pond to evaluate prey preferences. Lower feeding intensity was observed during the initial months which gradually increased as the fish's grew. The dominant phytoplankton groups in pond water according to the order of dominance were Chlorophyceae, Bacillariophyceae and Myxophyceae. Most abundant zooplankton group was Dinoflagellates followed by Copepods. The main phytoplankton groups found in the stomach according to the order of dominance were Bacillarophyceae, Chlorophyceae and Myxophyceae. Dominant zooplankton group in the stomachs were Dinoflagellets followed by copepods. Prey preference analysis revealed that *M. Cephalus* actively selected Bacillariophyceae as most preferred food material. Myxophyceae was also selected as second preference. Chlorophyceae was not at all selected and was probably swallowed mechanically during intake of other food stuffs. True positive selection of copepods during initial months of rearing indicates preference by *M. cephalus* juveniles.

Keywords: *Mugil cephalus*; Extensive farming; Feeding strategy; Prey preference; Sundarbans

Introduction

Feeding is the dominant activity in entire life cycle of fish; thus detailed knowledge on food and feeding habit of any fish is essential to get success in culture of that particular fish species. Food and feeding habits of a species of fish is intimately associated with the ecological niche that they occupy in the natural environment [1] and knowledge on this aspect is advantageous in their proper management and exploitation [2]. Quantitative and qualitative changes in fish food during the life span are useful tools to define the diet of a particular fish species [3,4]. Fish feed opportunistically yet are often selective in their diets [5-7]. The main factors that determine the type of prey are feeding preference [8-10], availability of prey, prey mobility and its distribution in the water column, catching efficiency of the predator, water temperature and turbidity [11]. Changes in feeding habits of a fish species are a function of the interactions among several environmental factors that influence the selection of food item [12]. Feeding behavior at the level of prey selection can have implications at the individual [13], population [14], and community levels [15].

Flathead grey mullet (*Mugil cephalus* L.) is an economically important euryhaline and eurythermal marine teleost contributing to sizable fisheries of estuarine and coastal regions in many countries [16] and has a cosmopolitan distribution between latitude 40°N and 40°S covering all the oceans [17-19]. The overall share of Aquaculture was 2.6% in the total production of Marine fishes and it was contributed substantially by flathead grey mullet (*M. cephalus*) as one of the species [20]. It is an excellent candidate species for both mono and polyculture as it feeds at the lower trophic level grazing on plant detritus and microflora [21,22]. This species is widely cultured in both brackish and freshwater semi-intensive fishponds [23] and can utilize both supplemental feed and/or natural food [24]. Studies on food and feeding habit of stripped grey mullet worldwide indicates that the food spectrum is mainly constituted with Diatoms, green and blue green algae, Dinoflagellates, Protozoa, macro plant detritus, Copepods, Nematodes and Foraminifera according to the order of dominance [25-31].

In India, *M. cephalus* occurs in the shallow coastal waters, lagoons, the coastal lakes and estuaries [32] and is mainly reared by polyculture method using seed collected from the wild and is restricted to traditional system [33] in extensive poly-farming impoundments. Studies on the food and feeding habits of *M. cephalus* in India were done earlier in Lake Pulicat [34], Thana creek [35], Gosthani estuary [36] and the East coast of Andhra Pradesh [37]. No such study has been reported from Hooghly-Matla estuarine complex, popularly known as 'Sundarbans' in spite of being a highly potential and ecologically important area where *M. cephalus* is an important component of extensive poly-farming. This study aimed to assess food and feeding ecology along with prey selectivity of grey mullet in extensive poly-culture system in Sundarbans as representative of natural environment.

Materials and Methods

The study was carried out at Gopalnagar Dakshin village (21.8029-21.8073°N, 88.2962-88.2985°E) of Patharpratima block in South 24 Parganas district of West Bengal, India from February to November, 2014. Three tide fed extensive ponds (0.7-0.9 ha) locally known as '*Bhery*' situated at the bank of a creek of Hatania-Doania river were

***Corresponding author:** Asish Mondal, Kakdwip Research Centre of Central Institute of Brackish water Aquaculture Kakdwip, South 24 Parganas, West Bengal-743 347, India, Tel: +9007806599; E-mail: asish177@gmail.com

selected. Ponds were dewatered and sun dried during December. During first week of January, unfiltered saline tidal water (18.2 ppt) was allowed to let in and the ponds were filled up to a depth of 110 cm and bamboo screen were fitted at the inlet. The traditional bamboo screen at the inlet allows entry of fry of different species during water exchanges but restricts exit of bigger individuals. No feed or fertilizer was applied following traditional practice. Entry of grey mullet fry along with other species was anticipated as seeds of *M. cephalus* remain available in north-east coast of India during January–March [16]. About 20-30% water was exchanged every lunar cycle following the common practice throughout the rearing period. Water and fish samples were collected from three ponds to eliminate any possible biasness. Both fish and water samples were frizzed in ice before those were carried to laboratory for subsequent analysis.

Water temperature, salinity, pH, dissolved oxygen (DO), nitrite-nitrogen (NO_2-N), nitrate-nitrogen (NO_3-N), ammonia-nitrogen (NH_3-N)and phosphate-phosphorus (PO_4-P) were analyzed following standard methods [38] from pond water samples collected between 09:00 and 10:00 hours at monthly interval. Salinity was recorded using a refractometer (ATAGO, Japan). Plankton samples were collected monthly by filtering 50 L of water through bolting silk plankton net (mesh size 64 μm). Plankton concentrates were immediately preserved in 5% buffered formalin for further qualitative and quantitative analysis.

From each three ponds, 10 fish were collected during middle of each month i.e., 30 fish in a month and total 300 fish were collected and analyzed throughout 10 months study period. Gravimetric data, namely, total length (TL, mm) was recorded with a slide caliper, while body weight (W, g) was measured using a digital electronic balance. The stomachs were removed intact by cutting above the cardiac and below the pyloric sphincters and preserved in a vial with 4% formalin. The stomach fullness degree was estimated and categorized as empty, ¼ full, ½ full, ¾ full, full and gorged [39]. Dissection entailed making an incision above the longitudinal axes of the stomach and intestine and stomach contents were transferred to a fixed volume of 4% formalin. Three 1ml sub samples of water and each stomach were then transferred to Sedgwick-rafter counting cell and planktonic constituents were counted and identified [40,41]. Food items of water and stomachs were grouped as Chlorophyceae (green algae), Bacillariophyceae (diatoms), Myxophyceae (blue-green algae), Dinoflagellates, Copepods and fish or shrimp parts and numeric percentages of each group were calculated. Additionally, organic matter and sand particles in stomach were also evaluated as major constituents. The percentage compositions of food items in the stomach falling under different groups were then compared with that of fish pond to evaluate prey preferences. Prey preferences were determined by the Ivlev Electivity Index [42] using the following formula:

$$E = \frac{r-p}{r+p}$$

Where, r=percentage of dietary item in ingested food, p= percentage of prey in the environment.

Results

Water quality parameters of the three studied ponds are presented in Table 1. Salinity showed wide variations in all three ponds throughout the culture duration and was highest during May and lowest during September. Temperature ranged between 18.7 and 33.1°C. Highest temperature was recorded during April and lowest during November. Dissolved oxygen (DO) and pH remained almost steady throughout the rearing period and ranged between 5.80 to 9.10 ppm and 7.92 to 8.72, respectively. Those were similar in pond 1 and 2, however, significantly ($p<0.05$) lower pH and DO were observed in pond 3. Concentrations of toxic nitrogenous metabolites like nitrite-nitrogen (NO_2-N) and total ammonia nitrogen (NH_3-N) varied between 9.33-24.47 and 21.83-44.08 μg/l, respectively considering all the ponds. Significantly ($p<0.05$) higher NH_3-N concentration was observed in pond 3 while it was similar in pond 1 and 2. Concentrations of nutrients like nitrate-nitrogen (NO_3-N) and phosphate-phosphorous (PO_4-P) ranged between 69.62-111.04 and 21.58-43.27 μg/l, respectively showing no significant difference among ponds. Both phytoplankton and zooplankton concentration showed marked difference ($p<0.05$) among ponds with highest value in pond 1 and lowest in pond 3 (Figure 1). The dominant phytoplankton groups in pond water according to the order of dominance were Chlorophyceae, Bacillariophyceae and Myxophyceae in all ponds. Most abundant zooplankton group was Dinoflagellates followed by Copepods, fish and shrimp parts were also present in the water column (Figure 2). Among Chlorophyceae, *Pediastrum, Chlorella* and *Tetraedron* were found to be the most abundant genera. In addition, *Ankistrodesmus, Coilastrum, Crucigenia, Scenedesmus* and *Pandorina* were recorded. Percentage composition of Chlorophyceae varied between 31.49-53.52% with highest and lowest value during May and September, respectively. The most abundant genera of Bacillariophyceae were *Navicula, Nitzschia, Cyclotella* and *Melosira*. Other genera of Bacillariophyceae were *Gyrosigma, Cymbella,* and *Synedra*. Bacillarophyceae constituted 4.06-15.36% of planktonic forms. *Anabaena* and *Oscillatoria* were the dominant genera among

Water parameters	P_1	P_2	P_3
Water temperature (°C)	29.9 ± 1.7	29.9 ± 1.7	29.7 ± 1.9
pH	8.04 ± 0.23^a	7.96 ± 0.25^a	7.78 ± 0.31^b
DO (mg L^{-1})	6.06 ± 0.42^a	5.99 ± 0.52^a	5.69 ± 0.52^b
Salinity (ppt)	12.87 ± 5.34	12.74 ± 5.32	12.89 ± 5.19
NO_2-N (μg L^{-1})	16.35 ± 5.83	15.91 ± 5.62	16.11 ± 6.63
NO_3-N (μg L^{-1})	93.12 ± 15.41	92.66 ± 11.14	92.97 ± 8.94
NH_4-N (μg L^{-1})	30.96 ± 5.61^b	31.19 ± 7.91^b	34.89 ± 6.27^a
PO_4-P (μg L^{-1})	32.07 ± 13.43	31.91 ± 11.98	31.89 ± 12.74
Phytoplankton (numbers/L^{-1} ×10^3)	15.38 ± 1.62^a	15.12 ± 1.94^b	14.95 ± 1.73^c
Zooplankton (numbers/L^{-1}×10^3)	3.05 ± 0.25^a	2.91 ± 0.23^b	2.83 ± 0.17^c

Means bearing different superscripts indicate statistically significant differences in a row ($p < 0.05$); Values are expressed as mean ± SE (n=10 for each impoundments every month).

Table 1: Water quality parameters of three extensive brackish water farming impoundments used for grey mullet feeding ecology study.

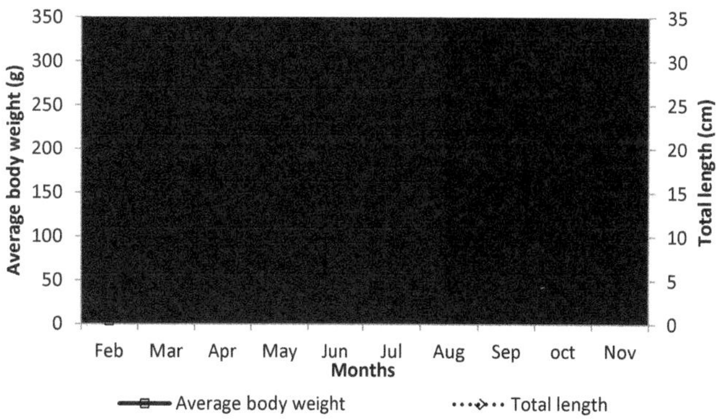

Figure 1: Growth of grey mullet (*Mugil cephalus*) in extensive farming system.

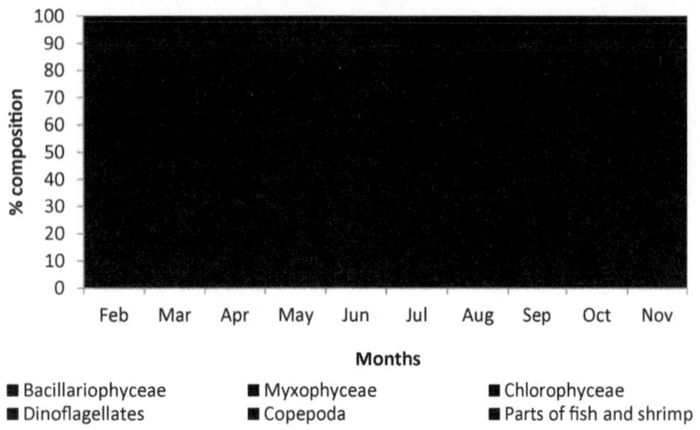

Figure 2: Percentage occurrence of suspended food materials of grey mullets (*Mugil cephalus*) in the ambient water of extensive farming system.

Myxiphyceae. Other genera such as *Chroococcus, Gleocapsa*, and *Merismopedia* were also present in the pond water. Composition of Myxophyceae was 3.88-8.09% throughout the study period. Common zooplankton groups were Dinoflagellates and Copepods. Rotifers and Cladocera represented other less abundant zooplankton groups. Maximum abundance of Dinoflagellates were observed during April (42.56%) and minimum during November (34.62%) whereas Copepoda constituted 0.22-4.35% of planktonic constituents. Common Dinoflagellate genera were *Ceratium* and *Peridinium* whereas; *Calanus* was the most common Copepod genera.

Growth of grey mullet in terms of weight and length over time is presented in Figure 1. Fishes were grown from 3.36 ± 0.15 g (6.37 ± 0.18 cm) as measured during 30 DOC to 307.20 ± 16.55 g (31.97 ± 1.14 cm) in 300 days. Feeding intensity of grey mullets in terms of stomach fullness is presented in Table 2. Lower feeding intensity with highest number of empty stomachs was observed during the initial months of rearing. Feeding intensity gradually increased as the fishes grew, indicated by more number of gorged and full stomachs in the later phase of rearing. Constituents of stomach contents throughout the study period are presented in Figure 3. The main phytoplankton groups found in the stomach according to the order of dominance were Bacillarophyceae (23.04-40.10%), Chlorophyceae (9.59-20.39%) and Myxophyceae (4.45-14.36%). Dominant zooplankton group in the stomachs were Dinoflagellets (1.60-6.58%) followed by copepods (0.20-2.97%). Among non-planktonic stomach constituents, parts of fish and shrimp constituted 0.12-0.98%, organic matter composed mainly of scraps of macrophytes and organic particles constituted 8.37-15.36%; sand and mud constituted 23.52-36.38% of stomach content. Organic matter and sand particles were not considered for prey preference analysis.

Electivity Index (E) for Bacillariophyceae varied between 0.50-0.85 with highest value during the month of May whereas E for Myxophycese ranged between 0.22-0.59 showing highest value during October and November (Figure 4). Chlorophyceae showed negative value throughout the study period and ranged between -0.18 to -0.31. Among zooplanktonic stomach constituents, Copepods showed higher values during early months of rearing which gradually decreased as the fishes were grown. Dinoflagellets and fish or shrimp parts showed negative values almost throughout the rearing period.

Discussion

Recorded water quality parameters were within optimum ranges for brackish water aquaculture [43] and differed significantly ($P<0.05$) with time. Concentrations of toxic gases like nitrite-nitrogen (NO_2-N) and ammonia-nitrogen (NH_4-N) remained lower than the critical level and concentrations of nutrients like nitrate-nitrogen (NO_3-N) and phosphate-phosphorous (PO_4-P) was much lower than fertilized ponds reported from Sundarbans [16,44]. Lower nutrient concentrations in the studied extensive system may be attributed to complete dependence on natural productivity without any additional input. Order of dominance of the planktonic groups in the ambient water in the present study was corroborated with that reported from the Hooghly estuary [45].

Months	Gorged	Full	3/4 Full	1/2 Full	1/4 Full	Little	Empty
Feb	0	6.66	9.99	16.65	20	33.3	13.32
Mar	0	3.33	9.99	9.99	26.6	40	9.99
Apr	3.33	6.66	6.66	16.65	23.3	30	13.32
May	3.33	9.99	9.99	19.98	23.3	26.6	6.66
Jun	6.66	9.99	13.32	16.65	26.6	20	6.66
Jul	9.99	13.3	13.32	19.98	26.6	13.3	3.33
Aug	6.66	13.3	19.98	26.64	23.3	9.99	0
Sep	13.32	20	23.31	13.32	16.7	13.3	0
Oct	16.65	9.99	23.31	19.98	20	6.66	3.33
Nov	13.32	13.3	16.65	23.31	26.6	6.66	0

n=30 for each month, 10 fishes were taken monthly from three impoundments under study.

Table 2: Feeding intensity of grey mullet in extensive brackish water farming impoundments used for grey mullet feeding ecology study.

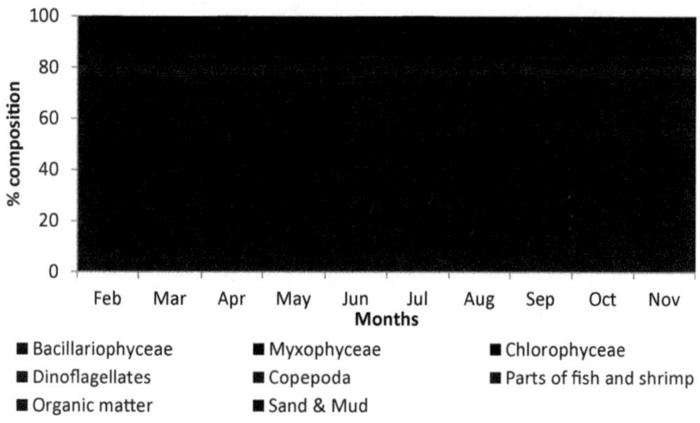

Figure 3: Percentage occurrence of food constituents in the grey mullet (*Mugil cephalus*) stomach collected from extensive farming system.

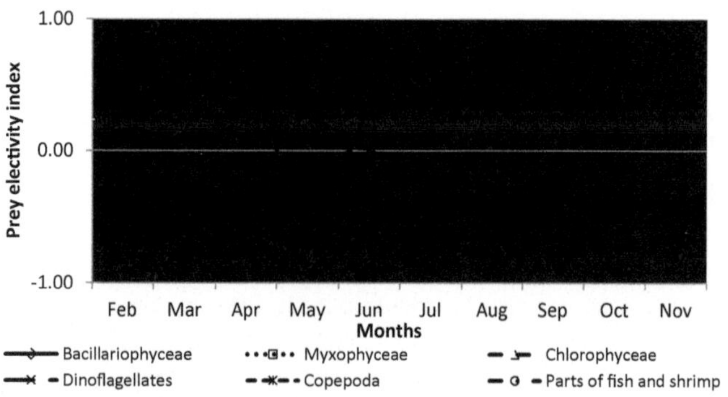

Figure 4: Electivity Index indicating prey preference of grey mullet (*Mugil cephalus*) reared in extensive farming system.

M. cephalus is a diurnal and opportunistic feeder and feeds almost continuously throughout the day [22]. The higher feeding frequency in bigger fishes than smaller ones may be attributed to the fear of potential predators by the smaller fishes while feeding as they are more vulnerable and would rather feed more cautiously than their bigger counterpart [46]. Larger fish may require more food to obtain the necessary energy for reproductive activity than smaller ones require for growth. Moreover, a wider mouth opening in larger fish helps to ingest relatively larger quantity food items at a time [47].

Bacillariophyceae followed by chlorophyceae and myxophyceae as most dominant food constituents of *M. cephalus* in brackish water environments has been reported from various parts of the world like East coast of Andhra Pradesh, India [37], coastal waters of Bangladesh [48], Northern Pacific coast of Ecuador [49], along Bulgarian Black Sea coast [31], Senegal River estuary [50] and earthen fishponds in Egypt [51]. Similar order of dominance of prey groups in different environments with varied planktonic constituents indicates that *M. cephalus* does not feed on whatever available, but have some preference. Chlorophyceae, in spite of being dominant group in water column ranked second in order of dominance in the stomach content in the present study further supports this view. Considering the complex nature of grey mullet feeding ecology, electivity index (E) analysis is necessary to throw some light on food preferences. As per Ivlev's equation, E ranges from -1 to +1, where -1 to 0 stands for negative selection, while values 0 to +1 can be interpreted as positive selection of that prey item. Subsequent investigation [52] suggested that a true positive or negative prey selection can be interpreted only at values >0.3 or <-0.3 respectively. Grey mullets in the present study actively selected bacillariophyceae as most preferred food material. In spite of ranking third in the order of dominance in the stomach content, myxophyceae was selected as second preference except during the month of May characterized with high temperature. Although being second dominant planktonic constituent in the stomach content, chlorophyceae was not at all selected and was probably swallowed mechanically during intake of other food stuffs. True positive selection of Copepods during initial months of rearing indicates preference by *M. cephalus* juveniles. Studies on the prey preference of *M. cephalus* is rare, however, it has been suggested that the organization of the alimentary system of a particular species, as for example in the relative concentrations of its digestive enzymes, may be such as to obtain maximum advantage for only a limited part of the range of material which the animal is actually capable of ingesting [53]. Feeding ecology study along Bulgarian Black Sea coast revealed that *M. cephalus* is able to utilize either the direct grazing or plant detritus food chains as an energy source depending upon which is the easiest to exploit. When both food sources are present in abundance, the mullet exhibits a preference for living microalgae. Such an algal diet has a higher caloric content than a macro plant detritus diet [31]. These might be the guiding force behind prey selection. Based on the present observations, it may be inferred that grey mullet (*M. cephalus*) does not feed at random but prefers diatoms (bacillariophycae) followed by blue green algae (myxophyceae) and higher production can be expected in extensive brackish water farming systems which are rich in those phytoplankton groups.

Acknowledgement

The authors are indebted to the authorities of Techno India University, Salt Lake, Kolkata and the Officer-in Charge, KRC of CIBA, Kakdwip for providing infrastructural support. Authors are thankful to the faculty members and scientists of both the organizations for their valuable guidance. Help from the supporting staffs and continuous encouragement by colleagues are thankfully remembered.

References

1. Oren OH (1981) Aquaculture of Grey mullets (International Biological program No. 26). Cambridge University press, Cambridge, England. 507.
2. Khan AA, Fatima M (1994) Feeding ecology of the Grey Mullet, Rhinomugil corsula (Hamilton) from the River Yamuna, North India. Asian Fish Sc 7: 256-266.
3. Shamsi M1K, Niamat R, Jafri AK (1995) Planktonic biota and food selection of freshwater fish puntius sophore in a perennial and sewage-fed tropical fish pond from Northern India. Egypt J Appl Sc 10: 217-224.
4. A1-Ake1 AS, A1-Kahem HF, Shamsi M1K, Ahmed Z (1996) Selection of food in various size groups of Nile Tilapia, Oreochromis niloticus (L.) in Wadi Haneefah stream, Saudi Arabia. Pak J Zool 27: 271-275.
5. Ehlinger TJ, Wilson DS (1988) Complex foraging polymorphism in bluegill sunfish. Proc Nat Ac Sc 85: 1878-1882.
6. Gerking SD (1994) Feeding Ecology of Fish, San Diego: Academic Press Inc 416.
7. Fry B, Mumford PL, Tam F, Fox DD, Warren GL, et al. (1999) Trophic position and individual feeding histories of fish from Lake Okeechobee Florida. Canadian J Fish Aquatic Sc 56: 590-600.
8. Shaw GW, Pankhurst PM, Purser GJ (2003) Prey selection by greenback flounder Rhombo solea (Günther) larvae. Aquaculture 228: 249-265.
9. Hagiwara A, Suga K, kazawa A, Kotani T, Sakakura Y (2007) Development of rotifer strains with useful traits for rearing fish larvae. Aquaculture 268: 44-52.
10. Nunn AD, Harvey JP, Cowx IG (2007) The food and feeding relationships of larval and 0+ year juvenile fishes in lowland rivers and connected water bodies II Prey selection and the influence of gape. J Fish Biol 70: 743-757.
11. Moore JW, Moore IA (1976) The basis of food selection in flounders, Platichthys flesus in the Sever Estuary. J Fish Biol 9: 139-156.
12. Ribeiro DFO, Nuñer APO (2008) Feed preferences of Salminus brasiliensis (Pisces, Characidae) larvae in fish ponds. Aquaculture 274: 65 -71.
13. Fraser D, Huntingford FA, Adams CE (2007) Foraging specialisms, prey size and life-history patterns a test of predictions using sympatric polymorphic Arctic charr (Salvelinus alpinus). Ecol Freshwater Fish published online June 7.
14. Herwig BR, Zimmer KD (2007) Population ecology and prey consumption by fathead minnows in prairie wetlands importance of detritus and larval fish. Ecol Freshwater Fish 16: 282-294.
15. Schleuter D, Eckmann R (2007) Generalist versus specialist the performances of perch and ruffe in a lake of low productivity. Ecol Freshwater Fish published online August 16.
16. Biswas G, De D, Thirunavukkarasu AR, Natarajan M, Sundaray JK, Kailasam M, et al. (2012) Effects of stocking density, feeding, fertilization and combined fertilization-feeding on the performances of striped grey mullet (Mugil cephalus L.) fingerlings in brackish water pond rearing systems. Aquaculture 338-341: 284–292
17. De Silva SS (1980) The biology of juvenile grey mullet a short review. Aquaculture 19: 21-36.
18. Thomson JM (1966) The grey mullet. Barnes H (Ed.) Ocean and Mar Biol-an Annual Review. Allen and Unwin, London. 4: 301-355.
19. Well RDS (1984) The food of the grey mullet (Mugil cephalus L.) in Lake Waahi and the Waikato River at Huntly. New Zealand J Mar and Freshwater Res 18: 13-19.
20. FAO (2010) The State of the World Fisheries and Aquaculture Food and Agricultural Organization of the United Nations Rome Italy.
21. Moriarty DJW (1976) Quantitative studies on bacteria and algae in the food of the mullet (Mugil cephalus L) and the prawn Metapenaeus bennettae (Recek & Dall). J Exp Mar Biol 22: 131-143.
22. Odum W (1970) Utilization of the direct grazing and plant detritus food chains by the striped mullet (Mugil cephalus L.): Steele, JJ (Eds) Marine Food Chains. Oliver and Boyd, Edinburgh 222-240.
23. Lupatsch I, Katz T, Angel DL (2003) Assessment of the removal efficiency of fish farm effluents by grey mullets a nutritional approach. Aquacult Res 34: 1367-1377.

24. Abdel-Tawwab M, Abdel-Hamid ME, Ali EA, El-Marakby HI (2005) The Assessment of Water Quality and Primary Productivity in Earthen Fishponds Stocked with Stripped Mullet (Mugil cephalus) and Subjected to Different Feeding Regimes. Turk J Fish and Aquatic Sc 5: 1-10.

25. Thompson J (1954) The organs of feeding and the food of some Australian mullet. Aust J mar and Freshwater Res 5: 469-485.

26. Wood E (1964) Study on the microbial ecology of Australian region Nova Hedwigia 8: 461-568.

27. Odum W (1968) Mullet grazing on a dinoflagellate bloom. Chesapeake Center for Sc 9: 202-204.

28. Brusle J (1981) Food and feeding in grey mullet: OH Oren (Ed.) Aquaculture of Grey Mullet. Cambridge University Press Cambridge UK: 185-217.

29. Porter CB, Krost P, Gordin, H, Angel DL (1996) Preliminary assessment of grey mullet (Mugil cephalus) as a forager of organically enriched sediments below marine fish farms. Israel J. Aquaculture 48: 47-55.

30. Jana SN, Garg SK, Patra BC (2004) Effect of periphyton on growth performance of grey mullet Mugil cephalus (Linn.) in inland saline ground water ponds. J. Appl. Ichthyol 20: 110-117.

31. Bekova R, Raikova-Petrova G, Gerdzhikov D, Petrova E, Vachkova V, et al. (2013) Food spectrum of grey mullet (Mugil cephalus L.) along the Bulgarian Black Sea coast. Agril Sc and Tech 5: 173-178.

32. Rheman S, Islam ML, Shah MMR, Mondal S, Alan MJ (2002) Observation on the fecundity and Gonadosomatic Index (GSI) of Grey mullet Liza parsia (Ham.). Online J. Biol. Sci. 2: 690-693.

33. Abraham M, Kailasam M, Kishore Chandra P, Sihranee P, Rajendran KV et al. (2000) Development of captive broodstock of the grey mullet, Mugil cephalus (L). Ind J Fish 47: 91-96.

34. Rangaswamy CP (1973) Studies on the age and growth and food habits of the grey mullet Mugil cephalus Linnaeus of the Lake Pulicat. J Inland Fish Soc India 5: 9-22.

35. Tandel SS, Athalye RP, Gokhale KS (1986) On the seasonal changes in food habit of Mugil cephalus of the Thana Creek. Ind J Fish 33: 270-276.

36. Rao LM, Sivani G (1996) The food preferences of five commercially important fishes of Gosthani estuary. Ind J Fish 43: 199-202.

37. Rao RK, Babu KR (2013) Studies on food and feeding habits of Mugil cephalus (linnaeus, 1758) east coast off Andhra Pradesh, India. Can J Basic and Appl Sc 7: 2499-2504.

38. APHA (1998) Standard Methods for the Examination of Water and Wastewater, 20th ed, American Public Health Association Washington DC USA.

39. Abdelghany AE (1993) Food and feeding habits of Nile tilapia from the Nile River at Cairo, Egypt: L Reintertsen, A Dhale, C Jorgensen, R Twinnereim (Eds) Fish Farm Technology AA Balkema, Rotterdam, The Neatherland 447-453.

40. Jhingran VG, Natarajan AV, Banerjee SM, David A (1969) Methodology on reservoir fisheries investigation in India. Central Inland Fisheries Research Institute, Barrackpore, India Bulletin No. 12. 109.

41. Prescott GW (1961) Algae of the Western Great Lakes Area. WMC Brown Company Publishers Dubuque IA USA 990.

42. Ivlev VS (1961) Experimental ecology of the feeding of fisheries. Yale University Press, New Haven CT, USA 322.

43. Chakraborti RK, Sundaray JK, Ghoshal TK (2002) Production of Penaeus monodon in the tide fed ponds of Sundarban. Ind J Fish 49: 419-426.

44. Saha SB, Bhattacharyya SB, Mitra A and Choudhury A (2001) Quality of shrimp culture farm effluents and its impact on the receiving environment. Bangladesh J Zool 29: 139-149.

45. Dutta S, Maity S, Bhattacharyya SB, Sundaray JK, Hazra S (2013) Diet composition and intensity of feeding of Tenualosa ilisha (Hamilton, 1822) occurring in the Northern Bay of Bengal, India. Proc Zool Soc Ind. 67: 33-37.

46. Akpan AW, Isangedighi IA (2004) Aspects of the Feeding Ecology of three Species of Pseudotolithus (Sciaenidae) in the Inshore waters of Southeastern Nigeria, East of the Niger Delta, Nigeria. J Aquat Sc 19: 51-58.

47. Isangedighi I, Uudo PJ, Ekpo IE (2009) Diet composition of Mugil cephalus (pisces: mugilidae) in the Cross River estuary, Niger delta, Nigeria. Nigerian J of Agril Food and Env 5: 10-15.

48. Islam R, Hossain MB, Das NG, Rafi RUN (2009) Food and feeding behaviour of grey mullet, Mugil cephalus of Bangladesh coastal water. Bangladesh J Pro Sc & Tech 7: 56-61.

49. Rramirez LV, Navia AF, Rubio EA (2008) Food habits and feeding ecology of an estuarine fish assemblage of northern Pacific Coast of Ecuador. Pan-Am J Aquatic Sc 3: 361-372.

50. Modou SS, Mouhameth C, Tinkoudgou KJA (2014) Seasonal feeding variation of the yellow mule (Mugil cephalus, Linnaeus 1758, Mugilidae) in Senegal River estuary fishery. Int J Agril Pol and Res 2: 125-131.

51. El-Marakby HI, Eid AM, Abdelghany AE, Abdel-Tawwab M (2006) The impact of stripped mullet Mugil cephalus on natural food and phytoplankton selectivity at different feeding regimes in earthen ponds. J Fish Aquatic Sci 1: 87-96.

52. Lazzaro X (1987) A review of planktivorous fishes: Their evolution, feeding behaviours, selectivities, and impacts. Hydrobiologia 146: 97-167.

53. Hartley P (1947) The natural history of some British freshwater fish. Proceedings of the Zoological Society of London 117: 129-206.

31

Optimization of Fertilization Success in the Bivalve Mollusk *Tivela mactroides* under Laboratory Conditions

Yajaira García de Severeyn[1]*, Ana Villasmil[1], Héctor Severeyn[2], Félix Morales[3] and Marynes Montiel[4]

[1]Universidad del Zulia, Facultad Experimental de Ciencias, Dpto. de Biología, Laboratorio de Cultivo de Invertebrados Acuático, Venezuela
[2]Universidad del Zulia, Facultad Experimental de Ciencias, Dpto. de Biología Laboratorio de Sistemática de Invertebrados Acuáticos, Venezuela
[3]Universidad del Zulia, Facultad Experimental de Ciencias, Dpto. de Biología Laboratorio de Oceanografía, Venezuela
[4]Universidad del Zulia, Facultad Experimental de Ciencias, Dpto. de Biología, Unidad de investigaciones de Microbiología Ambiental Maracaibo, Zulia, Venezuela

Abstract

With the objective to produce seed in an aquaculture facility it was studied under laboratory conditions, the effect of sperm concentration and contact time between gametes to optimized fertilization success of the bivalve mollusk *Tivela mactroides*. Mature ovocites and spermatozoids gotten through gonads dissection of adult mature male and female clams were artificially combined to produce fertilization. Solutions of 5.5×10^6 ovocites were fertilized with 10 different sperm concentrations. Results indicate that the percentage of succeed fertilization was low, highly variable and relying on the sperm concentration. The maximum mean percentage of fertilization was 22% between 1.3×10^7 and 1.7×10^7 spermatozoid/ml. A strong fall in fertilization success was observed at sperm concentrations lower than 10^6 spermatozoid/ml. The time lapsed after the first contact for fertilization that produced the highest fertilization rate was three hours, at sperm concentrations higher than 107 spermatozoids/ml. The highest fertilization percentage found in this study falls between the ranges reported for other bivalve mollusk species.

Keywords: Bivalve; Mollusks; Tivela; Veneridae; Fertilization; Sperm concentration; Oocytes; Contact time; Gamete

Introduction

The success of fecundation of the mature ovocites is a key factor to maximize the reproduction. This process in aquatic invertebrates depends on several factors such as the distribution, density and the individuals' behavior, the necessary density of sperms to fecundate the eggs, the longevity and the behavior of the gametes [1,2]. There are not only complex interactions among these biological factors but also with the physical properties and the dynamics of the aquatic environment [3]. Before a wide interval of sperm concentrations, the fecundation rates can diminish quickly when the sperms are diluted in the waster column and because they also become less active with time. To counteract the dilution of the gametes and to maximize the fecundation rates, the aquatic animals, after a natural spawn, have developed strategies which increase the probabilities of a successful fertilization [1].

However, the artificial fecundation in aquaculture facilities allows the control of the concentration of sperms therefore the ovocites can be inseminated to an appropriate concentration to maximize the fecundation rate. This procedure avoids the problems associated with dropping fecundation rates and the development of polispermia [4].

Until the publication of the classic study of Levitan [1], the literature on the study of the factors that could maximize the fecundation in aquatic invertebrates was scarce and dispersed. Nonetheless it is surprising that his paper did not mention any mollusk species when almost 30 years before started this type of studies in pectinid bivalve mollusks [5]. These authors studied the best sperm and ovocites concentrations to get viable larvae of *Pecten maximus.*

In the present investigation we studied the marine clam *Tivela mactroides*, a marine bivalve mollusk, and candidate to be cultivated in order to recover its natural banks in the southwest of the Venezuelan Gulf. These banks were decimated by a spill of petroleum in 1997 [6]. The objective was to optimize the production of embryos using, as high-priority factors, the concentration of sperms and the time lapsed after having been induced the contact among the gametes. The final objective of this and following investigations is to obtain viable larvae that can be used to repopulate the area affected by the oil spill.

Materials and Methods

Collection of adults of *Tivela mactroides*

The clams were gathered in the beach Miramar, an area of about two Km^2 of coast which occupies an arid-coastal plain. It is a high energy beach and strong surf located to the north in the capital of the Municipality Buchivacoa, Capatárida Falcon. (11°13 `53 ``North and 70°36 `51 `` West). This population of *T. mactroides* is located to 300 Km. east of the area affected by the spill and it still remains as a pristine area, only being exploited by artisanal fishermen.

The animals were collected by means of a trail with a mesh diameter of 20 mm. In some occasions the fishing was made manually. During the samplings, the quantity and size of the gathered clams was made at random, since the size of the animal is independent of the sexual maturity [7]. Of the gathered clams, 100 were placed into 20 L. plastic containers, with oxygen and water of the collecting size cooled with pack of ice. This was done to avoid as much as possible stress during their transfer by car to the laboratory thus avoiding that clams may have spontaneous spawning.

***Corresponding author:** Yajaira García de Severeyn, Universidad del Zulia, Facultad Experimental de Ciencias, Dpto. de Biología, Laboratorio de Cultivo de Invertebrados Acuático, Venezuela, E-mail: ygsevereyn@yahoo.com

Obtaining of the gametes

To obtain the gametes, the clams were opened up by dissection, and their gonads observed under the microscope to identify the sex. Only mature male and female gametes were used to produce fertilization. For this procedure, gonads were carefully perforated with a sterile micro-scissor to liberate the gametes. Them gonads were washed out and collected into Erlenmeyer. Ovocites were filled with 50 ml and sperm with 180 ml of artificial seawater (Instant Ocean) prepared to 40 ups to create the stock solution.

Obtaining of the experimental concentrations of the gametes

To estimate the concentration of sperms and ovocites in the stock solutions there were taken, 3 times, 5 ml of each solution, which were preserved with five drops of 10% formalin solution. Later on, with a camera of New bauer, the gametes were counted. The mean of the three replicates was taken as the concentration of each solution: 5.8×10^{9} spermatozoids/ml, and 5.5×10^{6} ovocites/ml. Starting from the stock solutions, there were carried out the dilutions to obtain the concentration series of sperms to study. In order to do them, there were prepared 10 dilutions adding to each rehearsal tube from 9 to 1 ml of artificial seawater to 40 ups and a series from 1 to 9 ml of the solution stock of sperms. Two controls of 1 ml of ovocites solution were set, the first one without sperms to discard that this species does not present parthenogenesis, and the second one with sperms previously inactivates through freezing at 0°C during 4 hours, with the purpose of observing if the sperm contains some protein that intervenes or stimulates the fecundation process.

Induction of the fecundation

After carrying out the dilutions of sperms, 1 ml of the ovocites solution stock was placed in each one of the rehearsal tubes with the respective concentrations to induce the fecundation. The experiments stayed to 28 ± 1°C in an environmental cabinet, during five hours starting from the mixture of the gametes. Each experiment was carried out with three replicates. Once united the gametes, there were taken samples of 0.05 ml at 30 minutes, three and five hours, placed in an excavated slide and examined under microscope. In this observations there were recorded the total number of ovocites, fecundated ovocites, those not fecundated and the embryos in division.

Results

Effect of the concentration of sperms

A notable result of this investigation was the high variability that was observed in the percentage of fecundation among the studied concentrations. Indeed, although the observed maximum value was of 50% in two single replicates, 33% of all experimental concentrations there was not fecundation of all. For this reason, the non-parametric analysis of variant (Kruskal-Wallis) did not detect significant differences among the studied concentrations ($p = 0.8854$). The maximum mean fertilization percentage, 22%, was observed between 1.3×10^{7} and 1.7×10^{7} spermatozoids/ml (Figure 1). Above or below these values, the fecundation decayed notably.

Effect of the time lapsed over fecundation

The time lapsed after the contact of gametes affected the percentage of fecundated eggs. Indeed, the percentage of fecundation after three hours was significantly smaller ($p=0.0210$) that during the first half hour or after five hours (Figure 1).

Discussion

In aquatic invertebrates, the fecundation success depends on many factors both internal as external to the gametes. To this respect, Levitan [1] points out four groups of factors: environmental, poblacional, individual and specifically internal (genetics) to the gametes. Products of the interaction of the first three categories are generated three conditions that have to do directly with the gametes: the concentration of these in the aquatic means, their dispersion rate (ability to swim longer distances) and the grade of mixture of the gametes of different animals. To these three conditions, the same author adds the intrinsic factors to the gametes: variation in the individual acting (for example longevity), the variation in the size and the form of the eggs, the morphology of the sperms and the rate of collisions among gametes.

However, it is of consensus that the concentration of sperms is one of the most critical factors. The success of the fecundation and therefore the percentage of fecundated ovocites it is highly dependent of the good concentration of sperms. [4,8-11]. When the fecundation takes place with excess of sperms, polispermia increase, generating a decrease of viable embryos [12,13].

In the present study, only two of the mentioned factors were investigated: The concentration of spermatozoids and the time lapsed after having induced the contact among the gametes (an indirect way to measure the effect of sperm longevity).

With regard to the first factor, it was already indicated that the results showed a lot of variability, so statistically speaking, no significant differences was detected among the studied sperm concentrations. These results coincide with other studies in several aspects.

First, high concentrations of sperms are required in the case of clams. Some authors [8] have reported that concentrations of more than 10 million spermatozoids/ml are necessary to successful fecundation. Our successful sperm concentrations were between 13 and 17 million. However, it is clear that the optimal concentration is a factor related to the species, so it is not surprising to find successful ranges very below respect to the reported here. Indeed, *Spisula sollidisima* [14] has optimal sperm concentrations from 800.000 up to 4.000.000 spermatozoids/ml.

The second aspect evaluated in this investigation with regard to the success of the fecundation was the necessary time to obtain the maximum fecundation percentage. As demonstrated above, three hours was the necessary time to maximize the quantity of fecundated ovocites. This time of maximum fecundation coincides with studied in other bivalve where the range goes from 10 minutes to six hours [2,14-16], but it is evident that this wide range should be product of other factors not evaluated in these researches.

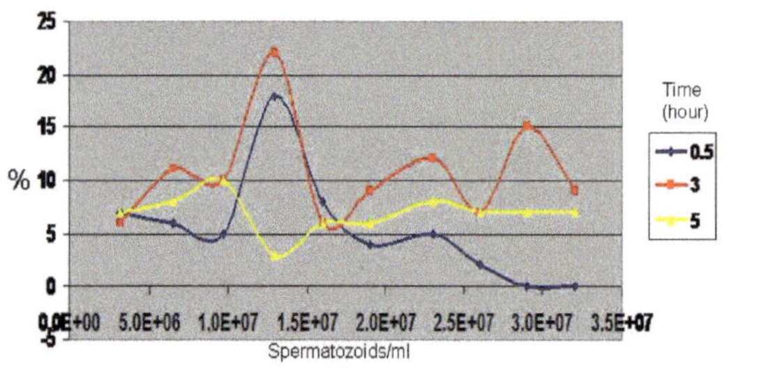

Figure 1: Effect of spermatozoids concentration and time lapse of contact after fertilization over the fertilization succed (porcentaje) in the marine clam *Tivela mactroides*.

Conclusions

Considering the main objective of the present study, is not conclusive which is the optimal concentration of sperms to maximize embryos production. However it is also clear that it is not the only necessary factor to achieve high percentages of fecundation in *Tivela mactroides*. Time of contact (longevity of spermatozoid) is also important as we saw in the present results. Both factors together but they do not work alone to affect the success of the fecundation measured as formation of viable embryos. Additional research is necessary to clear this situation.

Considering the necessity to produce the biggest possible number of viable embryos, the range of concentrations of sperms that maximized the fecundation in this investigation was between 1.3 and 1.7×107 spermatozoids/ml after three hours of contact among gametes. Half hour turned out to be little time, but five hours they did not improve the success of fecundation above three hours.

References

1. Levitan D (1998) Sperm Limitation, Gamete Competition, and Sexual Selection 175-202. In External Fertilizers, (1stedn), Academic Press, New York, USA.
2. Song Y, Suquet M, Quéau I, Lebrun I (2009) Setting of a procedure for experimental fertilization of Pacific oyster (*Crassostrea gigas*) oocytes. Aquaculture 287: 311–314.
3. Hodgson A, Le Quesne W, Hawkins S, Bishop J (2007) Factors affecting fertilization success in two species of patellid limpet (Mollusca: Gastropoda) and development of fertilization kinetics model. Mar Biol 150: 415-426.
4. Liu W, Alabi A, Pearce C (2008) Fertilization and embryonic development in the basket cockle, *Clinocardium nuttallii*. J Shellf Res 27: 393-397.
5. Gruffydd L, Beaumont A (1970) Determination of the optimum concentration of eggs and spermatozoa for the production of normal larvae in *Pecten maximus* (Mollusca, lamellibranchia). Helgolander wiss Moeresunters 20 486-497.
6. Severeyn H, Delgado J, Godoy A, García de Severeyn Y (2003) Efecto del derrame de petróleo del buque Nissos Amorgos sobre la fauna macro invertebrada bentónica del Golfo de Venezuela: cinco años después. Ecotrópicos 16: 83-90.
7. Reverol Y, Delgado J, García de Severeyn Y, Severeyn H (2004) Embrionary and larval development of the marine clam Tivela mactroides (Bivalvia:Veneridae)in Zulia State,Venezuela. Revista de Biología Tropical.
8. Powell D, Tyler P, Peck LL (2001) Effect of sperm concentration and sperm ageing on fertilization success in the Antarctic soft-shelled clam *Laternula elliptica* and the Antarctic limpet *Nacella concinna*. Mar Ecol Prog Ser 215: 191-200.
9. Luttikhuizen P, Drent J (2003) Fertilization in broadcast spawning marine bivalves is a small scale process: the Baltic clam *Macoma balthica*. J Evol Biol 119-126.
10. Luttikhuizen P, Honkoop P, Drent J, Van Der Meer J (2003) optimal egg size for external fertilization: theory on the role of sperm limitation and field observations for a marine bivalve. J Evol Biol 83-102.
11. Kupriyanova E (2006) Fertilization success in *Galeolaria caespitosa* (Polychaeta: Serpulidae): gamete characteristics, role of sperm dilution, gamete age, and contact time. Scientia Marina 70: 309-317.
12. Narvate M, Pascual M (2003) Fertilization, larval rearing and post-larval growth of the Tehuelche scallop *Aequipecten tehuelchus* D'Orb., 1846. Aquaculture 217: 259-274.
13. FARIAS J (2006) Cultivo de moluscos. 1era Edic.. Editorial Alfaomega. México, D.F., México.
14. Clotteu G, Dube F (1993) Optimization of fertilization parameters for rearing surf clams (Spisula *solidissima*). Aquaculture 114: 339-353.
15. Dos Santos A, Nascimiento I (1985) Influence of gamete density, salinity and temperature on the normal embryonic development of the mangrove oyster *Crassostrea rhizophorae* Guilding, 1828. Aquaculture 47:335-352.
16. O'connor W, Heasman M (1995) Spawning induction and fertilization in the doughboy scallop *Chlamys (Mimachlamys) asperrima*. Aquaculture 136: 117-129.

Impact of Different Colors of Artificial Light on Pigmentation and Growth Performance of Hybrid Red Tilapia (*Oreochromis mosambicus* × *O. hornorum*) Reared in Saline Well Water

Hadir A Aly, Mohamed M Abdel-Rahim*, Ayman M Lotfy and Basem S Abdelaty

Fish Rearing Lab, Aquaculture division, National Institute of Oceanography and Fisheries (NIOF), Alexandria, Egypt

Abstract

Four light colors (red, white, green and blue) were tested to evaluate their effects on both body color enhancement and growth performance of Florida red tilapia (*Oreochromis mosambicus* × *O. hornorum*). Fish were stocked in 12 fiberglass tanks, (each of 2 m3 water volume), at a stocking rate of 100 fish per tank with average initial weight of 2.66 g/fish, three replicates for each treatment. Fish were fed on a commercial diet containing 25% protein, two meals per day with a daily feeding rate of 7% in the first two weeks, then reduced to 5% until the end of the experimental period (6 weeks). The results of this study reveals that there are no significant differences (p>0.05) in growth performance indexes between treatments. However, fish exposed to red color showed highest growth values. With the same trend, results of survival percent and condition factor showed no significant differences between treatments. Significant differences (p≤0.05) are observed in the whole body chemical composition between treatments. Fingerlings that are exposed to blue color has the highest value of dry matter content, while, the lowest value is observed in fingerlings exposed to red color. Feed utilization (FI, FCR and PER) do not significantly affect the lighting color, while protein and energy utilization (PPV and EU) are significantly affected. Concerning the red color accumulated in the fish body, the fingerlings which are exposed to the blue color achieved the best β-carotene value (211.25 IU/100 g fish) with highly significant differences compared with other treatments. The content of β-carotene in blue color treatment is 8, 9, and 16 folds comparing with green, white, and red colors, respectively. Also, Red tilapia fish with black spots in the blue treatment were 12%, compared with 29% in green, 56% in white and 52% in red colors.

It could be concluded that cultivating Red tilapia in a full saline well water using the artificial blue color will significantly improve not only the pigmentation of Red tilapia, the quality of fish in terms of the content of dry matter, but also the protein and the energy utilization of the consumed feed.

Keywords: Red tilapia; Artificial light color; Pigmentation; B-Carotene; Growth; Feed utilization

Introduction

Tilapia is one of the most widely cultivated fishes in the world [1]. Nevertheless, under high salinity water, the culturists must face its sensitivity to handling, disease infections [2], remarked retardation in growth, increased feed conversion ratio, up to high mortality ratio [3,4]. Florida red tilapia strain was developed in the 1970s by crossing a normal colored *O. hornorum* female with a red-gold male *O. mossambicus* [5]. This strain was introduced to Egypt in 1993. Florida red tilapia strain grows well in high saline water [6] and can tolerate salinity up to 36.2 ppt, and the optimum limit is 17.8 ppt [7], while Nile tilapia does not tolerate salinities above 20 ppt [8]. Therefore, world is moving towards expanding the cultivation of Red tilapia, the most resistant strain to high salt water [3].

The red color of hybrid tilapia is one of the most favorite traits, in addition to its tolerance to salinity. There are many factors influencing the chromatic state of fish, including nutritional, environmental, genetic and neuro-hormonal factors [9]. Dietary carotenoids play an important role in regulating fish color [10]. Because fish, like other animals, are unable to synthesize carotenoids, the skin color in fish is highly dependent on carotenoids present in the diet [11]. Therefore, dietary sources of carotenoids (natural or artificial) should be enhanced the artificial diets for colorful fishes, such as Red tilapia. Otherwise, fish will suffer from pale colors. Dietary β-carotene suppressed skin β-carotene, and pigments of False Clownfish, *Amphiprion ocellaris* [10].

In addition, the environmental conditions, including background [12,13], light color, light intensity [14], external and/or internal stresses etc. [15,16]. can alter the coloration of cultivated fishes. This phenomenon causes technical, and commercial problems and may reduce the marketability of the cultured fish [15]. Changing light intensity or background color could temporarily change the coloration of Clownfish. According to the opinion of [10], manipulating environmental conditions is unlikely to permanently alter the pigmentation of fish. However, this rapid, simple and inexpensive intervention will improve the color of farmed fish and increase the profitability of aquatic organisms.

Many studies were interested in investigating the genetic system of body color inheritance in red tilapia, where it is reported to be single in complete dominance inheritance [17,18]. It has been stated that there are three different genetic mechanisms responsible for the body coloration system in tilapia [19]. A single gene with incomplete dominance for pink, red and black colors, two genes with epistatic interaction and two independent genes, one for producing melanin and the other for producing red pigmentation. In vertebrates, the skin color is organized by enzymes stimulation, but aquaculturists have to consider that

***Corresponding author:** Mohamed M Abdel-Rahim, Fish Rearing Lab, Aquaculture Division, National Institute of Oceanography and Fisheries (NIOF), Alexandria, Egypt, E-mail: mohamed_m_ar@yahoo.com

environmental colors are determining factors of skin color intensity and pattern overall the fish body. One of the environmental characteristic that affects the fish physiology is environmental color, which modulate several parameters such as; feeding [20], growth [21], reproduction [22], sex determination [23], aggression [24], larval jaw malformation [25], and stress response [26]. Moreover, fish can alter their color in response to environmental conditions, physiological challenges, stressful stimuli [27] and cultural condition (such as background color in red porgy, *Pagrus pagrus* tanks) [28]. However, fish could adapt to the background color by changing the skin color [29,30] and this phenomenon causes commercial problems in areas where this species is traditionally fished [28] and leads to a reduced marketability of the cultured fish [30].

A common problem with the Florida red tilapia is the color of the skin, which can revert to a black coloration phenotype like those of the Mozambique tilapia. This change can subsequently reduce market acceptability and demand, but change of the light coloration may mitigate this. The aim of this study was to evaluate different light colors on the growth, feeding efficiencies, biochemical composition and skin colorations of red tilapia reared in saline well water.

Material and Methods

Fish origin

Hybrid red tilapia (*O. mossambicus* × *O. hornorum*) was obtained from a GAFRD governmental marine fish hatchery located in Maryout Company for Fish Farms, west Alexandria, K21, Egypt. The fish were placed in 4 m^3 fiberglass tanks and acclimated for a week on marine well water with a salinity of 32 ppt, by gradual increasing from 8 ppt to 32 ppt.

Experimental procedures

This experiment is performed in El-Max Research Station for Applied Research (NIOF). After fish acclimation and settlement, 1200 fish with an average weight of 2.66 g are randomly selected and transferred to 12 experimental tanks each of 2 m^3 water volume at a stocking density of 50 fish/m^3. Four lightning colors; red, blue, green and white, were tested as an artificial source of lighting, with three tanks for each treatment (color) as shown in Figure 1. Black barriers were placed between each treatment to avoid color interference The fish are fed on a commercial diet purchased from Aller Aqua Egypt Company (www.alleraquaegypt.com), contains (25% Crude Protein, 5.6% Crude fat, 7.15% Crude fibers, 403 Kcal/100 gm DM feed Gross energy, and 62.03 N:C ratio (mg CP: Kcal)); 2 times daily with feeding rate 7% of fish biomass during the first two weeks, and then reduced to 5% starting from week 3 up to the end of the experiment (after 6 weeks). The onset of the experiment is actually registered after 3 days of stocking (acclimatization period), where dead fish are removed and replaced with the same number and weight.

Studied Traits

Water quality analyses

Water temperature, salinity, pH, total ammonia-nitrogen (TAN), and un-ionized ammonia are monitored weekly throughout the experimental period. Temperature and pH were measured using portable pH Meter (pH-8424) (HANNA Instrument). Dissolved oxygen was measured by HI-9142 (HANNA Instrument). Water salinity is measured by a refractrometer (http://www.atago.net).The concentration of total ammonia nitrogen (TAN) was analyzed using

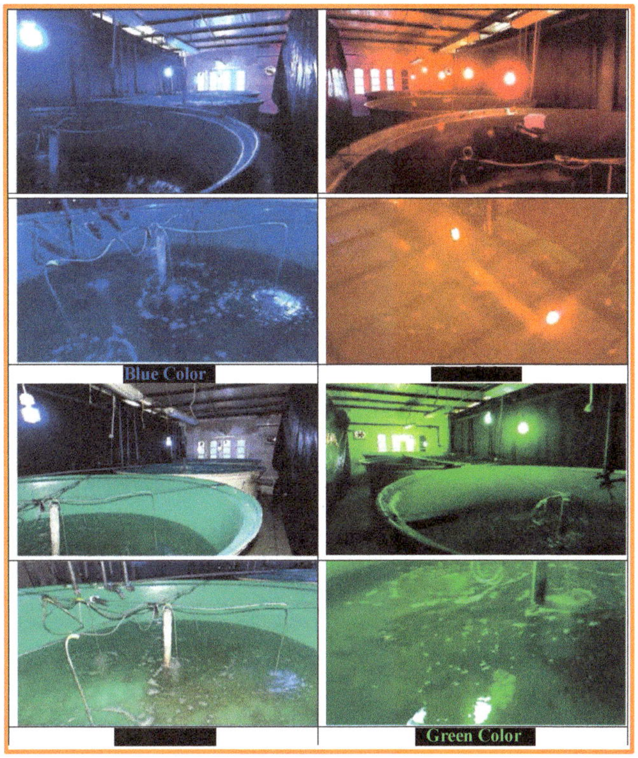

Figure 1: Photos of the experimental tanks supported with four artificial light colors (blue, red, white and green) during the experimental period.

YSI 9300 photometer and YSI Professional Plus. The concentration of un-ionized ammonia was calculated as a percentage of TAN according to U.S. Environmental Protection Agency [31].

Growth performance, condition factor, and survival

About 40 fish were randomly selected bi-weekly to estimate the gain in weight and length. The growth performance parameters are calculated according to the following equations:

Specific growth rate (%/day): $SGR = 100 \times (\ln Wt - \ln W0) / n$ [32]

Where: W_0: initial mean weight of fish (grams).

Wt: final mean weight of fish (grams).

n: Experimental period (days).

ln: natural logarithm.

After 6 weeks, the fish in each tank are weighed individually to determine survival and final weight gain.

Survival (%)=100 × (final number of fish/ initial number of fish) [33,34].

Condition factor $(K) = 100 \times (BW / L^3)$ [24].

Where: BW is fish body weight (gram); L is total fish length (cm)

Whole-body fish chemical analysis

At the beginning of the experiment, a representative fish sample is analyzed for body chemical composition (moisture, protein, lipid, ash). Also, at the end of the experimental period, ten fishes are randomly selected from each tank. The selected fishes are analyzed for body moisture, protein (Kjeldahl method), lipid (Soxhlet method), and ash (at 550°C for 16 hours) according to the AOAC (2000) [35].

Gross energy estimation (Kcal/100g)

Gross energy (GE) content of the experimental diet was estimated according to the following equation:

Gross energy (GE) was calculated as 5.64, 9.44 and 4.11 kcal GE/g for protein, lipid, and NFE, respectively.

Gross energy (GE) in Feed (kcal/100 g DM)=(protein content (%) × 5.64) + (Lipid content (%) × 9.44) + (carbohydrate content (%) × 4.1).

Gross energy (GE) in Fish Carcass (kcal/100g DM)=(protein content (%) × 5.64) + (Lipid content (%) × 9.44)).

Feed utilization measurements:

Feed utilization parameters are calculated according to the following equations [36]:

Feed Intake (FI, g/fish): This is the amount of feed given or supplied during the experimental period for each fish per gram.

Food conversion ratio (FCR)=feed intake (as DM in g)/body weight gain (g) [37].

Protein efficiency ratio (PER)=fish weight gain (g)/protein intake (g).

Protein productive value (PPV, %)=100 × (Pt –P0/protein intake (g)).

Where: P0: Protein content in fish carcass at the start.

Pt: Protein content in fish carcass at the end.

Energy gain, EG (Kcal): EG=Et-E_0

Where: E_0: energy content in fish carcass (Kcal) at the start

Et: energy content in fish carcass (Kcal) at the end

Energy utilization, EU (%): EU=100 × (Energy gain (Kcal)/ Energy intake (Kcal)].

Color measurement:

At the end of the experiment, a representative sample of fish from each tank is collected and sent to a specialized laboratory to measure the total content of B-carotene. It was measured by Reversed-Phase Liquid Chromatography (LC) method according to [38]. Also, a sample from each treatment was photographed with a high-resolution camera as another way to gauge the clear differences in fish body color. Also, the percentage of Red tilapia fish with black spots (FBS) was measured as a percentage of the total number of fish at the end of this experiment. Any fish with remarked black spots was counted in this trait.

Statistical Analysis

Data are transferred to excel sheet to calculate the mean and standard error for final weight, SGR, survival, condition factor, FCR, PER, PPV, EU, and b-carotene content. Then, these data are analyzed with a one-way analysis of variance (ANOVA) using SPSS 20 statistical package to evaluate the differences between the tested treatments. The differences within each experimental treatment are evaluated using Duncan's Multiple Range Test at 0.05 Probabilities [39].

Results

Water quality, growth performance and survival

Average values of water quality parameters in the experimental tanks are; (temperature 18.6 ± 1.3°C; salinity 32.27 ± 0.76 ppt; pH 7.92 ± 0.34; dissolved oxygen 5.74 ± 0.63 ppm; total ammonia nitrogen (TAN) 0.186 ± 0.034 ppm, and 0.004 ± 0.00ppm un-ionized ammonia, NH_3). No significant (p>0.05) differences are observed in water quality measurements between the tested artificial light colors.

Results of bi-weekly growth performance of hybrid red tilapia indicate a clear tendency towards red and white colors (Figure 2). The final data of growth performance indexes (SGR) are presented in Table 1. The data reveal that fish exposed to red and white light attain the higher values of growth but without any significant (p>0.05) differences with the other tested colors. However, the fish exposed to blue light color have the lowest values. Also, condition factor (K), which is another consideration of evaluating fish growth, shows no significant (p>0.05) differences between the four tested treatments (Table 1). The obtained values of K ranged between 1.45-1.61. The results of survival (%) shows that all treatments had low percent varied between 59.33-61% without any significant differences (p>0.05) between treatments (Table 1).

Whole-body proximate composition

Carcass composition of the Red tilapia at the end of this experiment is presented in Table 2. Moisture content is the highest in fingerlings which were tested with red light (68.91%), while the lowest value is

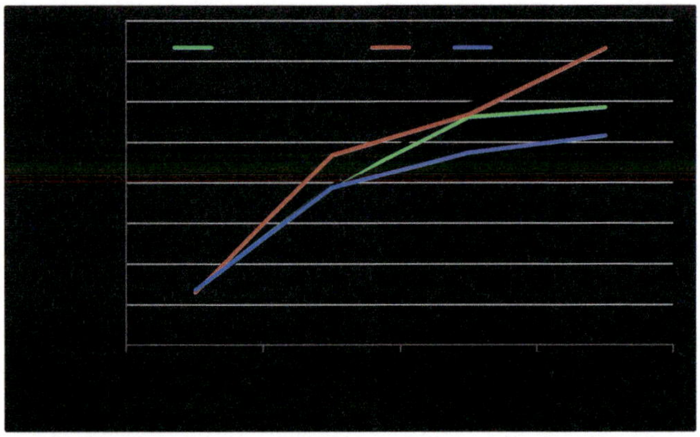

Figure 2: Periodical growth performance of hybrid red tilapia reared under different artificial light colors.

Treatments	Initial Weight, gm	Final Weight, gm	SGR,%/day	Condition factor (K)	Survival, %
Green	2.68 ± 0.03	4.94 ± 0.41	1.45 ± 0.19	1.57 ± 0.18	59.33 ± 1.86
White	2.64 ± 0.01	5.23 ± 0.08	1.63 ± 0.04	1.61 ± 0.28	61.00 ± 1.16
Red	2.65 ± 0.01	5.66 ± 0.52	1.81 ± 0.21	1.53 ± 0.05	60.00 ± 3.06
Blue	2.68 ± 0.04	4.58 ± 0.50	1.28 ± 0.25	1.45 ± 0.12	59.33 ± 3.18
Level of Significance	0.541	0.365	0.277	0.935	0.956

Means in the same column with different superscript are significantly different (p ≤ 0.05).

Table 1: Mean ± SEM of growth performance, condition factor, and survival of Red tilapia fingerlings tested with different artificial light colors.

obtained by fingerlings which were tested with the blue light (56.50%) with significant differences ($p \leq 0.05$) between treatments. The dry matter content in blue light treatment is the highest (43.50%) with significant differences ($p \leq 0.05$) when comparing with the other treatments.

The results of CP manifested that, the highest value is obtained with green treatment (49.83%), followed by the white one (49.4%) without any significant differences ($p>0.05$), while the poorest CP value is obtained by fingerlings which stocked in red light (46.93%) with significant differences ($p \leq 0.05$). Concerning the EE values, the highest values are achieved by fingerlings which are stocked in red and blue color (35.9 and 35.26%, respectively). Whereas, the poorest values were achieved by fingerlings which were stocked in white color (32.77%) with significant differences ($p \leq 0.05$). Regarding ash content, the lowest value is obtained by fingerlings which were stocked in red color followed by blue color (16.6 and 16.8% DM, respectively). Higher significant contents of carcass energy (Kcal/100 gm) were measured in tilapia fish treated with red and blue colors compared with green and white colors.

Feed utilization

Feed utilization (FI, FCR, PER, PPV, EG, and EU) are presented in Table 3. The differences between treatments regarding FI, FCR, PER, and EG are insignificant ($p>0.05$). Feed intake (FI) was affected by the lighting color, with 35.9% higher in FI under Red color (6.51 g/fish) compared with blue color (4.79 g/fish). The lowest value of FCR is obtained from fingerlings exposed to blue color (2.76), while the highest FCR (2.24) is achieved with red color treatment. Contrarily to the results of FCR, and PER, the PPV, and EU values are significantly in favor of blue color compared with the other tested colors. PPV (%) under blue color treatment is 26.9, 22.3 and 67.7% higher than green, white, and red color treatments, respectively. With the same trend, EU (%) under blue color treatment is 39.6, 39.6 and 60.9% higher than green, white, and red color treatments, respectively.

β-Carotene content

The data of β-carotene for Red tilapia fingerlings are presented in Figures 2-4. Schematic pictures and final data (Mean ± SEM) where β-carotene (IU/ 100 g) are clearly identified under wavelength 325 nm manifested the differences in body color between the tested treatments. Also, real photos for red tilapia at the end of the experiment can clearly confirm the results of the chemical analysis by the chromatograph as shown in Figure 5. The data illustrated that, the highest amount of carotene is achieved by fingerlings which were exposed to blue artificial light, the where β-carotene value is 211.25 IU/100g with highly significant differences ($p \leq 0.05$) in comparing with the other treatments. The blue color is followed by the green color (24.88 IU/100 g), whereas, the lowest value (13.03 IU/100 g) is recorded by red tilapia fingerlings exposed to red artificial light. This means that the content of β-carotene in blue color treatment was 8, 9, and 16 folds comparing with green, white, and red colors, respectively. Figure 5 clearly shows that the Red tilapia fish with black spots (representing retardation in genetic traits) in the blue treatment were 12%, compared with 29% in green, 56% in white and 52% in red colors.

Discussion

Growth performance and feed utilization

Red tilapia has higher market value than Nile tilapia, mainly due to its beautiful color that needs continuous selection to retain and pass their red color from generation to generation [40]. The available data discussing the relationship between illumination color and growth performance still very few. The obtained results of the present study shows that the blue color represented the lowest growth performance and it may due to the reduced vision, which prevents fish to detect the feed well, while red color has better growth performance in spite of the insignificance between the tested colors. This result is congruent with Volpato *et al.* [41]. Who found that red color stimulates feeding intake and improves growth performance more than yellow, green, blue, and white colors. This is attributed to that fish under stress environment such as red color is more aggressive and grew almost many times as much in the less stressful conditions (blue color) [26].

The effect of indirect or background color originated from the reflected color of fish rearing tank on fish growth has been investigated [12,13,42]. Steel *et al.* [39] notices that Summer flounder *Paralichthys dentatus* held in red tanks expresses the greatest weight increase. But, the fish held in dark blue tanks performed the poorest weight gain. However, no significant differences are observed between tank

Treatments	Moisture%	Dry matter, %	CP, as %DM	EE, as %DM	Ash, as %DM	Carcass Energy (Kcal/100gm)
Green	63.50 ± 0.18^{c}	36.50 ± 0.18^{b}	49.83 ± 0.13^{a}	32.77 ± 0.19^{c}	17.49 ± 0.13^{b}	590.44 ± 2.52^{b}
White	64.50 ± 0.09^{b}	35.50 ± 0.09^{c}	49.40 ± 0.14^{b}	32.60 ± 0.08^{c}	17.90 ± 0.06^{a}	586.36 ± 0.09^{b}
Red	68.91 ± 0.15^{a}	31.09 ± 0.15^{d}	46.93 ± 0.12^{d}	35.90 ± 0.08^{a}	16.60 ± 0.13^{c}	603.60 ± 0.47^{a}
Blue	56.50 ± 0.16^{d}	43.50 ± 0.16^{a}	47.33 ± 0.08^{c}	35.20 ± 0.10^{b}	16.80 ± 0.09^{c}	599.25 ± 0.98^{a}
Level of Significance	0.001	0.001	0.001	0.001	0.001	0.001

Means in the same column with different superscript are significantly different ($p \leq 0.05$).

Table 2: Mean ± SEM of carcass chemical composition of Red tilapia fingerlings tested with different artificial light colors

Treatments	FI, gm/fish	FCR	PER, g	PPV, %	EG, kcal	EU, %
Green	5.82 ± 0.99	2.61 ± 0.26	1.57 ± 0.19	37.66 ± 3.63^{b}	6.87 ± 0.84	30.00 ± 2.83^{b}
White	5.93 ± 0.10	2.29 ± 0.06	1.75 ± 0.06	37.85 ± 1.32^{b}	7.17 ± 0.19	30.01 ± 0.69^{b}
Red	6.51 ± 0.36	2.24 ± 0.24	1.84 ± 0.23	28.49 ± 2.99^{c}	6.89 ± 0.93	26.04 ± 2.14^{b}
Blue	4.79 ± 0.57	2.76 ± 0.49	1.53 ± 0.23	47.79 ± 2.33^{a}	8.24 ± 1.15	41.89 ± 1.45^{a}
Level of Significance	0.309	0.588	0.630	0.007	0.649	0.002

Means in the same column with different superscript are significantly different ($p \leq 0.05$).

Table 3: Mean ± SEM of feed intake (FI), feed conversion ratio (FCR), protein efficiency ratio (PER), protein productive value (PPV), energy gain (EG), and energy utilization (EU) of Red tilapia fingerlings tested with different artificial light colors.

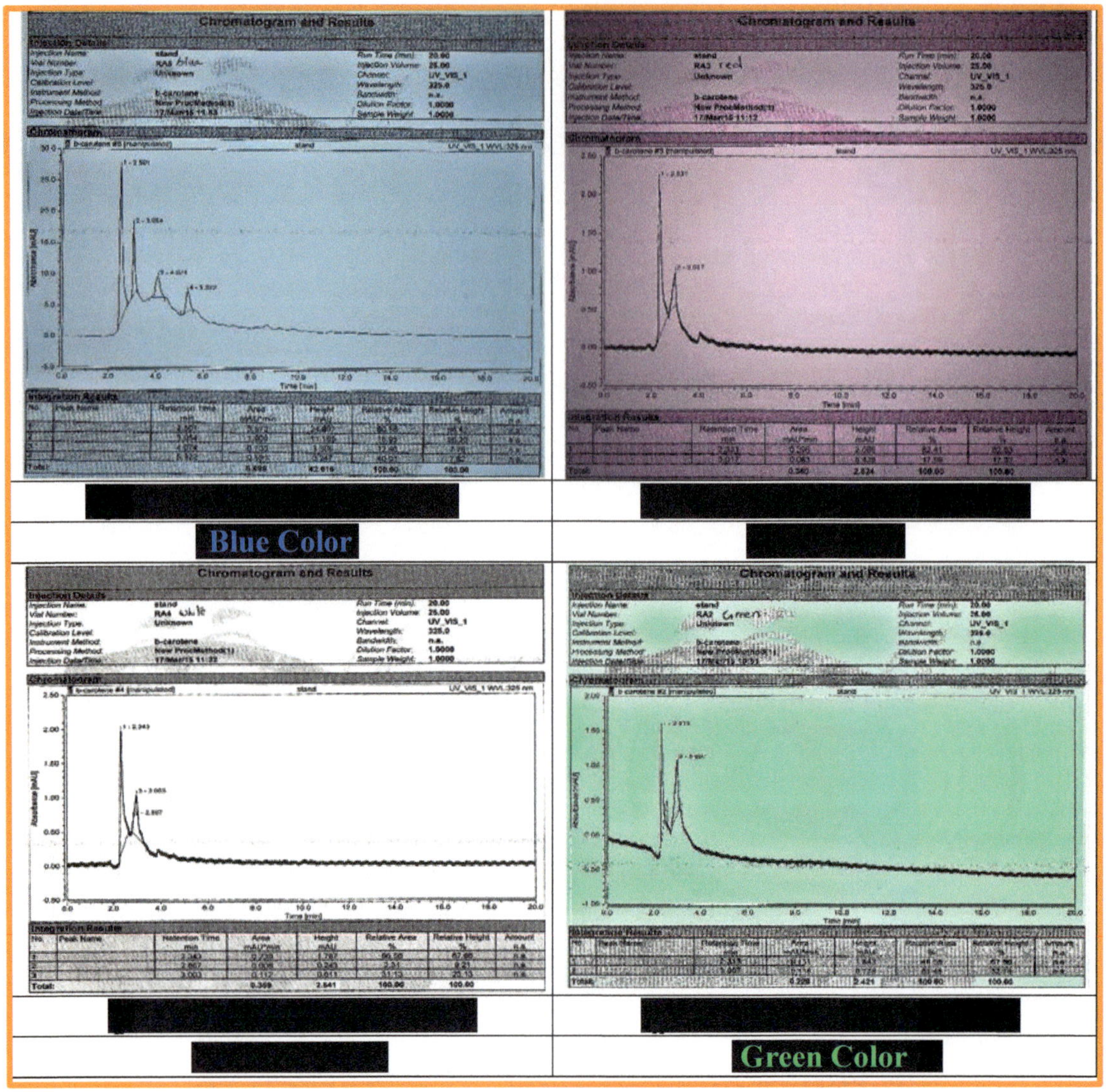

Figure 3: Schematic pictures and final data (Mean ± SEM) of a chromatogram for β-Carotene (IU/ 100 g) accumulated in Red tilapia whole body tested with different artificial lighting colors over fish rearing tanks for 42 days' experimental period. Means with different superscript are significantly different ($p \leq 0.05$).

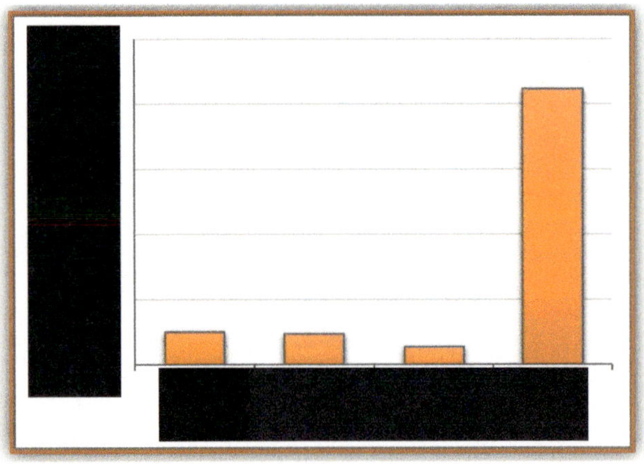

Figure 4: The content of β-Carotene (IU/100 g) accumulated in Red tilapia whole body tested with different artificial lighting colors over fish rearing tanks for 42 days' experimental period.

colors. Similar results are obtained by Karakatsouli *et al.* [43], who finds out that rainbow trout *Oncorhynchus mykiss* shows faster growth and improved physiological condition under red light, with reduced growth under blue light. The growth and skin color of Caspian kutum Rutilus frisii have been examined with five tank colors (white, red, blue, yellow and black) [12]. The results show that there are no significant differences in growth performance between the different experimental groups, but the final body weight tended to be higher in the yellow color tank. On the other hand, in the case of gilthead seabream *Sparus aura*, red light significantly increases brain dopaminergic activity, while a tendency towards reduced growth is also observed. Also, some other studies demonstrated that there are no significant differences in growth performance when comparing different environmental colors such as on Seahorses *Hippocampus abdominal* [44], on scaled carp *Cyprinus carpio* L. [45] and on Atlantic salmon *Salmo salar* L [46]. Nevertheless, in a study to evaluate the effect of tank wall color and light level, treatments affect the growth and survival of Eurasian perch larvae *Perca fluviatilis* L. [42]. The greatest growth in weight and length is observed in tanks

Figure 5: Photos of fish pigmentation and percentage of fish with black spots (FBS) of Red tilapia subjected to different artificial lighting colors at the end of the Experiment.

with light gray and white walls, while the lowest growth is recorded in the tank with the black wall.

Condition factor is an indicator of the general fish condition. It is used to assess the status of the aquatic ecosystem in which fish live [47,48]. When the value of condition factor is higher, this means that the fish has reared in a better condition, such as less stress, better water quality, availability of well-balanced feeds, etc. Condition factor is influenced by both biotic and abiotic environmental conditions. In the present study, it seems that the effect of artificial light color on the condition factor was very limited. The values of condition factor 'K' recorded in the present study are of greater than one under different lighting colors, indicating a good heath status [42]. K values of Red tilapia are similar with the findings of Ayoade *et al.* [49] with values 1.63-165 and less than 3.4 for females and 3.3 for males of Red tilapia [50,51].

Results of the present study for Feed utilization (FI, FCR, and PER) show insignificant differences between treatments with high values of FCR in all groups. This might be attributed to low water temperature. Lighting color definitely affects feed utilization in the present study. The lower value of FCR and the higher values of FI for tilapia fish tested with Red lighting color agree with the findings of [41], who finds out that red light stimulates feeding motivation and improves feed conversion in Nile tilapia, but without extra feeding. This is most likely by affecting central control centers. Also, PER values reflect all the previous results of growth performance parameters.

The importance of light reflection of rearing tank color on feed utilization has been confirmed. Similar results of the present study, obtained by Imanpoo *et al.* [12], detected no significant differences in growth between the different tested tank color, but feeding rate (FR) in the yellow color was lower than that in the black color with no significant differences. Furthermore, Strand *et al.* [52] observed that observed that there was significantly lower feeding rate in Eurasian Perch *Perca fluviatilis* held in yellow tanks when compare with black tanks. Also, McLean *et al.* [13] observed significantly lower FCR for Summer Flounder (*Paralichthys dentatus*) held in red tanks when compared against dark and light blue and green Aquaria. In the same study, the highest FCR (poorest) were observed in tilapia held in dark blue-colored aquaria.

In our experiment, PPV and EU values were significantly higher under blue color treatment, indicating that the fish health is much better than the other tested colors. The best health condition of tilapia fish under blue color is due to lack of secretion of cortisol associated with the inappropriate conditions or stressors [26,16] and other stress hormones like glucocorticoids [53,54].

Whole-body proximate composition

Based on the results of the present study, significant effects of the artificial light color on the chemical composition of Red tilapia can be clearly detected. There is an evident correlation between the blue light color and the quality of red tilapia. Volpato *et al.* [26] stated that blue light color can prevent or reduce stresses, and improves the welfare of the Nile tilapia, while red color has negative effects on fish welfare [51]. The positive effects of blue color on the quality of Red tilapia meat in the present study can be confirmed through the higher content of dry matter, and the lower content of ash. These results are incongruent with the findings of Sharaf, Volpato, Moharram, Zaki, *et al.* [3,26,55,56]. The main explanation is that plasma cortisol levels may be at low levels in fish submitted to a blue light as stated by Schierle *et al.* [38], while one putative explanation is that fish exposing to red light color may have higher levels of plasma cortisol. There was also a positive correlation between ventilation rate [VR] resulted from social stressors, and plasma cortisol concentration in Red tilapia [16]. There is a positive correlation between stressors and plasma cortisol levels in Red tilapia [16]. Increasing Plasma cortisol might effect on body composition leading to higher levels of moisture and lower levels of the protein [57].

The darker lights results in higher EE values and lower protein and ash content than the lighter colors. A similar trend was observed by many authors. In the same context [58], reported that Nile tilapia long-term exposure to green, red and darkness colors showed a significant decrease in total protein compared to the yellow color (control). On the other hand, total protein shows an insignificant decrease in the group exposed to long-term blue light.

Different background colors may also influence the whole body proximate composition of fish. Also, Imanpoo *et al.* [12] illustrated that total protein of the Caspian kutum *Rtilus frisii* is the highest in black-colored tanks and the poorest value was in yellow color treatment, without any significant differences. Nevertheless, lipid content in the yellow tanks treatment was significantly higher than black, but the body moisture was inverse. McLean *et al.* [13] demonstrated that there are no significant differences in dry matter and moisture between groups (dark blue, light blue, black, green and red tank colors) on tilapia fillet. Furthermore, the highest fillet lipid level is in fish maintained in red-colored tanks and the lowest is in tilapia held in black-colored tanks, without significant differences. Similarly, fillet protein level didn't vary across the various groups.

B-Carotene content

The purity and strength of color in red tilapia is one of the most important characteristics that help to increase the demand and at the same time affect the selling price. The influence of light coloration on the fish color and beta-carotene content was profound of the present study. The explanation of these values might be attributed to the stress-releasing hormones. The positive effects of blue light color prevented or reduced the confinement-induced cortisol response [26]. Under blue color, the ability of red tilapia to deposit β-Carotene present in the offered feed with higher rates might be improved than other colors. The result of the present study is consistent with Yasir *et al.* [59] who demonstrated that a blue or green background color could give strength to the orange color, while a white color made fish less color saturated but brighter. In another viewpoint, under a stressful condition, fish are responding by the secretion of stress hormones especially glucocorticoids, which have a remarked positive correlation with the value of carotenoid coloration in fish [53]. Only fish with good health can tolerate high glucocorticoid levels, and thus advertise this ability through better deposition of carotenoids in its body [54]. Higher plasma cortisol levels, faster ventilation rate, and darker skin color were intensified more by Red tilapia fish subjected to confinement and social stresses [16]. There is no conflict between the results of the present experiment and both points of view. Based on Figure 5, there is no conflict between the results of the present experiment and both points of view. To clarify this point, the degree of Red color intensity and Black color retardation rate in the tested fish, showed that the fish that were exposed to a lower degree of light stress (blue color), displayed a lower percent of fish containing the black spots (12%) and higher content of carotene in its body. However, those fish exposed to higher levels of stress, especially red color showed a higher percent of fish containing black spots (more than 4-folds) and lower percent of carotene (less than 16-folds) compared with blue color.

Rapid color change in fishes is controlled by the brain through the pigment called chromatophore [60]. The hormonal control of color alteration in fishes involves two peptide hormones released from the pituitary gland, namely α-MSH and MCH. When aquatic organisms are placed on a stressful dark background, the MSH cells are activated and release a higher concentration of α-MSH into the blood, causing a dispersion of pigments in the dermal melanophores of the skin [12]. Also, the higher level of α-MSH induces cortisol release from the internal tissue as demonstrated for *Oreochromis mossambicus* [61].

The tank color has a clear and documented effect on fish coloration. Imanpoo *et al.* [12] reported that tank color significantly affected on all body color parameters of Juvenile Caspian Kutum *Rtilus frisii*. The values of (lightness) and (whiteness) are significantly lower in black and red tank color than other color treatments. Redness and yellowness are higher in red and yellow, respectively. However, the blue and white colors do not significantly affect the body color. Also Sabri *et al.* [58]. reported that, there are no changes in skin color of Nile Tilapia in different environmental color (blue, green, red, yellow and darkness). With the exception of darkness treatment, fish skin color changed from gray to dark, and finally to the black color when the darkness effect is removed. It seems that the effect of lighting color is species-specific. The effects of substrate color and light intensity on skin color change of the juvenile seahorse, *Hippocampus Erectus* were investigated [62]. The juveniles cultured with substrates of mixed colors (green and orange, or green and red) had higher color change rates and survival than those reared in a single color substrate (green, orange, red, or black). Also, increasing light intensity increases color change rate of the juveniles. In this context Yasir *et al.* [14] found that ambient light intensity could regulate fish color performance, but cannot change the pigment dominance by β - carotene.

The results of β-Carotene obtained in this experiment are conclusive and very important economically and technically. However, many studies need to be done to clarify the scientific explanation for this color change phenomenon in Red tilapia.

Conclusion

It could be concluded that cultivating Red tilapia under a full saline well water using artificial blue color will significantly improve the pigmentation of red tilapia, and the quality of fish in terms of the content of dry matter, but also protein and energy utilization of the consumed feed. Despite the positive effect of red light on growth, but it had negative effects on the content of carotene, the whole-body chemical composition and the increased percentage of fish with black-color spots. In conclusion, our results might be useful in improving fish coloration with simple, economic and sustainable way.

Acknowledgment

We would like to show our gratitude to the "anonymous" reviewer for his so-called constructive insights, careful and conscientious reading and sharing his pearls of wisdom with us. We are also immensely grateful to Prof. Dr. Mohamed Essa, Fish Rearing Lab, NIOF, for his earlier comments that greatly improved the manuscript and Prof. Dr. Elsayed Elebiary for assistance with his papers about Red tilapia and continuous support and encouragements.

References

1. FAO (2014) The State of World Fisheries and Aquaculture 2014. Food and Agriculture Organization of the United Nations. Rome p: 223.
2. Jaspe CJ, Caipang CMA (2011) Small-scale hatchery and larval rearing techniques for local strains of saline-tolerant tilapia, Oreochromis spp. ABAH Bioflux 3: 71-77.
3. Sharaf SM, Mohamed KA, Eldanasoury MA, Nafea RR (2013) Growth and Physiological Responses of Nile Tilapia *Oreochromis niloticus* and Hybrid Red Tilapia (*O. mossambicus* ♂ X *O. niloticus*♀) by Changes in Salinity of Rearing Water. Journal of Animal, Poultry & Fish Production, Suez Canal University 1:23-31.
4. Lawson EO, Anetekhai MA (2011) Salinity Tolerance and Preference of Hatchery Reared Nile Tilapia, *Oreochromis niloticus* (Linneaus 1758). Asian Journal of Agricultural Sciences 3: 104-110.
5. Behrends LL, Nelson RG, Smitherman RO, Stone NM (1982) Breeding and Culture of Red-Gold Color Phase of Tilapia. *J. World Maricul. Soc.* 13: 210-220.
6. Watanabe WO, Fitzimmons K, Yi Y (2006) Farming tilapia in saline waters. In: C. Lim, C.D. Webster (eds.). Tilapia : Biology, Culture, and Nutrition. Food Products Press/Haworth Press, New York, London, Oxford. pp 347-447.
7. El-Sayed AM (2006) Tilapia Culture in Salt Water. Environmental Requirements, Nutritional Implications and Economic Potentials. In: Elizabeth Cruz Suárez L, Denis Ricque Marie, Mireya Tapia Sa lazar, Martha G Nieto López, David AV illarreal Cavazos, Ana C Puello Cruz y Armando García Ortega (eds.). Advances in Aquaculture Nutrition. VIII International Symposium on Aquaculture Nutrition. 15 - 17 November. Autonomous University of Nuevo León, Monterrey, Nuevo León, Mexico ISBN 970-694-333-5.
8. Baroiller JF, Clota F, Cotta HD, Derivaz M, Lazard J, Vergent A (2000) Sea water adaptability of two tilapia species (*S. melanotheron* and *O. niloticus*) and their reciprocal F1 hybrids. p: 303.
9. Fujii R (1993) Coloration and chromatophores. In: Evans DH (ed.) The Physiology of Fishes, Boca Raton, pp: 535-562.
10. Yasir I, Qin JG (2010) Effect of dietary carotenoids on skin color and pigments of false clownfish, *Amphiprion ocellaris*, Cuvier. Journal of the World Aquaculture Society 41: 308-318.
11. Torrissen OJ, Hardy RW, Shearer KD, Scott TM, Stone FE (1990) Effects of dietary canthaxanthin level and lipid level on apparent digestibility coefficients for canthaxanthin in rainbow trout (*Oncorhynchus mykiss*). Aquaculture 88: 351-362.
12. Imanpoo MR, Abdollahi M (2011) Effects of Tank Color on Growth, Stress Response and Skin Color of Juvenile Caspian Kutum *Rtilus frisii Kutum*. Global Veterinaria 6: 118-125.
13. McLean E, Cotter P, Thain C, King N (2008) Tank color impacts performance of cultured fish. Ribarstvo 66: 43-54.
14. Yasir I, Qin JG (2009a) Effect of Light Intensity on Color Performance of False Clownfish, *Amphiprion ocellaris* Cuvier. Journal of the World Aquaculture Society 40: 337-350.
15. Han D, Xie SH, Lei W, Zhu X, Yang Y (2005) Effect of light intensity on growth, survival and skin color of juvenile Chinese longsnout Catfish (*Leiocassis longirostris* Günther). Aquaculture 248: 299-306.
16. Templonuevo RMC, Vera Cruz EM (2016) Responses of Red Nile Tilapia (*Oreochromis niloticus* L.) subjected to Social and Confinement Stresses. The CLSU International J of Sci & Tec 1: 7-14.
17. Huang CM, Chang SL, Cheng HJ, Liao IC (1988) Single gene inheritance of red body coloration in Taiwanese red tilapia. Aquaculture 74: 227-232.
18. Hulata G, Karplus I, Harpaz S (1995) Evaluation of some red tilapia strains for aquaculture : growth and color segregation in hybrid progeny, Aquaculture Res 26: 765-771.
19. Tave D (1987) Genetics and breeding of tilapia: A review, Proc. of The Second International Symposium on Tilapia in Aquaculture, Bangkok, Thailand, p: 285- 294.
20. Duray MN, Estudillo CB, Alpasan LG (1996) The effect of background color and rotifer density on rotifer intake, growth and survival of the grouper (*Epinephelus suillus*) larvae. Aquaculture 146: 217-224.
21. Dowing G, Litvak MK (2000) The effect of photoperiod, tank colour and light intensity on growth of larval haddock. Aquaculture International 7: 369-382.
22. Volpato GL, Duarte CRA, Luchiari AC (2004) Environmental color affect Nile tilapia reproduction. Brazilian J. Medical and Biological Res 37: 479-483.
23. Turner PM (2008) Effects of light intensity and tank background color on sex determination in southern flounder (*Paralichthys lethostigma*). A thesis submitted to the Graduateb Faculty of North Carolina State University In partial fulfillment of the requirements for the Degree of Master of Science Zoology Raleigh, North Carolina p: 71.
24. Hoglund E, Balm PHM, Winberg S (2002) Behavioral and neuroendocrine effects of environmental background colour and social interaction in arctic char (*Salvelinus alpinus*). J. Experimental Biol. 205: 2535-2543.
25. Cobcroft JM, Battaglene SC (2009) Jaw malformation in striped trumpeter *Latris lineata* larvae linked to walling behaviour and tank colour. Aquaculture 289: 247-282.
26. Volpato GL, Barreto RE (2001) Environmental blue light prevents stress in the fish Nile tilapia. Brazilian J. Medical and Biological Res 34: 1041-1045.
27. Szisch V, Van der Salm AL, Wendelaar Bonga SEM, Pavlidis M (2002) Physiological colour changes in the red porgy, *Pagrus pagrus*, following adaptation to blue lighting spectrum. Fish Physiology and Biochemistry 27: 1-8.
28. Rotllant J, Tort L, Montero D, Pavlidis M, Martinez M, Wendelaar Bonga SE Balm PHM (2003) Background colour influence on the stress response in cultured red porgy *Pargus pargus*. Aquaculture 223: 129-139.
29. Fernandez PJ, Bagnara JT (1991) Effect of background color and low temperature on skin color and circulating alpha-MSH in two species of leopard frog. General and Comparative Endocrinol. 83: 132-141.
30. Han D, Xie SH, Lei W, Zhu X, Yang Y (2005) Effect of light intensity on growth, survival and skin color of juvenile Chinese longsnout catfish (*Leiocassis longirostris* Günther). Aquaculture 248: 299-306.
31. Environmental Protection Agency (1989) Ambient water quality criteria for ammonia (Saltwater). EPA 440/5-88-004, National Technical Information Service (NTIS), Springfield, VA.
32. Castell JD, Tiews K (1980) Report on the EIFAC, IUNS and ICES working group on the standardization of methodology in fish nutrition research. Hamburg, Federal Republic of Germany, EIFAC Technical Paper.
33. Newman RM, Martin FB (1983) Estimation of fish production rates and associated variances. Canadian J of Fisheries and Aquatic Sci 40: 1729-1936.
34. Ricker WE (1975) Computation and interpretation of biological statistics of fish populations. Fish Res Board Can Bull 191: 1-382.
35. AOAC International (2000) Official Methods of Analysis of AOAC International, 17th ed. AOAC, Arlington, VA, USA.
36. Brody S (1954) Bioenergetics and Growth. Reinhold Publishing, New York.
37. Jhingran, VG (1991) Fish and Fisheries of India (2nd edn.) India, p : 666.
38. Schierle J, Pietsch B, Ceresa A, Fizet C, Waysek EH (2004) Method for the Determination of β-Carotene in Supplements and Raw Materials by Reversed-Phase Liquid Chromatography: Single Laboratory Validation. Journal of AOAC International 87: 1070-1082.
39. Steel RGD, Torrie JH (1980) Principles and Procedures of Statistics: A Biometrical Approach (2nd edn.) McGraw Hill Book Co.: New York.
40. Lovshin LL (2000) Criteria for Selecting Nile Tilapia and Red Tilapia for Culture. CRSP Research Report p: 157.
41. Volpato GL, Bovi TS, de Freitas RHA, da Sliva DF, Delicio HC, *et al.* (2013) Red Light Stimulates Feeding Motivation in Fish but Does Not Improve Growth. PLoS ONE 8: e59134.
42. Tamazouzt L, Béatrice C, Pascal F (2000) Tank wall colour and light level affect growth and survival of Eurasian perch larvae (*Perca fluviatilis* L). Aquaculture 182: 85-90.
43. Karakatsouli N, Papoutsoglou SE, Pizzonia G, Tsatsos G, Tsopelakos A, *et al.* (2007) Effects of light spectrum on growth and physiological status of gilthead seabream *Sparus aurata* and rainbow trout *Oncorhynchus mykiss* reared under recirculating system conditions. Aquacult. Eng 36: 302-309.
44. Martinez-Cardenas L, Purser GJ (2007) Effect of tank colour on Artemia ingestion, growth and survival in cultured early juvenile pot-bellied seahorses

(*Hippocampus abdominalis*). Aquaculture 264: 92-100.

45. Papoutsoglou SE, Mylonakis G, Miliou H, Karakatsouli NP, Chadio S (2000) Effects of background color on growth performances and physiological responses of scaled carp (*Cyprinus carpio* L.) reared in a closed circulated system. Aquaculture Engineering 22: 309-318.

46. Stefansson SO, Hansen T (1989) Effects of tank colour on growth and smoltification of Atlantic salmon (*Salmo salar* L). Aquaculture 81: 379-386.

47. Anene A. (2005) Condition Factor of Four Cichlid Species of a Man-made Lake in Imo State, Southeastern Nigeria. Turk J Fish Aquat Sci 5: 43-47.

48. Datta SN, Kaur VI, Dhawan A, Jassal G. (2013) Estimation of length-weight relationship and condition factor of spotted snakehead *Channa punctata* (Bloch) under different feeding regimes. Springer Plus. 2:436.

49. Ayoade AA, Ikulala AOO (2007) Length weight relationship, condition factor and stomach contents of *Hemichromis bimaculatus*, *Sarotherodon melanotheron* and *Chromidotilapia guentheri* (Perciformes: Cichlidae) in Eleiyele Lake, Southwestern Nigeria. Revista de Biología Tropical 55: 969-977.

50. Malik A, Abbas G, Soomro MA, Shah SSA, Asadullah A, Bhutto AH, Jamali GQ, Roonjho Z (2017) Length-weight Relationship and Condition Factor of Red Tilapia (Hybrid) Reared in Cemented Tanks of Sun-bright Red Tilapia and Ornamental Hatchery-Karachi, Sindh-Pakistan. Sindh Univ Res Jour 49:159-162.

51. Volpato GL, Bovi TS, de Freitas RHA, da Sliva DF, Delicio HC, *et al.* (2013) Red Light Stimulates Feeding Motivation in Fish but Does Not Improve Growth. PLoS ONE 8: e59134.

52. Strand A, Alanara A, Staffan F, Magnhagen C (2007) Effects of tank colour and light intensity on feed intake, growth rate and energy expenditure of juvenile Eurasian perch, *Perca fluviatilis* L. Aquaculture 272: 312-318.

53. Backström T, Brännäs E, Nilsson J, Magnhagen C (2014) Behaviour, physiology and carotenoid pigmentation in Arctic Charr *Salvelinus alpinus*. J. Fish Biol. 84: 1-9.

54. Sefc KM, Brown AC, Clotfelter ED (2014) Carotenoid-based coloration in cichlid fishes- a review. Comparative Biochemistry and Physiology, Part A 173: 42-51.

55. Moharram SG, El-Ebiary SHE (1999) Effect of dietary lipid level on puberty and oocyte development in Florida red tilapia Oreochromis species (family: Cichlidae) reared under different salinity levels. Bulletin of the National Institute of Oceanography and Fisheries 25: 253-270.

56. Zaki MA, El-Ebiary EH (1997) The use of aquatic plant proteins in completed diets for Florida Red tilapia. Bull Nat Inst Oceanogr and Fish 23: 459-472.

57. Fatemeh A1, Sanaz G, Shahla J (2008) Plasma cortisol changes and body composition in Stizostedion lucioperca exposed to handling stress. Pak J Biol Sci. 11:623-7.

58. Sabri DM, Elnwishy N, Nwonwu F (2012) Effect of Environmental Color on the Behavioral and Physiological Response of Nile Tilapia, *Oreochromis Niloticuss*. Global J Science Frontier Research Biological Sciences 12:11-20.

59. Yasir I, Qin JG (2009b) Impact of background on color performance of false clownfish, *Amphiprion ocellaris*, Cuvier. The Journal of World Aquaculture Society 40:724-736.

60. Skold HN, Aspengren S, Wallin M (2012) Rapid Color Change in Fish and Amphibians - Function, Regulation, and Emerging Applications. Pigment Cell Melanoma Research 26:29-38.

61. Lamers AE, Flik G, Atsma W, Wendelaar Bonga SE (1992) A role for di-acetyl alpha-melanocyte stimulating-hormone in the control of cortisol release in the teleost *Oreochromis mossambicus*. J Endocrinol. 135: 285-292.

62. Lin Q, Lin J, Huang L (2009) Effects of substrate color, light intensity and temperature on survival and skin color change on juvenile seahorses, *Hippocampus erectus* Perry, 1810. Aquaculture 298: 157-161.

Seed Production of Commercially Important Portunid Crab, *Charybdis Feriata* (Linnaeus)

P Soundarapandian[1]*, N Ilavarasan[2], D Varadharajan[1] and K Gangatharan[1]

[1]*Faculty of Marine Sciences, Centre of Advanced Study in Marine Biology,Annamalai University, Parangipettai-608 502. Tamil Nadu, India*
[2]*Department of Zoology, Government Arts College, Karur, Tamil Nadu, India*

Abstract

To develop aquaculture activities mass seed production technology is essential. Hence in the present study an attempt was made to produce seeds in controlled condition. The complete larval development took a span of 22-26 days. According to Duncan's multiple range test the larval duration is more or less same for I and III zoeal stages. Similarly II, IV, V and VI zoeal stages are also similar. However, I and III and II, IV, V and VI zoeal stages are significantly varied between each other. A maximum of 95% of survival was observed during the first zoeal stage and thereafter the survival was gradually decreased. The survival percentage shows that mortality was high in all zoeal and megalopa stages. The survival rate of zoeal and megalopal stages are significantly varied each other.

Keywords: Survival; Zoeal; Megalopa; Crucifix crab; Modern aquaculture

Introduction

The crab *Charybdis feriata is* a portunid crab formerly classified as *Charybdis cruciata* and commonly known as the crucifix crab. It is a commercially important species but is not being cultured commercially. *Charybdis feriata* is a good potential aquaculture species because of its meat quality, taste and size. However in recent years it is noticed that the catch of the crucifix crab is decreasing and for future development and conservation of the species, aquaculture is a right choice to increase the production. In India very few studies have been attempted concern with reproductive biology and population characteristic of crucifix crab [1,2]. No studies have been reported on the life cycle and seed production of the species. Few attempts have been made in the West Asian countries on the larval stages and seed production [3-7]. Knowledge on the life history of the species is very important for captive seed production. Considering the potential of the species seed production experiment was initiated in the present study on the crucifix crab, *Charybdis feriata*.

Materials and Methods

Broodstock maintenance

Healthy gravid females of *Charybdis feriata* with all the appendages intact and the characteristic yellow colured eggs were collected from the Parangipettai coastal waters. They were disinfected with 200 ppm formalin for 30 minutes. Crabs were kept in holding tanks at a salinity of 35 ± 1%, pH 8.2 ± 0.1, temperature 27 ± 2°C and photophase 14 hours light and 10 hours darkness with continuous aeration. Filtered seawater was used for the entire operation and 50% of water was exchanged every day. Mussel and clam meat were given as food. Females about to hatch their eggs were identified by their egg colouration, absence of yolk and vigorous limb movements in the embryo. Broodstock with grey eggs were transferred to the hatching tanks. Eggs hatched during the early hours of the day by jerking movements of the abdomen which is probably to disperse the newly hatched larvae.

Structure of larval rearing tanks

Circular fiberglass tanks (flat bottom) with the capacity of 250 litres were used (size 3 feet height × 2 feet in diameter).

Water quality management

Seawater was brought to the laboratory and filtered by a 0.5mm mesh size cloth sock. The seawater was allowed to settle in a sedimentation tank for 24 hours and passed through sand filters. The water was disinfected by adding calcium hypochlorite (10.8 grams for 500 litres) and allowed to stand for another 24 hours [7]. The water was vigorously aerated for the next 24 hours and passed through cotton filters at the rate of five litres per hour. Antibiotics such as oxytetracyclin and ciproflaxin were added to the rearing water. During the experimental period, salinity, dissolved oxygen, pH and temperature were recorded by using a Centuary Water Analyzer Kit Model CK 711.

Larval culture

Immediately after hatching the aeration was stopped for few minutes for settling the hatching waste and the weak larvae. The newly hatched active photopositive zoeae I congregate along water interfaces. They were siphoned into glass beakers and counted. The number of larvae was estimated (at the rate of 50/litre,) and introduced into rearing tanks.The entire larval cycle (zoeae I to megalopa) was carried out at 35% filtered seawater.

Live feed culture

Cultures of algae, *C. marina* and rotifer, *B. plicatilis* were raised simultaneously in the wet laboratory. The culture details of *C. marina*, rotifer and also *Artemia* hatching were already mentioned earlier.

Feeding

Newly hatched zoea I was fed three hours after stocking. Larvae

***Corresponding author:** P Soundarapandian, Faculty of Marine Sciences, Centre of Advanced Study in Marine Biology, Annamalai University, Parangipettai-608502, Tamil Nadu, India, E-mail: soundsuma@gmail.com

were fed twice daily in the morning (8:00 AM) and evening (5:00 PM). All the zoeal stages and megalops were fed with rotifers (*B. plicatilis*) and *Artemia* nauplii daily. In the morning larvae were fed with *B. plicatilis* at the rate of 5-15 per ml (zoea Iand II), 15-25 per ml (zoea III), 25-40 per ml (zoeae IV-VI) and 70-80 per ml (megalopa). In the evening thawed *Artemia* nauplii, 2-20 per larvae for zoeae I- III and 20-50 per larvae for zoeae IV to megalopa were provided. The zoeae VI on reaching the megalopa stage, were provided with pebbles as substrate, oyster shells were suspended by nylon ropes as hide in the rearing tanks.

Daily the rearing water was exchanged up to three fourths of the tank capacity. Dead larvae and exuviae were siphoned out during this time to prevent contamination. A cloth sock was used to prevent the loss of live zoeae. Mild and continuous aeration was provided (10 to 15 bubbles per minute) to the rearing tanks by an air compressor. Care was taken to prevent irregular aeration using a generator.

Larval numbers were estimated daily by counting 8 replicates taken in 500 ml beakers from the rearing tanks. Assessment of each zoeal stage was done at completion of different levels of metamorphosis to determine feeding and survival rates. Thus mass rearing from different broods was carried out simultaneously in 5 rearing tanks in order to replicate the experiment.

Statistical Analysis

The data were subjected to One- way analysis of variance (ANOVA) and difference between means were determined by Duncan's multiple range tests ($P<0.05$) using SPSS version 17.0.

Results

The regular monitoring of water quality parameters in the culture medium did not show much variation. Parameters like salinity-35 ± 1ppt, dissolved oxygen-5 ± 1 ppm, temperature-30 ± 1°C and pH-7.3 to 8.2 were recorded during the study period. The larval development of *Charybdis feriata* includes six zoeal stages and a megalopa stage. Each zoea has a long rostrum, a dorsal spine and a pair of short lateral spines on the carapace. Zoea resemble the typical portunid larva in morphology and are very active and photopositive. The total duration of the zoeal phase varied between 23-26 days during different trails. The stages were idenitified based on the setation of the telson, number of abdominal segments and pleopods of the abdominal somites.

Identification of larval stages and intermoult duration

Zoea-I: Eyes are sessile. Abdomen is five segmentated plus the telson. Telson forked and each fork bearing one inner and dorsal spine. Inner margin of each fork bears three, long serrated setae (3+3). It took 4-5 days to reach next stage.

Zoea-II: Development of stalked eyes. Abdomen is five segmented as in the previous stage. Abdominal somites 3-5 bears more distinct lateral spines. Telson bears a pair of short, plumose setae on median margin of clept part (4+4). It took 3-4 days to reach next stage.

Zoea-III: Abdomen develops 6 segments and lateral spines on 3-5 somites are longer. Telson is similar to that of previous stage (4+4). It took 3-4 days to reach next stage.

Zoea-IV: Abdomen is as in the previous stage. Pleopod buds just started appearing at the ventral posterior end of somites 2-5. Telson adds one additional short setae on inner margin (4+1+4). It took 3-4 days to reach next stage.

Zoea-V: Pleopod buds on the abdominal somite second to sixth developed. Telson has developed additional short setation inner margin (5+5). It took 3-4 days to reach next stage.

Zoea-VI: Abdominal somites and telson are similar to previous stage. Pleopod buds well developed, biramous on somites 2-5 and uniramous on somite 6. Telson setation same as in previous stage (5+5). It took 3-4 days to reach next stage.

The complete larval development took a span of 22-26 days. According to Duncans multiple range test the larval duration is more or less same for I and III zoeal stages. Similarly II, IV, V and VI zoeal stages are also similar. However, I and III and II, IV,V and VI zoeal stages are significantly varied between each other. The details of the intermoult duration of different larval stages are given in table 1.

Survival

The details of the survival rate of different larval stages are presented in table 2. A maximum of 95% of survival was observed during the first zoeal stage and thereafter the survival was gradually decreased. The survival percentage shows that mortality was high in all zoeal and megalopa stages. The survival rate of zoeal and megalopal stages are significantly varied each other.

Discussion

In order to reduce the gap between supply and increasing demand through the commercialization of captive raised organisms, one special constraint must be overcome – larval mass rearing [8]. To date broodstock development and hatchery seed production of crabs (in terms of percentage of survival) have been experimental, though the technology has developed for the production of crab seeds in many countries. Several studies related to the survival of the commercial portunid crab larvae have used brine shrimp, rotifers and algae as food, since the nutrition turns to be vital to the larval survival. Apart from the live feed, the water quality parameters such as salinity and temperature will also play an important role in the larval growth and survival during mass culture experiments.

Larval Stage	Intermoult Duration (Days)
I Zoea	$4.52^{b} \pm 0.41$
II Zoea	$3.72^{c} \pm 0.36$
III Zoea	$4.01^{b} \pm 0.28$
IV Zoea	$3.97^{c} \pm 0.58$
V Zoea	$3.48^{c} \pm 0.83$
VI Zoea	$3.49^{c} \pm 0.48$
Megalopa	$5.20^{a} \pm 0.12$
Total days	**23-26**

Means with different superscript are significantly different ($P<0.05$; Duncan's multiple range test).

Table 1: Intermoult Duration of Different Larval Stages.

Larval stage	Survival (%)
I Zoea	$95.00^{a} \pm 1.46$
II Zoea	$71.75^{b} \pm 2.36$
III Zoea	$44.0^{c} \pm 3.36$
IV Zoea	$34.5^{d} \pm 4.79$
V Zoea	$25.5^{e} \pm 0.50$
VI Zoea	$12.0^{f} \pm 0.50$
Megalopa	$5.25^{g} \pm 0.50$

Means with different superscript are significantly different ($P<0.05$; Duncan's multiple range test).

Table 2: Survival (%) of Different Larval Stages.

Seawater appears to be an excellent medium for bacterial survival and the microbiological safety of all sea- and freshwater used must be assured to make it the first line of defense against harmful bacterial contamination from other sources [9]. Water quality (temperature, salinity, nutrient and hygiene) is a significant factor in larval survival [10]. Since the nutritional aspects of the hatcheries have been standardized much, it is hypothesized that the microbiological environment in the cultures is now the most significant constraint on the achievement of consistently high levels of survival through the larval cycle at production scale [11]. So, now a days the role of antibiotics is much felt in the mass scale culture and they are found to be enhancing the premetamorphic survival of zoeae while rearing rate of zoeal development and success of metamorphosis to megalopa unaltered [2,12,13]. In the present study, the antibiotics such as oxytetracyclin and ciproflaxin were used to control the microbial load in the rearing tanks and also to provide a healthy environment to the larvae to grow and metamorphose successfully with less intermoult duration. The survival and the health of the larvae were improved after applying these two antibiotics to the rearing medium [14] made a preliminary study on antibiotics such as penicillin-G, streptomycin and polymycin-B, individually and in combination, on survival and development of the crab larvae of *S. serrata.* [12,13] made a similar study in *S. tranquebarica* and *P.sanguinolentus* by treating water with ciproflaxin and oxytetracyclin to control the microbial load in the larval rearing medium.

Survival and longevity of marine invertebrate larvae are influenced by abiotic factors such as water temperature and salinity, and by biotic factors such as food availability, food quality and predation. In the present study the water salinity was maintained at 35 ppt since the spawning, embryogenesis and hatching of eggs generally takes place in coastal regions as in *P. pelagicus* [15] and *P.sanguinolentus* [13]. The results of the present study indicate that the most suitable range of temperature for crab larvae was found to be 30 to 32.5°C. The higher mortality rate of the zoea I, especially during the first three days of culture might be due to fluctuations in water temperature in brooder's tank and the mass culture tanks [16] have noticed that the temperature shock caused larval stress and mortality once surmised when there is an unintentional temperature fluctuation of 5°C's due to equipment failures that lead to abnormally high mortality rates. The better survival evidenced in the present study with the larvae of *Charybdis feriata* might also be due to the higher temperature (30 ± 1°C). The previous studies on the effect of temperature on larval rearing revealed that the larvae could not survive in lower temperatures. The lower survival rate was evident when the larvae of *S. serrata* reared at low mean temperatures, *i.e.*, at 24°C by Du Plessis [17], at 22°C by Warner [18] and at 27°C.

The quantity and quality of food supply are the chief factors regulating the duration of larval development [19,20]. Insufficient food supply will prolong larval development, thus, increasing the risks of larval mortality due to predation and starvation. At first feeding, larvae usually restrict the size of the food particles that can be ingested. Providing prey of a suitable size is one of the more important feeding strategy aspects of crustacean larvae, which hatch with little or no yolk reserve. Suitable prey for larvae should meet three general criteria: namely, they should be an appropriate size for easy capture and consumption, they should be present at an adequate concentration, and they should contain essential dietary nutrients [20]. The seed production of aquatic species almost entirely depends on the successful production of live food organisms, principally rotifers, followed by *Artemia* and that too enriched live food organisms of improved quality. The superiority of the live food organisms in larval nutrition over existing compounded diet is partly due to the availability of exogenous enzymes through the live food, which in combination with endogenous enzymes of the animal lead to efficient digestibility [21]. Young animals with less developed digestive system benefit more from exogenous enzymes than do adults. The exact quantity of food required at each stage cannot be prescribed as it depends on the utilization of feed by the larvae and must be judged visually by the operator.

In the present study the zoea were initially fed with *B. plicatilis*, since the small size of first zoea refused to feed on *Artemia* nauplii. *B. plicatilis* is small in size and can be ingested completely by small decapod crustacean larvae. Rotifer gut is usually filled with bacteria and algae, which could provide additional nutrition for the larval forms of decapods. It has a short life cycle with simple dietary requirements can be cultured in high densities and has a favorable nutritional content. Chen and Lin HF and Lovett and Fielder [22,23] reported that caloric content of rotifers per gram ash dry weight is not significantly different from that of *Artemia* nauplii. The larvae of *Charybdis feriata* were provided with *Artemia* nauplii only later stages especially from III zoea onwards. The studies on the crab larvae showed that the absence of small prey during the early zoeal stage of *C. sapidus* resulted in high mortalities [24] reported that the smaller size and slower swimming speed of *B. plicatilis* apparently allow their capture and manipulation by small crab zoea. He also found out that newly hatched larvae of *C. sapidus* couldn't pass to the next stage when fed with *Artemia* nauplii [25] observed that the *M. malcolmsonii* early larval stages apparently graze on the appendages of *Artemia* nauplii but could consume entire rotifers. The swimming crab *P. trituberculatus* was fed with *Artemia* nauplii from third zoea stage to avoid cannibalism [26].

Combination of *Artemia* nauplii and rotifer obtained mixed results when fed to the crab larvae by different authors like Takeuchi et al. [27] showed that mud crab larvae fed on *Artemia* nauplii alone had a higher survival rate than those fed on rotifers. He suggested that the addition of rotifers might have contributed to the deterioration of the culture medium, through oxygen consumption or release of metabolites, without providing any nutritional benefit for the larvae [20] reported that *Rithropanopeus harrisii* larvae fed on rotifer could not metamorphose due to low lipid content and low feeding efficiency. Hill and Maheswarudu et al. [28,29] demonstrated that the rotifers are more important than *Artemia* nauplii for maintaining the survival rate of the first and second zoeal stages, whereas supplying *Artemia* or rotifers as the sole prey failed to maintain the survival rate of mud crab. In most of the previous studies, successful seed productions obtained when rotifer and *Artemia* nauplii were used as feed [15]. Successful seed production was reported in *P. trituberculatus* offered with rotifer and *Artemia* nauplii [30,31]. Minagawa and Murano [32] recommended mixed diets (*Artemia* nauplii + rotifer) for mass seed production of frog crab, *R. ranina*. In the present study, both rotifer and *Artemia* nauplii have been offered to the larvae of *Charybdis feriata* as experimented in the previous study. However, the survival rate is not encouraging. Various reasons are attributed for the lower survival even though standard live foods were used.

Information on larval nutritional requirements is important for the establishment of successful seed production technology. Improving the food value of *Artemia* through enrichment prior to feeding of fish or crustacean larvae is now a common practice among marine hatcheries for improved survival, growth and stress resistance [33,34]. The advantage of using *Artemia* nauplii for feeding the larval crab is that it could have contribute to the lipid and energy resulting in a high feeding efficiency. In general live food is lack of n-3 HUFA, without which the growth of the developing larvae will not be optimized. Although data on the nutritional requirements of brachyuran crabs in the larval stage is limited, the essential fatty acid requirements has been revealed for several species [12,13,15,16,26,27]. Earlier study showed that feeding mud crab larvae with live food containing a low nutrition value, especially n-3 HUFA resulted in lower survival and longer intermoult period. emphasized that as the amount of n-3 highly unsaturated fatty acids (n-3HUFA) increases in the feed, the larval survival, growth and velocity of development were also been improved in the larvae of swimming crab, *P. trituberculatus*. The swimming crab larvae fed with *Artemia* containing n-3 HUFA from the 3rd stage to obtain high survival rate [31]. All the *Artemia* do not possess all the essential fatty acids in required concentrations, particularly 22:6 n-3 [34]. Similarly Watanabe et al. [35] reported that rotifer cultured with the yeast were quiet low in n-3 unsaturated fatty acids. The larvae fed with cuttle fish liver oil enriched *Artemia* nauplii and rotifer showed accelerated growth, and survival [36]. Larval growth and morphogenesis might be controlled by the nutritional conditions of the prey and the larva itself. Although it has not been shown for brachyuran larvae that the morphogenesis is affected by nutritional factors, there are a few reports that dietary n-3HUFA improves the growth which was represented by the carapace width of the first crab stage in the larval rearing of *S. serrata* [37], *S. paramamosain* [38,39] and the swimming crab, *P. trituberculatus* [27,39].

The larval development of *Charybdis feriata* resembles other portunid crabs with variation in number of zoeal stages [40-42]. *Charybdis feriata* crab had six zoeal stages, the maximum number recorded for a portunid crab in the Indo-Pacific region. The larval stages recorded in the present study are found to be similar to the larval stages of *Charybdis feriata*, reported from other regions [2,3,5,6]. The final survival rate of *Charybdis feriata* larvae in the present study was 5.25% from VI zoea to megalopa. The survival rate is comparatively higher than that reported for *S. serrata* by Heasman et al. [19]. Almost similar report obtained in *P.sanguinolentus* [13] and *Charybdis feriata* [3]. The reasons could be that the feeding schedules, incorporated combinations of B. *plicatilis* and *Artemia* nauplii. Similarly, Anuar et al. [43] pointed out that the combination of *B. plicatilis* and *Artemia* nauplii served as feed gave better survival [44]. The survival rate in the zoeal stages (I to IV zoeae, 65 to 35 per cent) achieved by Heasman et al. [19] in the larvae of *S. serrata* was due to high concentration of *Artemia* nauplii (5 to 30 nos. per ml) used once in a day. Larvae of the related family *P. trituberculatus* have been reared with success (over 60%) on the same species of rotifer in combination with *Artemia* by Hue et al. [30]. Minagawa [32] recommend a combination of diets to mass rear the larvae of *R. ranina*.

In the present study mortality in all zoeal stages of *Charybdis feriata* regardless of the feeding regimes is quite comparable with the results obtained by Josileen and Menon [44] in *Charybdis feriata* and Soundarapandian et al. [15] in *P. pelagicus* and Nunnam et al. [13] in *P. sanguinolentus*. They pointed out from their experiments that initial mortality during the first two days of the experiment occurred often and relatively high regardless of feeding regimes and are rather unpredictable. They related the mortality to the low viability of the individual larva. Similarly mortality during the zoea VI and megalopa stages was either before moulting, during moulting (includes incomplete moulting) in *Charybdis feriata* larvae. The possible reason cited by Anger [45] is that mortality due to depletion of reserves resulting in larval inability to catch the prey. Similarly, Rosenberg and Costlow [46] suggested that the majority of the larval population is preparing for the premetamorphic moult to megalopa. Costlow and Bookhout and Christiansen and Costlow [47,48] have observed high mortalities in the larvae of *R. harrisii* at the premetamorphic stage. They attribute two reasons for such mortality – 1. The larvae at this premetamorphic stage are extremely susceptible to unfavorable environmental conditions at this time of life cycle and 2. The metabolic cost of metamorphosis is very high and appears to decrease the capacity of larvae to counteract these unfavorable conditions. Cannibalistic tendency was observed in all zoeal and megalopal stages and it was the main reason for the higher mortality from Ist zoea to megalopa. The shelter provided to this larval stage was found effective and reduced the cannibalism to some extent.

The maintenance of good water quality and hygiene during the larval culture results in higher survival percentages. The hygiene begins with the preparation of the broodstock and continues up to the metamorphosis of megalopa. The aim is to restrict the growth of potential pathogens, including bacteria, fungi, viruses and protozoa in the culture system. It can be safely assumed that all inputs into a culture tank are potential sources of infection that may reduce rates of larval survival and metamorphosis. All tanks and equipment used in the culture must first be effectively sterilized following standard methods before use as a simple precautionary measure [49,50].

In the present study larval mortality was spread throughout the six zoeal and megalopa stages and was not confined to any one particular stage of development. However, highest mortality occurred during the transition from first zoea to second and sixth zoea to megalopa. This may be due to the inability of the larvae to break completely away from their casts. Similar observations were already made in *P.sanguinolentus* [13] and *Charybdis feriata*. Many workers have reported that the mortality was high during the first zoeal moult to second zoea in *S. serrata* [19,42,49].

[50] have observed high larval mortality in the first zoeal stage in *Panopeus herbstii*. [42] reported 40% mortality in the first zoeal stage in *S. oceanica*. [51] in *C. sapidus* and [52] in *P. pelagicus* have reported high larval mortalities in the first two zoeal stages. [53] found more mortality in later zoeal stages of *Cyclograpsus cinereus*.

The crab *Charybdis feriata* has been listed one of the six suitable species for stock enhancement and culture in the international forum on the culture of portunid crabs [54]. The present study also found that *Charybdis feriata* suitable for seed production. Hence, using the present study as database, it is clear that the larvae of *Charybdis feriata* can be cultured on a large scale by adding more importance to the water quality to prevent mortality by bacterial infection and with the combination of *B. plicatilis* and *Artemia* nauplii as feed. The rotifer and the enriched *Artemia* nauplii were found to be supporting the larval development very much and hence better survival could be observed. No ciliate and bacterial infection was noted during the study period.

References

1. Padayatti PS (1990) Notes on population characteristics and reproductive biology of the portunid crab charybdis (charybdis) feriatus (linnaeus) at cochin. Indian J Fish 37: 155-158.

2. Josileen J (2011) Captive spawning, hatching and larval development of crucifix crab, Charybdis feriatus (Linnaeus, 1758). J Mar Biol Ass India 53: 35-40.

3. Motoh H, Villaluz DK (1976) Larvae of Decapod Crustacea of the Philippines–I. The zoeal stages of a swimming crab, *Charybdis cruciata* (Herbst) reared in the laboratory. Bulletin of The Japanese Society Of Scientific Fisheries 42: 523-531.

4. Shinkarenko L (1979) Development of the Larval Stages of the Blue Swimming Crab Portunus pelagicus L. (Portunidae : Decapoda : Crustacea). Australian Journal of Marine and Freshwater Research 30: 485-503

5. Parado-Estepa, Emilia TQ, Eduard MR (2002) Seed Production of the Crucifix Crab *Charybdis feriatus*. Aquacult Asia 3: 37.

6. Parado-Estepa, Emilia TQ, Eduard MR (2007) Seed Production of *Charybdis feriatus* (Linnaeus). Aquac Res 38: 1452-58.

7. Lagoc J (1990) Disease prevention in shrimp hatcheries. SEAFDEC, Aqua Farm News VIII 2.

8. Calado R, Narciso L, Morais S, Rhyne AL, Lin J (2003) A rearing system for the culture of ornamental decapod crustacean larvae. Aquacult 218: 329-339.

9. Blackshaw AW (2001) Larval culture of *Scylla serrata*: maintenance of hygiene and concepts of experimental design. Asian Fish Sci 14: 239-242.

10. Motoh H, de la Pena D, Tampos E (1977) Laboratory breeding of the mud crab *Scylla serrata* (Forskal) through the zoea and megalopa stages to the crab stage. Res Rep Aqucult Dep SEAFDEC 1: 14-18.

11. Mann D (2001) The influence of microbiology on the success of mud crab larval culture. ACIAR Proceedings No 78: 153-158.

12. Thirunavukkarasu N (2005) Biology, nutritional evaluation and utilization of mud crab *Scylla tranquebarica* (Fabricius, 1798). Ph.D. Thesis, Annamalai University, India. 1- 127.

13. Nunnam John Samuel, Soundarapandian Peyail, Anand Thananjayan (2011) Seed production of commercially valuable portunid crab *Portunus sanguinolentus* (Herbst). European J Exp Biol 1: 23-32.

14. Brick RW (1974) Effects of water quality, antibiotics, phytoplankton and food on survival and development of larvae of Scylla serrata (Crustacea: Portunidae). Aquaculture 3: 231-244.

15. Soundarapandian P, Thamizhazhagan E, Samuel JN (2007) Seed production of commercially important blue swimming crab *Portunus pelagicus* (Linnaeus). J Fish Aqua Sci 2: 302-309.

16. TRIÑO1 AT, MILLAMENA OM, KEENAN CP (2001) PondCulture ofMud Crab Scylla serrata (Forskal) Fed Formulated Diet With or Without Vitamin and Mineral Supplements Asian Fisheries Science 14: 191-200.

17. Du Plessis A (1971) A preliminary investigation into the morphological characteristics, feeding, growth, reproduction and larval rearing of *Scylla serrata* Forskal, held in captivity. Unpublished report of the Fisheries Development Corporation of South Africa 24.

18. Warner GF (1977) Food and feeding. In: The biology of crabs. Gresham Press Old Working Surrey UK 85-94.

19. Heasman MP, Fielder DR (1983) Laboratory spawning and mass rearing of the mangrove crab, Scylla serrata (Forskal), from first zoea to first crab stage. Aquaculture 34: 303-316.

20. Roberts MH Jr. (1974) Larval development of pagurus longicarpus say reared in the laboratory. V. Effect of diet on survival and molting. Biol Bull 146: 67-77.

21. McConaugha, Jr (1985) Nutrition and larval growth. In: Wenner, AM (ed.). Crustacean Issues. 2: Larval growth. AA Balkema, Rotterdam 127-159.

22. Chen HY, Lin HF (1992) Effects of different artemia diets on the growth and digestive enzyme activities of early post-larval *Penaeus monodon*. Asian Fish Sci 5: 73-81.

23. Lovett DL, Fielder DL (1988) Evaluation of the rotifer Brachionus plicatilis as a substitute for Artemia in feeding larvae of Macrobrachium rosenbergii. Aquaculture 71: 331-338.

24. Emmerson WD (1984) Predation and energetics of Penaeus indicus (Decapoda: Penaeidae) larvae feeding on Brachionus plicatilis and Artemia nauplii. Aquaculture 38: 201-209.

25. Sulkin SD (1975) The significance of diet in the growth and development of larvae of the blue crab, Callinectes sapidus rathbun, under laboratory conditions. J Exp Mar Biol Ecol 20: 119-135.

26. Soundarapandian P, Kannupandi T, John Samuel M (1998) Effects of feeds on digestive enzymes of juveniles of *Macrobrachium malcolmsonii* (H. Milne Edwards). Ind J Exp Biol 36: 720-723.

27. Takeuchi T, Nobukazu Satoh, Sachio Sekiya, Tomohito Shimizu, Takeshi Watanabe (1999) Influence of the different feeding schedule of food organisms for larval swimming crab *Portunus trituberculatus*. Nippon Suisan Gakkaishi 65: 804-809.

28. Hill BJ (1974) Salinity and temperature tolerance of zoeae of the portunid crab Scylla serrata. Mari Biol 25:21-24.

29. Maheswarudu G, Josileen Jose, Manmadhan Nair, Arputharaj MR, A Ramakrishnan, et al. (2007) Larval rearing of mud crab, Scylla tranquebarica (Fabricius, 1798) and feeding requirements of its zoea1. nng of mud crab. J Mar Biol Ass India 49: 41-46.

30. Hue JS, Bang KS, Rho YK (1972) Studies on growth and artificial rearing of the larval blue crab, *Portunus trituberculatus*. Bull Korean Fish Res Dev Agency 9: 55-70.

31. Takeuchi T (2001) A review of studies on the effect of dietary N-3 highly unsaturated fattyacids on larval swimming crab, *Portunus trituberculatus* and mudcrab, *Scylla tranquebarica*. Proc JSPS Int Symp Fisheries Sci Tropical Area 244-247.

32. Minagawa M, Murano M (1993) Effects of prey density on survival, feeding rate and development of zoeas of the red frog crab, *Ranina ranina* (Crustacea: Decapoda: Raninidae). Aquaculture 113: 91-100.

33. Sorgeloos P, Leger PH (1992) Improved larviculture outputs of marine fish, shrimp and prawn. J World Aquacult Soc 23: 251-264.

34. Izquierdo MS (1996) Essential fatty acid requirements of cultured marine fish larvae. Aquaculture Nutrition 2: 183-191.

35. Watanabe T, Kitajima C, Fujita S (1983) Nutritional values of live organisms used in Japan for mass propagation of fish: a review. Aquacult 34: 115-143.

36. Kannupandi T, Veera Ravi A, Soundarapandian P (2003) Efficacy of enriched diets on the larval development and survival of an edible crab, *Charybdis lucifera* (Fabricius). Indian J Fish 50: 21-23.

37. Kobayashi T, Takeuchi T, Arai D, Sekiya S (2000) Suitable dietary levels of EPA and DHA for larval mud crab during *Artemia* feeding period. Nippon Suisan Gakkaishi 66 : 1006-1013.

38. Suprayudi MA, Takeuchi T, Hamasaki K, Hirokawa J (2002) The effect of N-3 HUFA content in rotifers on the development and survival of mud crab, *Scylla serrata*, larvae. Suisanzoshoku 50: 205-212.

39. Hamasaki K, Takeuchi T, Sekiya S (1998) Dietary value for larval swimming crab *Portunus trituberculatus* of marine rotifer *Brachionus rotundiformis* cultured with several feeds. Bulletin of the Japanese Society of Scientific Fisheries 64: 841-846.

40. Greenwood JG, Fielder DR (1983) Description of the later zoeal stages and megalopa of Charybdis truncata (Fabricius, 1798) (Crustacea: Portunidae). J Nat Hist 17: 15-30.

41. Fielder DR, Greenwood JG, Campbell G (1984) The Megalopa of Charybdis feriata (Linnaeus) with Additions to the Zoeal Larvae Descriptions (Decapoda, Portunidae). Crust 46: 160-165.

42. Anil MK (1997) Studies on the fishery and culture prospects of mud crabs (genus *Scylla* De Haan) along the Kerala coast. Ph.D. Thesis, Cochin University of Science and Technology 1-213.

43. Anuar H, Hai TN, Anil C, Mithun S (2011) Preliminary Study on the Feeding Regime of Laboratory Reared Mud Crab Larva, Scylla serrata (Forsskal, 1775). World Applied Sciences Journal 14: 1651-1654.

44. Josileen J, Menon NG (2004) Larval stages of the blue swimmer crab, *Portunus pelagicus* (Linnaeus, 1758) (Decapoda, Brachyura). Crust 77: 785-803.

45. Anger K (1985) Influence of salinity on larval development of the spider crab, Hyas araneus, reared in the laboratory. Marine biology of polar regions and effects of stress on marine organisms. Wiley, Chichester.

46. Rosenberg R, Costlow JD Jr (1979) Delayed response to irreversible non-genetic adaptation to salinity in early development of the brachyuran crab Rhithropanopeus harrisii, and some notes on adaptation to temperature. Ophelia 18: 97-114.

47. Costlow JD, Jr, Bookhout CG (1971) The effect of cyclic temperatures on larval development in the mud crab *Rithropanopeus harrisii*. In: Crisp JD, D.J. (Ed.). Fourth European Marine Biology Symposium, Cambridge University Press 211-220.

48. Christiansen ME, Costlow JD Jr (1975) The effect of salinity and cyclic temperature on larval development of the mud-crab Rhithropanopeus harrisii (Brachyura: Xanthidae) reared in the laboratory. Mar Biol 32: 215-221.

49. Ong KS (1966) The early developmental stages of *Scylla serrata* Forskal (Crustacea, Portunidae), reared in the laboratory. Proc Indo-Pacific Fish Coun 11: 135-146.

50. Costlow JD, Bookhout CG Jr (1962) The larval development of *Hepatus epheliticus* (L.) under laboratory conditions. J Elisha Mitchell Scient Soc 78: 113-125.

51. Costlow JD Jr, Bookhout CG (1959) The larval development of *Callinectes sapidus* Rathbun reared in the laboratory. Biol Bull Mar Biol Lab Woods Hole 116: 373-396.

52. Raman K, Srinivasagam S, Rangaswamy CP, Krishnan S, Joseph KO, Sultana M (1987) A note on larval rearing of the edible crab, portunus pelagicus linnaeus, at ennore hatchery, Madras. Indian J Fish 34: 128-131.

53. Costlow JD, Jr, Fagetti E (1967) The larval development of the crab, *Cyclograpsus cinereus* Dana, under laboratory conditions. Pacif Sci 21: 166-177.

54. Costlow JD, Bookhout CG Jr, Monroe R (1966) Studies on the Larval Development of the Crab, Rhithropanopeus harrisii (Gould). I. The Effect of Salinity and Temperature on Larval Development. Physiol Zool 39: 81-100.

The Levels of Toxic Metals in Blue Crab *Portunus segnis* from Persian Gulf

Mehdi Hosseini[1]*, Seyed Mohammad Bagher Nabavi[2] Jamileh Pazooki[1] and Yaghoob Parsa[2]

[1]*Department of Marine Biology, Faculty of Biological Science, Shahid Beheshti University, Tehran, Iran*

[2]*Department of Marine Biology, Faculty of Marine Science, Khoramshahr University of Marine Science and Technology, Iran*

Abstract

Levels of heavy metals) Ni, Co, Cd and Pb(in muscle, gill and hepatopancreas of blue swimming crab *P. segnis* and relationship with its food source from Persian Gulf, were investigated. In this study, there was a direct relationship between heavy metals levels in tissues crabs with organisms that feeding on them. Female crabs feed on plant and detritus and are close to bottom sediment and receive more sediment associated metals. Also, there was a positive correlation between metal concentrations in tissues with size of food items. Heavy metal concentrations were highest in female hepatopancreas whereas lowest in the muscle of all crab species. From the human consumption point of view, Heavy metals concentrations were below the admissible limits except for hepatopancreas of female crabs. Thus, precautions should be taken on account of higher content of metals as well as in other organs that could be affected by industrial pollution.

Keywords: Heavy metal; Monitoring; *Portunus segnis*; Persian Gulf

Introduction

Heavy metals can be classified as potentially toxic elements such as cadmium, lead and mercury and essential elements such as copper, zinc and iron. Toxic metals naturally occur in aquatic environments in very low concentrations, but their concentration levels have increased due to anthropogenic pollutants over time. Industrial activities as well as agriculture and mining create a potential source of metals pollution in aquatic environment.

Accumulation of metals in the organisms depends on various biological and environmental factors such as size, age, feeding habit, temperature and dissolved oxygen [1]. However, feeding habit plays a significant role in the accumulation of metals in organism's tissues [2]. Because metals have tendency to be biomagnified through food chains [3]. Most aquatic food chains begin from invertebrates. However, ability of invertebrates to accumulate contaminants from water varies from species to species. Thus biomagnification of contaminants entered aquatic food chains is, significantly affected by invertebrate species.

It is known that certain forms of metals can readily accumulate within crustacean tissues at much higher levels than those in the water column and in sediment [1]. The blue swimming crab *Portunus segnis* belongs to the phylum; arthropoda which make up about three quarters of living animal species. For the fauna, the species most studied are the benthic macroinvertebrates, particularly those with low mobility, which accumulate larger concentrations of metals compared to animals that live in open water. Among the most-studied are the crab Portunidae which has important characteristics that allow the study of bioaccumulation: they feed mainly on a wide variety of fish, bivalves, plant, crustaceans and benthic animals [4], as well as the sediment itself [5] and they have a slow growth rate and long life cycle [4]. Therefore, this crab species is especially appropriate for use in studies of environmental impact by metals from an ecosystemic.

The Persian Gulf is a shallow and semi-enclosed sea that its environment is changing rapidly [6]. The discovery of oil in this sea led to a massive increase in anthropogenic activities in the area. In general, the agricultural use of fertilizers, herbicides, pesticides, petrochemical and oil industries are the major sources of pollution in this area [6]. Therefore, levels of Ni, Co, Cd and Pb in tissues (muscle, hepatopancreas and gills) of blue swimming crab *Portunus segnis*, from northwest of the Persian Gulf, were studied.

Materials and Methods

Sampling area was selected along the Persian Gulf (Figure 1). The Persian Gulf lies on the south Iran, between longitudes 48°25' and 56°25" East, and latitudes 24°30" and 30°30' North. It has an estimated area of 260 km^2 and extends 600 km offshore to a depth average of about

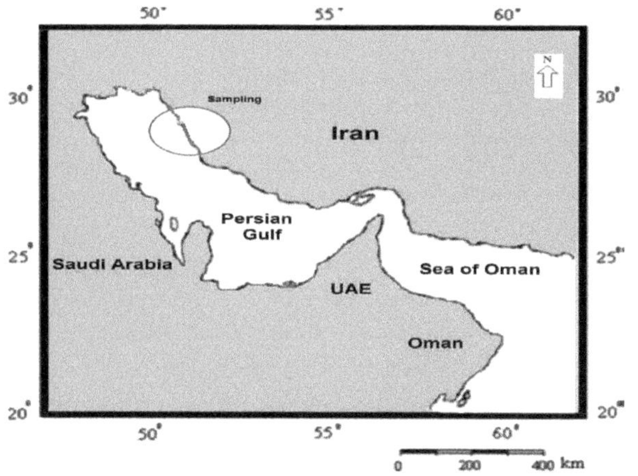

Figure 1: Map of Persian Gulf showing sampling sites and the study area.

***Corresponding author:** Mehdi Hosseini, Department of Marine Biology, Faculty of Biological Science, Shahid Beheshti University, Tehran, Iran
E-mail: smhbio@yahoo.com

30-40 m. This area is a sandy-muddy habitat that provides a suitable ecosystem for a lot of marine organisms such as crab, shrimp and fish [7]. Major sources of contamination in this area include the agricultural use of fertilizers, herbicides, pesticides, hazardous substance spills from various refineries, petrochemical and oil industries. Sampling was performed by trawl net. The samples placed on ice, immediately transported to the laboratory on the same day and stored at -20°C until analysis [8].

Each species was properly cleaned by rinsing with distilled water to remove debris, planktons and other external adherent, and the sex of each individual was determined according to morphological characteristics. Male crabs are bright blue and their abdomens (womb area) are narrow and in the form of a spear, while female crabs are green–brown and have round abdomens [9]. The samples were dissected with sterilized scissors and tweezers to remove samples of two tissues (muscle, gill and hepatopancreas), in standardized locations: (i) muscle of the chelar propodus, due to higher metal accumulation verified by Chen et al. [10]; (ii) hepatopancreas tissue, which has a particularly high metabolic rate [11] and (iii) gills, because of their osmoregulatory function [11]. It was then drained under folds of filter, weighed, wrapped in aluminum foil and then frozen at 10ºC prior to analysis. The tissues were placed in clean watch glasses and were oven dried at 105ºC for 1 hour and later cooled in the desiccators. Each sample of crab was homogenized in an acid-cleaned mortar and 2 g were digested in triplicate in a water bath at 60ºC for 6 h after adding 2.5 mL each of concentrated HNO_3 and H_2SO_4 [12].

Each sample was analyzed for metals by the mineralization method with HNO3 at 65 percent, according to Basset et al. [13]. Analyses were optimized by hollow cathode lamps (LCO), according to the metallic element analyzed, and samples were read using a GBC-932 AA atomic absorption spectrophotometer [8]. The equipment was calibrated using metal stock solutions (1000 ppm). The recovery means for Ni, Co, Cd and Pb was 99%, 103%, 99% and 102% respectively.

All data were tested for normal distribution with Shapiro-wilk normality test. The comparisons of metals levels between muscle, gill and hepatopancreas of crab were carried out by t-test. All concentrations are reported in μg g^{-1} dry weight and a probability of $p<0.01$ was set to indicate statistical significance.

Results

Table 1 shows scientific name, sex, the mean body weight and food type for the blue swimming crab *Portunus segnis*. Metals concentrations were calculated in microgram per gram wet basis (μgg-1). In order to check the validity of the measurements, reference material (Multi-4, Merck) was used. The mean concentration and standard error of studied heavy metals are given in Table 2.

Ni

Nickel concentration in the hepatopancreas tissue of crabs ranged from (09.2 μg g^{-1}) to (20.5 μg g^{-1}) for male and from (11.2 μg g^{-1}) to (31.1 μg g^{-1}) for female, while concentrations in the gill were in the range of (07.5 μg g-1) to (15.2 μg g^{-1}) for male and from (10.4 μg g^{-1}) to (20.6 μg g^{-1}) for female crabs. The nickel concentrations in the muscle of the male and the female crabs were observed to be between (7.6 μg g^{-1}) to (14.1 μg g^{-1}) and between (8.1 μg g^{-1}) to (17.5 μg g^{-1}), respectively.

Co

Cobalt concentration in the hepatopancreas tissue of crabs ranged from (0.41 μg g^{-1}) to (0.9 μg g^{-1}) for male and from (0.45 μg g^{-1}) to (1.2 μg g^{-1}) for female, while concentrations in the gill were in the range of (0.19 μg g^{-1}) to (0.6 μg g^{-1}) for male and from (0.24 μg g^{-1} 2) to (0.7 μg g^{-1}) for female crabs. The nickel concentrations in the muscle of the male and the female crabs were observed to be between (0.16 μg g^{-1}) to (0.5 μg g^{-1}) and between (0.21 μg g^{-1}) to (0.6 μg g^{-1}), respectively.

Scientific name	Sex	Mean Weight (g)	Food type
Portunussegnis	Female (n = 122)	170 ± 0.10	Shrimp; Fish; Mollusca; Plant; Detritus
	Male (n = 107)	151 ± 0.02	

Table 1: Sex, food type, and weight (Mean ± SE) of the blue swimming crab *Portunussegnis*.

Metal	Sex	Gill	Hepatopancreas	Muscle
Ni	Male	15.4 ± 0.05	20.4 ± 0.02	14.5 ± 0.06
	Female	20.2 ± 0.01	31.9 ± 0.01	17.3 ± 0.05
Pb	Male	1.1 ± 0.02	2.2 ± 0.02	0.43 ± 0.03
	Female	1.5 ± 0.05	2.5 ± 0.05	0.71 ± 0.02
Co	Male	0.63 ± 0.06	0.91 ± 0.01	0.53 ± 0.06
	Female	0.71 ± 0.02	1.2 ± 0.03	0.62 ± 0.01
Cd	Male	0.43 ± 0.01	0.61 ± 0.01	0.15 ± 0.04
	Female	0.51 ± 0.03	0.88 ± 0.02	0.38 ± 0.02

Table 2: Metal concentrations (μg g^{-1}) in the gill, hepatopancreas and muscle of male and female of blue swimming crab *P. segnis*.

Pb

Lead concentration in the hepatopancreas tissue of crabs ranged from (0.75 μg g^{-1}) to (2.2 μg g^{-1}) for male and from (0.93 μg g^{-1}) to (2.5 μg g^{-1}) for female, while concentrations in the gill were in the range of (0.65 μg g^{-1}) to (1.1 μg g^{-1}) for male and from (0.74 μg g^{-1}) to (1.5 μg g^{-1}) for female crabs. The nickel concentrations in the muscle of the male and the female crabs were observed to be between (0.36 μg g^{-1}) to (0.54 μg g^{-1}) and between (0.53 μg g^{-1}) to (0.81 μg g^{-1}), respectively.

Cd

Cadmium concentration in the hepatopancreas tissue of crabs ranged from (0.31 μgg^{-1}) to (0.61 μg g^{-1}) for male and from (0.48 μg g^{-1}) to (0.88 μg g^{-1}) for female, while concentrations in the gill were in the range of (0.19 μg g^{-1}) to (0.43 μg g^{-1}) for male and from (0.24 μg g^{-1} 2) to (0.55 μg g^{-1}) for female crabs. The nickel concentrations in the muscle of the male and the female crabs were observed to be between (0.11 μg g^{-1}) to (0.25 μg g^{-1}) and between (0.15 μgg^{-1}) to (0.38 μg g^{-1}), respectively.

Correlation

Correlation analysis also showed relationship between individual elements including Co and Ni (r=0.65) and between Cd and Co (r=0.76). Pearson correlation showed that there was no significant correlation among these metals ($p>0.05$).

In our samples, hepatopancreas cadmium concentrations were positively correlated with those in gill of the blue swimming crab ($r = 0.80$ $P<0.002$, Figure 2a). Also, there was a positive correlation between lead concentrations in hepatopancreas with those in muscle ($r = 0.75$ $P<0.001$, Figure 2b)

Discussion

This area is surrounded by many petrochemical units and petroleum refinery. Also, metals concentration may be due to discharge of sewage and urban effluents and related to the oil tankers traffic in the area. The fact that total concentration of some petrochemical-related metals such as Ni, Co and Cd in the northwest of Persian Gulf

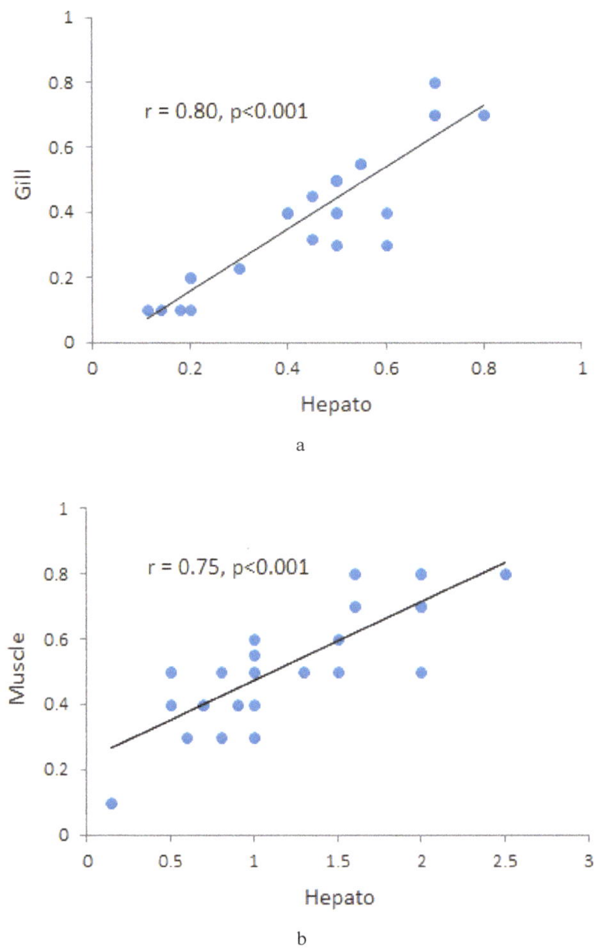

Figure 2: Relationship of cadmium concentrations (µg g^{-1}) between hepatopancreas and gill (a) and lead concentrations between hepatopancreas and muscle (b) in blue swimming crab *P. segnis*.

decreased with distance from Boushehr coasts is strong evidence reflects that the metals in this area were sourced from petrochemical activities, and not from background geological sources. Besides, some main creeks of this estuary flow into two crowded and industrialized cities and consequently receive huge amounts of domestic effluents and urban wastewaters [8].

It is likely that this high concentration of nickel was related to the oil tankers traffic and petrochemical complex in this area [8]. This area is surrounded by some petrochemical complex and petroleum refinery. The high concentration of lead was situated at the boat stations where high boating activities takes place and in located where the pH of the sediment was the lowest. the highest concentration of lead was found in the Turlok Lake when there were high boating activities.

The result demonstrated that metals concentrations in the different organs followed the hierarchical pattern H > G > M (Figure 3). Heavy metal accumulation in shrimp organs depend on the physiological role of the organs [14]. Some tissues such as hepatopancreas are considered as target organs for metals accumulation [15]. The very high levels metals in the hepatopancreas in comparison to other tissues may be related to the content of metallothionein protein in hepatopancreas tissue. Metallothionein protein that plays a significant role in the

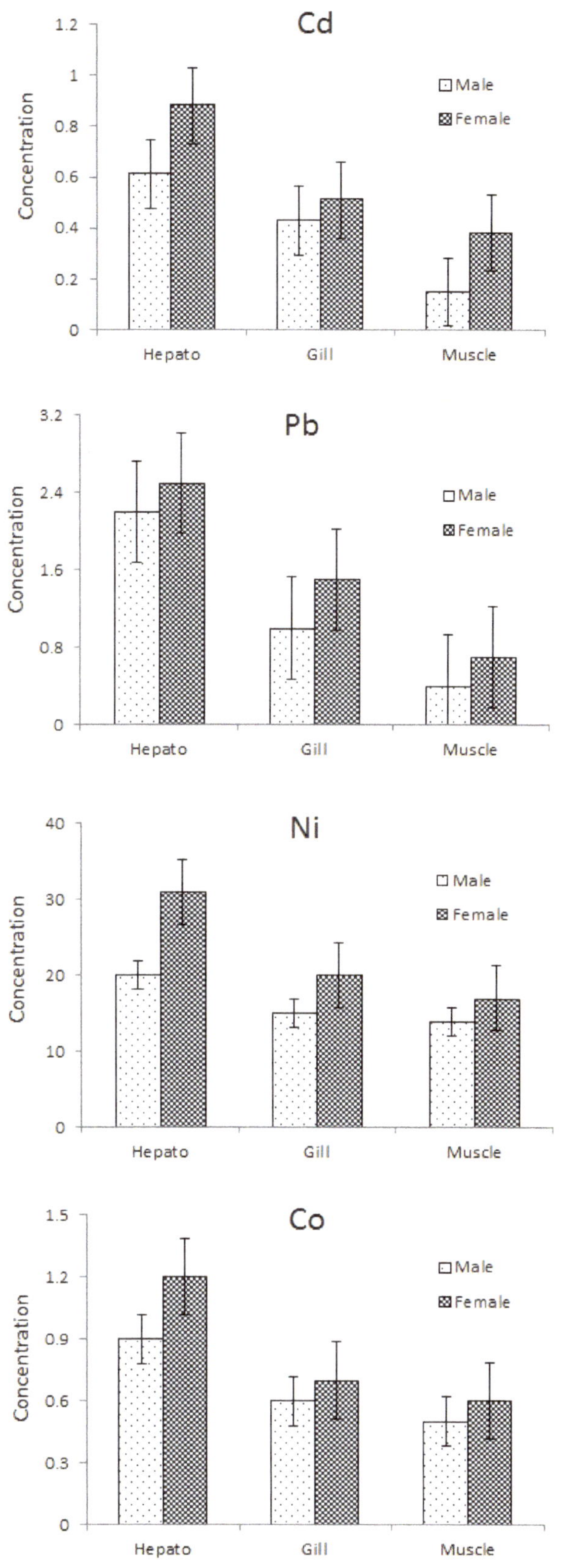

Figure 3: Comparison of Ni, Cd, Co and Pb concentrations in hepatopancreas, gill and muscle of male and female of blue swimming crab *P. segnis*.

regulation and detoxification of metals is produced in high levels in hepatopancreas tissue [16]. This protein contains a high percentage of amino group, nitrogen and sulphur that sequester metals in stable complexes. In general, the accumulation of metals in the hepatopancreas could be resulted from the abundance of metallothioneins proteins in these tissues in comparison to gill and muscle.

Gills usually reflect the concentrations of metals in surrounding water. This organ is directly in contact with water and suspended materials thus could absorb different substances from the surrounding environment. They also serve a variety of physiological functions such as osmoregulation and gas exchange. Due to these functions, gills have remarkable influences on the exchange of toxic metals between a fish and its environment. However, the muscle tended to accumulate less mercury in comparison to the liver and gills. This finding may reflect the low concentration of metallothioneins in the muscle tissue.

The result also showed that there are differences in metal concentration in the samples collected from different station. The indicated variability of metal concentration in the same species depends on their habitats [15] because chemical reactions, bioavailability of metals, and sources of pollution that potentially could affect the accumulation of metals are different between stations.

There have been several studies on accumulation of heavy metals in crustacean species, yet few studies have taken into account the effect of sexual changes with respect to the metal accumulation and distribution among tissues. We found that metals values were larger in tissues of female of the species than the males. Differences in accumulation between the genders have been mainly attributed to differences in diet or differences in habitat [17].

The male crab more feed on fish and bivalvia and females crab more feed on shrimp, plant and detritus [4]. Plants have relationship with sediment and receive more sediment associated metals. The roots of plants have an important role in depurating the water and the sediment, retaining large quantities of organic material and trace metals brought by the tides [17]. Metals are closely bound to the plant cell wall, slowing its translocation from roots to buds. Because plants are salt-excluder, it hinders the entry of metals through its root system [18].

It is known that certain forms of metals can readily accumulate within crustacean tissues at much higher levels. Shrimp have been reported as a vector of the transfer of mercury element to top marine predators of the food chains. Therefore, female crabs more feed on shrimp and plant and receive high levels of metals (Figure 3).

Since larger organisms generally exhibit higher contaminant level in their bodies [8] and crabs that are higher on the food chain also accumulate more contaminants when comparing to crabs that eat a range of different foods or eat smaller organisms. We expected to see higher metals levels in tissues of female crabs because they are larger and can eat larger food items. In general metal levels have been shown to increase with size and age of the ingested crab and it tends to be higher in species that occupy higher trophic levels [19], based on this logic we predicted that there should be higher levels of metals in the larger predators. Gewurtz et al. [20] have shown that higher metals levels in female fish were due to the increased consumption of food, relative to males, to meet the increasing demands of reproduction.

Conclusions

The chief purposes of current investigation were to determine to be biomagnification of Ni, Co, Cd and Pb in three tissues (muscle, hepatopancreas and gills) of blue swimming crab *Portunus segnis* and relationship with its food source. There was a positive correlation between metals concentrations in tissues with size of food items. Therefore, we expected to see higher metals levels in tissues of female crabs because they are larger and can eat larger food items. The results of this study show that highest mean metals level were found in the herbivorous crabs, followed by detritivorous, carnivorous and omnivorous. Plant-eating crab can be used to develop a sensitive bioindicator for ecological and human health parameters. One advantage of studies such as this is that they provide baseline data which can be used in future to evaluate differences within and across specific geographical areas. Without such a database it will be difficult to evaluate and interpret future results from the region, or to identify place with disturbing trends in pollution levels. Finally, there is a need to develop and refine herbivorous crab model to serve as a sentinel of ecosystem health, to help provide early warning indications for possible human exposure.

Acknowledgement

Financial support was carried out by Shahid Beheshti University and Environment Protection institute and Environmental project, Iran.

References

1. Beltrame MO, Marco SGD (2010) Influences of sex, habitat, and seasonality on heavy-metal concentrations in the burrowing crab (*Neohelice granulate*) from a coastal lagoon in Argentina. Arch Environ Contam Toxicol 58: 746-756.
2. Karadede HI, Oymak SA, Unlu E (2004) Heavy metals in mullet, Liza abu, and catfish, *Silurustriostegus*, from the Ataturk Dam Lake (Euphrates), Turkey. Environ Int 30: 183-188.
3. Dalman O, Demirak A, Balci A (2006) Determination of heavy metals (Cd, Pb) and trace elements (Cu, Zn) in sediments and fish of the Southeastern Aegean Sea (Turkey) by atomic absorption spectrometry. Food Chemistry 95: 157-162.
4. Williams MJ (1981) Methods for analysis of natural diet in portunid crabs (Crustacea: Decapoda: Portunidae). J Experimental Marine Biol Ecol 52: 103-113.
5. Prasad PN, Neelakantan N (1988) Food and feeding of the mud crab *Scylla serrata* (Forskal) (Decapoda: Portunidae) from Karwar waters. Indian Journal of Fisheries 35: 164-170.
6. Sheppard C, Al-Husiani M, Al-Jamali F, Al-Yamani F, Baldwin R, et al. (2010) The Gulf: A young sea in decline. Mar Pollut Bull 60: 13-38.
7. ROPME (1999) Regional report of the state of the marine environment. Regional Organization for the Protection of the Marine Environment (ROPME), Kuwait 220.
8. Abdolahpur MF,PeeryS, Karami O, Hosseini M, Bastami AA, et al. (2012) Distribution of Metals in the Tissues of Benthic, *Euryglossaorientalis* and *Cynoglossusarel.*, and Bentho-Pelagic, *Johniusbelangerii.*, Fish from Three Estuaries, Persian Gulf. Bull Environ Contam Toxicol 89: 489-494.
9. Potter IC, Lestang SD (2000) Biology of the blue swimmer crab *Portunuspelagicus* in Leschenault Estuary and Koombana Bay, south-western Australia. J R Soc West Aust 83: 443-458.
10. Fu CT, Wu SC (2005) Bioaccumulation of polychlorinated biphenyls in mullet fish in a former ship dismantling harbour, a contaminated estuary, and nearby coastal fish farms. Mar Pollut Bull 51: 932-939.
11. MourenteG (1996) In vitro metabolism of ^{14}C-polyunsaturated fatty acids in midgut gland and ovary cells from *Penaeuskerathurus* Forskal at the beginning of sexual maturation. Comp Biochem Physiol Part B Biochem Mol Biol 115: 255-266.
12. Athanasopoulos N (1993) Flame methods manual for atomic absorption. GBC Scientific Equipment PTY Ltd, Victoria.
13. Basset J, Denney RC, Jeffery GH, Mendhan J (1981) Vogel: AnaliseInorganica Quantitativa, fourthed. Guanabara SA, Rio de Janeiro.
14. Uysal K, Emre Y, Köse E (2008) The determination of heavy metal accumulation ratios in muscle, skin and gills of some migratory fish species by inductively coupled plasma-optical emission spectrometry (ICP-OES) in Beymelek Lagoon (Antalya/Turkey). Microchemical J 90: 67-70.

15. Yilmaz AB, Yilmaz L (2007) Influences of sex and seasons on levels of heavy metals in tissues of green tiger shrimp (*Penaeussemisulcatus*). Food Chem 101: 1664-1669.

16. Sen A, Semiz A (2007) Effects of metals and detergents on biotransformation and detoxification enzymes of leaping mullet (*Liza saliens*). Ecotoxicol Environ Saf 68: 405-411.

17. Beckvar N, Field J, Salazar S, Hoff R (1996) Contaminants in aquatic habitats at hazardous waste sites: mercury. NOAA, Seattle.

18. Bernini E, Silva MAB, Carmo TMS (2006) Composicaoquımica do sediment e de folhas das especies do manguezal do estuario do Rio Sao Mateus, Espırito Santo, Brasil. Rev Bras Bot 29: 689-699.

19. Phillips GR, Lenhart TE, Gregory RW (1980) Relation between trophic position and mercury accumulation among fishes from the Tongue River reservoir. Environ Res 22: 73-80.

20. Gewurtz SB, Bhavsar SP, Fletcher R (2011) Influence of fish size and sex on mercury/PCB concentration: importance for fish consumption advisories. Environ Int 37: 425-434.

Size and Growth of *Cardiosoma armatum* and *Cardiosoma guanhumi* as Ecological Parameters for Mangrove Ecosystem

Elegbede Isa Olalekan[1]*, Lawal-Are AO[2] and Titilade PR[2]

[1]*Department of Environmental Planning, Brandenburg University of Technology, Cottbus-Senftenberg, Germany*

[2]*Department of Marine Sciences, University of Lagos, Akoka, Lagos, Nigeria*

Abstract

The Samples of *Cardiosoma armatum* and *Cardiosoma guanhumi* were collected from the Lagos Lagoon mangrove area of the University of Lagos. They were studied for their size composition and influence on growth pattern. A comparative analysis was done on both crabs. The carapace length of *Cardiosoma armatum* and *Cardiosoma guanhumi* examined ranges from 2.50 cm to 9.30 cm and 2.50 cm to 9.20 cm respectively. The carapace-width examined ranges from 2.70 cm to 9.40 cm and 2.80 cm to 9.40 cm respectively. The total weight of the *Cardiosoma armatum* ranges between 96.00 g and 290.00 g while *Cardiosoma guanhumi* ranges between 4.70 g and 295.00 g. The length- weight relationship of the two crabs showed low correlation value of 0.3378 and 0.2113 respectively. The Statistical T-test of the right and left chelipeds, the carapace length and carapace width of *Cardiosoma armatum* and *Cardiosoma guanhumi* showed that there were no statistical significances ($p>0.05$) between carapace length of both crabs. There are statistical significance ($p<0.05$) between carapace weight of both crabs collected in February, March, June and July. The statistical significance ($p<0.05$) between the right and left chelipeds of both crabs are highly recognized. This research study indicates almost similar biological features for both species and evidently shows that both crabs are important spotlight for mangrove habitat.

Keywords: Crustacean; Size composition; Length-weight relationship; Mangrove ecosystem; Lagoon systems; Ecological values

Introduction

Crabs are decapods in the brachyuran infra order. The family comprises more than 14,750 species known for their ten legged creature (decapod). They are physiologically and structurally diverse across marine and freshwater environment. Their importance are diverse, this type of decapods are used as biological models and are highly edible for consumptions [1,2]. Crabs have flourish to be a predominant icon in the invertebrate fauna because of their ubiquitous in some part of the world including marine, freshwater and terrestrial environment [3]. The crab fisheries are also predominant in the marine, coastal, freshwater and lagoon fishery [4]. According to Geist et al., [5] crabs are known as ecosystem engineers due to their ability to burrow into sediment structures. These enhance sediment aeration and improve the biodiversity of surrounding vegetation associated to the mangrove environment. The size and growth of crabs are important ecological indicators, which can monitor the abundance of the crab species in the mangrove ecosystem. The ranges of their significant sizes expose the richness of the habitat with nutrient [6]. In Nigeria, *Geryon maritae* (Deep water crab), *Ocypode africana* (Ghost crab), *Goniopsis pelii,* and *Sesarma sp.* (Mangrove crabs), *Uca tangerii* (Fiddler crabs), *Callinectes latimanus, C. amnicola, C. pallidus* and *C. marginatus* (Swimming crabs), *Cardiosoma armatum* and *Gecarcinus weileri* (Land crabs) are common crab species found in brackish and marine environments [7]. The mangrove crabs have been found in mangrove habitat of the Lagos Lagoon. They have been subjected to reclamation by anthropogenic activities. However, this has played ecological role and significantly modify the mangrove ecosystem. The crabs have also been instrumental to clean up the environment by feeding on the fallen leaves [8]. Their environmental cleaning qualities are important with the omnivorous feeding attribute, depending on plant and animal [9]. According to Turner et al., [10] these mangrove land crabs developed burrows, which have strength for at least five years with temperature stability and high humid medium. The burrows are great soil nutrient improvement agents and enhance plant seedlings production. *Cardiosoma* is a genus of Land crabs. Young individuals are often very colorful with a purple-blue Carapace and orange-red legs, exhibit color change as old age is reached [11]. There have been difficulties in the classification of these two crabs. Measurement of the size-frequency is one of most widely used method for growth pattern especially in the wild. This method gives more information about the ecological status of the species in the mangrove habitat [2]. The mangrove of the Lagos lagoon plays an important role on the diversity of the crab species, though various impacts have pose alarm on the environment which also affects the distribution of other important fauna and flora community of the ecosystem. The aim of this research is to provide baseline information on the relevance of using size and growth of *Cardiosoma armatum* and *Cardiosoma guanhumi* as ecological parameter of mangrove ecosystem and make comparison of the population differences based on morphological analyses of the two crab's species: *Cardiosoma armatum* and *Cardiosoma guanhumi*.

Materials and Methods

The study site for this project is the coastal and mangrove area of University of Lagos Lagoon front, located opposite the Lagos Lagoon, with the geographical platform of 6°26'N and 6°39'N and longitude 3°29'E and 3°50'E (Figure 1). The lagoon is the largest of the four lagoon

***Corresponding author:** Isa Olalekan E, Department of Environmental Planning, Brandenburg University of Technology, Cottbus-Senftenberg, Germany
E-mail: isaelegbede@gmail.com

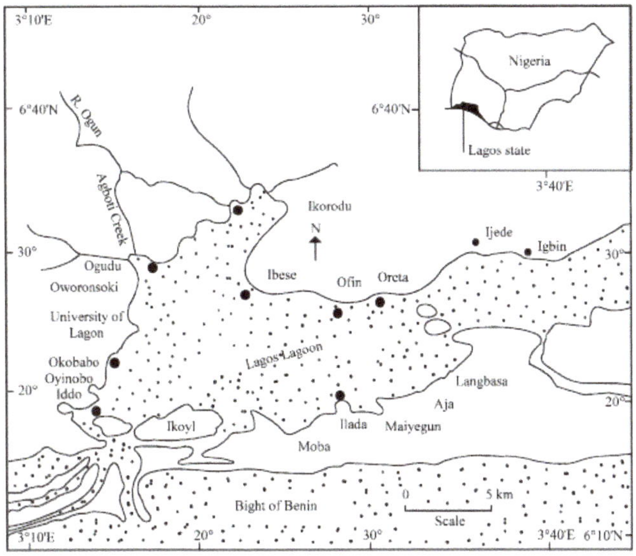

Figure 1: Map Showing the sampling location (Joseph and Juliet, 2010).

systems of the Gulf of Guinea and is located at South Western Nigeria. The mangrove swamp connects to the Lagos lagoon by tidal creek [12]. The Crab species (Plates 1 and 2) were collected at the mangrove part of the Lagos Lagoon along University of Lagos using the removal method [5]. They were caught between 7 pm and 11 pm. The collection was done randomly over a period of six months between February and July, 2012. The crabs were collected at two different stations within the mangrove swamp. A total of 858 crabs were collected from the site and were preserved immediately in a deep freezer in the laboratory prior to examination. The crabs were removed from the freezer and allowed to thaw. Excess water was removed from the specimens using filter paper. The carapace length and width of the two crabs was measured using a Simple vernier caliper. Total weight, weight of left and right chelipeds were measured to the nearest tenth of a gram using Sartorious Top Loading Balance (Model 1106); the results were recorded in a proformer for each specimen before dissection. The relationship between the carapace length-frequency distributions was established for each month and the cumulative (summary for the six months) worked out. Specimens were collected randomly in each of the six months. The crabs were examined to obtain their size composition and abundance. For the growth pattern, data of the length-weight relationship and width-weight relationship were compiled. The carapace length-weight relationship was expressed based on Rickter, [13] with the following equation:

$W=aL^b$

Where W=weight of crabs in grams.

L=length of the carapace in cm.

a=regression constant.

b=regression coefficient.

The equation was transformed to a linear relationship as

$\text{Log Wt}=\text{Log a} + \text{b Log L}$

Scatter diagram of log weight - log length were plotted to illustrate these relationships.

Plate 1: *Cardiosoma armatum* (Self-Created).

Plate 2: *Cardiosoma guanhumi* (Self-Created).

Result

Size composition of *Cardiosoma armatum* and *Cardiosoma guahunmi*

418 and 440 specimens of *Cardiosoma armatum* (Plate 1) and *Cardiosoma guanhumi* (Plate 2) were studied respectively making a total of 858 species of crabs collected and studied. The specimens were studied for the length and width frequency distributions between the months of February to July, 2012 (Figures 2 and 3). The carapace length of *Cardiosoma armatum* and *Cardiosoma guanhumi* examined ranged from 2.50 cm to 9.30 cm and 2.50 cm to 9.20 cm respectively while their carapace width were examined ranging from 2.70 cm to 9.40 cm and 2.80 cm to 9.40 cm respectively. The largest specimen of *Cardiosoma armatum* and *Cardiosoma guanhumi* weighed 290.00 g and 295.00 g respectively. The smallest size range collected in the same month weighed 9.60 g and 4.70 g. The two crabs were studied for length and width frequency distributions between February-July, 2012 (Tables 1 and 2). The carapace length frequency polygon of *Cardiosoma armatum*

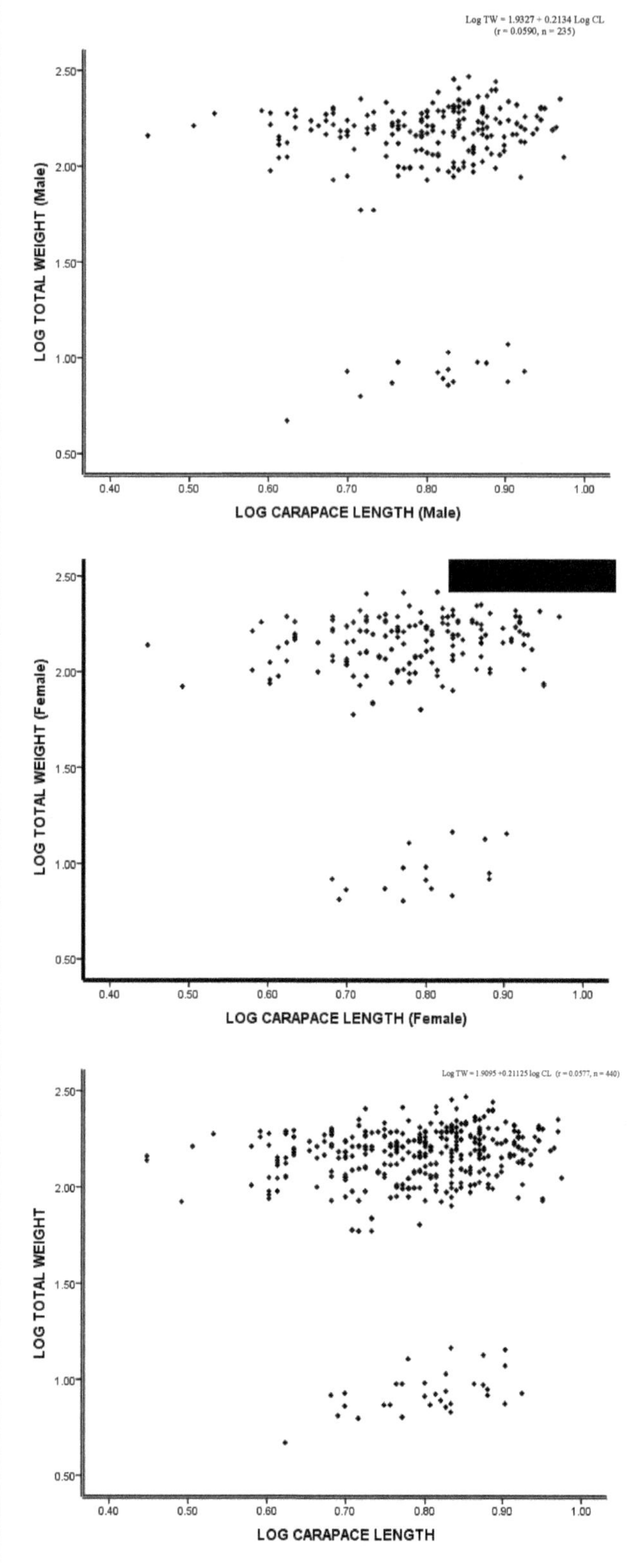

Figure 2: Log Total Weight against Log Carapace Length of Male, Female and combined sex for *Cardiosoma guanhumi* from Lagos Lagoon mangrove swamps in Nigeria (February- July, 2012).

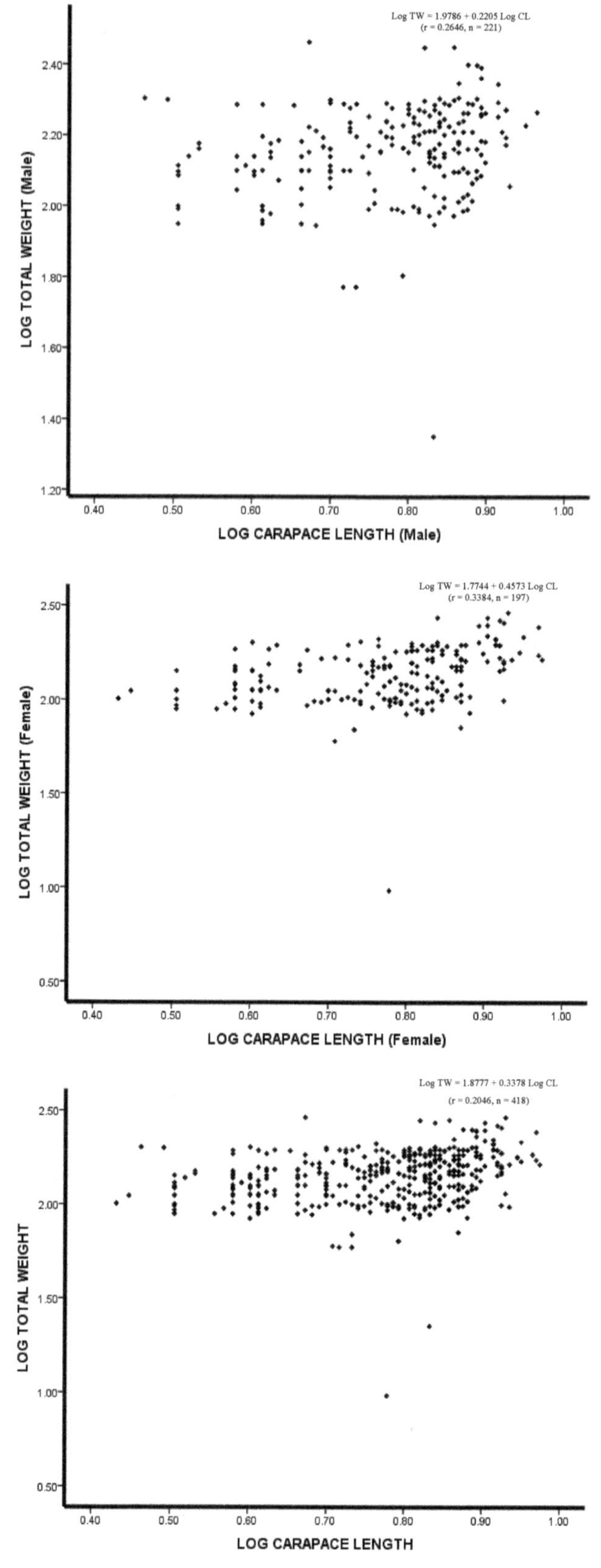

Figure 3: Log Total Weight against Log Carapace Length of Male, Female and combined sex for *Cardiosoma armatum* from Lagos Lagoon mangrove swamps in Nigeria (February- July, 2012).

Month	Number Collected					
	Cardiosoma armatum			*Cardiosoma guanhumi*		
	Female	Male	Total	Female	Male	Total
February	26	30	56	24	36	60
March	29	33	62	34	41	75
April	33	37	70	27	34	61
May	31	38	69	41	37	78
June	39	39	78	40	40	80
July	39	44	83	39	47	86
Total	197	221	418	205	235	440

Table 1: Monthly collection of *Cardiosoma armatum* and *Cardiosoma guahunmi* from Lagos Lagoon mangrove swamp in Nigeria (February- July, 2012).

	Length				Width			
Carace length	*Cardiosoma armatum*		*Cardiosoma guanhumi*		*Cardiosoma armatum*		*Cardiosoma guanhumi*	
	Frequency	%	Frequency	%	Frequency	%	Frequency	%
2.5-3.4	20	4.8	5	1.1	43	10.3	12	2.7
3.5-4.4	51	12.2	34	7.7	49	11.7	39	8.9
4.5-5.4	59	14.1	78	17.7	55	13.2	87	19.8
5.5-6.4	93	22.2	115	26.1	111	26.6	126	28.6
6.5-7.4	120	28.7	118	26.8	108	25.8	110	25.0
7.5-8.4	63	15.1	68	15.5	44	10.5	48	10.9
8.5-9.4	12	2.9	22	5.0	8	1.9	18	4.1
Total	418	100.0	440	100.0	418	100.0	440	100.0

Table 2: The frequency distribution of Carapace length and width of *Cardiosoma armatum* and *Cardiosoma guahunmi* from Lagos lagoon mangrove swamps in Nigeria (February to July, 2012).

and *Cardiosoma guanhumi* showed distinct size groups. The size group 6.5-7.4 cm was abundant with 25% and 28.7% for *Cardiosoma armatum* and *Cardiosoma guanhumi* respectively.

Growth pattern of *Cardiosoma armatum* and *Cardiosoma guahunmi*

The total weight of the *Cardiosoma armatum* ranged between 96.00 g and 290.00 g while *Cardiosoma guanhumi* 4.70 g and 295.00 g were for the combined sex. The carapace length of *Cardiosoma guanhumi* ranges between 2.80 cm to 9.40 cm and carapace length of the *Cardiosoma armatum* ranges from 2.70 cm and 9.40 cm for the combined sex. This result showed increase in length with increase in weight (Figures 4 and 5). The carapace length-total weight of the two crabs was transformed into a logarithmic form. The Log length-weight relationship showed a linear relationship between length and weight of the crab. The carapace length-weight relationship was determined using the formula below:

Log TW=a + b Log CL

Where, W=Total weight of crab in grams (g)

CL=Carapace length of crab in grams (cm)

a=regression constant

b=regression coefficient

The value of length-weight relationship for C. *armatum* and C. *guanhumi* are given as follows;

The Total length-weight relationship for *Cardiosoma armatum* of the least square common fit of the transformed data got the following linear equation.

Female: Log TW=1.7744 + 0.4573 Log CL (r=0.3384, n=197)

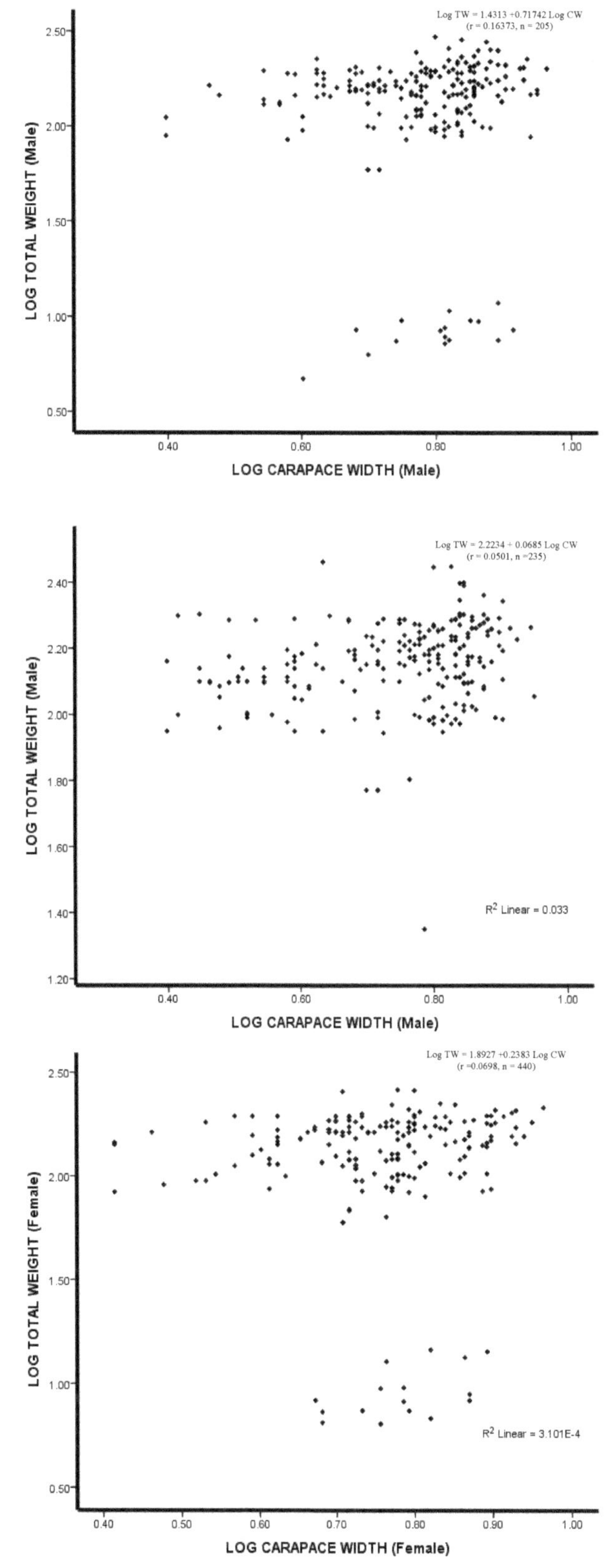

Figure 4: Log Total Weight against Log Carapace Width of Male, Female and combined sex for *Cardiosoma guanhumi* from Lagos Lagoon mangrove swamps in Nigeria (February- July, 2012).

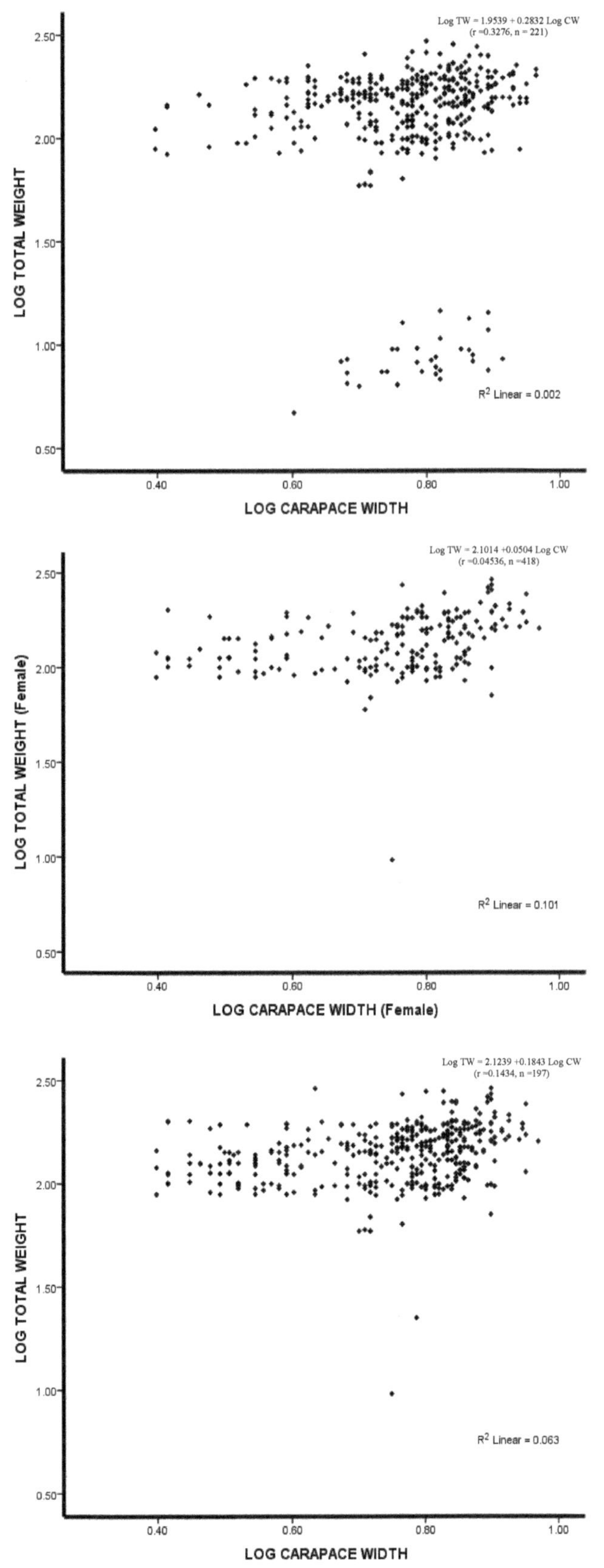

Figure 5: Log Total weight against Log Carapace Width of Male, Female and combined sex for *Cardiosoma armatum* from Lagos Lagoon mangrove swamps in Nigeria (February- July, 2012).

Male: Log TW=1.9786 + 0.2205 Log CL (r=0.2646, n=221)

Combined (sex): Log TW=1.8777 + 0.3378 Log CL (r=0.2046, n=418)

The values of b were less than 3 in both sexes of *Cardiosoma armatum* which also indicated a positive isometric growth. The values were 0.4573, 0.2205 and 0.3378 for male, females and combined sexes respectively.

The carapace length-weight relationship for *Cardiosoma guanhumi* of the least square common fit of the transformed data gave the following linear equation;

Female: Log TW=1.9999 + 0.1038 log CL (r=0.0266, n=205)

Male=Log TW=1.9327 + 0.2134 log CL (r=0.0590, n=235)

Combined sex=Log TW=1.9095 +0.21125 log CL (r=0.0577, n=440)

Statistical analysis of *Cardiosoma armatum* and *Cardiosoma auahunmi*

T-test statistical analysis was conducted; there is no statistical significance for carapace weight of *Cardiosoma armatum* and *Cardiosoma guahunmi* collected in April and May respectively (Tables 3 and 4). No statistical significance was observed for the left chelae of *Cardiosoma armatum* and *Cardiosoma guahunmi* for the month of February. The statistical T-test analysis of the right chelae of *Cardiosoma armatum* and *Cardiosoma guahunmi* for the month of March and May do not show statistical significant.

Discussions

The assessment of the size and growth pattern of the Hairy Mangrove Crab, *Sersema huzardii* was examined by Lawal-Are and Nwankwo [4] in the Lagos Lagoon mangrove area. The length of the

	Weight		Length of Carapace	
Month	***Cardiosoma Guanhumi***	***Cardiosoma armatum***	***Cardiosoma armatum***	***Cardiosoma guanhumi***
Feb	159.34 ± 7.65 ᵃ	141.58 ± 6.12 ᵇ	3.16 ± 0.17 ᵃ	3.06 ± 0.26 ᵃ
Mar	153.76 ± 5.96 ᵃ	118.58 ± 5.60 ᵇ	4.01 ± 0.18 ᵃ	4.12 ± 0.16 ᵃ
April	155.48 ± 4.42 ᵃ	149.19 ± 5.83 ᵃ	4.98 ± 0.28 ᵃ	5.05 ± 0.24 ᵃ
May	161.83 ± 3.39 ᵃ	167.53 ± 4.44 ᵃ	6.01 ± 0.29 ᵃ	5.99 ± 0.26 ᵃ
June	97.22 ± 8.81 ᵃ	164.56 ± 3.46 ᵇ	6.96 ± 0.29 ᵃ	6.94 ± 0.29 ᵃ
July	141.96 ± 4.78 ᵃ	124.79 ± 3.33 ᵇ	7.92 ± 0.32 ᵃ	7.96 ± 0.31 ᵃ

In each row, means with a common letter are not significantly different (P > 0.05).

Table 3: T-test of weight *Cardiosoma armatum* and *Cardiosoma guahunmi* and length of Carapace of *Cardiosoma armatum* and *Cardiosoma guahunmi* from the Lagos Lagoon mangrove swamps (February- July, 2012).

	Right Chelae		Left Chelae	
Period	***Cardiosoma Guanhumi***	***Cardiosoma armatum***	***Cardiosoma guanhumi***	***Cardiosoma armatum***
Feb	9.84 ± 0.18 ᵃ	9.91 0.21 ᵇ	10.22 ± 0.24 ᵃ	9.35 ± 6.08 ᵇ
Mar	9.80 ± 0.22 ᵃ	9.76 ± 0.28 ᵃ	10.45 ± 0.311 ᵃ	10.32 ± 0.28 ᵃ
April	7.94 ± 035 ᵃ	10.28 ± 0.29 ᵇ	9.21 ± 0.32 ᵃ	9.73 ± 0.31 ᵃ
May	6.98 ± 0.23 ᵃ	7.17 ± 0.23 ᵃ	8.32 ± 0.23 ᵃ	8.22 ± 0.24 ᵃ
June	7.04 ± 0.26 a	8.19 ± 0.25 b	8.75 ± 0.24 a	9.20 ± 0.22 a
July	8.39 ± 0.25 a	10.13 ± 0.21b	9.15 ± 0.26 a	9.24 ± 0.05 a

Table 4: T-test of Right and Left Chelae for *Cardiosoma armatum* and *Cardiosoma guahunmi* from the Lagos Lagoon mangrove swamps in Nigeria (February- July, 2012).

carapace ranged from 1.5 cm to 4.7 cm which is low when compared to 2.5 to 9.5 cm of the present study for both species under consideration. Both *Sersema* spp and *Cardiosoma* spp are both mangrove crabs; however *Sersema spp* are smaller in size than *Cardiosoma spp*. The biology and distribution of some Mangrove Crabs in the Wouri River Estuary of Douala in Cameroon was studied by Ngo-Massou et al., [14] some *Cardiosoma* species were also studied, carapace length and width ranges from 3.6 -9.79 cm and 4.49 -13.13 cm respectively. These values are smaller than the *Cardiosoma* spp studied in this research, which is due to the abundance of trees and the richness of the mangrove ecosystem of University of Lagos Lagoon front of the Lagos Lagoon. The two crabs show high chela weight, which is in accordance with the body weight. This is in relation to Sneddon, Huntingford and Taylor, [15] that there is a great relationship between the chela size and the body weight, the chelae influences the contest between crabs in the same habitat. Akin-oriola et al. [3] also reported the maximum carapace length of 6.0 cm for *Cardiosoma armatum* and 7.0 cm for *Callinectes pallidus* respectively from Badagry creek. There was only one predominant generation of crabs sampled and they belong to the same year of class in their first year of life. It was observed that there was very low distribution in the 8.5-9.4 cm (1.9%) of *Cardiosoma armatum*, which is in contrast to Hartnoll et al. [16] of the mangrove crab of *Johngarthria lagostoma* with the size 90-80 cm size group. The ranges in the data obtained were due to the collection methods of the crabs, differences in the habitat terrain and topography of the area. The monthly distribution of the two crabs *Cardiosoma armatum* and *Cardiosoma guanhumi* for the month of July showed greater distribution pattern of 83 in July, 2012 for *Cardiosoma armatum* and 86 in July, 2012 for *Cardiosoma armatum* respectively. These values are in conformity with the research of Akin-oriola et al. [3] which showed the increase in the number of crabs caught in June and July including the size group of 4.0-4.9 cm showed the highest distribution pattern in May, September and November. *Cardiosoma armatum* in the Gulf of Guinea showed a maximum carapace length of 9.5 cm, Akin-oriola et al. [3] this is in agreement with the present study which also conform with the work of Atar and Secer, [17] and Lawal-Are and Nwankwo [4]. The logarithmic form of length-weight relationship of both crab species show low b value. *Cardiosoma armatum* shows a positive isometric growth with values 0.4573, 0.2205 and 0.3378 for male, female and combined sexes respectively. The length-weight relationship for *Cardiosoma guanhumi* was 0.1038, 0.2134 and 0.2113 for male, female and combined sexes respectively, this value are supported by Turner et al. [10]. However, females of both crabs showed a lower b value of 0.2205 and 0.2113 for *Cardiosoma armatum* and *Cardiosoma guanhumi* respectively, Which is scientifically attributed to the slow rate of growth of female crabs and the great amount of energy invested in the reproductive process at the expenses of growth by females [10]. Observations of higher b value of *Cardiosoma guanhumi* over *Cardiosoma armatum* were linked to the higher population and standing stock biomass and condition indices [17]. The overall low symmetrical or isometric growth of b values was less than 3 and this is due to the recruitment stock in biomass. Which is due to the peculiarity of coastal dwelling land crabs to show irregular recruitment pattern with uncertainty of returning to a small land mass after the planktonic laval phase [18]. T-test statistical analysis between the two crabs *Cardiosoma armatum* and *Callinectes pallidus* for carapace length, weight and chelae were reported by Akin-oriola et al. [3] the result shows concurrence with the present research.

Conclusion

The analysis of the two crabs, *Cardiosoma armatum* and *Cardiosoma guanhumi* for the Size and growth parameter indicate the ecological richness of the mangrove ecosystem. The significant amount of size and the abundance including the frequency of the two crabs greatly expose that the crabs have influenced the vegetation which is attributed to the lagoon ecosystem. This research has served as a baseline for the use of crustaceans to monitor the ecological significance of the lagos lagoon habitat.

Acknowledgements

This study was a research to contribute to the scholarly attribute of the mangrove area of the Lagos Lagoon around the University of Lagos. We appreciate the support and contributions of the Department of Marine Sciences, University of Lagos, Akoka, Lagos, most especially for the permission to use their facilities. We also thank the people who assisted with the fieldwork, laboratory activities and the data analysis.

References

1. Hosseini M, Vazirizade A, Parsa Y, Mansori A (2012) Sex Ratio, Size Distribution and Seasonal Abundance of Blue Swimming Crab, *Portunus pelagicus* (Linnaeus, 1758) in Persian Gulf Coasts, Iran. World Applied Sciences Journal 17: 919-925.
2. Vogt G (2012) Ageing and longevity in the Decapoda (Crustacea): A review. Zoologischer Anzeiger - A Journal of Comparative Zoology 251: 1-25.
3. Akin-Oriola G, Anetekhai M, Olowonirejuaro K (2005) Morphometric and Meristic Studies in Two Crabs: *Cardiosoma armatum* and *Callinectes pallidus*. Turkish Journal of Fisheries and Aquatic Sciences 5: 85-89.
4. Lawal-Are AO, Nwankwo H (2011) Biology of the Hairy Mangrove Crab, *Sersema huzardii* (Decapoda: Graspidae) from a Tropical Estuarine Lagoon. Journal of American Science 7: 45-48.
5. Geist SJ, Nordhaus I, Hinrichs S (2012) Occurrence of species-rich crab fauna in a human-impacted mangrove forest questions the application of community analysis as an environmental assessment tool. Estuarine, Coastal and Shelf Science 96: 69-80.
6. Schlacher TA, De Jager R, Nielsen T (2011) Vegetation and ghost crabs in coastal dunes as indicators of putative stressors from tourism. Ecological Indicators 11: 284-294.
7. Oluwatoyin A, Akintade A, Clarke E, Kusemiju V (2013) A study of length-weight relationship and condition factor of west African blue crab (*Callinectes pallidus*) from Ojo Creek, Lagos, Nigeria. American Journal of Research Communication 1: 102-114.
8. Olafsson E, Buchmayer S, Skov MW (2002) The East African decapod crab, *Neosarmatium meinerti* (de Man) sweeps mangrove floors clean of leaf litter. Ambio 31: 569-573.
9. Hostetler ME, Mazzotti FJ, Taylor AK (2003) Blue Land Crab (*Cardisoma guanhumi*) Dept of Wildlife Ecology and Conservation, Florida Cooperative Extension Service, Institute of Food and Agricultural Sciences, University of Florida.
10. Turner LM, Hallas JP, Morris S (2011) Population Structure of the Christmas Island Blue Crab, Disc Geist SJ, Nordhaus I, Hinrichs S (2012) Occurrence of species-rich crab fauna in a human-impacted mangrove forest questions the application of community analysis as an environmental assessment tool. Estuarine, Coastal and Shelf Science 96: 69-80.
11. Ng PKL, Guinot D, Davie PJF (2008) Systema Brachyurorum. Part I. An annotated checklist of extant brachyuran crabs of the world. Raffles Bulletin of Zoology 17: 1-286.
12. Osman C (2013) Nutrients to close Tropical Lagun for Primary Production.3.
13. Ricker WE (1973) Linear regressions in fisheries research. Journal of Fisheries Research Board of Canada 30: 409-434.
14. Ngo-Massou VM, Essomè-Koum GL, Kottè-Mapoko E, Din N (2014) Biology and Distribution of Mangrove Crabs in the Wouri River Estuary, Douala, Cameroon. Journal of Water Resource and Protection.
15. Sneddon LU, Huntingford FA, Taylor AC (1997) Weapon size versus body size as a predictor of winning in fights between shore crabs, *Carcinus maenas* (L.). Behavioral Ecology and Sociobiology 41: 237-242.
16. Hartnoll RG, Broderick AC, Godley BJ, Saunders KE (2009) Population

structure of the land crab *Johngarthia lagostoma*on Ascension Island. J Crus Biol 29: 57-61.

17. Atar HH, Secer S (2003) Width/length-weight relationships of the blue crab (*Callinectes sapidus*, Rathbun, 1986) population living in Beymelek lagoon lake. Turk. J. Vet. Anim. Sci 27: 443-447.

18. Hartnoll RG, Clark PF (2006) A mass recruitment event in the land crab *Gecarcinus ruricola* (Linnaeus, 1758) (Brachyura: Grapsoidea: Gecarcinidae), and description of the megalop. Zoological Journal of the Linnean Society 146: 149-164.

36

Size at Maturity, Sex Ratio and Variant Morphometrics of Blue Swimming Crab *Portunus segnis* (Forskal, 1775) from Boushehr Coast (Persian Gulf)

Mehdi Hosseini*, Jamileh Pazooki and Mohsen Safaei

Department of Marine Biology, Faculty of Biological Science, Shahid Beheshti University, Tehran, Iran

Abstract

Size at maturity, sex ratio and variant morphometrics of blue swimming crab *Portunus segnis* was studied for one year along the Persian Gulf coasts, Iran. Sex ratio during a year of study was M:F=0.92:1 which should be the relative frequency of females. Male blue crabs have a V-shaped abdomen and female have an abdomen broad and rounded. The carapace width/length- weight relationship was studied in both sexes of crab. The allometric relationships between the characters of this set suggest that most relationships are positive and highly significant. An analysis of covariance indicates that there is a significant difference between sexes with respect to the carapace width-weight relationship. The crab's carapace widths ranged were from 75 to175 mm for males and 70 to170 mm for females. The crab's carapace lengths were from 35 to 80 mm for male and 30 to 80 mm for female. The results of this study show that at the same size, weight and the parameters of carapace length and carapace width male of *Portunus segnis* are higher than females in Persian Gulf coasts.

Keywords: Size at maturity; Sex ratio; *Portunus segnis*; Persian Gulf

Introduction

Crabs belong to a group of animals knows as brachyura crustaceans. Most of the marine crabs occurring along the Persian Gulf coasts belong to the family Portunidae. The blue swimming crab *Portunus segnis* is distributed from the eastern Mediterranean to east Africa in the Indian Ocean, and to Pakistan, Red Sea and Persian Gulf [1]. Also live in a wide range of inshore and continental shelf areas, including sandy, muddy or algal and sea grass habitats, from the intertidal zone to at least 50 m depth. They are usually found in large numbers in shallow bays with sandy bottom and are important commercial species is Iran and South East Asia [2-4].

This species supports substantial commercial fishery in the Persian Gulf and is an important component of many recreational fisheries in Iran and other parts of the world and the major species crab caught in trawl nets. They have five paired legs (ten legs), some of the legs are pointed and others flat. The pointed ones are used for crawling and the flat ones are used like paddles for swimming. The first pair of legs, which are modified as claws, called chelipeds. The cheliped are long, elongated ridged, catch and hold food and bring it to the mouth, digging, cracking open shells and warning off would be attackers. At the same size, males are heavier than females in the coasts of Persian Gulf.

In crustaceans, as growth progresses, certain dimensions of the animal's body may grow much more than others, resulting in the phenomenon known as relative growth [5]. Studies of relative growth are often used to determine changes in the form and size of the abdomen, pleopods, or chelipeds during ontogeny. Knowledge of these distinguishing characters and size relationships in sexually mature individuals is of particular importance in the study of commercially valuable crustaceans. Such knowledge can be useful for further studies on the life history of the species and in the development of its fishery, resource management, and culture. The mathematical length-weight relationship thus yields information on the general well-being of individuals, variation in growth according to sex, size at first maturity, gonadal development, and breeding season. Study of the length-weight relationship in aquatic animals has wide application in delineating the growth patterns during their developmental pathways [6].

In population studies, morphometric analysis provides a powerful complement to genetic and environmental stock identification approaches and length-weight relationships allow the conversion of growth-in-length equations to growth-in-weight for use in a stock assessment model [6]. Information about individual body weight-length/width relationships in populations is important for estimating the population size of a stock, specifically for the purpose of its exploitation. The length-width/weight relationships are regarded as more suitable for evaluating crustacean populations [7-12]. The interrelationships between various morphometric characters, carapace width and length and chelar propodus length, and abdominal width in females and males, were analysed and are here presented. The results will be useful in comparing the different stocks of the same species at different geographical locations.

Materials and Methods

Study area

The study was carried out in coasts of Boushehr, which is one of the several adjoining coasts in the Persian Gulf (Figure 1). The Persian Gulf lies on the south Iran, between longitudes 48°25' and 56°25' East, and latitudes 24°30' and 30°30' North. It has an estimated area of 260 km^2 and extends 909 km long to a depth average of about 30-40 m. Specifically, the Boushehr lies between Longitudes 50°6' and 52°58'East and Latitudes 27°14' and 30°16' North and it is about 60 km long [1].

Sample collection

Sampling was performed in Persian Gulf coast (Boushehr province)

***Corresponding author:** Mehdi Hosseini, Department of Marine Biology, Faculty of Biological Science, Shahid Beheshti University, Tehran, Iran
E-mail: smhbio@yahoo.com

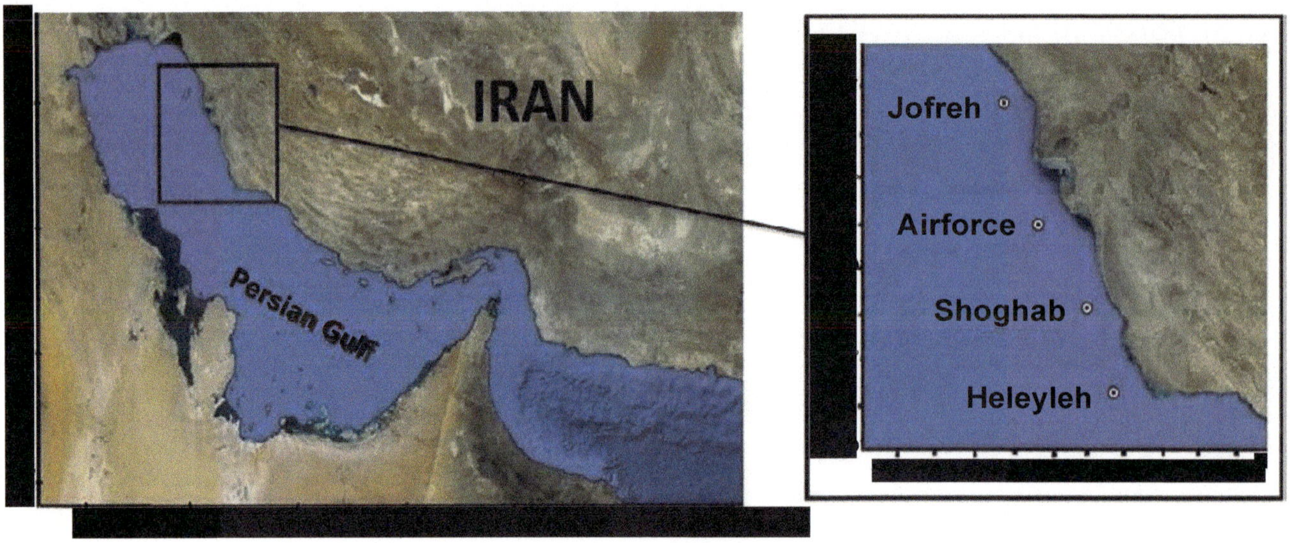

Figure 1: Map of Persian Gulf showing sampling stations and the study site.

including stations four along the Buoshehr coasts .The crabs for study were collected seasonal for one year from January 2011 to December 2011. Sampling was performed with several methods such as trawl net, research ship. After sampling, samples were taken to the laboratory in a cooler and stored in a deep freezer for further analysis. Then the crabs are washed for removing the mud and algae's and barnacles sticked to the external skeleton. Crabs were identified to species level carried out using photo cards and available identification keys [7,8]. Therefore each crab was sorted into species. The first sex and their morphological characteristics (Wet weight, carapace width and carapace length, width and length of the abdomen and chelar (propodus) length) were taken. Male blue swimming crabs have a V-shaped abdomen and female crab have an abdomen broad and rounded [8]. Vernier callipers with an accuracy of 0.5 mm were used for length measurements, and the total weight of the crab was determined to the nearest gram using a digital balance (1 g).

Carapace width (CW) was taken as the distance between the tips of the posteriormost lateral carapace spines. Carapace length (CL) was measured dorsally along the midline, between the frontal notch and the posterior margin of the carapace. Right chelar propodus length (Ch L) was measured from the tip of the propodus' fixed finger to the base of the propodus. When the right chela was missing or damaged, the left chelar propodus length and height were taken, as in *P. segnis* both chelae are equally well developed. Abdominal width was measured at the maximum width, i.e., across the fourth somite. Abdominal length was measured along the midline from the anterior margin of the first somite to the posterior margin of the telson.

The carapace width-weight relationship was estimated using the log form of the allometric growth equation $W=aL^b$ [7], where W=expected weight, L=total carapace width, 'a'=y-intercept or the initial growth coefficient, and 'b'=the slope or growth coefficient. The values of constants of 'a' and 'b' were calculated by the least squares method. The differences in the carapace width-weight relationship between sexes were tested by ANOVA (MS Excel).

Data were presented as mean ± standard deviation. Appropriate measure of the size of maturity (CW^{50}) in both males and females crabs size of maturity was determined when 50% of the crabs were sexually

Characteristics	Male		Female	
	Min	Max	Min	Max
Carapace width (cm)	6	17.5	5.5	17
Carapace Length (cm)	4	8	3.5	7
Total Weight (gr)	40.0	275.0	32.50	242.2

Table 1: The morphological characteristics (carapace width, carapace length and total weight) of *P. segnis*.

matured. CW_{50} was determined by using the linear regression at CW range that produced percentage of maturity at 50%.

Results

Sex ratio during a year of study was M: F=0.92: 1 which should be the relative frequency of females. The crabs CW ranged were from 6.0 cm to 17.5 cm for males and 5.5 cm to 17.0 cm for females. The crabs CL were from 4.0 cm to 8.0 cm for male and 3.5 cm to 7.0 cm for female. The BW ranged from 40.0 to 275.0 gr for males and 32.50 to 242.2 gr for females (Table 1). The results of this study show that at the same size, BW and the parameters of CL, CW and CPL male *P. segnis* are higher than females in Persian Gulf coasts.

A total of 448 female crabs of *P. segnis* of various sizes were randomly sampled and only 278 crabs of *P. segnis* had mature abdomen. The largest immature female encountered was 16.5 cm CW and the smallest mature female was 7.40 cm CW. Using the abdominal flap changes for female as an indicator of crab maturity, the percentage of female crabs that mature was calculated for each 0.5 cm CW interval from CW range of 6.0 cm (the smallest size range of the crab sampled) to CW range of 17.0 cm (the largest size range of the crab sampled). Using the linear regression at the CW ranges that shows percentage of maturity, CW^{50} occurred at7.5 cm CW in the female crabs with y=30.5x+10.5 and R2=0.9368 (Figure 2).

Out of 418 male crabs sampled, only 242 crabs of *P. segnis* were mature. Male crab maturity was determined based on the presence of enlarged and white vas deferens. The largest immature male encountered was 17.5 cm CW and the smallest mature female was 7.5 cm CW. Using the presence of vas deferens for male as an indicator of crab maturity, the percentage of male crabs that mature was calculated

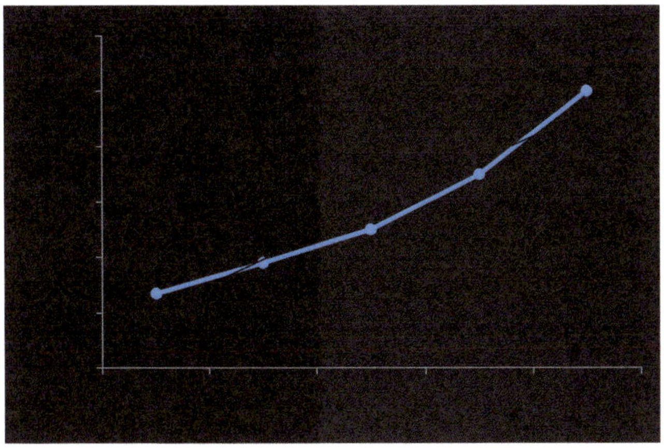

Figure 2: Linear regression of morphometric relationship between CW range and percentage of maturity of female *P. segnis.*

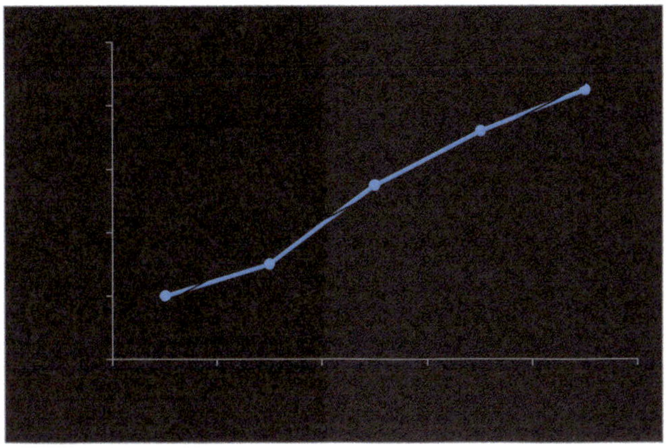

Figure 3: Linear regression of morphometric relationship between CW range and percentage of maturity of male *P. segnis.*

for each 0.5 cm CW interval from CW range of 6.0 cm (the smallest size range of the crab sampled) to CW range of 17.5 cm (the largest size range of the crab sampled). Linear regression analysis at the CW ranges that shows percentage of maturity, shows that the size at maturity, CW^{50} occurred at 8.24 cm CW in the male crabs with y=17.2x+0.8 and R^2=0.9844 (Figure 3).

Allometric equations with respect to males and females of *P. segnis* are indicated in Tables 2 and 3. The allometric relation between the set of characters studied suggested that in most cases the relationship was positive and highly significant. The study has shown that males are marginally heavier than females up to 120-125 mm carapace width, and there after males are heavier than females. The scatter diagram for males and females was obtained by plotting weight against carapace width/length of individual crabs. From the data presented, a distinct relationship was found between width and total weight, as judged from the closeness of the scatter dots, as well as from the parabolic nature of the plot.

The exponential values (b) for the carapace width-weight relationship in males and females (2.334 and 2.554, respectively) show that there is marked deviation from the isometric growth pattern (Table 4). The 't ' test confirmed that 'b' significantly differs from 3, in both sexes. The exponential values (b) for carapace length-weight in males and females (2.112 and 3.443, respectively) indicate that the significant departure from isometric growth is only evident in females. In males, the curve follows an isometric growth pattern. The 't' values are given in Table 5.

The relationship between female abdomen width and CW. is shown in (Figures 4 and 5). The abdomen width shows positive allometry in the first phase (crabs <8.7 cm CW) and the allometry becomes more positive in the second phase (crabs >7.5 cm CW). Thus the size at which puberty molt occurs is between 7.1 and 8.0 cm CW.

The relationship between chelar propodus length and CW is shown in (Figures 6 and 7). for male crabs. In male crabs chela allometry is slightly positive in first phase (crabs having <7.5 cm CW) and then allometry changes and chela grows more positively in second phase (crabs having >8.7 cm CW). This shows that the puberty molt occurs between 7.5 and 8.7 cm CW in male crabs.

Discussion

Sexual maturity is defined the ability to successful mate and results

Independent variable (*x*)	Dependent variable (*y*)	Allometric growth equation (*y*=a+bx)	'r^2' value
Carapace width	Abdomen width	AW=−18.228+0.4563 CW	0.5567*
Carapace width	Abdomen length	AL=−16.442+0.7664 CW	0.3343*
Carapace Length	Abdomen width	AW=−15.232+0.6673 CL	0.7675*
Carapace Length	Abdomen length	AL=−6.664+0.5579 CL	0.4454*
Abdomen width	Abdomen length	AL=−7.886+0.7764 AW	0.7677*
Carapace width	Chelar propodus length	CPL=−69.334+1.4676 CW	0.6777*
Carapace length	Chelar propodus length	CPL=−58.665+1.7655 CL	0.3343*

Table 2: Allometric equations and correlation coefficient (r) values between different variables in males of *P. segnis.*

Independent variable (*x*)	Dependent variable (*y*)	Allometric growth equation (*y*=a+bx)	'r^2' value
Carapace width	Abdomen width	AW=−17.008+0.7656 CW	0.6754*
Carapace width	Abdomen length	AL=−15.332+0.7334 CW	0.6884*
Carapace Length	Abdomen width	AW=−12.552+0.8873 CL	0.7332*
Carapace Length	Abdomen length	AL=−6.7764+0.8979 CL	0.6112*
Abdomen width	Abdomen length	AL=−5.6554+0.9884 AW	0.6442*
Carapace width	Chelar propodus length	CPL=−59.556+1.5556CW	0.5221*
Carapace length	Chelar propodus length	CPL=−51.225+1.4443CL	0.5546*

Table 3: Allometric equations and correlation coefficient (r) values between different variables in females of *P. segnis.*

Measurements	Sex	Logarithmic equation	Parabolic equation
Carapace Width-Total Weight	Male	Log=−16.532+2.334 log L	$W=0.00006469L^{3.552}$
	Female	Log=−15.278+2.554 log L	$W=0.00002233L^{2.443}$
Carapace Length-Total Weight	Male	Log=−6.551+2.112 log L	$W=0.00005445L^{3.033}$
	Female	Log=−5.720+3.443 log L	$W=0.00003442L^{2.332}$

Table 4: Carapace width/length-total weight relationship in males and females of *P. segnis.*

Relationship	Sex	'*t*' value	Remarks
Carapace Width-Total Weight	Male	18.22	Significant at 1% level
	Female	5.45	Not significant
Carapace Width-Total Weight	Male	2.33	Significant at 1% level
	Female	6.51	Significant at 1% level

Table 5: The '*t*' values for the carapace width/length and total weight relationship in males and females of *P. segnis.*

Figure 4: This shows relationship between female abdomen width and CW.

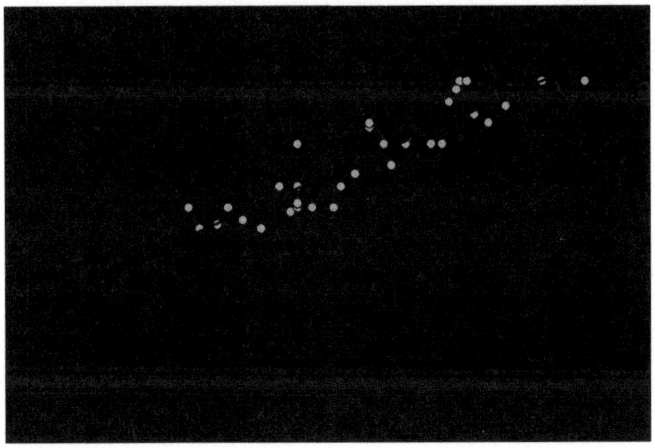

Figure 5: Relationship between carapace width (CW) and abdomen width (AW) of mature female crab.

in the extrusion of fertilized egg [8]. The pubertal molt stage in *P. segnis* is said to be a reliable indicator of the ability of female crab to mate. Studies have also shows that the female *Portunus* spp. molt again after the pubertal molt [12-14]. The result from the present study shows that the *P. segnis* male mature at a smaller size than female crab for both morphological appearances and relative growth.

Knowledge of the distinguishing characteristics and size relations of sexually mature individuals is of particular importance in the study of commercially valuable crustaceans. Crabs of the family Portunidae are distinguished by having the dactylus of their fifth pereiopods enlarged and flattened, which facilitates swimming. In the Indo-Pacific region, the genus *segnis* is represented by a member that has nine antero-lateral spines at each side of the carapace, the posteriormost one of which is enlarged. All these characters are in agreement with the descriptions of *Portunus segnis*. In addition, in this species sexes are easily differentiated by the colour pattern of their exoskeleton. Males are brilliantly coloured with bright blue, and females have a dull brown colour. This unique feature is not seen in other common portunids. Gross morphological differences in external anatomy between sexes are similar to those of other portunid crabs [11-15].

The results of the length-weight relationship analysis in *P. segnis* indicate that in juveniles and pre-adult crabs, weight gain is almost uniform; males are slightly heavier than females until they attain 10-12 cm carapace width. Thereafter males are heavier than females at any given length [6]. Sukumaran and Neelakantan [16] found a weight increase was evident above a carapace width of 115 mm in P. pelagicus. The tendency of males to be heavier than females in portunids is in accordance with the observations of Potter et al. [17] in P. pelagicus and P. sanguinolentus (Herbst, 1783), from Cochin [17], in P. sanguinolentus from Mangalore of Prasad et al. [18], in *P. sanguinolentus, P. pelagicus,* and *Scylla serrate* from Karwar and of Sukumaran, and of Neelakantan [18] in *P. pelagicus* and *P. sanguinolentus* from Mangalore [19]. However, from the reports of these workers it is understood that there are marked variations in the results of carapace width-weight relationship in portunids from different places, and also within various regions of the same coast.

In contrast to the result and reports of the above workers, found that females of *P. pelagicus* are heavier than males at a given length. There have been a large number of studies on morphometric in marine crabs along coasts of Persian Gulf [13-15]. Ghorbani et al. [15] reported that male of *P. pelagicus* are heavier than female in coasts of Persian Gulf (Boushehr Province), but Jazayeri et al. [13] found that females of *P. pelagicus* are heavier than males in coasts of Persian Gulf (khuzestan Province). Tadi et al. [20] reported that male of *P. segnis* are heavier than female in coasts of Persian Gulf (Hormozgan Province) [20]. Same studies showed that different size with carapace width ranging

Figure 6: Relationship between carapace width (CW) and chelar propodus (CPL) of immature male crab.

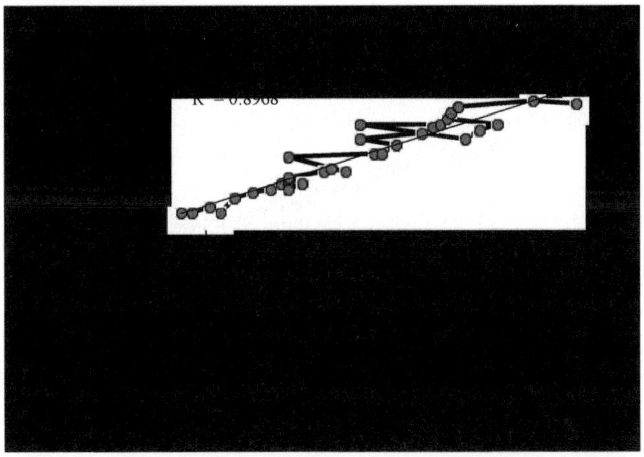

Figure 7: Relationship between carapace width (CW) and chelar propodus (CPL) of mature male crab.

from 8 to 15.9 cm for female and 7.8 to 13.7 cm for male and mean weight of males was considerably more than females from Moreton Bay, Eastern Australia [21,22], also Razek [23] showed that carapace width ranging from 4.5 to 16.5 cm, carapace length from 1.9 to 7 cm for the same species from Bardawil lagoon, Northern Egypt.

In brachyuran crabs chela in males and abdomen in females are considered as secondary sexual characters because of their functions in reproduction. The male crab uses its chela for territorial defense, combat, mating and courtship as well as in carrying and holding the female during copulation. The abdomen in adult females forms an incubation chamber for the developing eggs, which are attached to the setose pleopods. The increase in relative growth of the male chela and female abdomen at the puberty molt brings these structures to full functional size at sexual maturity. Hence the relative growth of chela in males and abdomenin females has been used to determine size at which puberty molt occurs or functional maturity attained.

The result of size at maturity obtained using morphological appearences (male CW_{50}=8.24 cm CW; female CW_{50}=7.44 cm CW) are within the value obtained using relative growth (male=7.5-8.7 cm CW; female=7.1-8.0 cm CW). Thus, this shows that both methods can be used to determine the CW_{50}.

The sex ratio obtained for *P. segnis* was M: F=0.92:1 (male/female) which showed a significantly more abundant female population. Similar results were reported for *Portunus pelagicus, Portunus pelagicus* by Jazayeri et al. [13] and *Portunus segnis* by Tadi et al. [20].

Acknowledgment

We thank Dr. Amir Vazirizadeh and Mr. Ali Mansoori for insightful comments; and for field assistance. Financial support was carried out by Shahid Bheshti University, Iran.

References

1. ROPME (1999) Overview on Land-based Sources and Activities Affecting the Marine Environment in the ROPME Sea Area. UNEP/GPA Coordination Office & ROPME 127.
2. Willams AB (1974) The swimming crabs of the genus Callinectes (Decapoda: portunidae). United States Fishery Bull 72: 685-798.
3. Atar HH, Sector S (2003) Width/length-weight relationships of the blue crab (Callinectes sapidus Rathbun, 1896) population living in Beymelek lagoon lake Turkish. Anim Sci 27: 443-447.
4. Batoy CB, Sarmago JF, Pilapil BC (1987) Breeding season, sexual maturity and fecundity of the blue crab, Portunus pelagicus (L.) in selected coastal waters in Leyte and vicinity, Philippines. Annals of Tropical Research 9: 157-177.
5. Kardavani P (1995) Iran waters ecosystems, Persian Gulf and Oman Sea. Tehran, Semat pree 680.
6. Carmona-Suarez C, Conde FE (2002) Local distribution and abundance of swimming crabs (Callinectes spp. and Arenaeus cribrarius) on a tropical arid beach, Ensenada de la vela, Venezuela. Fisheries Bull 100: 11-25.
7. Davis, Gregory Bruce (1988) The biology of the blue manna crab - (Portunus pelagicus) - in estuaries of Southwest Western Australia / by Greg Davis 1-6.
8. Dai A, Yang S (1991) Crabs of the China Seas. China Ocean Press, Beijing, 682.
9. Dittel A, Epifanio CE (1982) Seasonal abundance and vertical distribution of crab larvae in Delaware Bay. Estuaries 5: 197-202.
10. Edgar GJ (1990) Predator-prey interactions in sea grass beds. II. Distribution and diet of the blue manna crab Potunus pelagicus Linnaeus at Cliff Head, Western Australia. Journal of Experimental Marine Biology and Ecology 139: 23-32.
11. Fielder DR, Eales AJ (1972) Observations on courtship, mating and sexual maturity in Portunus pelagicus (L., 1766) (Crustaceas, Portunidae). Journal of Natural History 6: 273-277.
12. Gaddes SW, Sumpton WD (2004) Distribution of barnacle epizoites of the crab Portunus pelagicus in the Moreton Bay region, Eastern Australia. Marine and Freshwater Research 55: 241-248.
13. Jazayeri A, Papan F, Savari A, Sakinejad T (2008) Biological Investigation of Persian Gulf blue swimmer crab (Portunus pelagicus) In Khuzestan coasts, Iran. Journal of American Science 7: 7-1.
14. Hosseini M, Vazirizadeh A, Parsa Y, Mansoori A (2011) Sex ratio, size distribution and seasonal abundance of blue swimming crab, Portunus pelagicus (Linnaeus, 1758) in Persian Gulf coasts, Iran. World Applied Sciences Journal 17: 919-925.
15. Ghorbani N, Sayfabadi JS, Oufi F, Abtahi B (2002) Relationships carapace width-lentgh with weigh of blue swimmer, crab Portunus pelagicus in coasts Boushehr waters. Journal of Fisheries 2: 59-66.
16. Sukumaran KK, and Neelahantan B (1998) Food and feeding of Portunus sanguinolentus (herbst) and Portunus pelagicus (Linnaeus) (Branchyuran: Portunidae) along Karanataka coast (India). Journal of Marine Science 26: 35-38.
17. Potter IC, Chrystal PJ, Loneragan NR (1983) The biology of the blue manna crab Portenus pelagicus in an Australian estuary. Marine Biology 78: 75-85.
18. Prasad PN, Reeoy J, Kusuma N, Neelakantan B (1989) Width-W eight and Length-Weight Relationships in Three Portunid Crab Species Uttar Pradesh. Journal Zoology 9: 116-120.
19. Stephenson W, Campbell B (19590 The Australian portunids (Crustacea: Portunidae). The genus Portunus. Australia Journal of marine and Freshwater Research 10: 84-124.
20. Romano N, Zeng C (2006) The effects of salinity on the survival, growth and haemolymph osmolality of early juvenile blue swimmer crabs, Portunus pelagicus. Aquaculture 260: 151-162.
21. Tadi F, Pazooki J, Safaei M (2012) Food and feeding habits of blue swimming crab Portunus segnis along Persian Gulf (Hormozgan coast). International Conference the Biology of Iran, Press Iran.
22. Sagun LC, Tureli E, Duysak O (2009) Width/length-weight and width-length relationships for 8 crab species from north-Mediterranean coast of Turkey. Journal of Animal and Veterinary Advances 8: 75-79.
23. Razek FAA (2006) Population biology of the blue crab from lagoon northern sina Egypt. National institute of oceanography and fisheries Egypt.

Socio Economic Status of Fisher Folks Engaged in Mussel Fishery at Kanyakumari District of Tamil Nadu, India

M. Dalin Mary[1], Saritha Kailasam[2]*, M. Jansi[3] and Jamila Patterson[2]

[1]St. Jude's College, Thoothoor, Tamil Nadu, India
[2]Suganthi Devadason Marine Research Institute, Tuticorin, Tamil Nadu, India
[3]St. Hindu College, Nagercoil, Tamil Nadu, India

Abstract

This paper attempts to explore the extent of improvement due to mussel fishing in the ten sites of Kanyakumari district. As per our study the data on the percentage of fishermen involved in mussel fishery and their marital status, age status, income status, educational status, family status and housing pattern were collected through personal interview. Data regarding marital status show that 80.7% were married, 16.7% unmarried and 2.6% widower among fishermen involved in mussel fishery in Kanyakumari district. In case of age group, the fishermen involved in mussel fishery in Kanyakumari district revealed that 46% of them were in the age group of 40-50 followed by 50-60(24%) and 30-40(18%).The income status of the sample respondent varied and 55% of them earned Rs 5,000-15,000. Only 24% of them earn Rs 15,000-25,000 while other remains in other categories. In Kanyakumari district most of the fishermen (68.7%) involved in mussel fishery are illiterate and 26.6% are educated to primary level. Most of the fishermen involved in mussel fishery in Kanyakumari district have 2-6 members in their family (92.8%). When compared with four types of houses used by the fisher folks most of them (70%) were in semi pucca colony house. Based on our results it shows that the literacy level and awareness about education was very low around the fisherman. So it may be suggested that more attention should be given to enhance the literacy level of the coastal population.

Keywords: Mussel fishery; Marital status; Age status; Income status; Educational status; Family status; Housing pattern

Introduction

Kanyakumari district is approximately having 68 km of coastal area with nearly 1 lakh fishermen involved in fishing. Kanyakumari coastal area covers about 26 percent of total fishermen of Tamil Nadu. Marine mussel fisheries are very important for the economy and well-being of coastal communities, providing food security, job opportunities, income and livelihood as well as traditional cultural identity. Maintaining the long term prosperity and sustainability of marine mussel fisheries is not only a political and social significance but also of economic and ecological importance. Details of the mussel fishery in Kanyakumari coast during the earlier years were given by Jones, Jones and Alagarswami, Appukuttan and Nair, Appukuttan et al. and Joel and Ebenezer [1-5]. Fishing is carried out throughout the year, except during the fishing holidays from first May to middle of June. The fisher folks are much aware about the different types of fishing, so they change the fishing according to the variation in the fishing season. Different forms of fishing like finfish and shell fish fishing were done in the kanyakumari coast. One group of fishermen knows skin diving and do mussel fishing in the rocky coastal areas during October to April. Present study was carried out to analyze the age groups, marital status, income status family status, educational status and housing pattern of the fishermen who are doing mussel fishing for their livelihood. Besides increasing fish production to the maximum, the fisheries development should aim at raising nutritional level of the people and improving the socio economic conditions of the fisherman. The socio economic condition of the marine fishermen and their attitude towards development schemes will serve as background information which is vital for implementing new technologies and politics.

Materials and Methods

Kanyakumari, district of Tamilnadu, forms the southern tip of India where the Bay of Bengal, Indian Ocean and Arabian Sea meet. Mussels are available plenty in rocky coastal areas of this district. Two species of mussels such as green mussel, *P. viridis* and the brown mussel *P. indica* are available more in the coastal areas of Enayamputhenthurai, Enayam, Midalam, Kurumpanai, Vaniyakudi, Kodimunai, Colachal, Kadiapatinam, Muttom and Kanyakumari (Figure1). Among these two species *P. indica* is dominant in Kanyakumari coast which forms food for the coastal people. Mussel fishing is normally carried out as an off time occupation by the fishermen. They collect the mussels from intertidal rocks using iron scalpel with wooden handle during low tide. The craft and gears used for mussel fishing were observed and recorded from the fishermen. Socio-economic characteristics viz., marital status, age, monthly income, educational status, family status and Housing facility level were selected for the study. The data were collected from 1170 respondents through personal interview with the help of pre-tested survey schedule and the data were analyzed using the Excel XP 2007 software [6].

Results

Marital status

Marital Status of the fishermen involved in mussel fishing was analyzed and the result was presented in (Figure 2). The results revealed

***Corresponding author:** Saritha Kailasam, Suganthi Devadason Marine Research Institute, Tuticorin, Tamil Nadu, India, E-mail: saritha23kailasam@gmail.com

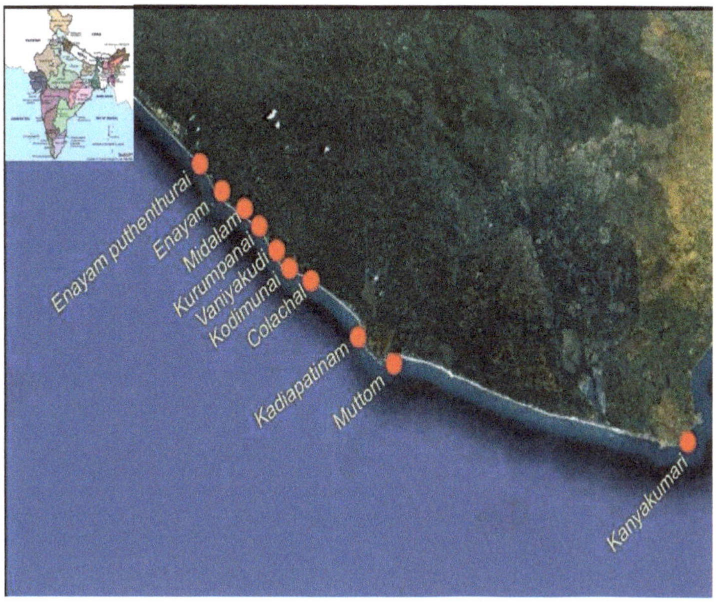

Figure 1: Map showing the study areas.

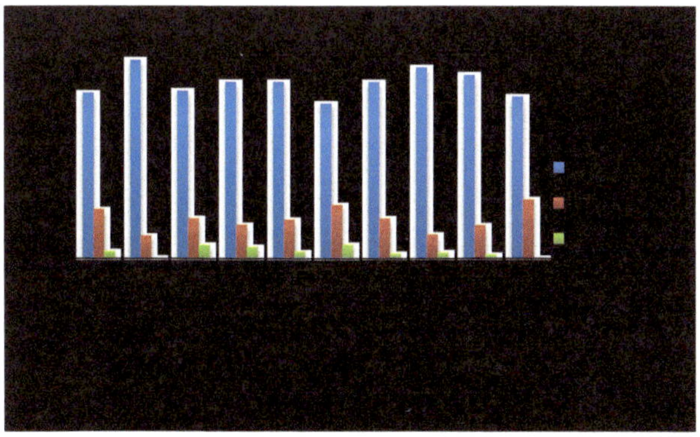

Figure 2: Marital Status of the fishermen involved in mussel fishing.

that in Enayamputhenthurai 90% of the mussel fishing fishermen are married and this was highest compared to the other coastal areas. The lowest percentage (70%) of married persons involved in mussel fishing was from Kurumpanai coastal area. In Kanyakumari coastal area 26% of fishermen involved in mussel fishing were unmarried, which was the highest when compared with other coastal areas. The lowest percentage of unmarried persons involved in mussel fishing was from Enayamputhenthurai (10%). The widower status of the persons involved in mussel fishing was analyzed during the study period in the ten coastal areas. It was found that in Colachel and Kurumpanai 6% of the fishermen involved in mussel fishing were widowers, compared to the other coastal areas. The lowest percentages of widower involved in mussel fishing were from Vaniyakudi (1.7%) coastal area. As a whole 80.7% were married, 16.7% unmarried and 2.6% widower among fishermen involved in mussel fishery in Kanyakumari district.

Age status

The status of the age group of the fishermen involved in mussel fishing was analyzed during the study period and the results were presented in (Figure 3). It was found that 15% of fishermen doing mussel fishery are below the age of 30 in Enayam, which was highest result when compared with the other coastal areas. In colachel coastal area 6% of these age group persons were involved in mussel fishing. In Colachel and Midalam coastal areas 30% of fishermen involved in mussel fishery were above 30-40 age group and this result was the highest compared to the other coastal areas and 8.3% of t persons of this age group were involved in mussel collection in Vaniyakudi .In Vaniyakudi coastal area 58.3% of fishermen in the age group of 40-50 were involved in mussel fishery and it was highest result compared to the other coastal areas. But in Kanyakumari costal area 33.3% of these age group fishermen were involved in mussel fishery. In this region 46.7% of fishermen engaged in mussel fishery were above the age group 50-60. Lower percentage (12.5%) of this age group persons involved in mussel fishery were observed in Enayamputhenthurai. In Midalam coastal area 10% of old men were involved in mussel fishing and only 2% of old men were involved in mussel fishing in Colachel and Kurumpanai coastal areas. As a whole the age status of the fishermen involved in mussel fishery in Kanyakumari district revealed that 46% of them were in the age group of 40-50 followed by 50-60 (24%) and 30-40(18%).

Income status

The income due to mussel fishing in different coastal areas was analyzed and results were presented in (Figure 4). The income status of the persons involved in mussel collection was analyzed during the study period 2010-2012 in all the ten coastal areas. It was found that in Kanyakumari 56.7% of persons involved in mussel fishing earns below Rs 5000 and 43.3% of person earns above Rs 5,000-10,000. The lower percentage of this income of Rs 5000-10000 was noted among the fishermen of Enayamputhenthurai (22.5%). In Kadiapatinam station 33.3% of the fishermen earn Rs 10, 000-15,000 by mussel fishing and this was the highest when compared with the other coastal areas and the lowest percentage (15%) was observed in Colachel. In Enayam, 20% of fishermen involved in mussel fishing earns above Rs. 15,000-20,000 and this was the highest when compared with other coastal areas and the lowest percentages of this income was from the station

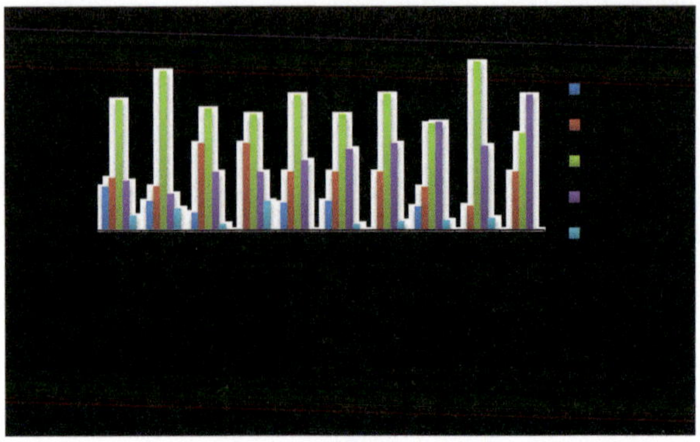

Figure 3: Age range of fishermen involved in mussel fishing.

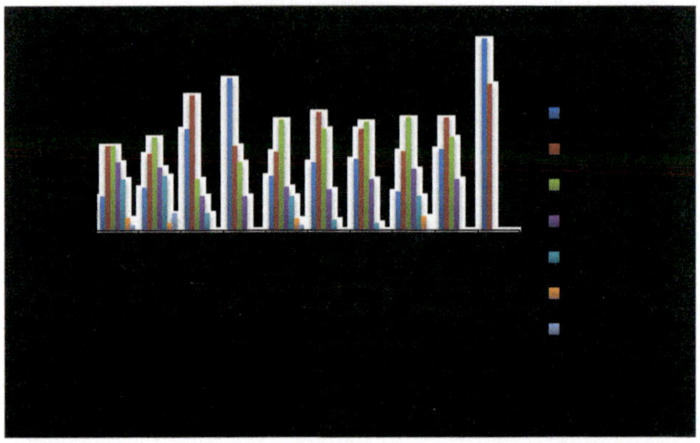

Figure 4: Monthly Income Status of fishermen involved in mussel fishing.

Colachel and Midalam (10%). In station Enayamputhenthurai, 16% of persons involved in the collection of mussels were earning above Rs 20000-25,000 which was highest when compared with other stations and the lowest percentage of this income was observed from the station Kodimunai (2%). In station Kadiapatinam 4% of persons involved in the collection of mussels were earning above Rs 25,000-30,000, which was highest and the lowest percentage of this income status was with fishermen of Enayamputhenthurai (2.5%). The highest income status of Rs 30,000 and above were earned by fishermen of Enayam (5%) followed by Muttom and Enayamputhenthurai (1.5%) As a whole, in these ten sites, the income status of the sample respondent varied and 55% of them earned Rs 5000-15000. Only 24% of them earn Rs 15000-25000 while other remains in other categories.

Educational status

The educational status due to mussel fishing in different coastal areas was analyzed and results were showed in (Figure 5). The Educational status of the persons involved in mussel collection were analyzed during the study period in the all the coastal areas. The percentage of the educational status of persons involved in mussel fishing varied between different stations. It was found that 93.3% of persons involved in the collection of mussels were illiterate in Kanyakumari which was the highest when compared to other coastal areas and 6.7% of persons were educated up to primary level. In station Enayam 50% of the fishermen involved in mussel fishing were with primary level education and 47.5% were illiterate. In Muttom 20% of persons involved in the mussel fishing were with educational status of primary level, 8% at middle school level, 1.3% at High school level and 0.7% at Higher Secondary level. In station Colachel only 2% of persons involved in mussel collection were educated to High school level, 3% to middle school level, 23% to primary level and the remains of 72% were illiterate. In Kanyakumari district most of the fishermen (68.7%) involved in mussel fishery are illiterate and 26.6% are educated to primary level.

Family status

The family status due to mussel fishing in different coastal areas was analyzed and results were represented in (Figure 6). The family status of the persons involved in mussel fishing was analyzed during the study period. It was found that in Kanyakumari, coastal region 3.3% of persons involved in mussel fishing coming under the family status of below 2 members. In Kurumpanai 72% of fishermen had 2-4 members in their family and this was the highest and the lowest percentage of 40% were observed in Enayam. In Vaniyakudi 55% of fishermen had 4-6 members in their families, and 24% of this size family was observed in Kurumpanai. In Enayam 11% of the families had 6-8 members where as 0.8% of families had this number of member's in Vaniyakudi. In Enayam and Kadiapatinam (2%) of the fishermen family had 8-10 members in their families which was the highest followed by Vaniyakudi (1.7%) and Muttom (0.6%) Most of the fishermen involved in mussel fishery in Kanyakumari district have 2-6 members in their family (92.8%).

House pattern

The house types of fishermen involved in the mussel fishing at different coastal areas were analyzed and result was presented in (Figure 7). Housing type is one of the most important indicators of the economic status of the family. The mussel fishermen in Kanyakumari district lived in four types of house namely Pucca house, Semi-pucca house, Mud-tiled house and Rental house. It was observed from the above table that only 3% of fishermen have their own Pucca houses (i.e., a dwelling with tiled or concrete roof and brick wall) Most of the fishermen have their own semi-pucca houses (Government free colony houses) i.e., 70%. Nearly 25% of the fishermen were accompanied by mud-tiled house .It is apt to mention here that some of the fishermen don't have their own house. Only 2% of the fishermen lived in rental houses. When compared with four types of houses, most of them (70%) were in semi pucca colony houses. So it could be informed that the economic status of the households of sample respondents of fishermen, as revealed by their housing conditions is poor even though most of the fishermen have their own house.

Discussion

Fisheries in India support the livelihood of about 10 million people. Fisheries development program are implemented to improve the socio economic status of the fishermen in terms of education, health, income and standard of living. In the present study, the analysis of the age status of fishermen involved in mussel fishing in Kanyakumari district revealed that 46.3% of fishermen are in the age group of 40-50, 24.1% of fishermen are in 50- 60 age group, 17.9% of fishermen are 30-40 age group, 7.9% of fishermen are below 30 age group and 3.8% of fishermen are above 60 age group, 70% of fishermen in the age range of 40-60 years were involved in mussel fishery. The family status of the person involved in mussel collection revealed that 52% of the fishermen family with 2-4 members, 41% of the family with 4-6 members, 4.4% of the family with 6-8 members and 2% of the family with below 2 members and 2% of the family with 8-10 members. The average family size in

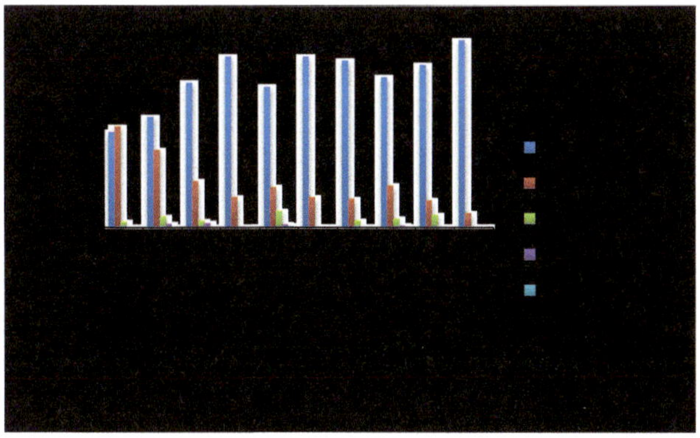

Figure 5: Educational Status of fishermen involved in mussel fishing.

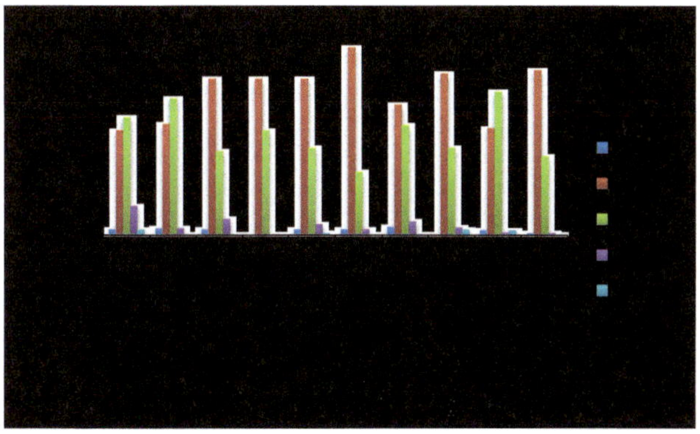

Figure 6: Family Status of persons involved in mussel fishing.

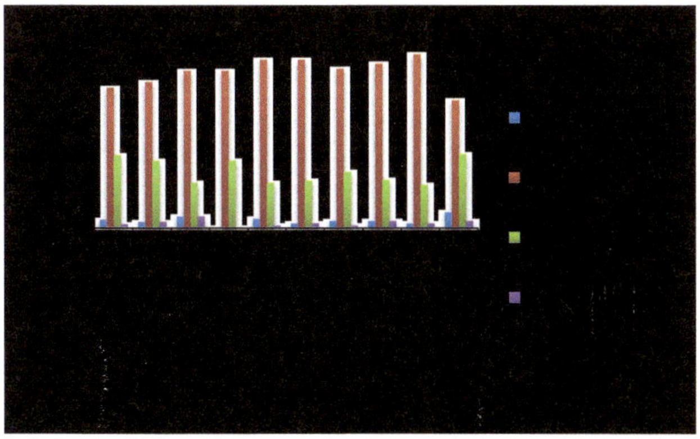

Figure 7: House Pattern of Fisherman involved in mussel fishing.

Poonthurai village near Thiruvananthapuram in 1985 was 6.3 and about 69% were living in huts [7]. The illiteracy rate was 46% and 89% of the literates had education up to primary level only. The educational status of person involved in mussel collection in Kanyakumari district revealed that the illiterate level of the mussel fishermen were higher (68.7%) than that of the fishermen at Poonthurai village [7] 26.6% are at primary level, 3.9% of them are at middle school level, 0.7% at high school level. During 1983, the economic condition of the fishermen in some selected villages of Maharashtra and Gujarat revealed that the illiteracy rate ranged from 48-75% and among the literates majority had primary education only [7] similar to the present observation. The present study reported that the income status of the person was increased as the mussel cost increased from Rs 90 to 120 for 100 mussels 35% of fishermen were coming under the income status of Rs 5,000- 10,000, 21% of the fishermen earn about were Rs 10,000-15,000 income 15% of the fishermen are below Rs 5000. Similarly 15% of the fishermen earn about Rs 15,000-20000 income, 10% of fishermen earn about 20,000-25,000 income, 3% of the fishermen earn about Rs 25,000-30,000 and 1% of the fishermen were earning above Rs 30,000 per month. It was reported that the fishermen received an annual income between Rs 3000 and Rs 10000 in Puthiappa-puthiangadi area of calcutta during 1980 [7]. In this area the size of the fishermen family was large (9.0) as compared to non-fisherman family. Average family size in 1992 in Ernakulam district of Kerala was estimated to be 7.7, 5.5 and 5.6 for marginal small and large fish farmers respectively. The literacy level of the marginal farmers was less than high school qualification while 6% of the small farmers were graduates and post graduates. The housing pattern of the fishermen involved in mussel fishing in the Kanyakumari district revealed that most of them (70%) have their own semi-pucca houss (government free colony house) and 25% of them lived in mud tiled houses. Socio economic study of fishermen along the Madras coast showed that most of the fishermen (90%) in Thiruvettyurkuppam and 6% in Pudimanikuppam were lived in huts. The study conducted during 1989-90 to evaluate the socio economic condition of traditional fishermen living along the Tanjavur coast of Tamil Nadu indicated that most of fishermen were living in temporary structures like Kutcha house and Huts [7]. John Kurien [8] reported that most of the fishermen were living in huts (71.8%) at Poonthura village of Triuvandrum District.But in our study 70% of fishermen owned their own pucca houses.

Conclusion

Based on the studies conducted on the field of socioeconomics, it is suggested that more attention should be given to enhance the literacy level of the coastal population. Since the literacy rate is low, So small family norms should be popularized among the fishermen communities .because he bigger number of dependents creates poverty due to low economic status So from our study we suggest government should take necessary steps to improve the literacy rate, provide loan/ subsidies and create awareness about mussel processing techniques among the fisher communities to improve their economic status.

Acknowledgement

The authors are thankful to Dr. J.K. Patterson Edward, Director, Suganthi Devadason Marine Research Institute, Tuticorin, India for providing us the facilities to carry out the work.

References

1. Jones S (1950) Observations on the bionomics and fishery of the brown mussel (Mytilus sp.) of the Cape Comorin region of peninsular India. J Bombay Nat Hist Sac 49: 519-528.
2. Jones S, Alagarswami K (1973) Mussel fishery resources of India. Proc. Symp. Living Resources of the Seas around India, Central Marine Fisheries Research Institute, Cochin 641-647.
3. Appukuttan KK, Nair TP, Thomas KT (1989) Spat settlement of brown mussel Perna indica Kuriakose and Nair in the southwest coast of India. J mar biol Ass India 31: 266-275.
4. Appukuttan KK, Nair TP, Mathew Joseph, Thomas KT (1988) Brown mussel (Perna indica) resources of the southwest coast of India and the results of farming experiments at Vizhinjam. CMFRI Bull 42: 257-263.
5. Joel JJ, Ebenezer IP (1989) The current trend of the mussel fishery in the

Kanyakumari district of Tamil Nadu. Mar Fish Infer Seru T & E Ser 100: 9-13.

6. Singh JY, Santhakumar R, Pandey DK, Bharati H, DebRoy P (2012) Adoption of Hygienic Fish Handling Practices by Fishermen. Indian Res J Ext Edu 12.

7. Narayanakumar R, Panikkar KKP, Sekara DBS, Sathyadas R (2000) In: Marine Fisheries Research and Management (Menon, Pillai (Eds)) Socioeconomic analysis of marine fisherman in India, CMFRI publication, 60: 895-906.

8. John kurien (1980) Fishermen's cooperatives in Kerala-A critique (BOBP/MISC/1, Bay of Bengal programme, Madras).

Separate and Joint Effects of Polycyclic Aromatic Hydrocarbons (PAH) and Polychlorinated Biphenyls (PCB) on Aromatase CYP19A Transcription Level in Atlantic Tomcod (*Microgradus tomcod*)

Adam Tulu[1], Ali Ishaque[1]*, Egbe Egiebor[1], Christopher Chambers[2] and Rosemary Jagus[3]

[1]*Department of Natural Science, University of Maryland, Eastern Shore, Princess Anne, MD 21853, USA*
[2]*NOAA-Fisheries, 74 Magruder Rd., Highlands, NJ 07732, USA*
[3]*Center of Marine Biotechnology, University of Maryland Biotechnology Institute.701 E Pratt st. Baltimore, MD 21202, USA*

Abstract

Ovarian cytochrome 19A (CYP19A) expression is recognized as a useful biomarker for exposure of fish to environmental contaminants such as PAHs and PCBs. In this study, a laboratory approach using Atlantic tomcod (*Microgradus tomcod*) from the Hudson River was used to evaluate the additive and interactive effects of a PAH (benzo[a]pyrine) and a PCB mixture (Aroclor 1242) with respect to their effects on various metrics of reproduction. The experimental design was a two-way factorial with each treatment at 0, 0.1 and 1 ppm and replicated three times. Fish embryos were subjected to a short term aqueous exposure (DMSO as vehicle) whereas larvae were exposed via repeated feedings of contaminated prey (*Artemia*). Quantitative reverse transcription-polymerase chain reaction (RT-qPCR) was used to evaluate the mRNA expression level of the ovarian aromatase CYP19A as a biomarker to assess reproductive stress in Microgadus tomcod. When tested alone, expression of aromatase CYP19A was significantly up regulated at the higher level of PCBs but no effect was observed from benzo-a-pyrene (B[a]P). In the PCB/PAH combined treatment, both low-PCB with both low/high levels of PAH and high-PCB with both low/high levels of PAH treatment groups had no significant effect on aromatase CYP19A transcript levels compared with the control group. Gonadosomatic Index (GSI) of reproductively mature females showed that only the high-PCB treatment group exhibited significant gonadal loss. The results confirm the expectation that transcription of both PCB and PAH responsive genes are upregulated since they both exert their toxic effects through the aryl hydrocarbons receptor (AHR) pathway.

Keywords: Aromatase; Biomarkers; Biotransformation; Endocrine disruption pollutants

Introduction

The potential adverse effects of environmental contaminants that function as endocrine disruptors have raised concerns in recent years. Endocrine disrupting chemicals (EDCs) in the environment are associated with adverse reproductive and developmental effects in wildlife by mimicing the actions of the female sex hormone estradiol [1]. Polychlorinated biphenyls (PCBs) and polycyclic aromatic hydrocarbons (PAHs) in the environment are associated with potential disruptive effects on vertebrate and invertebrate reproductive activity [2,3]. Many studies have shown that PCBs and PAHs are toxic to fish and other aquatic organisms [4-7]. PCBs and PAHs are thought to have contributed to diseases, deformities, and decline in some wild populations, including fish and birds of the Hudson River [8,9].

PCBs and PAHs can act through both receptor-dependent and receptor-independent (i.e., nonhormonal) pathways. Compared to PCBs, PAHs are biodegraded easily and accumulate in fish tissues in quantities that reflect recent exposure. PAH toxicity depends highly on the chemical structure of the PAH [10-12], and even PAH isomers with the same number of benzene rings may vary from nontoxic to extremely toxic depending on different steric positioning of these benzene rings [13,14]. As PCBs and PAHs are hydrophobic, they are found bound to the organic content of the sediments where they become bioavailable to the benthic organisms and bioaccumulate in the food chain [15]. The principal target organs for PAHs and PCBs are those that produce, regulate, or respond to estrogenic hormones such as: brain, gonad, liver, uterus, mammary, adrenals, prostate, and placenta, as well as organs in the developing embryo or fetus [16]. PCBs and PAHs are known to adversely affect estrogen receptor (ER) function either through direct binding to the receptor or by the activation of other receptor pathways that modulate ER activity [17-19].

Aromatase cytochrome P450 (CYP19A, aromatase), is the terminal enzyme in the steroidogenic pathway [20,21], and is responsible for the conversion of androgens to estrogens [22,23]. The product of the CYP19 gene catalyzes the conversion of androgens to estrogens (specifically, C19-androstenedione into C18-estrone and testosterone into E2) [17]. Its activity is an important modulator of plasma β-estradiol (E2) concentrations and is critical to the regulation of reproductive processes controlled by E2 [24]. Both aromatase CYP19A and E2 concentrations increase simultaneously with seasonal gonadal development [25-27].

In lower vertebrates, aromatase CYP19A is predominantly expressed in the ovary and plays important roles in sex-differentiation and oocyte growth, [28-31]. This enzyme complex is localized in the endoplasmic reticulum of cells in which it is expressed [29,32]. Studies

***Corresponding author:** Ali Ishaque, Department of Natural Science, University of Maryland, Eastern Shore, Princess Anne, MD 21853, USA
E-mail: Abishaque@umes.edu Abishaque@umes.edu

have shown that transcript levels of gonadal aromatase (CYP19A) are increased in association with aromatase enzyme activity during vitellogenesis in Atlantic cod (*Microgadus tomcod*) [33], rainbow trout [34,35], tilapia [20], and red seabream [36]. Aromatase gene expression has been proposed as a key step of estrogen synthesis for ovarian differentiation [37]. Moreover, plasma E2 is directly related with the transcript levels and enzyme activity of aromatase CYP19A during oocyte development [2]. Studies have shown that activated aryl hydrocarbon receptor (AHR) induces the recruitment of estrogen receptor- (ERa) to AHR-regulated genes and that AHR is recruited to ERα-regulated genes [38-43].

Fish such as Microgadus tomcod are an appropriate model for studying the effects of EDCs because Microgadus tomcod have been exposed to multiple sources of EDCs in their environment including sewage, industrial effluent, and urban and agricultural runoff. The value of using local species and exposing them to two abundant xenobiotics in their natural environment and at ecologically relevant doses is that this allows inspection of possible interactive effects between PCBs and PAHs. Our logic in using two classes of contaminants – both are problematic in the Hudson River – is that such an approach permits a determination of whether the primary effect on the Microgadus tomcod population was mediated by bioaccumulation of PCB, PAH, or both. The laboratory-based study spans most of Microgadus tomcod life cycle, i.e., from embryos to feeding larvae to juveniles to reproductive maturation of adults.

Hence, by using both PCB and PAH on local species, we formulated a picture of the uptake and toxicity of contaminants within this ecosystem. We hypothesized that the effect of PCB and PAH together will be higher than their individual effects on CYP19A transcription and that both chemicals will cause reproductive stress. This study has two main objectives pertaining to the evaluation of the additive and/or interactive sub lethal effects of early-life exposures of fish to PAHs and PCBs. First, to determine the interactive effects of PCB and PAH exposure on hepatic CYP19A transcript levels of Microgadus tomcod. Second, to quantify the toxic and morphological effects of PAHs and PCBs on the tissue, organ (ovary mass) and whole body size of female adult Microgadus tomcod.

Materials and Methods

Exposure of Microgadus tomcod

All female adult Microgadus tomcod (F_0generation, *N*=6), were collected on the same day in mid-February of 2008 from the Hudson River just prior to their spawning. Fish were transferred to James J. Howard Marine Sciences Laboratory at Sandy Hook where they were maintained at 4°C in clean laboratory water at salinity of 5 ppt for 21 days. The F_1 generation (fertilized eggs) and larvae were relocated to the grow-out static water tanks (40-L) held at 8°C. Salinity was increased incrementally from 5 ppt to 25 ppt in order to mimic the downstream movement of tomcod larvae in nature.

All exposures were conducted in a temperature and light controlled cold room at 6 to 7°C. During 12 wk post-hatching (wph) period, larvae were exposed to the PAH (benzo[a]pyrene), and the PCB (Aroclor 1242) at 0.1 (low), and 10 (high) times the observed concentration in the Hudson River. For embryo exposures, 100 µl of the high or low stocks was mixed with 100 ml of 5 ppt seawater in 150 ml glass beakers just prior to the addition of embryos. Thus, the final high and low exposures represented concentrations of 359 and 3.59 ng/ml (parts per billion), respectively. When both PAH and PCB were used, the PAH was administered to embryos for 48 h before the exposure to the PCB. The exposure medium was Dimethyl sulfoxide(DMSO) for the embryos and via a contaminated food source (*Artemia*) for the larvae for 10 sequential feedings. Vehicle control groups were exposed to 0.1% (v/v) acetone (100 µl/100 ml 5 ppt seawater).

At 14 wph, juveniles was transferred to 120-L flow-through tanks, weaned onto frozen foods, and tank temperatures were adjusted weekly to mimic seasonally changes in nature. Lastly, as juveniles achieved a size at which implant tags could be applied (~4 to 6 cm at 16 wph), they were marked with a subcutaneous injection of one of two colors of elastomer at one of six location associated with the dorsal fins (left or right side of anterior, medial, or posterior fins) according to treatment and pooled by replicate into 600-L tanks. The tomcod were fed *ad libitum* on thawed frozen bloodworms (chironomids), adult *Artemia*, and squid, and maintained at ambient temperature and salinity until they matured. The experiment was terminated at 36 wpf and maturing adults were sampled. During the rear-out period larvae and juveniles were checked daily for mortalities.

Toxic- interaction scale

A two parameter additive index approach model was used to determine the joint toxic action of PCB and PAH chemicals in a mixture. A Toxic Interaction Scale (S), was defined as $(S)=[(A_m/A_i)+(B_m/B_i)]$, where, A_m and B_m= toxicity of PCB and PAH when present in a mixture, and A_i and B_i = toxicity of PCB and PAH when they are tested separately. The values of S was measured in toxic unit (TU), and categorized into three groups: S=1 represents an additive effect, S<1 represent antagonistic effect, and S>1 represents synergistic effect.

Gonadosomatic Index (GSI)

Gonadosomatic index (GSI) quantifies the morphological effects of PAH and PCB on the ovary. GIS is calculated by the ratio of ovary weight to whole body weight; it is indicative of critical reproductive success in the life cycle of Microgadus tomcod population to determine its survival [44]. GSI was used for this study to serve as a time-integrated indicator of the general wellbeing of Microgadus tomcod gonad development.

Condition Factor (CF)

Condition factor (also known as coefficient of condition, or length-weight factor) was computed to assess the relative robustness, or degree of well-being, of a fish. *Condition Factor (CF)=[weight/(length)] X 100%*, where length was measured in centimeters (cm) and weight was measured in grams (g). Growth represents the integration of feeding, assimilation and energy expenditure over a period of time. Poor growth means less energy was available for reproduction, which will in turn reduce the species fitness and lead to a decline in the population. A high condition factor, therefore, reflects good environmental quality; while a low condition factor reflects poor environmental quality.

Aromatase CYP19A Transcription as an Indirect Assessment of E2 levels

In order to measure the mean difference of aromatase CYP19A transcription as indirect assessment of E2 level in Microgadus tomcod ovaries, three distinct phases of adult female Microgadus tomcod reproductive activity were evaluated. First, spawning stage was defined as fish whose ovaries are well developed and near peak mass, in which eggs are extruded from the body under slight pressure or when ovary sac is broken, and eggs were easily separated from each other. Second, matured stage was defined as fish whose ovaries had passed the peak spawning stage and begun to regress. At this stage, the ovaries contain less than 50% scattered clear eggs, and the eggs cannot be easily separated from each other. Third, spent stage was defined as fish whose the ovary had fully regressed and contained less than 5% white dry ova and reddish unspawned scattered eggs.

Isolation of Microgadus Tomcod CYP19A Complementary DNA (cDNA)

Approximately 100 mg of ovarian tissue were used as starting material for total RNA isolation with Ambion Pure Link total RNA extraction mini kit (Cat # 12183-018A) following the manufacture's protocol. All RNA preparations were adjusted to 0.5 µg/µl and used as template for reverse transcription (RT). In all reactions, cDNA was synthesized using 1 µg of total RNA with 100 ng random hexamer primers(10 µl total volume), following the manufacturer's instructions (Clontech). Reactions were terminated by heating at 70°C for 10min, then diluted 1:5 with DNase-free water and used as template for polymerase chain reactions (PCR).

Degenerate primers for Microgadus tomcod CYP19A PCR amplification were designed based on CYP19A cDNA sequences from the closely related *Gadusmorhua* along with six other most closely related teleost species (Table 1). *Gadusmorhua, Mugilcephalus, Epinepheluscoioides, Epinephelusakaara, Latescalcarifer, Larimichthyscrocea, and Micropogoniasundulatus* (Accession #s: DQ402370, AY859425, AY510711, AY547354, AY684256, FJ800566, and DQ184486 respectively) were selected from the NCBI database. Standard group codes were used to represent degenerate primers (W=A+C; S=G+C; R=A+G; K=G+T; V=A+G+C; Y=C+T).

Cloning and Sequencing of CYP19A Partial cDNA

The PCR amplification cycles were varied based on expected product size and the primer melting temperatures. A major RT-PCR product of approximately 1000 bp was isolated from 1.5% agarose/ethidium bromide (EtBr) gels and ligated into pGEM-T vector (Invitrogen). Sequencing was arranged using 250 to 400 ng concentration of plasmid DNA per 8 µl total volume on the ABI 3730xl 96-capillary automated DNA sequencers (Brisbane, Australia). Both strands of plasmid DNA containing the cDNA insert were sequenced in both directions using T7 (forward) and SP6 (reverse) sequencing primers. Thereafter, all positive sequence results were aligned with the CLUSTAL W method to generate a contig sequence using Mega4 software (http://www.megasoftware.net/ mega4/.html).

Rapid Amplification of cDNA Ends (RACE)

Rapid amplifcation of cDNA ends (RACE)-ready cDNA libraries were generated using Microgadus tomcod ovarian total RNA and a commercial kit Clonetech SMARTer RACE cDNA Amplification Kit (Cat. Nos. 634923, CA, USA), following the manufacturer's protocol. Microgadus tomcod CYP19A specific primers for 3' RACE, outer-

Primers	Sequence	Tm	GC%
MtAro268F	5'-GCCAGYAACTACTACAACRACAAG-3'	62-65	42-50
MtAro698F	5'-TTTACACRTGGCAGACTGTR-3'	57-61	43-52
MtAro1384R	5'-GGATGGAYTTCATCATCACCATG-3'	61-63	43-48
MtAro1476R	5'-GGCTGCTGGASAGGTTGTTG-3'	61-62	43-44

Table 1: List of degenerate primers designed for tomcod CYP19A partial sequence.

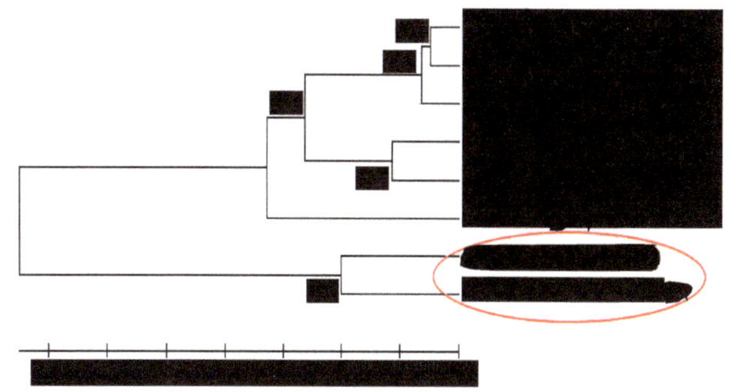

Figure 1: Phylogenetic analysis of *Microgadus tomcod* CYP19A amino acid sequence. The circle shows that *Microgadus tomcod* CYP19A is most closely related to CYP19A of *Gadusmorhua*.

primer: 5'-TGGTGGACGTCTCCAACACCCTCTTC -3' and inner-primer: 5'- CTCAGGAGCTGCAGGACGCCATAGA -3' were used. For nested PCR and for 5' RACE, outer primer 5'-CGGTGCATGC-GTCCCACGTT - 3' and inner primer 5'- CTCAGGAGCTGCAG-GACGCCATAGA -3' were used. Primers were adjusted to 10 pmol for both 5' and 3' RACE reactions. Thermal parameters for RACE touch-down PCR were: 95 °C for 2 min; 5 cycles of 95°C for 5 s and 72°C for 180 sec; 5 cycles of 95°C for 5 sec, 70°C for 10 sec and 72°C for 180 sec; 25 cycles of 95°C for 5 sec, 68°C for 10 sec and 72°C for 180 sec using the Biorad i-cycler thermal cycler. Thermal parameters for RACE Nested PCR were: 95°C for 2 min; 35 cycles of 95° C for 30 sec, 55°C for 30 sec, and 72 min, followed by 72°C for 10 min final extension using the Biorad i-Cyclear thermal cycler. PCR products were gel purified as stated above and ligated into pGEM-T vector (Invitrogen) per the manufacturer's protocol. Confirmation of the identified nucleotide sequence was achieved by comparing Microgadus tomcod CYP19 cDNA to *Gadusmorhua* aromatase CYP19A sequence using the Sequencer TM 4.0 Windows Version Package (Gene Codes Corporation).

Deduced amino acid comparison was performed using the CLUSTAL W multiple sequence alignment aligorithm and a molecular phylogenetic analysis was performed using the neighbor joining method in MEGA version 5. To determine its relation to other aromatase proteins, the deduced Microgadus tomcod CYP19A protein was included in a molecular phylogenetic analysis with six other teleost sequences (Accession #s: DQ402370, AY859425, AY510711, AY547354, AY684256, FJ800566, and DQ184486), (Figure 1).

Real Time Quantitative PCR (RT-qPCR) Primers

To prevent genomic DNA (gDNA) amplification in RT-qPCR applications, primers were designed to anneal across a unique cDNA regions of splice junction. Primers spanning exon-exon junctions were designed to cross exon 8 and exon 9 boundaries by viewing the detailed alignment of the Microgadus tomcod CYP19A cDNA query sequence against zebra fish CYP19A genomic sequence [22]. Since CYP19A is a highly conserved gene in all vertebrates, the exon-exon boundaries of tomcod and zebra fish can be expected to be the same. Forward: 5'- TCCGGCAGTGTGTGC TGGAGATG -3' and reverse: 5'- CCGGAGCTTCCTCTCGCCTATGA -3' oligonucleotides were used for RT-qPCR. The validity of results obtained using RT-qPCR was improved by including Microgadus tomcod S2 ribosomal protein (S2RP) as a reference gene to correct for sample to sample variations in, RT-qPCR efficiency and errors in sample quantification (NCBI database Accession # AY292477). Forward- 5'-AGGGCTGGTCAGCGTACA-3'- and reverse -5'-CTGGAACGATGGACAGCTTG-3'- primers for

S2RP were designed using primer3 software (Whitehead Institute, Cambridge, MA).

Real Time RT-PCR Analyses (TaqMan)

Real-time RT-PCR using Microgadus tomcod ovary was performed to determine abundance for aromatase CYP19A mRNA transcription. cDNA was produced using 500 ng total RNA primed with 100 ng random hexamer oligonucleotides and used as template for RT reactions. RT reactions were diluted (1:5) with DNase-free water and 2 μλ of each diluted RT was mixed with 23 μλ of a master mix containing a commercial qRT-PCR enzyme mix (Taqman). Each 25 μλ PCR reaction was set up as duplex with the threshold cycle (C_t) for the gene of interest (GOI) and the reference (REF) measured at the same time to minimize differences in reaction efficiency. The standard curve for CYP19A was generated using an *in vitro* transcribed Microgadus tomcod aromatase CYP19A plasmid DNA as described above. A serial dilution was prepared to be targeted in the range of 10^8 to 10^2 copies for the standard curve was prepared. Real Time RT-qPCR was performed on a BIORAD®iCycleriQ (Catalog number 170-8740). For each set of PCR analyses, the transcription levels were determined in a "standard sample" to determine interassay variability.

Statistical Analysis

The experimental design employed was a 3 x 3 factorial with 3 replicates per treatment combination. Equal variance and normality were checked and analyzed on the mean value by analysis of variance (ANOVA) with factorial design followed by Tukey's pairwise comparisons. Polynomial contrasts were used for the actual spacing of the treatment levels for determining the existence and nature of trends in the treatment level means. The sums of squares for linear (degree 1), quadratic (degree 2), and cubic (degree 3) trends were computed if needed. Statistical analysis was performed using the statistix package, analytical software 9.0 (Tallahassee, FL 32317).

Results

Isolation of Microgadus tomcod CYP19A cDNA

The deduced cDNA and amino acid (AA) sequences of the isolated Microgadus tomcod aromatase CYP19A cDNA are presented in (Figure 2) Initially, a central part of the mRNA was amplified from

```
1 ATATGTACGCCAGTA ACTTACTACAACGAC AAGTACGGGAGCATA GCCCGCGTATGGATC
1 AACGGGGAAGAGACC
76 I W T P V T Y Y N D K Y G S I A R V W I N G E E T
26 CTCATCCTCAGCAGC TCCTCTACAGTGCGC CACGTGCTGAGGAGC ACCCAGTACACCTCG
151 CGCTTCGGGAGCCAG
51 L I L S S S S T V R H V L R S T Q Y T S R F G S Q
226 CGGGGCCTGGGCTGC CTGGGCATGGACGGG CGAGGCATCATCTTC AACAACGACGTGGCC
76 CTGTGGAGGAAGACT
301 R G L G C L G M D G R G I I F N N D V A L W R K T
101 CGCAGCTACTTCGCC AGAGCTCTGAACGGT CCGGGTGTGCTGCAG ACGGTGGACGTGTGC
376 ACCGCCTCCACCCAG
126 R S Y F A R A L N G P G V L Q T V D V C T A S T Q
451 ACCCACCTGGACAGC TTGAGGGCGGACCGC CCCGGCAGGGGACCC GGGGACAGCCTGCG
151 G CAGGTGGACGTGCTG
526 T H L D S L R A D R P G R G P G D S L R Q V D V L
176 GGCGCGCTGCGGGGC ATCGTGGTGGACGTC TCCAACACCCTCTTC CTGGGAGCGCCGTTC
601 GATGAGAAAGATCTG
201 G A L R G I V V D V S N T L F L G A P F D E K D L
676 CTTGAGAAGATCCTG AAGTATTTTGACACG TGGCAGACTGTACTC ATTAAACCAAACCTT T
226 ACTTCAAGTTGGAC
751 L E K I L K Y F D T W Q T V L I K P N L Y F K L D
251 TGGATCCACCGGAGG CAGAGGGATGCAGCT CAGGAGCTGCAGGAC GCCATAGAGGTTCTG
826 ATCGATCAGAAGAGG
276 W I H R R Q R D A A Q E L Q D A I E V L I D Q K R
901 AGAGAACTGCAGGAG GCGGATAAACTGGAC TGCATCAACTTCACG GCGGACCTCATATTT
301 GCACAGGGTCACGGG
976 R E L Q E A D K L D C I N F T A D L I F A Q G H G
326 GAGTTGTCAGCCGAC CACGTCCGGCAGTGT GTGCTGGAGATGGTG ATCGCGGCCCCGGAC
105 ACCCTGTCTGTTAGT
1 E L S A D H V R Q C V L E M V I A A P D T L S V S
351 CTGCTCTTCATGCTG CTGCTGCTCAAGCAG CACCCAGAGGTGGAG CTGAGGCTCCTCAAA
112 GAGATCGACTCTGTC
```

351 CTGCTCTTCATGCTG CTGCTGCTCAAGCAG CACCCAGAGGTGGAG CTGAGGCTCCTCAAA
112 GAGATCGACTCTGTC
6 L L F M L L L L K Q H P E V E L R L L K E I D S V
376 ATAGGCGAGAGGAAG CTCCGGAACGAGGAT CTGCCCCGGCTCTGT GTCCTGGAGCACTTC
120 ATCAACGAGTCCTTA
1 I G E R K L R N E D L P R L C V L E H F I N E S L
401 CGCTTCCACCCCGTG GTGGACTTCACCATG CGCAGGGCCCTGTCC GACGACGTCATCGAA
127 GGCTACAGGGTCCCC
6 R F H P V V D F T M R R A L S D D V I E G Y R V P
426 AGGGGCACCAACGTC ATCCTGAACGTGGGA CGCATGCACCGCACC GAGTTCTACCCCAAA
135 CCCGACGAATTCAGC
1 R G T N V I L N V G R M H R T E F Y P K P D E F S
142 CTGGACAACTTTGAG AAAAGCGTCCCCAGC CGCTACTTCCAGCCG TTTGGCTCGGGGCCG
6 CGCTCCTGCGTGGGC
150 L D N F E K S V P S R Y F Q P F G S G P R S C V G
1 AAGCACATCGCCATG GTGATGATGAAGTCC ATCCTGGTGACGCTG CTGTCATGCTACACG
CTGTCTCCCCACGAG
K H I A M V M M K S I L V T L L S C Y T L S P H E
GGTCTGACGCTGGCC AGCATGCCCCAGACC AACAACCTCTCCCAG CAGCCCATGGAGAGC
GAGCAGGACCCCCAC
G L T L A S M P Q T N N L S Q Q P M E S E Q D P H
TACCTCGCCATGACC CTCACACCCAGACGA AGAACAGCAACAGGA AGCTACAGCGGAGCA
GGACTCTGAGAACCC
Y L A M T L T P R R R T A T G S Y S G A G L *
CCCCCCCGTGTGCGC GTGGCTGTGTGTGTG TGTGTGTGTGTGTGT GTGTGTGTGTGTGTG
TGTGTGTGTGTGTGT
GTGTGTGTGTGTGTG TGTGTGTGTGTGTGT GTAAATGATTTGTTT ATTTTTTCTTATATA A
AATTTGTATTCTAA
AAAAAAAAAAAAAAA AAA

Figure 2: The nucleotide sequence and its predicted amino acid sewuence *Microgadus tomcod aromatase* CYP19A. A microsatellite region was underlined. The asterisk (*) represent the stop codon. Numbers to the left represent nucleotide base or amino acid residue position.

degenerate primers and an ovarian total RNA. The sequence of this 1020 bp product was similar to other teleost CYP19A cDNA sequences (BLAST analysis, http://www.ncbi.nlm.nih.gov/blast/; data not shown). It is known that some genes are easily cloned as full-length forms and some are not. RACE reaction using Microgadus tomcod aromatase specific primers allowed identification of a 1415 bpcDNA, which encodes a gene product with a predicted molecular weight of 52.5, representing an almost complete CYP19A sequence. The deduced amino acid sequence of *Microgadus tomcod* CYP19A was aligned with *G. moruha* (DQ402370), *E. coioides* (AY859425), *L. calcarifer* (AY510711), and *M.cephalus* (AY547354) (Figure 2). The 1342 bp ORF encodes a protein of 521 amino acids which covers 96% of the ORF size compared with the cDNA for Atlantic cod and fully aligns with the other teleost CYP19A proteins. The sequence shares 86% identity with Atlantic cod and 65–78% identity with the other selected fish species CYP19A sequence. Figure 3 represent multiple alignments of *Microgadus tomcod* CYP19A mRNA with other teleost aroamatase CYP19A mRNA sequences.

Comparison of aromatase transcription levels in Microgadus tomcod reproductive development cycle

Changes in CYP19A mRNA as an indirect measure of E2 level in the control group and on PCB/ PAH separately treated groups were evaluated on Figures 4 and 5, respectively. Both the control group and PCB / PAH separately treated groups showed the highest aromatase CYP19A mRNA level at spawing, followed by matured and the lowest at the spent stage of the reproduction cycle. Neither PCB nor PAH affected the expected CYP19A transcription pattern, (Spawing>matured>spent). This finding suggested that the level of E2 steroid for PCB or PAH individually treated groups was at the highest seasonal values during pre-spawing and spawning stage. After spawning, the ovaries were regressed and the E2 level in the plasma continued to fall to very low levels.

Joint Effect of PCB and PAH on Aromatase CYP19A Transcript Level

Treatment groups exposed to PCB and PAH in combination showed the highest level of aromatase CYP19A transcript at matured stage, followed by spawning stage and lastly spent stage of reproduction cycle (matured>spawning>spent). Due to the additive affect of PCB and PAH on the combined treatment groups, the level of CYP19A mRNA and E2 production were disrupted (Figure 6).

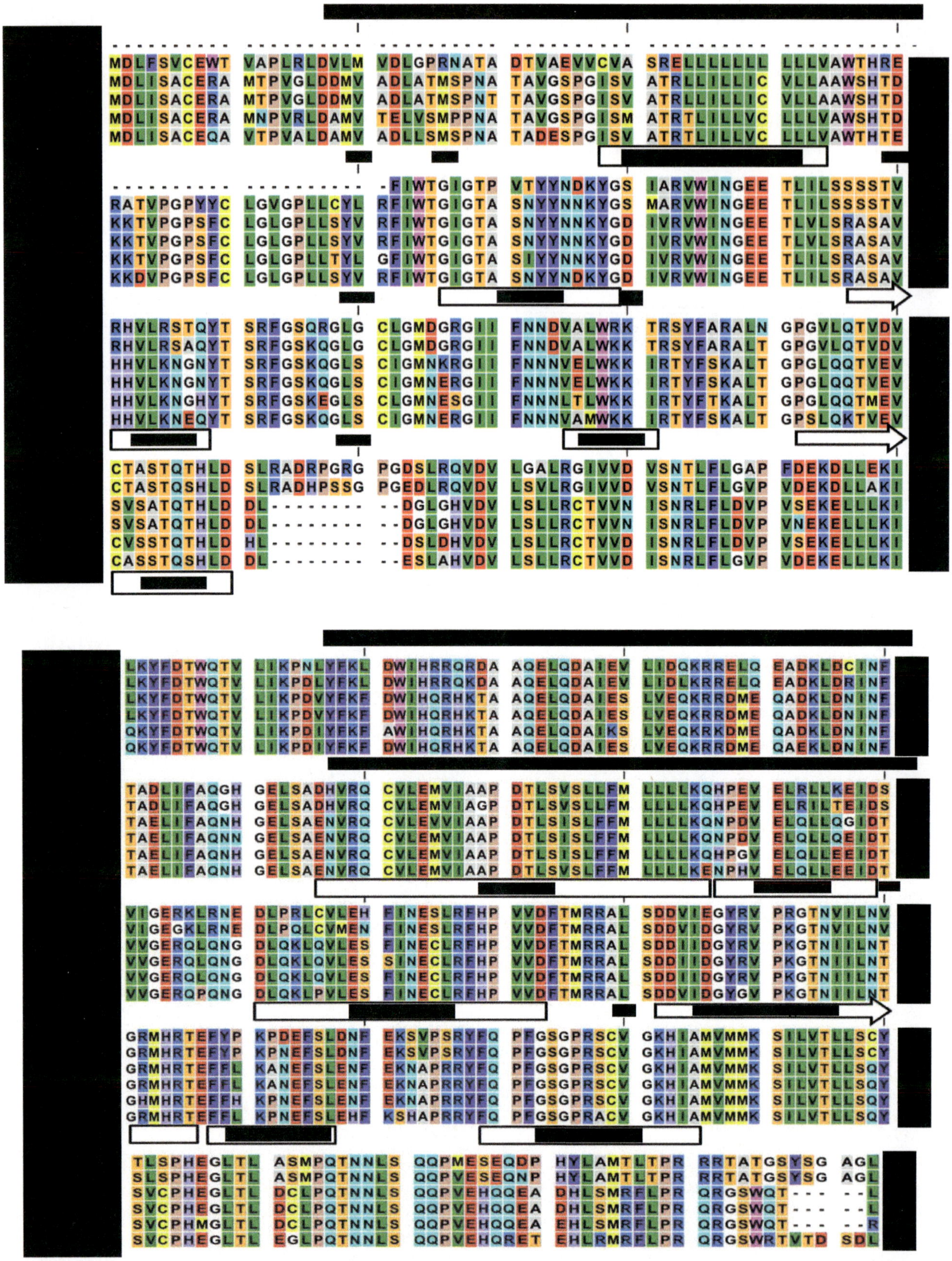

Figure 3: Multiple alignment of *Microgadus tomcod* CYP19A with other teleost aroamatase CYP19A sequences. The deduced amino acid sequence of Microgadus tomcod CYP19A was aligned with *G. moruha* (DQ402370), *E. coioides* (AY859425), *L. calcarifer* (AY510711), and *M.cephalus* (AY547354).

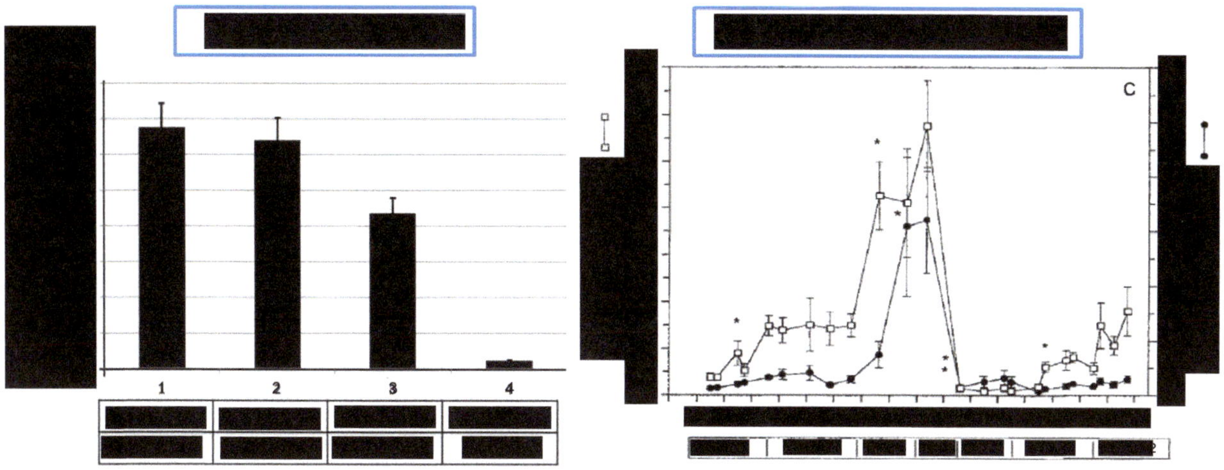

Figure 4: Changes in CYP19A transcript levels as an indicator of E2 level. A) Bars represent means +1S.E. of aromatase CYP19A transcript levels of female tomcod for control (DMSO) group. Spawning stages and sample size (N) are shown x-axis below the graph. B) Changes in plasma E2 and testosterone values for female winter flounder. Values are mean ± SEM; (PREP 1 initial rapid phase of gonadal growth; PREP 2 continuing phase of slower growth of gonads; PRES pre-spawning phase; SPW spawning; POST post spawned interval) (Figure 4B Modified from (Harmin *et al.*, 1995).

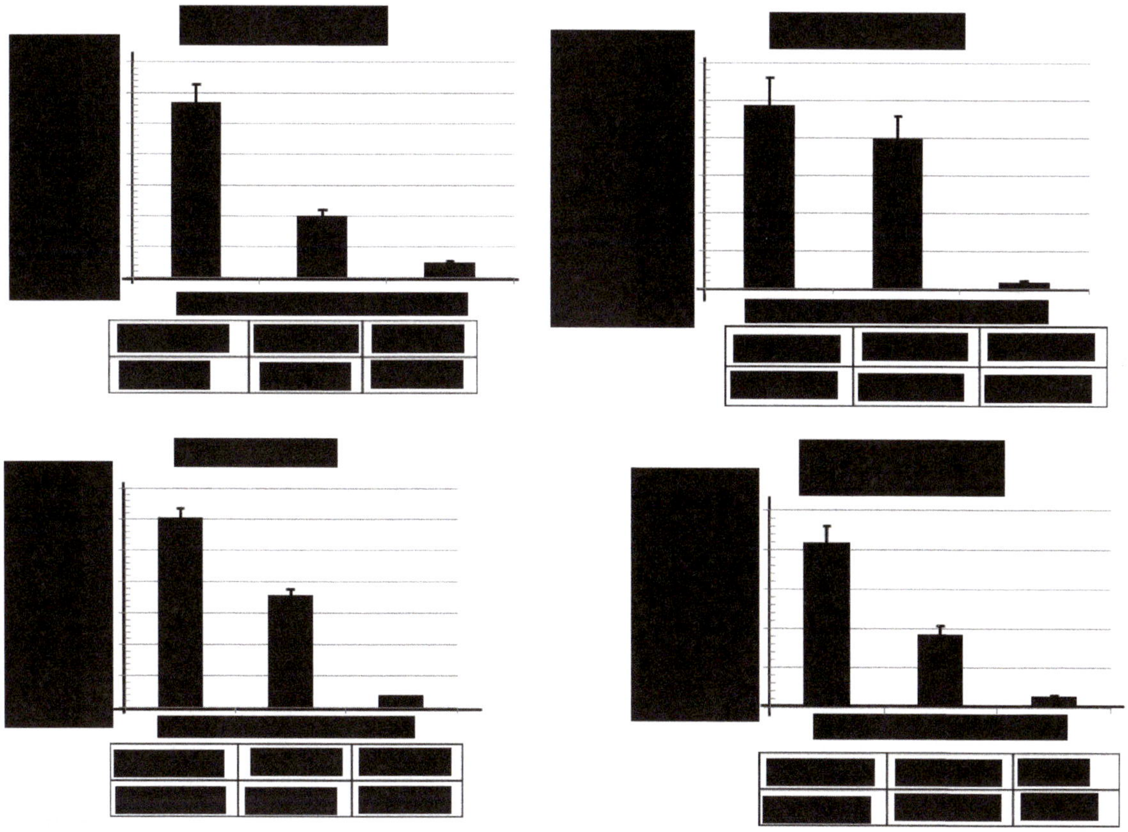

Figure 5: Effect of PCB (Figure A and B), and PAH (Figure C and D) alone on aromatase CYP19A transcript levels.Bars represent means ± 1S.E. of aromatase CYP19A transcript levels of female tomcod. Spawning stages and sample sizes are shown below x-axis. Well-developed and large ovaries at spawning stage showed the highest copy number compared to mature and spent developmental stages at all experimental groups.

Analysis of CYP19A Transcript Level Using Absolute Quantification

The interaction effect of PCB and PAH on CYP19A transcripts when reproductively mature female Microgadus tomcod aromatase CYP19A mRNA was analyzed using RT-qPCR assay (Figure 7). For individually treated groups, neither PCB nor PAH showed a significant effect on aromatase CYP19A mRNA level. For groups treated with both PCB and PAH, PCB significantly upregulated aromatase CYP19A transcript level at high-PCB/high-PAH (100 fold), and high-PCB/low-

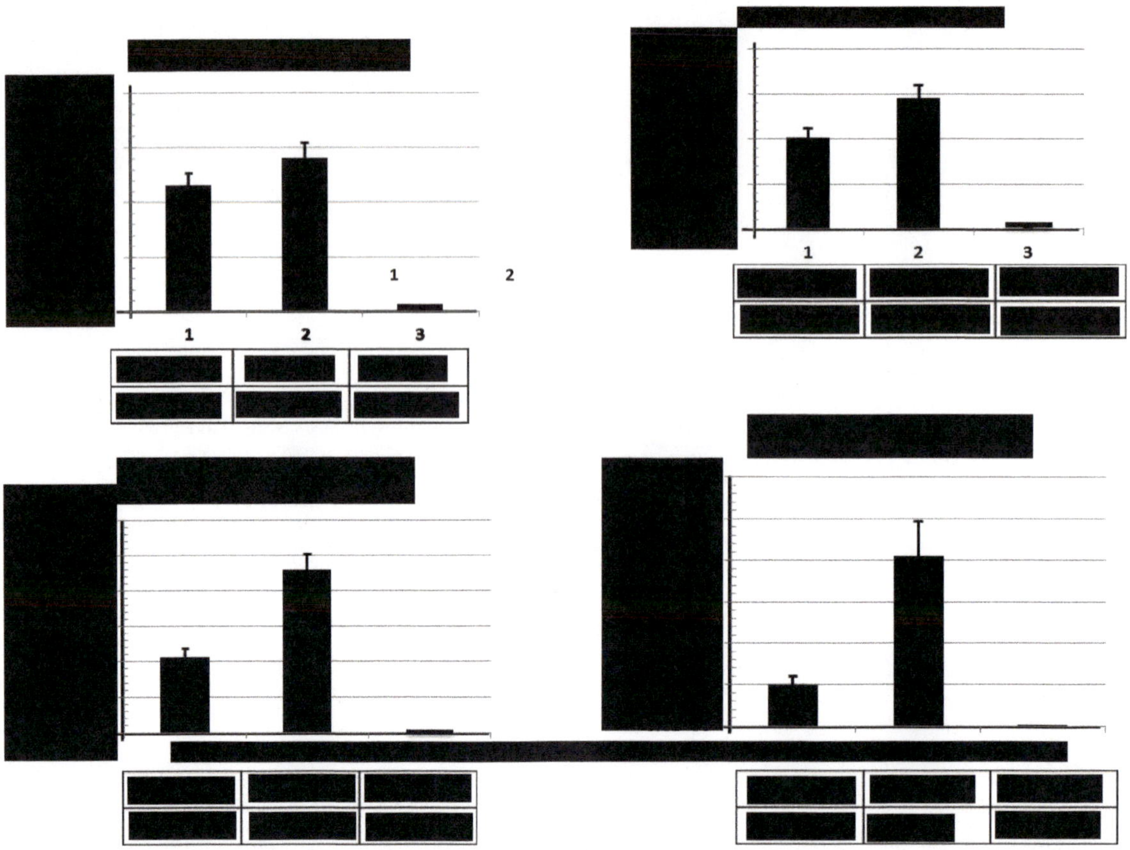

Figure 6: Effect of PCB and PAH in combination on aromatase CYP19A transcript levels. Bars represent means ± 1S.E. of aromatase CYP19A transcript levels of female tomcod. Sample size is shown below x-axis (see materials and methods for a description of the reproductive stages). The above pattern is different from the expected CYP19A transcription pattern, which is, matured>spawning>spent.

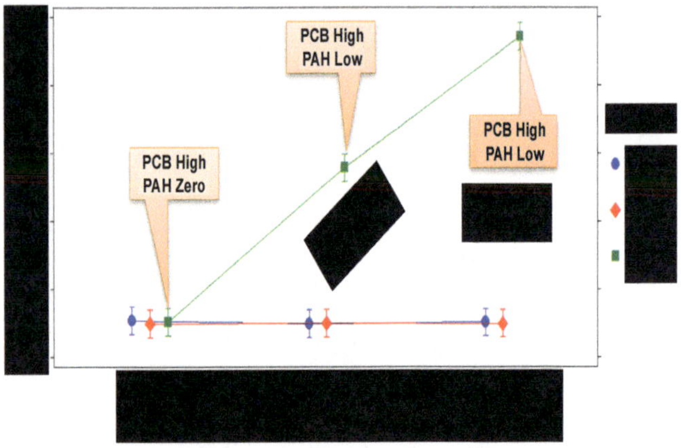

Figure 7: Effect of PCB and PAH alone and in combination on CYP19A expression. PAH level (Horizontal), and PCB level (vertical) of all treatment groups are shown. Zero, low (0.1 ppm), and high (1.0 ppm) levels are shown for both compounds. Copy number of CYP19A transcript is presented on the left vertical column.

PAH, (50 fold) treatment groups. At the highest level of PCB, increasing the levels of PAH increased CYP19A mRNA expression because the toxicity of PCB was enhanced in the presence of PAH. In contrast, at zero and low levels of PCB, increasing level of PAH had no significant effect on CYP19A mRNA expression level. The toxic interaction effects (S) of PCB and PAH mixtures were mainly synergistic (S=1.9 to S=755

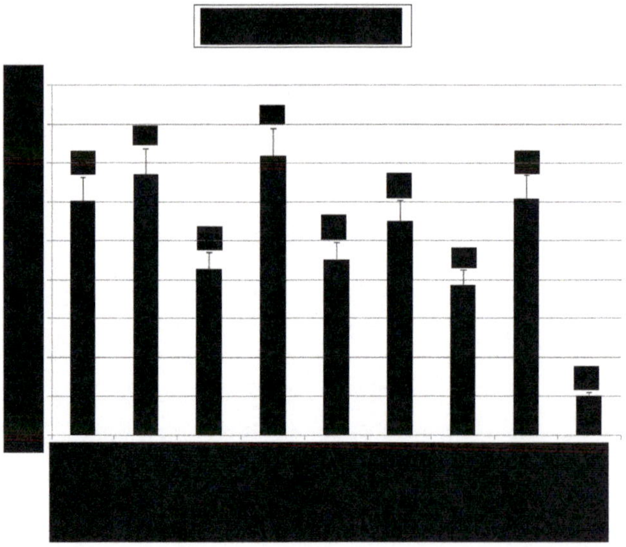

Figure 8: Individual and joint effect of PCB and PAH for the analysis of GSI. Bars with different letters are statistically different ($p<0.05$).

TU). The toxic interaction effects of the low-PCB/low-PAH and low-PCB/high-PAH treatment groups (4.5 and 1.9 TU, respectively) were not significantly high compared to those of the high-PCB/low-PAH,

(437.9 TU), and high-PCB/high-PAH (754.5 TU) treatment groups.

Gonadosomatic Index (GSI)

For either PCB or PAH individually treated groups, a significant gonadal loss occurred only in the high-PCB/high-PAH level of treatment (Figure 8). GSI did not change with low-PCB, and low or high-PAH individually treated groups compared to the control group. Exposure to estrogens mimicking compounds such as PAH and PCB can have various detrimental effects in fish. It can reduce general viability, induce gonadal malformations or feminization of genetic males, or lead to sterilization [45]. Even though there was no significant GSI for most treatment groups; we observed reproductive impairment ranging from underdeveloped gonads to decrease egg production and abnormal larval offspring. Morphological analysis was not conducted since it was beyond the scope of this study. However, abnormal morphological variation such as aplasias, atrophy, fusions, degeneration and intersex due to environmental pollutants need further study.

Condition Factor (CF)

The result of this study showed that there was no statistically significant difference in CF due to PCB or PAH exposure in any treatment group. There was also no significant PCB/PAH interaction observed in any treatment group. CF and the length-weight relationship comparison conclude that neither PCB nor PAH influence the wellbeing of fish reproductive cycle.

Discussion

This study demonstrated major differences in the way Microgadus tomcod CYP19A mRNA response to the effects of PCB and PAH at molecular and organismal levels. The individual effect of either PCB or PAH on Microgadus tomcod ovaries are not significant, but PCB caused higher CYP19A transcript levels in the PCB/PAH joint treatment groups. Moreover, low levels of aromatase CYP19A mRNA and lower GSI during the spawning stage of reproductive cycle was resulting from the additive effect of high-PCB and high-PAH. Previous studies on various teleosts found E2, testosterone, GSI, and oocyte diameter to peak during pre-spawning and spawning stages, and to be low or undetectable at spent stage [25,46-48]. This finding suggested that the level of CYP19A mRNA was compromised by the additive effect of PCB and PAH at any treatment group.

The results of all CYP19A transcription level, GSI, and CF tests revealed little or no significant effect of PAH exposure, suggesting that Microgadus tomcod from the Hudson River might have developed a protective mechanism against PAH. This result may be explained by the fact that PCBs caused reproductive impairment by reducing aromatase expression whereas the effect of PAH was reduced or compromised in Microgadus tomcod from the Hudson River.

Both PCBs and PAHs are known to mimic sexual hormones and/or disrupt steroidogenic enzyme function, [44,49], including aromatase CYP19A [22,50-52], PCBs and PAHs binding to the Aryl hydrocarbon receptor (AhR) can result in synergestic effects due to cross-talk between the AhR and the ER [53,54]. PCBs have also been shown to alter or inhibit hormone production by steroidogenic tissue in reproductive organs [16,55].

AhR is known to be involved in modulating estrogen-dependent transcription [34,39,56,57], to control target-specific regulation of ER [37,58], and to have crosstalk with several other important signal transduction pathways [9,24]. Some species are inherently capable of tolerating a contaminant through existing physiological acclimation mechanisms [59-61]. Previous exposure of environmental contaminants most likely decreases the sensitivity of the AhR pathway in Hudson River Microgadus tomcod population. The recent finding by Wirgin et al., [3] that a six-base deletion in the AhRreseptor 2 (AhR2) of Hudson River Microgadus tomcod is ths basis of six-base deletion as the bases for resistance. The finding suggests that the Hudson River Microgadus tomcod population has undergone rapid evolution, probably due to contaminant exposure. This finding of the basis of resistance in a vertebrate population provides evidence of evolutionary change due to selective pressure at a single locus. There is a possibility that Microgadus tomcod may instead develop a tolerance defense mechanism by which Microgadus tomcod do not reduce PAH toxicity or its action, but instead alleviate the negative consequences of such contaminats and action [62,63]. Previous studies have shown that wild caught Microgadus tomcod from the Hudson River have been developing various stages of resistance compared to Atlantic tomcod from relatively uncontaminated tributaries [1,3,9,61,64]. Tolerance to contaminants within cellular organelles is possible by preventing toxicants from exerting their toxic effects despite their elevated presence in receptor tissues. Further investigations are required to determine if resistance may be due to genetic or epigenetic modifications on Microgadus tomcod from Hudson River.

When EPA banded the production of PCB in 1977, it was estimated that 1.3 million pounds of PCBs were discharged in the Hudson River from two General Electric capacitor manufacture plants located in the town of Fort Edward and Hudson Falls, New York. This study supports the idea that resistance to pollution is a recent phenotype in Microgadus tomcod from the Huddson River, suggesting that Microgadus tomcod have rapid evolutionary responses with respect to their environment. Besides, future study is needed to determine if resistance is a common or rare response of chronically exposed fish populations with different selective pressure mechanisms. An important question is, how much does resistance among Microgadus tomcod is matter to the ecology of the contaminated Hudson River? Is an adapted population necessarily an "unhealthy" one? Whether resistance evolves or the population is eliminated depends on the rapidity of onset and severity of the stress, and the capacity of the population to adapt to it. Answering these difficult questions will require to be addressed by moving resistance studies beyond the species-contaminant pairings method. The evolutionary costs of resistance to Hudson River populations or to their communities remain to be experimentally determined.

In general, the polluted sites of the Hudson River may become well known for PCBs and PAHs contaminants, but it is rare that these compounds were released into the environment by industry or municipalities without the release of many other compounds. This issue may be particularly critical for a species such as Microgadus tomcod that appears to serve a unique position in estuarine food chains. Thus, exposures to single or two contaminants can be made in the laboratory, but their results are often confounded by the unknown effects of pre-exposures to a great variety of other contaminants in wild population. It is believed that studies of two major contaminants are the appropriate departures for future resistance research. The possibility exists that resistance is fostered by synergies among contaminants, at least in some instances. Future analysis will address whether AhR is potentially responsible for conferring resistance in population adapted to chronic exposure to chemical pollutants in the different superfound sites.

Acknowledgement

This research was supported and funded by National Oceanic and Atmospheric Administration (NOAA) – Living Marine Resources Cooperative Science (LMRCS). We thank to all of those at The Institute of Marine and Environmental Technology (IMET) at University of Maryland Center of Environmental Science (UMCES) and the James J. Howard Marine Sciences Laboratory at Sandy Hook.

References

1. Wirgin I, Kreamer G, Grunwald C, Squibb K, Garte SJ, et al. (1992) Effects of prior exposure history on cytochrome P450IA mRNA induction by PCB congener 77 in Atlantic tomcod. Marine Environmental Research 34: 103-108.
2. Swedenborg E, Pongratz I (2010) AhR and ARNT modulate ER signaling. Toxicology 268: 132-138.
3. Wirgin I, Roy KN, Loftus M, Chambers RC, Franks D, et al. (2011) Mechanistic Basis of Resistance to PCBs in Atlantic Tomcod from the Hudson River. Science 331: 1322-1325.
4. Bunton TE (1996) Experimental chemical carcinogenesis in fish. ToxicolPathol 24: 603-618.
5. Chen S, Zhang F, Sherman MA, Kijima I, Cho M, et al. (2003) Structure–function studies of aromatase and its inhibitors: a progress report. J Steroid BiochemMolBiol 86: 231-237.
6. Anderson JW, Lee RF (2006) Use of biomarkers in oil spill risk assessment in themarine environment." Human and Ecological Risk Assessment 12: 1992-1222.
7. Buryskova B, Hilscherova K, Blaha L, Marsalek B, Holoubek I (2006) Toxicity and modulations of biomarkers in Xenopuslaevis embryos exposed to polycyclic aromatic hydrocarbons and their N-heterocyclic derivatives. Environ Toxicol21: 590-598.
8. Laukers JD (1986) Disposal and destruction of waste PCBs. CRC Press: 83-152.
9. Wirgin I, Chambers RC (2006) Atlantic tomcod Microgadus tomcod:A model of responses of Hudson River fish to toxicants. AFS Symposium 51: 331-365.
10. Oost Ron, van der b, Jonny Beyer, Nico Vermeulen (2003) Fish bioaccumulation and biomarkers in environmental risk assessment: a review. Environmental Toxicology and Pharmacology 13: 57-149.
11. Chavan H, Krishnamurthy P (2012) Polycyclic aromatic hydrocarbons (PAHs) mediate transcriptional activation of the ATP binding cassette transporter ABCB6 gene via the aryl hydrocarbon receptor (AhR)." J Biol Chem 287: 32054-32068.
12. Liao C, Lv J, Fu J, Zhao Z, Liu F, et al. (2012) Occurrence and profiles of polycyclic aromatic hydrocarbons (PAHs), polychlorinated biphenyls (PCBs) and organochlorine pesticides (OCPs) in soils from a typical e-waste recycling area in Southeast China. Int J Environ Health Res 22: 317-330.
13. Aas E, Beyer J, Jonsson G, Reichert L, Andersen O (2001) Evidence of uptake, biotransformation and DNA binding of polyaromatic hydrocarbons in Atlantic cod and corkwing wrasse caught in the vicinity of an aluminium works. Mar Environ Res 52: 213-229.
14. Wurl O, Obbard JP (2004) A review of pollutants in the sea-surface microlayer (SML): a unique habitat for marine organisms.Mar Pollut Bull 48: 1016-1030.
15. Lee J, Choe E (2009) Effects of phosphatidylcholine and phosphatidylethanolamine on thephotooxidation of canola oil. J Food Sci 74: C481-c486.
16. Yan X, Richard M, Yu K, Zhang X, Giesy Margaret JG, et al. (2006) Effects of PCBs and MeSO2-PCBs on adrenocortical steroidogenesis in H295R human adrenocortical carcinoma cells. Chemosphere 63: 772-784.
17. Simpson ER, Mahendroo MS, Means GD, Kilgore MW, Hinshelwood MM, et al. (1994) Aromatase cytochrome P450, the enzyme responsible for estrogen biosynthesis." Endocr Rev 15: 342-355.
18. Frye CA, Bo E, Calamandrei G, Calzà L, Dessì-Fulgheri F, et al. (2012) Endocrine disrupters: a review of some sources, effects, and mechanisms of actions on behaviour and neuroendocrine systems. J Neuroendocrinol 24: 144-159.
19. Stasinakis AS, Mermigka S, Samaras VG, Farmaki E, Thomaidis NS (2012) Occurrence of endocrine disrupters and selected pharmaceuticals in Aisonas River (Greece) and environmental risk assessment using hazard indexes. Environ SciPollut Res Int 19: 1574-1583.
20. Chang XT, Kobayashi T, Kajiura H, Nakamura M, Nagahama Y (1997) Isolation and characterization of the cDNA encoding the tilapia (Oreochromisniloticus) cytochrome P450 aromatase (P450arom) changes in P450arom mRNA, protein and enzyme activity in ovarian follicles during oogenesis. J MolEndocrinol18: 57-66.
21. Stocco C (2008) Aromatase expression in the ovary: Hormonal and molecular regulation. Steroids 73: 473-487.
22. Trant JM, Gavasso S, Ackers J, Chung BC, Place AR (2011) Developmental expression of cytochrome P450 aromatase genes (CYP19a and CYP19b) in zebrafish fry (Daniorerio). J ExpZool 290: 475-483.
23. Lara LA, Duarte AA, Reis RM, Vieira CS, Rosa-e-Silva AC (2012) Endocrine disrupters: potential risk factors affecting sexual function in both men and women. J Sex Med 9: 941-942.
24. Carlson AE, Roy KN, Wirgin II (2009) Microarray Analysis of Polychlorinated Biphenyl Mixture-Induced Changes in Gene Expression Among Atlantic Tomcod Populations Displaying Differential Sensitivity to Halogenated Aromatic Hydrocarbons. Environ ToxicolChem 28: 759-771.
25. Harmin SA, Crim LW, Wiegand DM (1995) Plasma sex steroid profiles and the seasonal reproductive cycle in male and female winter flounder, Pleuronectesamericanus. Marine Biology 21: 601-610.
26. Shozu M, Zhao Y, Bulun SE, Simpson ER (1998) Multiple splicing events involved in regulation of human aromatase expression by a novel promoter. Endocrinology 139: 1610-1617.
27. Maranghi F, Mantovani A (2012) Targeted toxicological testing to investigate the role of endocrine disrupters in puberty disorders. Reprod Toxicol 33: 290-296.
28. Young G, Kagawa H, Nagahama Y (1983) Evidence for a decrease in aromatase activity in the granulosa cells of amago salmon (Oncorhynchusrhodurus) associated with final oocyte maturation. Biol Reprod 29: 310-315.
29. Gelinas D, Pitoc GA, Callard GV (1998) Isolation of a goldfish brain cytochrome P450 aromatase cDNA: mRNA expression splicing events involved in regulation of human aromatase expression by a novel promoter. Endocrinology 139: 1610-1617.
30. Tchoudakova A, Kishida M, Wood E, Callard GV (2001) Promoter characteristics of two cyp19 genes differentially expressed in the brain and ovary of teleost fish. J Steroid Biochem MolBiol 78: 427-439.
31. Doering JA, Giesy JP, Wiseman S, Hecker M (2013) Predicting the sensitivity of fishes to dioxin-like compounds: possible role of the aryl hydrocarbon receptor (AhR) ligand binding domain. Environ SciPollut Res Int 20: 1219-1224.
32. KazetoK, Place A, Trant MJ (2004) Effects of endocrine disrupting chemicals on the expression of CYP19 genes in zebrafish (Daniorerio) juveniles. Aquat Toxicol 69: 25-34.
33. Kortner T, Arukwem A (2007) Effects of 17α-methyltestosterone exposure on steroidogenesis and cyclin-B mRNA expression in previtellogenic oocytes of Atlantic cod (Gadusmorhua). Comparative Biochemistry and Physiology 146: 569-580.
34. Tanaka M, Telecky TM,Fukada S, Adachi S, Chen S, et al. (1992) Cloning and sequence analysis of the cDNA encoding P-450 aromatase (P450arom) from a rainbow trout (Oncorhynchusmykiss) ovary; relationship between the amount of P450arom mRNA and the production of oestradiol-17ß in the ovary." J Mol Endocrinol 8: 53-61.
35. Fukada S, Tanaka M, Matsuyama M, Kobayashi D, Nagahama Y (1996) Isolation, characterization, and expression of cDNAs encoding the medaka (Oryziaslatipes) ovarian follicle cytochrome P-450 aromatase." Mol Reprod Dev 45: 285-290.
36. Gen K, Okuzawa K, Kumakura N, Yamaguchi S, Kagawa H (2001) Correlation between messenger RNA expression of cytochrome P450 aromatase and its enzyme activity during oocyte development in the red seabream (Pagrus major). Biol Reprod65: 1186–1194.
37. Heid SE, Pollenz RS, Swanson HI (2000) Role of heat shock protein 90 dissociation in mediating agonist-induced activation of the aryl hydrocarbon receptor. Mol Pharmacol 57: 82-92.
38. Tanguay RL, Abnet CC, Heideman W, Peterson RE (1999) Cloning and characterization of the zebrafish (Daniorerio) aryl hydrocarbon receptor.

Biochim Biophys Acta 1444: 35-48.

39. Barouki R, Aggerbeck M, Aggerbeck L, Coumoul X (2012) The aryl hydrocarbon receptor system. Drug Metabol Drug Interact 27: 3-8.

40. Kiss EA, Diefenbach A (2012) Role of the Aryl Hydrocarbon Receptor in Controlling Maintenance and Functional Programs of RORgammat(+) Innate Lymphoid Cells and Intraepithelial Lymphocytes." Front Immunol 3: 124.

41. MayatiA, Le Ferrec E, Lagadic-Gossmann D, Fardel O (2012) Aryl hydrocarbon receptor-independent up-regulation of intracellular calcium concentration by environmental polycyclic aromatic hydrocarbons in human endothelial HMEC-1 cells. Environ Toxicol 27: 556-562.

42. He J, Hu B, Shi X, Weidert ER, Lu P, et al. (2013) Activation of the aryl hydrocarbon receptor sensitizes mice to non-alcoholic steatohepatitis by deactivating the mitochondrial sirtuindeacetylase Sirt3. Mol Cell Biol 33: 2047-2055.

43. Mittelholzer C, Andersson E, Consten D, Hirai T, Nagahama Y, et al. (2007) 20b-hydroxysteroid dehydrogenase and CYP19A1 are differentially expressed during maturation in Atlantic cod (Gadusmorhua). J Mol Endocrinol39: 319-328.

44. Hoffmann LJ, Oris TJ (2006) Altered gene expression: A mechanism for reproductive toxicity in zebrafish exposed to benzo[a]pyrene. Aquat Toxicol78: 332-340.

45. Senior AM, Nakagawa S (2013) A comparative analysis of chemically induced sex reversal in teleosts: challenging conventional suppositions. Fish Fisheries 14:60-76.

46. Stegeman JJ, Elskus AA, Pruell R (1992) Endogenously-mediated, pretranslational suppression of cytochrome P4501A in PCB-contaminated flounder. Marine Environmental Research 34: 97-101.

47. McMaster ME, Munkittrick KR, Luxon PL, Van Der Kraak GJ (1994) Impact of low-level sampling stress on interpretation of physiological responses of white sucker exposed to effluent from a bleached kraft pulp mill. Ecotoxicol Environ Saf 27: 251-264.

48. Williams JP, Courtena CS, Wilson EC (1998) Annual sex steroid profiles and effects of gender and season on Cytochrome P450 mRNA induction in Atlantic tomcod (Microgadus tomcod). Environmental Toxicology and Chemistry 17: 1582-1588.

49. Jonny BJ, Johnson G, Porte C, Krahn MM, Ariese F (2010) Analytical methods for determining metabolites of polycyclic aromatic hydrocarbon (PAH) pollutants in fish bile: A review. Environ Toxicol Pharmacol30: 224-244.

50. Halm S, Rand-Weaver M, Sumpter JP, Tyler CR (2001) Cloning and molecular characterization of an ovarian-derived (brain-like) P450 aromatase cDNA and development of a competitive RT-PCR assay to quantify its expression in the fathead minnow (Pimephalespromelas). Fish Physiology and Biochemistry 24: 49-62.

51. Kishida M, Callard GV (2001) Distinct cytochrome P450 aromatase isoforms in zebrafish (Daniorerio) brain and ovary are differentially programmed and estrogen regulated during early development. Endocrinology 142: 740-750.

52. Greytak SR, Champlin D, Callard GV (2005) Isolation and characterization of two cytochrome P450 aromatase forms in killifish (Fundulusheteroclitus): differential expression in fish from polluted and unpolluted. Aquatic Toxicology 71: 371-389.

53. Bemanian V, Rune Male, Anders Goksøyr (2004) The aryl hydrocarbon receptor-mediated disruption of vitellogenin synthesis in the fish liver: Cross-talk between AHR- and ERα-signalling pathways. Comp Hepatol.

54. Vogel CF, Matsumura F (2009) A new cross-talk between the Aryl hydrocarbon receptor and RelB, a member of the NF-κB family.Biochem Pharmacol 77: 734-745.

55. Yuan Z (2003) Resistance of cytochrome P4501A1 mRNA to induction by halogenated aryl hydrocarbons in Hudson River Atlantic tomcod-its prevalence and mechanistic basis " Doctoral dissertation, Department of Environmental Medicine, New York University School of Medicine, Tuxedo, New York.

56. O'Donnell EF, Kopparapu PR, Koch DC, Jang HS, Phillips JL, et al. (2012) The aryl hydrocarbon receptor mediates leflunomide-induced growth inhibition of melanoma cells. PLoS One 7: e40926.

57. Rager JE, Fry RC (2012) The aryl hydrocarbon receptor pathway: a key component of the microRNA-mediated AML signalisome. Int J Environ Res Public Health 9: 1939-1953.

58. Oda Y, Nakajima M, Mohri T, Takamiya M, Aoki Y, et al. (2012) Aryl hydrocarbon receptor nuclear translocator in human liver is regulated by miR-24." Toxicol Appl Pharmacol 260: 222-231.

59. Klerks PL, Leberg PL, Lance RF, McMillin DJ, Means JC (1997) Lack of development of pollutant-resistance or genetic differentiation of darter gobies (Gobionellusboleosoma) inhabiting a produced-water discharge site. Marine Environmental Research 44: 377-395.

60. Weis JS, Mugue N, Weis P (1999) Mercury tolerance, population effects, and population genetics in the mummichog, Fundulusheteroclitus, in: V.E " Forbes (Ed.), Genetics and Ecotoxicology, Taylor & Francis, Philadelphia, 1999: 31-54.

61. Wirgin I, Waldman JR (2004) Resistance to contaminants in North American fish populations.Mutat Res 552: 73-100.

62. Lefèvre T, Williams AJ, de Roode JC (2011) Genetic variation in resistance, but not tolerance, to a protozoan parasite in the monarch butterfly. Proc Biol Sci 7: 751-759.

63. Shivanna B, Zhang W, Jiang W, Welty SE, Couroucli XI, et al. (2013) Functional deficiency of aryl hydrocarbon receptor augments oxygen toxicity-induced alveolar simplification in newborn mice. Toxicol Appl Pharmacol 267:209-217.

64. Yuan Z, Courtenay S, Wirgin I (2006) Comparison of hepatic and extra hepatic induction of cytochrome P4501A by graded doses of aryl hydrocarbon receptor agonists in Atlantic tomcod from two populations. Aquat Toxicol 76: 306-320.

Universal Primers for Exon-Priming Intron-Crossing (EPIC) PCR on Ribosomal Protein Genes in Marine Animals

Seinen Chow*, Takashi Yanagimoto and Yoji Nakamura

National Research Institute of Fisheries Science, 2-12-4 Fukuura, Kanazawa, Yokohama, Kanagawa 236-8648, Japan

Abstract

Relatively conserved exon sequences among distant aquatic animal taxa (Copepoda, Echinodermata, and Teleostei) were determined for 12 ribosomal protein genes, and 19 primer pairs were designed for an exon-primed intron-crossing polymerase chain reaction strategy. The universal utility of these primers was evaluated in distant animal species (sea urchin, squid, crab, shark, and tuna). Fragment amplification was confirmed for at least three species for all primer pairs, and single fragment amplification was observed for at least one species in 16 primer pairs. Primer sets presented here may serve as an initial step for isolating single-copy nuclear DNA sequences in a wide variety of marine animals.

Keywords: Universal primers; Ribosomal protein genes; Intron; Single copy nuclear loci; Marine animals

Introduction

Introns may be a good source for finding DNA polymorphisms in the eukaryotic genome as they rapidly accumulate mutations without greatly affecting gene function. Exon-primed intron-crossing (EPIC) polymerase chain reaction (PCR) has been developed and applied to find genetic polymorphisms in a wide variety of animals [1-5]. Primers designed to target conserved areas may have a wide utility across a broad range of animal taxa [6-10]. The universality of the primers for EPIC PCR depends on the conservation of exonic sequences, and the efficiency of PCR amplification may be primarily affected by gene copy number, including pseudogenes. Genes encoding ribosomal proteins (RPs) may be highly suitable for designing universal primers, as RPs are the most highly conserved genes among eukaryotes and are typically encoded by a single gene [11,12]. Indeed, the coding sequences of RP genes are reported to be highly conserved among humans, fruit flies, and nematodes [13]. In this study, we designed 19 primer pairs in exons of 12 RP genes and demonstrated the universal utility of these primer pairs using samples of distantly related aquatic animal species.

Materials and Methods

Eighty different ribosomal protein genes were mapped to human chromosomes [14], and cDNA data for the homologous genes of the copepod (*Lepeophtheirus salmonis, Acartia pacifica*, or *Calanus helgolandicus*), sea urchin (*Strongylocentrotus purpuratus* or *Paracentrotus lividus*), and fish (*Danio rerio* or *Ictalurus punctatus*) were obtained from the Gen Bank database. RP genes that were not found in all three of these reference animal taxa were not analyzed. BLAST searches [15] using these reference cDNA sequences against the draft genome sequences of the Pacific blue fin tuna [*Thunnus orientalis*; accession nos. BADN01000001-BADN01133062 in the DNA Data Bank of Japan (DDBJ)] [16] detected partial sequences of 31 RP genes among 35 bluefin tuna scaffolds. For each of the 31 genes detected, a multiple sequence alignment was constructed by ClustalX version 1.83 [17], and conserved regions suitable for designing universal primers were surveyed. Crude DNA extractions of the long-spined sea urchin (*Diadema savignyi*), neon flying squid (*Ommastrephes bartramii*), golden crab (*Chaceon granulatus*), mako shark (*Isurus oxyrinchus*), and skipjack tuna (*Katsuwonus pelamis*) derived from our laboratory collection were used for evaluating the universal utility of the primers. PCR amplifications were performed in a 12 µL final volume containing 1 µL of template DNA, 1.2 µL 10x buffer, 1 mM of each deoxynucleotide phosphate (dNTP), 0.4 µM of each primer, and 0.5 units of EX Taq polymerase (Takara, Japan). The same reaction conditions were applied for all primer pairs in which the reaction mixtures were preheated to 94°C for 4 min, followed by 35 amplification cycles (94°C for 30 sec, 55°C or 58°C for 30 sec, and 72°C for 50 sec), with a final extension at 72°C for 7 min. PCR products were electrophoresed on a 1.5% agarose gel (UltraPure Agarose, Invitrogen, Carlsbad CA), stained with ethidium bromide and photographed.

Results and Discussion

Relatively conserved exon sequences among distant reference animal taxa were determined for 12 genes in which 19 primer pairs (A–S) were designed for EPIC PCR (Table 1). To design the primers, we adopted a primer length of 20 bases, a moderate GC content (approximately 50%), and usually less than three degenerated bases, and because the conserved regions among the variety of taxa were not extensive, there were limited options for selecting primer positions.

At a lower annealing temperature (55°C), amplification of single to multiple fragments ranging from approximately 100 to 4,000 bp (fragments smaller than 100 bp were not counted) was observed in at least three taxa for each primer pair (Figure 1). The number of amplified fragments decreased at the higher annealing temperature (58°C), to the extent that no amplified fragments were obtained for three primer pairs (B, I, and K; Figure 2). Amplification failure may be attributable to primer mismatch, large numbers of processed pseudogenes [18], and/or non-specific annealing, which may be responsible for multiple fragment amplification. At the higher annealing temperature, the number of primer pairs which amplified a single fragment was 13 pairs in tuna, followed by 8 in sea urchin, 6 in shark, 3 in squid, and only 2

***Corresponding author:** Seinen Chow, National Research Institute of Fisheries Science, 2-12-4 Fukuura, Kanazawa, Yokohama, Kanagawa 236-8648, Japan
E-mail: chow@affrc.go.jp

Pair	Gene	Primer	Orientation	Primer sequence (5'-3')	Distance*
A	S2	RPS2exCF	forward	CCHGTNCCCAAGAAGCTGCT	70
	S2	RPS2exDR	reverse	GGTYTCCTTCCASAGRTCAG	43
B	S3	RPS3ex1F	forward	AAATCTCVAAGAAGAGRAAG	2
	S3	RPS3ex2R	reverse	ACRCCGGAGTANCCATCCTC	62
C	S3	RPS3ex2F	forward	GAGGATGGNTACTCCGGYGT	53
	S3	RPS3ex3R	reverse	CKCTTCTGVACMACAGCGGT	48
D	S3	RPS3ex3F	forward	ACCGCTGTKGTBCAGAAGMG	30
	S3	RPS3ex4R	reverse	GTCGCAACCTTCTCAGCRTA	24
E	S3	RPS3ex4F	forward	TAYGCTGAGAAGGTTGCGAC	74
	S3	RPS3ex5R	reverse	AASCKCAGRACACCATARCA	6
F	S14	RPS14exAF	forward	GCCTCMTTYAACGAYACCTT	29
	S14	RPS14exBR	reverse	CKGTCRGCCTTYACCTTCAT	30
G	S17	RPS17exBF	forward	AGWGGMATCTCCATCAARCT	49
	S17	RPS17exCR	reverse	GGRTCRACCTCGATGATSTC	18
H	S23	RPS23exAF	forward	TCACACGCCAARGGCATYGT	15
	S23	RPS23exBR	reverse	ATRGCAGAGTTRGGCTGCTT	14
I	S23	RPS23exBF	forward	AAGAACGGCAARAAGATCAC	42
	S23	RPS23exCR	reverse	TGACCYTTACGWCCRAATCC	26
J	S27	RPS27exAF	forward	GCTAARATYCARGAYAAGGA	4
	S27	RPS27exBR	reverse	GTGCGKCCRTCYTCCAGCTG	43
K	L3	RPL3exAF	forward	GGCTACAAGGCYGGYATGAC	39
	L3	RPL3exBR	reverse	CAGTTYTTGTAGAAKCGACG	149
L	L3	RPL3exBF	forward	GCGTCGMTTCTACAARAACT	2
	L3	RPL3exCR	reverse	GTGAARGCCTTCTTCTYGGA	9
M	L3	RPL3exDF	forward	CAAGGGWCACGGRTDCAAGG	2
	L3	RPL3exER	reverse	GTCTTRCGGGGMAGCTTCTT	25
N	L8	RPL8exBF	forward	AGTTCATCTACTGCGGCAAG	37
	L8	RPL8exCR	reverse	ATGAYGGTDCCYTCAGGCAT	5
O	L9	RPL9exCF	forward	RAACTTCYTGGGRGAGAAGT	110
	L9	RPL9exDR	reverse	ACDGTGGTGGCYTGCTGGAT	10
P	L11	RPL11exAF	forward	CARACGCCHGTCTTCTCCAA	5
	L11	RPL11exBR	reverse	AAGGATCKYACAGTGTAGCG	4
Q	L11	RPL11exBF	forward	CGCTACACTGTRMGATCCTT	87
	L11	RPL11exCR	reverse	CCCARRTCGATGTGYTCCTG	65
R	L17	RPL17exAF	forward	GAGWCMGCTCAGGCCATCAA	132
	L17	RPL17exBR	reverse	TCGGCRCTCTTCTTRGGCCA	8
S	L24	RPL24exBF	forward	AGMCGCAARCACAAGAAGGG	13
	L24	RPL24exCR	reverse	ACYTCAGGCTTCTGRTTCCT	85

*Number of nucleotides between 3'-end of the primers and start or end of the target introns

Table 1: Sequences of primers for amplifying 19 introns of 12 ribosomal protein genes.

in crab. Higher scores observed in tuna and sea urchin are reasonable, since RP gene data of teleost and sea urchin were used to design primers. The lowest result in crab may indicate distant relationship between copepod and decapod.

Two to three amplicons (13 in total) from five primer pairs (D, E, G, L, and Q) showing a single fragment amplification at the higher annealing temperature (Figure 2) were selected and subjected to direct nucleotide sequencing using PCR primers. Partial or complete nucleotide sequences were successfully determined for eight (one each from primer pairs D, E, and G; two from L; and three from Q) out of 13 amplicons. Conserved exon sequences and expected exon–intron motifs for the respective RP genes were confirmed in six amplicons from D, E, G, and Q, whereas these were not determined in two amplicons

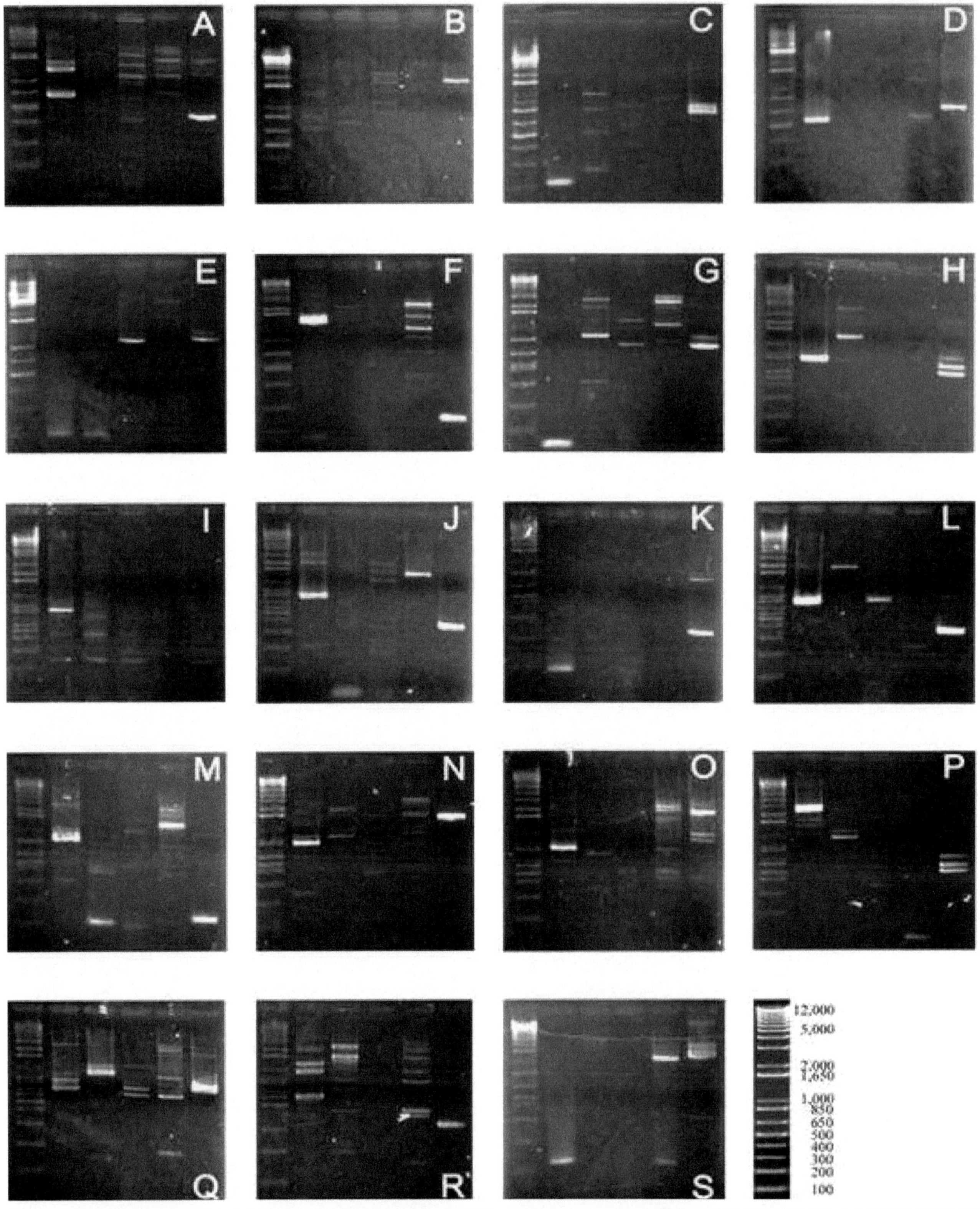

Figure 1: Agarose gel electrophoresis images of the PCR products obtained by 19 primer pairs using a 55°C annealing temperature. A-S correspond to the primer pairs shown in Table 1. The left-most lane (1st lane) of each gel image is the size marker (1 Kb Plus DNA Ladder, Invitrogen), and the 2nd-6th lanes correspond to long-spined sea urchin (*Diadema savignyi*), neon flying squid (*Ommastrephes bartramii*), golden crab (*Chaceon granulatus*), mako shark (*Isurus oxyrinchus*), and skipjack tuna (*Katsuwonus pelamis*), respectively.

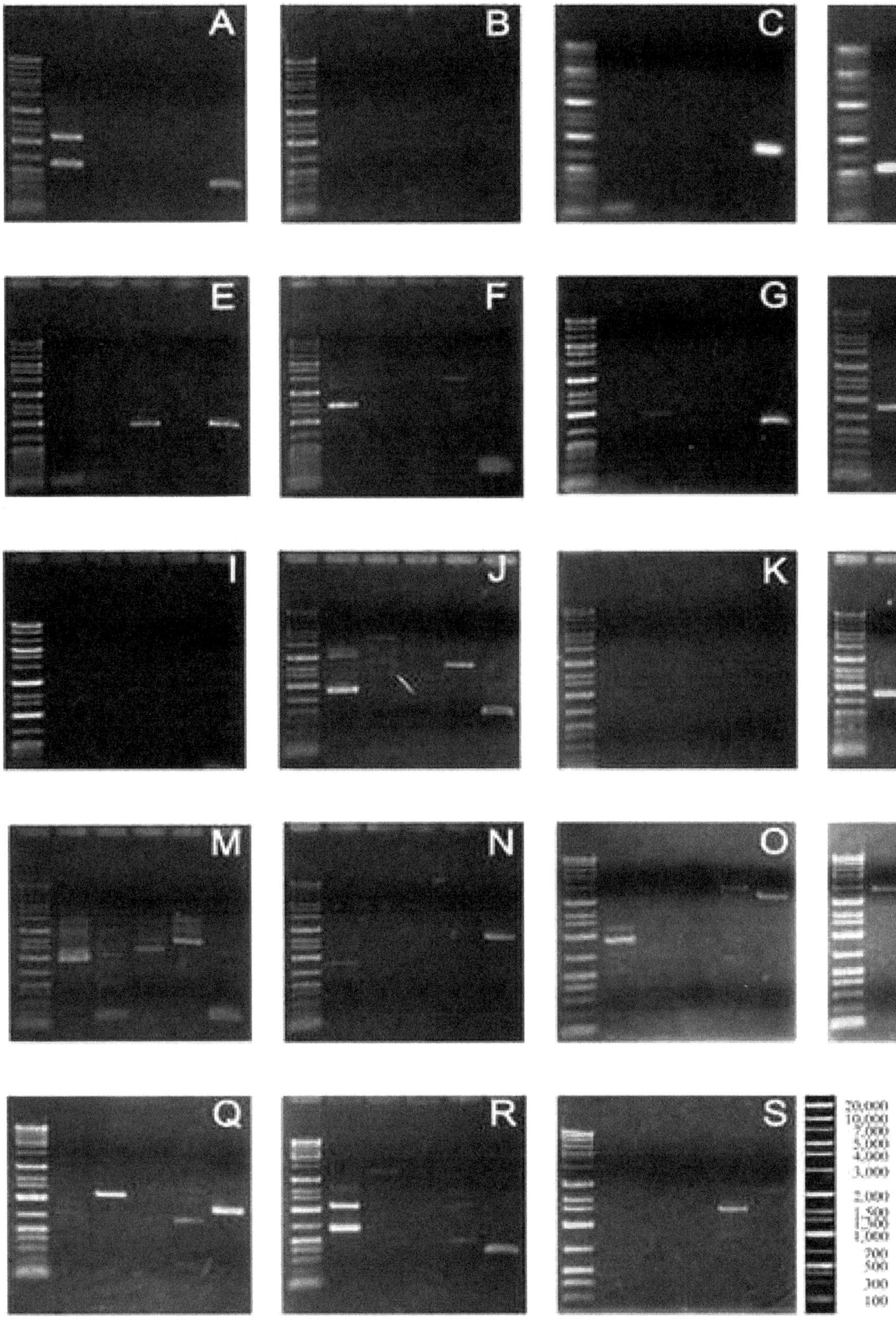

Figure 2: Agarose gel electrophoresis images of the PCR products obtained by 19 primer pairs at a 58°C annealing temperature. The left-most lane (1st lane) of each gel image is the size marker (Gene Ladder Wide 1, Wako), and the species used are the same as those in Figure 1.

from L due to the short distance between the primer annealing site and an intron (Table 1). All sequences are available in DDBJ and GenBank under the accession numbers LC055480–LC055485, LC055565, LC055566.

Although intensive investigation into the optimal PCR conditions was not performed, amplification failure and multiple fragment amplification may be improved by controlling the annealing temperature or modifying the primer sequences for taxa of interest.

Thus, the present study offers new nuclear primers potentially applicable to a wide variety of marine animal taxa, which may be an initial step toward isolating single-copy nuclear DNA sequences in a wide variety of marine animals. Subsequent polymorphism detection may also provide a more informative database for genetic species identification and population study.

Acknowledgments

This work was partially supported by a Grant-in-Aid for Scientific Research on Priority Areas (C) (No. 25450292) from the Ministry of Education, Science, Sports, and Culture, Japan.

References

1. Lessa EP (1992) Rapid survey of DNA sequence variation in natural populations. Mol Biol Evol 9: 323-330.
2. Palumbi SR, Baker CS (1994) Contrasting population structure from nuclear intron sequences and mtDNA of humpback whales. Mol Biol Evol 11: 426-435.
3. Corte-Real HBSM, Dixon DR, Holland PWH (1994) Intron-targeted PCR: a new approach to survey neutral DNA polymorphism in bivalve populations. Mar Biol 120: 407-413.
4. Chow S, Clarke S, Nakadate M, Okazaki M (2007) Boundary between the north and south Atlantic populations of the swordfish (*Xiphias gladius*) inferred by a single nucleotide polymorphism at calmodulin gene intron. Mar Biol 152: 87-93.
5. Daguin C, Bonhomme F, Borsa P (2001) The zone of sympatry and hybridization of *Mytilus edulis* and *M. galloprovincialis*, as described by intron length polymorphism at locus mac-1. Heredity 86: 342-354.
6. Chow S, Hazama K (1998) Universal PCR primers for S7 ribosomal protein gene introns in fish. Mol Ecol 7: 1255-1256.
7. Chow S, Nakadate M (2004) PCR primers for fish G6PD gene intron and characterization of intron length variation in the albacore *Thunnuns alalunga*. Mol Ecol Notes 4: 391-393.
8. Hassan M, Lemaire C, Fauvelot C, Bonhomme F (2002) Seventeen new exon-primed intron crossing polymerase chain reaction amplifiable introns in fish. Mol Ecol Notes 2: 334-340.
9. Pinho C, Rocha S, Carvalho BM, Lopes S, Mourão S, et al. (2010) New primers for the amplification and sequencing of nuclear loci in a taxonomically wide set of reptiles and amphibians. Cons Genet Res 2: 181-185.
10. Jennings BM, Etter RJ (2011) Exon-primed, intron-crossing (EPIC) loci for five nuclear genes in deep-sea protobranch bivalves: primer design, PCR protocols and locus utility. Mol Ecol Res 11: 1102-1112.
11. Wool IG (1979) The structure and function of eukaryotic ribosomes. Ann Rev Biochem 48: 719-54.
12. Wool IG, Chan Y-L, Gluck A (1995) Structure and evolution of mammalian ribosomal proteins. Biochem Cell Biol 73: 933-947.
13. Yoshihama M, Uechi T, Asakawa S, Kawasaki K, Kato S, et al. (2002) The human ribosoma protein genes: sequencing and comparative analysis of 73 genes. Genome Res 12: 379-390.
14. Uechi T, Tanaka T, Kenmochi N (2001) A complete map of the human ribosomal protein genes: assignment of 80 genes to the cytogenetic map and implications for human disorders. Genomics 72: 223-230.
15. Altschul SF, Gish W, Miller W, Myers EW, Lipman DJ (1990) Basic local alignment search tool. J Mol Biol 215: 403-410.
16. Nakamura Y, Mori K, Saitoh K, Oshima K, Mekuchi M, et al. (2013) Evolutionary changes of multiple visual pigment genes in the complete genome of Pacific bluefin tuna. PNAS 110: 11061-11066.
17. Thompson JD, Gibson TJ, Plewniak F, Jeanmougin F, Higgins DG (1997) The CLUSTALX windows interface: flexible strategies for multiple sequence alignment aided by quality analysis tools. Nucleic Acids Res 25: 4876-4882.
18. Duvov KP, Perry RP (1984) The gene family encoding the mouse ribosomal protein L32 contains a uniquely expressed intron-containing gene and an unmutated processed gene. Cell 37: 457-468.

PERMISSIONS

All chapters in this book were first published in JMSRD, by OMICS International; hereby published with permission under the Creative Commons Attribution License or equivalent. Every chapter published in this book has been scrutinized by our experts. Their significance has been extensively debated. The topics covered herein carry significant findings which will fuel the growth of the discipline. They may even be implemented as practical applications or may be referred to as a beginning point for another development.

The contributors of this book come from diverse backgrounds, making this book a truly international effort. This book will bring forth new frontiers with its revolutionizing research information and detailed analysis of the nascent developments around the world.

We would like to thank all the contributing authors for lending their expertise to make the book truly unique. They have played a crucial role in the development of this book. Without their invaluable contributions this book wouldn't have been possible. They have made vital efforts to compile up to date information on the varied aspects of this subject to make this book a valuable addition to the collection of many professionals and students.

This book was conceptualized with the vision of imparting up-to-date information and advanced data in this field. To ensure the same, a matchless editorial board was set up. Every individual on the board went through rigorous rounds of assessment to prove their worth. After which they invested a large part of their time researching and compiling the most relevant data for our readers.

The editorial board has been involved in producing this book since its inception. They have spent rigorous hours researching and exploring the diverse topics which have resulted in the successful publishing of this book. They have passed on their knowledge of decades through this book. To expedite this challenging task, the publisher supported the team at every step. A small team of assistant editors was also appointed to further simplify the editing procedure and attain best results for the readers.

Apart from the editorial board, the designing team has also invested a significant amount of their time in understanding the subject and creating the most relevant covers. They scrutinized every image to scout for the most suitable representation of the subject and create an appropriate cover for the book.

The publishing team has been an ardent support to the editorial, designing and production team. Their endless efforts to recruit the best for this project, has resulted in the accomplishment of this book. They are a veteran in the field of academics and their pool of knowledge is as vast as their experience in printing. Their expertise and guidance has proved useful at every step. Their uncompromising quality standards have made this book an exceptional effort. Their encouragement from time to time has been an inspiration for everyone.

The publisher and the editorial board hope that this book will prove to be a valuable piece of knowledge for researchers, students, practitioners and scholars across the globe.

LIST OF CONTRIBUTORS

Ghaeni M
Department of Fisheries, Science and Research Branch, Islamic Azad University, Tehran, Iran

Matinfar A
Department of Aquaculture, Iranian Fisheries Research Institute, Tehran, Iran

Varadharajan D, Soundarapandian P and Pushparajan N
Faculty of Marine Sciences, Centre of Advanced Study in Marine Biology, Annamalai University, Parangipettai-608502, Tamil Nadu, India

Soundarapandian P and Varadharajan D
Faculty of Marine Sciences, Centre of Advanced Study in Marine Biology, Annamalai University, Parangipettai - 608 502, Tamil Nadu, India

Ravichandran S
Department of Zoology, Government Arts College, Kumbakonam, Tamil Nadu, India

D Varadharajan and P Soundarapandian
Faculty of Marine Sciences, Centre of Advanced Study in Marine Biology, Annamalai University, Tamil Nadu, India

A Babu, P Sampathkumar, T Balasubramanian, D Varadharajan and T Manikandarajan
Faculty of Marine Sciences, Annamalai University, Parangipettai - 608 502, Tamil Nadu, India

Tahar Gharred, Azza Naija and Rawka Rayena Bouali
Research Unit of Genome, Immune Diagnosis and Valorization, Higher Institute of Biotechnology of Monastir, University of Monastir, Tunisia

Zohra Haouas
Laboratory of Histology Cytology and Genetics, Faculty of Medicine, Monastir 5019, Tunisia

Benoît Chénais
Laboratoire Mer Molécules Santé, EA2160, Université du Maine, Le Mans, France

Ensibi Cherif and Wifek Hannaoui
University of Carthage, Faculty of Sciences of Bizerte, Laboratory "Biodiversity and functioning of aquatic systems", 7021 Zarzouna, Bizerte, Tunisia

Olivier Pringault and Mohamed Nejib Daly Yahia
Laboratoire Ecosystèmes Lagunaires, UMR 5119 CNRS-UM2-IFREMERIRD ECOLAG, Station Méditerranéenne de l'environnement Littoral, Sète. Université de Montpellier II, France

Abdulsalam A Benkhayal and Khadija S El-turki
Chemistry Department, Faculty of Science, University of Benghaz, Libya

Ravinesh Ram and Paul C. Southgate
Centre for Sustainable Tropical Fisheries and Aquaculture, College of Marine & Environmental Sciences, James Cook University, Townsville, Queensland 4811, Australia

Roveena Vandana Chand
School of Biological and Chemical Sciences, Faculty of Science Technology and Environment, University of the South Pacific, Suva, Fiji Islands

D. Varadharajan and P. Soundarapandian
Faculty of Marine Sciences, Centre of Advanced Study in Marine Biology, Annamalai University, Parangipettai-608 502, Tamil Nadu, India

Olusola JO and Festus AA
Department of Chemistry, Federal University of Technology, Akure, Nigeria

Soundarapandian P, Varadharajan D, Jaideep Kumar and Suresh B
Faculty of Marine Sciences, Centre of Advanced Study in Marine Biology, Annamalai University, India

Ilavarasan N
Department of Zoology, Government Arts College, Karur, India

E Isa Olalekan
Environmental and Resource Management, Brandenburg University of Technology, Cottbus, Germany

Aderonke Omolara Lawal-Are
Department of Marine Sciences, University of Lagos, Lagos, Nigeria

Angsuman Chanda
Post Graduate Department of Zoology, Raja NL Khan Women's College, Gope Palace, Midnapur, Paschim Medinipur-721 102, WB, India

Yogesh Kumar and Pani Prasad Kurcheti
Division of Aquatic Environment and Health Management, Central Institute of Fisheries Education, (Deemed University), Mumbai-400 061, India

Lina M. Rotermund
Department of Physics & Atmospheric Science, Dalhousie University, Halifax, NS B3H 4R2, Canada

John Samson
Deep Inwater Imaging, 6589 Chebucto Rd, Halifax, NS B3L 1L9, Canada

Hans Jürgen Kreuzer
Department of Physics & Atmospheric Science, Dalhousie University, Halifax, NS B3H 4R2, Canada

Chien Tu and Shu-Ting Kuo
Animal Health Research Institute, Council of Agriculture, Taipei, Taiwan

Shao-Kuang Chang
Institute of Veterinary Medicine, School of Veterinary Medicine, National Taiwan University, Taipei, Taiwan

Sue-Min Huang
Animal Health Research Institute, Council of Agriculture, Taipei, Taiwan
Institute of Veterinary Medicine, School of Veterinary Medicine, National Taiwan University, Taipei, Taiwan

Hung-Chih Kuo
Department of Veterinary Medicine, National Chiayi University, Chiayi, Taiwan

Chi-Chung Chou
Department of Veterinary Medicine, National Chung-Hsing University, Taichung, Taiwan

D Varadharajan and P Soundarapandian
Faculty of Marine Science, Centre of Advanced Study in Marine Biology, Annamalai University, Parangipettai-608 502, Tamil Nadu, India

Awik Puji Dyah Nurhayati
Biology Department, Mathematic and Natural Science Faculty, Sepuluh November Institute of Technology Surabaya, Indonesia

Rarastoeti Pratiwi and Istriyati
Biology Faculty, GadjahMada University, Yogyakarta, Indonesia

Subagus Wahyuono
Department of Biological Pharmacy, Faculty of Pharmacy, Gadjah Mada University, Yogyakarta, Indonesia

Hari Purnomo
Department of Chemistry Pharmacy, Faculty of Pharmacy, Gadjah Mada University, Yogyakarta, Indonesia

Syamsudin Abdillah
Department of Pharmacology, Faculty of Pharmacy, Pancasila University, South Jakarta, Indonesia

Yaghoob Parsa
Department of Marine Biology, Faculty of Biological Science, Shahid Beheshti University, Tehran, Iran

Seyed Seyed Mohammad Bagher Nabavi, Seyedeh Narges Nabavi and Mehdi Hosseini
Department of Marine Biology, Faculty of Marine Science, Khoramshahr University of Marine Science and Technology, Iran

Kwaansa-Ansah EE, Agyemang D and Opoku F
Department of Chemistry, Kwame Nkrumah University of Science and Technology, Kumasi, Ghana

Mansoreh Ghaeni
Department of Fisheries, Ahvaz Branch, Islamic Azad University, Ahvaz, Iran

Khadijeh Nabi Ghahfarokhi and Ladan Zaheri
Fisheries Scientific Association, Islamic Azad University, Ahvaz, Iran

Soundarapandian P, Varadharajan D and Boopathi A
Faculty of Marine Sciences, Centre of Advanced Study in Marine Biology, Annamalai University, Tamil Nadu, India

Madenjian CP
US Geological Survey, Great Lakes Science Center, Ann Arbor, Michigan, USA

Jensen OP and Vastano AR
Department of Marine and Coastal Sciences, Rutgers University, New Brunswick, New Jersey, USA

Krabbenhoft DP, DeWild JF and Ogorek JM
US Geological Survey, Wisconsin Water Science Center, Middleton, Wisconsin, USA

Craig S Tepper
Department of Biology, Cornell College, Mt. Vernon, IA 52314, USA

Sophia C Gaynor
Department of Biology, Cornell College, Mt. Vernon, IA 52314, USA
Interdisciplinary Graduate Program in Genetics, University of Iowa, Iowa City, IA 52242, USA

Kefi AS, Chungu NP, Mupenda N and Mumba C
National Aquaculture Research and Develop-ment Centre, P. O. Box 22797, Kitwe, Zambia

Mahmoud M Kardousha, Abdulrahman Al-Muftah and Jassim A Al-Khayat
Department of Biological and Environmental Sciences, Qatar University, Doha, Qatar

Kunal Mondal
Faculty of Fishery Sciences, West Bengal University of Animal and Fishery Sciences, Kolkata-700094, India

Subhra Bikash Bhattacharyya and Abhijit Mitra
Department of Oceanography, Techno India University, Kolkata- 700091, India

A Fernández, E. Sierra, M. Méndez, S. Sacchinni, Y. Bernaldo de Quirós, M. Andrada, M. Rivero, O. Quesada and Arbelo M
Veterinary Histology and Pathology, Institute of Animal Health, Veterinary School, University of Las Palmas de Gran Canaria, Canary Islands, Spain

V. Martín and M. Tejedor
Society for the Study of Cetaceans in the Archipelago Canary (SECAC) - Canary Islands, Spain

Susmita Mandal
Kakdwip Research Centre of Central Institute of Brackish water Aquaculture Kakdwip, South 24 Parganas, West Bengal-743 347, India

Asish Mondal and Bhattacharyya SB
Kakdwip Research Centre of Central Institute of Brackish water Aquaculture Kakdwip, South 24 Parganas, West Bengal-743 347, India
Department of Oceanography, Techno India University, Salt Lake campus, Kolkata, India

Deepta Chakravortty and Abhijit Mitra
Department of Oceanography, Techno India University, Salt Lake campus, Kolkata, India

Yajaira García de Severeyn and Ana Villasmil
Universidad del Zulia, Facultad Experimental de Ciencias, Dpto. de Biología, Laboratorio de Cultivo de Invertebrados Acuático, Venezuela

Héctor Severeyn
Universidad del Zulia, Facultad Experimental de Ciencias, Dpto. de Biología Laboratorio de Sistemática de Invertebrados Acuáticos, Venezuela

Félix Morales
Universidad del Zulia, Facultad Experimental de Ciencias, Dpto. de Biología Laboratorio de Oceanografía, Venezuela

Marynes Montiel
Universidad del Zulia, Facultad Experimental de Ciencias, Dpto. de Biología, Unidad de investigaciones de Microbiología Ambiental Maracaibo, Zulia, Venezuela

Hadir A Aly, Mohamed M Abdel-Rahim, Ayman M Lotfy and Basem S Abdelaty
Fish Rearing Lab, Aquaculture division, National Institute of Oceanography and Fisheries (NIOF), Alexandria, Egypt

P Soundarapandian, D Varadharajan and K Gangatharan
Faculty of Marine Sciences, Centre of Advanced Study in Marine Biology,Annamalai University, Parangipettai-608 502. Tamil Nadu, India

N Ilavarasan
Department of Zoology, Government Arts College, Karur, Tamil Nadu, India

Mehdi Hosseini and Jamileh Pazooki
Department of Marine Biology, Faculty of Biological Science, Shahid Beheshti University, Tehran, Iran

Seyed Mohammad Bagher Nabavi and Yaghoob Parsa
Department of Marine Biology, Faculty of Marine Science, Khoramshahr University of Marine Science and Technology, Iran

Elegbede Isa Olalekan
Department of Environmental Planning, Brandenburg University of Technology, Cottbus-Senftenberg, Germany

Lawal-Are AO and Titilade PR
Department of Marine Sciences, University of Lagos, Akoka, Lagos, Nigeria

Mehdi Hosseini, Jamileh Pazooki and Mohsen Safaei
Department of Marine Biology, Faculty of Biological Science, Shahid Beheshti University, Tehran, Iran

M. Dalin Mary
St. Jude's College, Thoothoor, Tamil Nadu, India

Saritha Kailasam and Jamila Patterson
Suganthi Devadason Marine Research Institute, Tuticorin, Tamil Nadu, India

M. Jansi
St. Hindu College, Nagercoil, Tamil Nadu, India

Adam Tulu, Ali Ishaque and Egbe Egiebor
Department of Natural Science, University of Maryland, Eastern Shore, Princess Anne, MD 21853, USA

Christopher Chambers
NOAA-Fisheries, 74 Magruder Rd., Highlands, NJ 07732, USA

Rosemary Jagus
Center of Marine Biotechnology, University of Maryland Biotechnology Institute.701 E Pratt st. Baltimore, MD 21202, USA

Seinen Chow, Takashi Yanagimoto and Yoji Nakamura
National Research Institute of Fisheries Science, 2-12-4 Fukuura, Kanazawa, Yokohama, Kanagawa 236-8648, Japan

Index